6th Global Conference on Polymer and Composite Materials (PCM 2019)

IOP Conference Series: Materials Science and Engineering
Volume 634

Bangkok, Thailand
8 – 11 July 2019

ISBN: 978-1-5108-9836-3
ISSN: 1757-8981

Printed from e-media with permission by:

Curran Associates, Inc.
57 Morehouse Lane
Red Hook, NY 12571

CURRAN ASSOCIATES INC.
proceedings
.com

Some format issues inherent in the e-media version may also appear in this print version.

Institute of Physics
Dirac House, Temple Back
Bristol BS1 6BE UK

Phone: 44 1 17 929 7481
Fax: 44 1 17 920 0979

techtracking@iop.org

Additional copies of this publication are available from:

Curran Associates, Inc.
57 Morehouse Lane
Red Hook, NY 12571 USA
Phone: 845-758-0400
Fax: 845-758-2633
Email: curran@proceedings.com
Web: www.proceedings.com

Table of contents

Volume 634

The 6th Global Conference on Polymer and Composite Materials (PCM 2019)

8–11 July 2019, Bangkok, Thailand

Accepted papers received: 29 August 2019
Published online: 14 October 2019

Preface

Papers

IOP Conf. Series: Materials Science and Engineering **634** (2019) 011001 doi:10.1088/1757-899X/634/1/011001

PCM 2019 Preface

The 6th *Global Conference on Polymer and Composite Materials (PCM 2019)* was held from July 8 -11, 2019 in Bangkok, Thailand. The conference, with 165 participants, had a technical program consisting of 5 international keynote speakers, 87 oral presentations, and a poster session with 48 works. The conference had 7 oral parallel sessions with dedicated selected subjects related to Polymers and Composite Materials.

The present issue of IOP Conference Series: Materials Science and Engineering (MSE) records the proceedings of PCM 2019 and contains 48 specially selected manuscripts submitted to the conference. The electronic submission and handling of manuscripts via the conference website, including the selection of reviewers and evaluation of manuscripts, were identical to the procedures applied to manuscripts submitted as regular contributions for international per-reviewed publications.

The organization of this conference and the preparation of proceedings volume would have been impossible without the tremendous efforts and dedication of many individuals, especially from Ms. Yoko Ye, who oversaw the organization of the conference and the program, and a large team of reviewers with their timely submission of quality reports. We express our sincere thanks to all authors and presenters for their contributions. We also thank very much all our sponsors for their generous support, specially to Rangsit University.

The 7th Global Conference on Polymer and Composite Materials (PCM 2020) will be held during August 17th-20th, 2019, in the city of Cairo, Egypt. Located near the Nile Delta, the land composing the present-day Modern Cairo was the site of ancient national capitals whose remnants remain visible in parts of Old Cairo. The city's metropolitan area is one of the largest in Africa, and is associated with ancient Egypt, as the famous Giza pyramid complex and the ancient city of Memphis are located in its geographical area. Today, Greater Cairo has the largest number of educational schools, and higher learning institutes among other cities of Egypt. The great majority of publishing houses and media outlets and nearly all film studios are there, as are half of the nation's universities. Cairo accounts for 11% of Egypt's population and 22% of its economy. Cairo also has long been a center of the region's political and cultural life, and is titled "the city of a thousand minarets" for its preponderance of Islamic architecture. Some of landmarks of Cairo include Tahrir Square, the Museum of Egyptian Antiquities, the Cairo Tower, and the area of Old Cairo. All participants are welcome to the renewed and vibrant PCM 2020! More information will be available at the conference website: http://www.cpcmconf.org.

Prof. Dr. Esteban Broitman
General Chairman PCM2019
SKF Research & Technology Development (RTD), Netherlands, August 2019

IOP Conf. Series: Materials Science and Engineering **634** (2019) 011002 doi:10.1088/1757-899X/634/1/011002

Peer review statement

All papers published in this volume of *IOP Conference Series: Materials Science and Engineering* have been peer reviewed through processes administered by the proceedings Editors. Reviews were conducted by expert referees to the professional and scientific standards expected of a proceedings journal published by IOP Publishing.

IOP Conf. Series: Materials Science and Engineering **634** (2019) 012001 doi:10.1088/1757-899X/634/1/012001

Composites with graphene as reinforcement

A O Surendranathan[1,2]

[1]National Institute of Technology Karnataka, India

E-mail: aos_nathan54@au.yahoo.com

Abstract. In this paper, the developments in the area of graphene-reinforced composite have been reviewed. It also depicts the work guided by the author. In the work, WC powder with 6% cobalt (Co) and graphene (0.2%) in the form of graphene nanoplatelets (GPLs) was prepared by high energy rate ball milling and ultrasonification. The mixture was sintered using spark plasma sintering at 1250°C for 10 min. To study the effects of graphene reinforcement, the sample was analyzed for the physical (density), magnetic (coercivity, and magnetic saturation), microstructural (porosity) and mechanical properties (hardness, fracture toughness). Under Visual Inspection of the Spark Plasma Sintered (SPS) sample, the top surface showed more homogeneous distribution of the constituents than the bottom surface. The results of the surface measurements showed that the top and bottom surfaces have almost the same roughness. It was found that the SPS sample exceeded liquid phase sintered (LPS) specimens with respect to hardness and fracture toughness values. The coercive force was found to be higher for the LPS sample whereas the saturation magnetization value was found to be higher for the SPS sintered specimen. The density value is higher for the LPS sintered specimen. The presence of pores is seen in the back scattered electron (BSE) image. There was certain amount carbon pick up from the graphite die used in SPS. There was uniform distribution of the cobalt binder. The x-ray diffraction pattern for the SPS sintered sample indicates the presence of WC, Co and GNP.

1. Introduction

In the recent past, carbonaceous nanofillers play a promising role [1]. This is because they have better structural and functional properties. Also, they have broad range of applications in each field. Examples for such fillers are graphene and carbon nanotubes (CNTs).

2. Literature review

Garima *et al* [1] has reviewed the synthesis and properties of CNTs. Along with CNTS, the use of graphene as a novel substitution to the nanotubes as fillers was also reviewed. They resolved that CNTs and graphene have the power to alter properties of materials. The alteration can be exploited in the broad spectrum of applications. Xu and Gao [2] reported an efficient method to prepare nylon-6–(PA6–) graphene (NG) composites by *in situ* polymerization of caprolactam in the presence of graphene oxide (GO). During the polycondensation, GO was thermally reduced to graphene simultaneously. By adjusting the feed ratio of caprolactam to GO, various composites with 0.01–10 wt % content of graphene were obtained. The highly grafting nylon-6 arms on graphene sheets was confirmed by XPS, FTIR, TGA and AFM measurements, showing the grafting content up to 78 wt % and homogeneous 2D brush-like morphology from AFM observations. The efficient polymer-chain grafting makes the graphene homogeneously dispersed in PA6 matrix and depresses the crystallization of PA6 chains.

Barun *et al* [3] evaluated the mechanical properties of polyvinyl alcohol (PVA) and poly (methyl methacrylate) (PMMA)-matrix composites reinforced by functionalized a few-layers graphene (FG) using the nano-indentation technique. A significant increase in both the elastic modulus and hardness was observed with the addition of 0.6 wt.% of graphene. The crystallinity of PVA also increased with the addition of FG. This and the good mechanical interaction between the polymer and the FG, which provided better load transfer between the matrix and the fiber, are suggested to be responsible for the observed improvement in mechanical properties of the polymers.

Fan *et al* [4] dispersed a few-layered graphene sheets, synthesized by direct current arc-discharge method using NH_3 as one of the buffer gases, in chitosan/acetic acid solutions. FTIR and X-ray photoelectron spectroscopy showed the presence of oxygen-containing functional groups on the surface of graphene sheets that may assist the good dispersion of graphene in chitosan solution. Graphene/chitosan films were produced by solution casting method. The mechanical properties of composite films were tested by nanoindentation method. With the addition of a small amount of graphene in chitosan (0.1–0.3 wt%), the elastic modulus of chitosan increased over ~200%. The biocompatibility of graphene/chitosan composite films was checked by tetrazolium-based colorimetric assays in vitro. The cell adhesion result showed that the L929 cell can adhere to and develop on the graphene/chitosan composite films as well as on pure chitosan film, indicating that graphene/chitosan composites have good biocompatibility. Because there is no metallic impurity in graphene raw materials, the time-consuming purification process for removing metal nanoparticles entrapped in carbon nanotubes is thus avoided when graphene is used to prepare biomedical materials. Graphene/chitosan composites are potential candidates as scaffold materials in tissue engineering.

Wang *et al* [5] reinforced aluminum composites with graphene nanosheets (GNSs) through a feasible methodology based on flake powder metallurgy. The tensile strength of 249 MPa was achieved in the Al composite reinforced with only 0.3 wt.% GNSs, which is 62% enhancement over the unreinforced Al matrix.

Chatterjee *et al* [6] investigated the influence of reinforcements on the mechanical and thermal properties of graphene nanoplatelets/epoxy composites. Amine functionalized expanded graphene nanoplatelets (EGNPs) were dispersed within epoxy resins using high-pressure processor followed by three roll milling. Functionality on the EGNPs was confirmed with FTIR and micro-Raman spectroscopy. Bending and nano-mechanical testing was performed on the composites. Incorporation of EGNPs improved the flexural modulus and hardness of the composite and increased fracture toughness by up to 60%. Marked improvement was observed in thermal conductivity of the composites reaching 36% at 2 wt.% loading. Functionalized EGNPs exhibited significant improvements indicating favorable interaction at EGNPs/polymer interface.

Zhao *et al* [7] report that nanocomposites based on fully exfoliated graphene nanosheets and poly (vinyl alcohol) (PVA) were prepared via a facial aqueous solution. A significant enhancement of mechanical properties of the graphene/PVA composites was obtained at low graphene loading; that is, a 150% improvement of tensile strength and a nearly 10 times increase of Young's modulus were achieved at a graphene loading of 1.8 vol%.

Garima *et al* [8] reviewed the synthesis and properties of CNTs along with the use of graphene as a novel substitution to the nanotubes as fillers. A comparative analysis has been carried out between the importance of graphene and carbon nanotubes as fillers and the substantial changes in mechanical and electrical properties of matrix.

Lenka *et al* [9] had prepared silicon nitride + 1wt.% graphene platelet composites using various graphene platelets with the aim to improve the fracture toughness of Si_3N_4. The fracture toughness was significantly higher for all composites in comparison to the monolithic Si_3N_4, with the highest value of 9.92 $MPam^{0.5}$. The main toughening mechanisms originated from the presence of graphene platelets, and the increase in the fracture toughness values was attributable to crack deflection, crack branching and crack bridging.

Li *et al* [10] grew hybrid fillers composed of carbon nanotubes (CNTs) on graphene nanoplatelets (GNPs). They were dispersed into epoxy matrix to serve as promising reinforcements. And the CNT–

GNP/epoxy composite shows distinctive self-sensing behavior for *in situ* monitoring the onset of irreversibly permanent deformation. Here it has been established that the embedding of CNT–GNP hybrids into pristine epoxy endows optimum dispersion of CNTs and GNPs as well as better interfacial adhesion between the carbon fillers and matrix, which results in a significant improvement in load transfer effectiveness. Remarkably enhanced mechanical properties in the CNT–GNP/epoxy composite were achieved at ultralow hybrid loading (0.5 wt.%). The tensile modulus showed ~40% increase and the tensile strength was enhanced by ~36% with respect to the neat epoxy. The reinforcement efficiency of the CNT–GNP hybrids is found to outperform that of the CNT+GNP mixture predicted using the modified Halpin-Tsai modelling.

Xia *et al* [11] established the electrical behaviour of both uncured and oven cured GNPs/epoxy composites with loadings of up to 10 wt%. Above the critical loading found for oven cured materials (~8.5 wt%) the electrically conducting networks of GNPs formed in the matrix showed the ability to act as integrated nanoheaters when an electric current was passed through them,

successfully curing the composites by Joule heating. Composites prepared by this Out-of-Autoclave (OoA) curing method (as an alternative to the traditional oven based one) at 10 wt% loading of GNPs were also prepared and compared to the oven cured ones. They showed more compact composite structures, with less micro voids and a preferred orientation of the GNPs in the matrix relative to the oven cured material at identical loading.

Chu *et al* [12] has shown that approximately 2 wt. % of graphene in the matrix of a unidirectionally-reinforced carbon fiber epoxy composite leads to a significant enhancement in mechanical properties. Particularly, it is found that the axial stiffness of the composites is increased by ~10 GPa accompanied by an increase in axial strength of 200 MPa.

Young *et al* [13] have undertaken a detailed study of the mechanisms of stress transfer in polymeric matrices with different values of Young's modulus, E_m, reinforced by graphene nanoplatelets (GNPs). For each material, the Young's modulus of the graphene filler, E_f, has been determined using the rule of mixtures and it is found to scale with the value of E_m.

Li *et al* [14] prepared epoxy nanocomposites reinforced by graphite nanoplatelet (GNP) and nano-scale X-ray computed tomography was used to visualize the geometry, morphology and defects of the flakes, as well as the three-dimensional agglomerates that are normally difficult to characterise by other techniques. In combination with micromechanical analysis, the taxonomy of the nanoplatelets is shown to be of great importance in controlling the mechanical properties of nanocomposites, and this has been shown to explain the deviations of the predictions of micromechanical models from the measured values.

3. Our work [15]

3.1. Experimental

3.1.1. Characterisation of graphene. The graphene powder used was characterized using Raman spectroscopy.

3.1.2. Processing of WC-Co powder. Wax, which acted as the binder during compaction, was added to the composition of WC-6% Co. This mixture was analyzed for its magnetic properties. It was then subjected to high energy rate ball milling. The jar used was made up of WC. The balls used were having 1 cm diameter and was made from WC. The ball: powder ratio was 10:1 by weight. The speed selected was 200 rpm, and the milling duration was 1 h.

3.1.3. Ultrasonication. Different percentages of graphene were added. The mixture was ultrasonicated in N-Methyl Pyrrolidone for 1 h. The compositions of ultrasonicated WC-Co-graphene composite powders are given in table 1.

Table 1. Compositions of ultrasonicated WC-Co-graphene composite powders.

Grade Name	Composition
G0	WC + 6 w/o Co + 0 w/o graphene
G1	WC + 6 w/o Co + 0.2 w/o graphene

3.1.4. Vacuum drying. To separate liquid incorporated during ultrasonication, the mixture was dried in an oven at a temperature of 140ºC for 12 h.

3.1.5. Spark plasma sintering (SPS). The dried mixture was subjected to SPS. The die used was made of graphite and it was cylindrical in shape with 20 mm dia. The SPS temperature was 1250ºC for a time duration of 10 min. Uniaxial pressure of 50 MPa was applied.

3.1.6. Characterisation

- Coercive force (H_c)

The sintered specimen was fully magnetized in an electric field in a coil. Afterwards, it was demagnetized by applying electric field in the opposite direction. The coercive force (force required to reduce residual magnetism to zero) was determined.

- Magnetic saturation ($4\pi\sigma$)

In order to find the saturation magnetic field, the sintered sample was magnetized fully in permanent magnetic field. Then the sample was removed and the magnetic moment was measured. The obtained magnetic moment was divided by the weight of the sample to get the saturation magnetic field.

- Specific gravity

The specific gravity of the sintered sample was measured using Archimedes principle.

- Hardness

In order to check the hardness of the sintered specimen, it was ground with fine diamond grit. Later polishing was effected using diamond paste on rigidly supported laps. Then it was subjected to ultrasonic cleaning. The hardness was measured on the cleaned surface using Vickers indentation method. For the measurement, 30 kgf load was applied with a dwelling period of 10s at three sufficiently distant spots.

- Fracture toughness

The fracture toughness of the sintered specimen was evaluated by indentation method. For this, the lengths of the cracks emanating from the corners of Vickers hardness indentations were measured using optical microscope at 400X magnification. The obtained values were plotted as P (indentation load) vs. L (sum of the lengths of the four cracks). From the plot, W (slope) was found out. The fracture toughness (K_{IC}) was calculated using the formula:

$$K_{IC} = 0.0889(HW)^{½} \quad (1)$$

where, H is the Vickers hardness number.

- Porosity

The unetched specimen was analyzed for the presence, type and distribution of porosity and uncombined carbon using a metallurgical microscope (Olympus make). To observe the pores up to 10 µm, a magnification of 200X was used. To observe the pores in the range of 10 to 25 µm, a magnification of 100X was used. Uncombined carbon / graphite was assessed by comparing 100X micrographs with standard charts.

- Etching

The polished specimens were subjected to two levels of etching using Murakami's etchant.

➢ γ-Etching

The surface that was etched with Murakami's reagent was analyzed for γ-phase. The etching was done at a temperature just below 100ºC for 15 min [16]. Care was taken to photograph the same spot on the specimen with this etch. A light and careful retouching on the polishing cloth was necessary to remove a film formed on the surface of free tungsten carbide. Microscopic examinations and photomicrographs were made with a 4 mm apochromatic objective of numerical aperture 0.95 and with a green light filter. Etching was done in a solution held just below the boiling point.

- α-Etching

For obtaining the percentage of α-phase, the surface of the polished specimen was etched with Murakami's reagent in the presence of an electric current. Then the surface of the specimen was assessed for grain size and size distribution.

3.1.7. Grain size measurements. Two types of parameters were measured under grain size measurements.

- α_{max}

The first parameter was the maximum grain size of α-phase (α_{max}). For this, the α-etched specimen was scanned at 100X for the identification of the largest α grain. Its approximate size was measured.

- CαP

Coarse α patches in the specimen were analyzed for the size range that encapsulated the maximum number of coarse α grains. For this again the α-etched specimen was observed at 100X magnification.

3.1.8. X-Ray diffraction. The phases present in the SPS sample were identified using the X-Ray diffractometer (make-JOEL, model-JPX8). Nickel filter was used for the Cu K_α radiation (λ = 1.54 angstroms). Scanning rate of 2°/ min was used. The diffraction angle (2θ) range was 20°-90°. The diffraction spectrum was analyzed using Xpert Pro software.

4. Results and discussion

4.1. Characterization of graphene using Raman spectroscopy

The Raman spectrum obtained is shown in figure 1. High intensity G peak was obtained at 1585.7 cm^{-1}. The moderately intense D peak was seen at 1357.6 cm^{-1}. A fraction of graphene was present with a disordered structure. The lower intensity D peak indicates lower magnitude of defects. 2D peak (figure 2) at 2725 cm^{-1} indicates the presence of small amount of impurities.

Planar crystallite size was:

$$L_a = 4.4(I_G/I_D) = 5.9 \text{ nm} \quad (2)$$

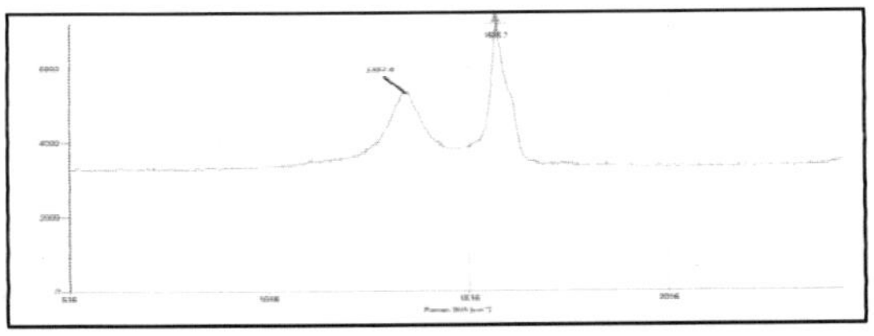

Figure 1. Raman spectrogram of graphene showing G and D peaks.

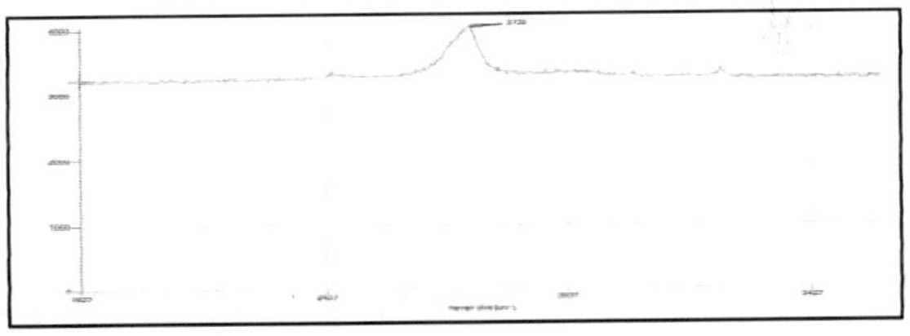

Figure 2. Raman spectrogram of graphene showing 2D peak.

4.2. Visual inspection of the spark plasma sintered sample

The top surface showed more homogeneous distribution of the constituents than the bottom surface. No contrast was seen on the side as it is as sintered without etching.

4.3. Surface roughness measurements

The results of the surface measurements are given in table 2. It is found that the top and bottom surfaces have almost the same roughness. The side is far smoother than the top and bottom surfaces. This is because the densification takes place along the side. It is the characteristic of uniaxial pressing. The table also gives the roughness values for the liquid phase (LP) sintered specimen. It is almost having the same values as that of the side of a SP sintered one. This is because the liquid formed during LP sintering spreads and fills all the open pores as well. On cooling after sintering, this results in a continuous solid phase without any pores. Surface roughness of SPS specimen is higher than that of LPS one because hot working gives rougher surface than cold working.

Table 2. Results of surface measurements.

Location	Ra (µm)	Rt (µm)	Rz (µm)
Top Surface	2.9-4.2	19.8-30.6	14.7-22.4
Bottom Surface	2.8-4.3	20.7-29.0	15.9-20.8
Side	1.0-1.8	9.6-18.0	5.8-11.3
WC-Co (LPS)	0.9-1.0	6.0-7.0	8.0-9.0

LPS-Liquid Phase Sintered.

4.4. Hardness and fracture toughness

The hardness and fracture toughness values of both the SPS and LPS specimens are given in table 3. Both the values exceed that for LPS sample.

Table 3. Hardness and fracture toughness values.

Location	Vickers Hardness, H_v	Palmqvist Fracture Toughness, K_{IC}
SPS	2000 (kgf/sq.mm)	11.8 (MPa$\sqrt{m}$)
LPS without graphene	1595 (kgf/sq.mm)	9.7 (MPa$\sqrt{m}$)

4.5. Magnetic properties & density

These properties are given in table 4. The coercive force was found to be higher for the LPS sample whereas the saturation magnetization value was found to be higher for the SPS sintered specimen. This shows that SPS sintering makes the material more ferromagnetic. The density value is higher for the LPS sintered specimen because of the presence of a liquid phase during sintering. The liquid phase fills the pores.

Table 4. Magnetic & physical properties of LPS & SPS samples.

Sample Type	Coercive Force, H_c	Magnetic Saturation, $4\pi\sigma$	Density
LPS	200 Oersted	11.3 μTm^3 /kg	14.89 g/cm^3
SPS	284 Oersted	5.2 μTm^3 /kg	15.20 g/cm^3

4.6. Microstructural characterisation

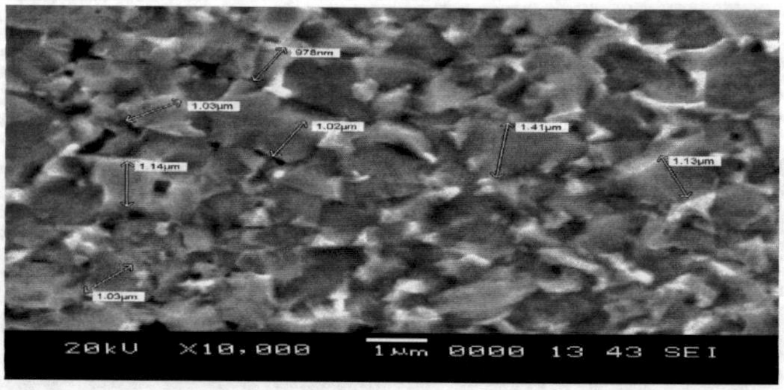

Figure 3. Scanning Electron photomicrograph of LPS sample.

The scanning electron photomicrograph of the LPS sample is shown in figure 3. The white areas in the figure show the presence of cobalt, whereas the grey areas represent the WC phase. The black areas contain the pores. The carbide particle size ranges from 978 nm to 1.41 µm. Figure 4 compares the secondary and back-scattered electron images of the SPS sample. The presence of pores is clearly seen in the back scattered electron (BSE) image.

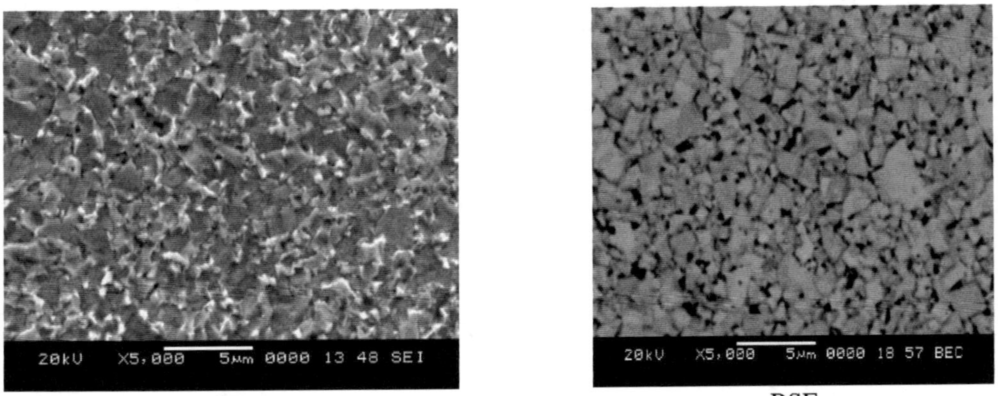

SE BSE

Figure 4. Scanning electron photomicrographs under secondary and back-scattered electron modes.

4.7. Carbon pickup & binder distribution

It is found from the photomicrograph of the top surface of the SPS sample (figure 5) that there was certain amount carbon pick up from the graphite die used in SPS. The same figure also shows that there was uniform distribution of the cobalt binder.

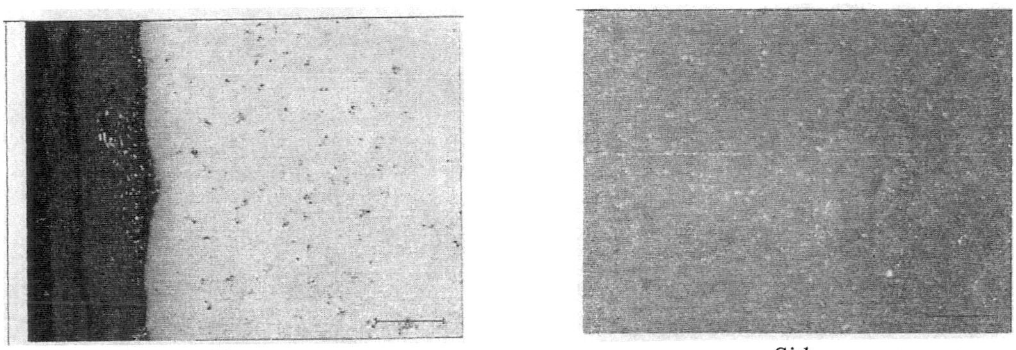

Top Side

Figure 5. Optical photomicrographs of SPS sample showing carbon pickup (top) and binder distribution (side).

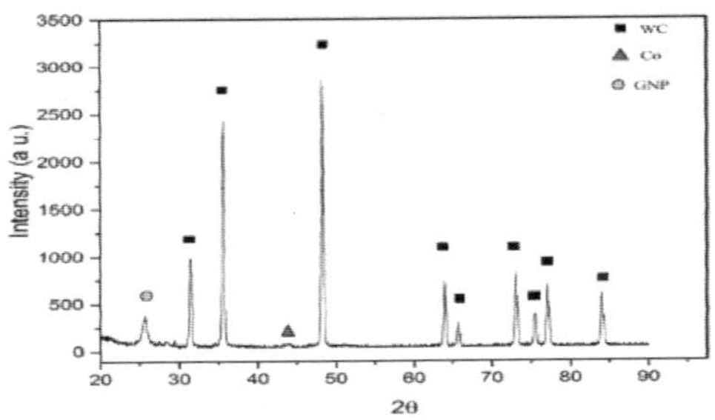

Figure 6. X-ray diffractograph of SPS sample.

4.8. X-ray diffraction

The x-ray diffraction pattern for the SPS sintered sample is shown in figure 6. It indicates the presence of WC, Co and GNP.

5. Conclusion

Composites with graphene as reimbursement has become a hot topic of the day as evidenced by the number of papers getting published. Spark plasma sintering of GNP-reinforced WC-Co composite resulted in significant increase in toughness. SPS gave higher hardness (~400 H_v). SPS makes the grain size distribution narrower. Reaction of graphene is limited in SPS. Graphene acts as a protective coating against oxidation.

References

[1] Garima M, Vivek D, Kyong Y R, Soo-Jin P and Wi R L 2015 A review on carbon nanotubes and graphene as fillers in reinforced polymer nanocomposites *Journal of Industrial and Engineering Chemistry* **21** 11-25

[2] Xu Z and Gao C 2010 *Macromolecules* **43** 6716-23

[3] Barun D, Eswar Prasad K, Ramamurty U and Rao C N R 2009 Nano-indentation studies on polymer matrix composites reinforced by few-layer graphene *Nanotechnology* **20** 12

[4] Fan H L, Wang L L, Zhao K K, Li N, Shi Z J, Ge Z G and Jin Z X 2010 Fabrication, mechanical properties, and biocompatibility of graphene-reinforced chitosan composites *Biomacromolecules* **11** 2345-51

[5] Wang J Y, Li Z Q, Genlian Fan, Huanhuan P, Zhixin C and Di Z 2012 Reinforcement with graphene nanosheets in aluminum matrix composites *Scripta Materialia* **66** 594-7

[6] Chatterjee S, Wang J W, Kuo W S, Tai N H, Salzmann C, Li W L, Hollertz R, Nüesch F A and Chu B T T 2012 Mechanical reinforcement and thermal conductivity in expanded graphene nanoplatelets reinforced epoxy composites *Chem. Phy. Letters* **531** 6-10

[7] Xin Z, Qinghua Z, Dajun C and Ping L 2010 Enhanced mechanical properties of graphene-based poly(vinyl alcohol) composites *Macromolecules* **43** 2357-63

[8] Garima M, Vivek D, Kyong Y R, Soo-Jin P, Wi R L 2015 A review on carbon nanotubes and graphene as fillers in reinforced polymer nanocomposites *J. of Industrial and Engineering Chem.* **21** 11-25

[9] Lenka K, Annamária D, Pavol H, Ján D, Péter K and Csaba B 2012 Fracture toughness and toughening mechanisms in graphene platelet reinforced Si_3N_4 composites *Scripta Materialia* **66** 793-6

[10] Weikang L, Anthony D and Jinbo B 2013 Carbon nanotube–graphene nanoplatelet hybrids as high-performance multifunctional reinforcements in epoxy composites *Composites Science and Tech.* **74** 221-7

[11] Tian X, Desen Z, Zheling L, Robert J Y, Cristina V and Ian A K 2018 Electrically conductive GNP/epoxy composites for out-of-autoclave thermoset curing through Joule heating *Composites Science and Techn.* **164** 304-12

[12] Jingwen C, Robert J Y, Thomas J A S, Timothy L B, Broderick C, Ludovic C, Aurèle V and Zheling L 2018 Realizing the theoretical stiffness of graphene in composites through confinement between carbon fibers *Composites Part A* **113** 311-7

[13] Robert J Y, Mufeng L, Ian A K, Suhao L, Xin Z, Cristina V and Dimitrios G P 2018 The mechanics of reinforcement of polymers by graphene nanoplatelets *Composites Science and Technology* **154** 110-6

[14] Zheling Li, Thomas J A S, Xinyu M, Yingjie Y, Robert J Y, and Timothy L B 2019 The taxonomy of graphite nanoplatelets and the influence of nanocomposite processing *Carbon* **142** 99-106

[15] Gorti V K, Golla S, Bhavani S B, Kantubhukta J, Kokkiligadda J, Ramesh S R and Surendranathan A O 2018 Fabrication and characterization of graphene reinforced tungsten

carbide-cobalt composite *The Journal of Advances in Mechanical and Materials Engineering* **1** 12-22
[16] Watertown Arsenal 1945 Report No. 710/412

IOP Conf. Series: Materials Science and Engineering **634** (2019) 012002 doi:10.1088/1757-899X/634/1/012002

Flexible hydrogen sensor based on Pd/TiO_2 nanofilm with fast response

S H Wu[1,2], Z Y Li[1], S Mao[1], H Y Wang[1] and X R Wang[1,2]

[1]School of Optoelectronic Science and Engineering, University of Electronic Science and Technology of China, Chengdu 610054, China

E-mail: shwu@uestc.edu.cn/xiangruwang@uestc.edu.cn

Abstract. Flexible hydrogen sensors based on palladium (Pd) and titanium dioxide (TiO_2) nanofilm on poly(dimethylsiloxane) (PDMS) substrate have been demonstrated. We have prepared PDMS using anodic aluminum oxide template and researched the performance of Pd/TiO_2 nanofilm based hydrogen sensors with different thickness of TiO_2. The sensor with 6nm thickness of TiO_2 has demonstrated the best performance with a response/recovery time of 1.6 s/6 s at 0.2% hydrogen concentration, respectively.

1. Introduction

Clean energy has become a global hot topic due to the increasing environmental pollution, non-renewability of fossil fuels and greenhouse effect. Hydrogen (H_2) is good clean energy for its renewable and abundant energy with zero pollutant emission [1]. However, its inflammability above a concentration of 4% restricts the extensive utilization of H_2 energy, which motivates a lot of researches on the high performance H_2 sensors. In recent years, researchers have focused on the new types of hydrogen sensors such as semiconductors, thermoelectric, optics and so on [2]. Especially, metal palladium (Pd) plays an important role in hydrogen sensors due to its special ability of absorbing hydrogen gas at a normal temperature and pressure. Our group have been working on high efficient Pd based hydrogen sensors [3-7]. We tried to use anodic aluminum oxide (AAO) as a substrate to prepare Pd film with porous nano-network structure, which shows good hydrogen detection performance partly due to the increased specific surface area [5,6]. Meanwhile, flexible H_2 gas sensors have attracted general attention because of their light-weight, good mechanical property, various applications and so on [8-12]. Lim *et al* reported a flexible H_2 sensor using Pd nanotubes with large surface area and well-interconnected structure exhibiting extremely high sensitivity [8]. Krško *et al* reported TiO_2 thin film based flexible H_2 sensor, which demonstrated a response (R_0/R_{H2}) of about 10^4 for 1% H_2 concentration and a capability of sensing H_2 concentration of 30 ppm [9].

Poly(dimethylsiloxane) (PDMS) was chosen as an elastomeric substrate for H_2 sensors because of its compatibility with Pd, chemical inertness to H_2, and mechanical flexibility. In this work, we have fabricated flexible H_2 sensors based on Pd/TiO_2 nanofilm using PDMS substrate and researched the relationship between the thickness of TiO_2 and the performance of hydrogen sensors at room temperature. The sensor with 6nm thick TiO_2 achieves a fast response/recovery time of 1.6 s/6 s at 0.2% hydrogen concentration, respectively.

2. Material and methods

AAO template was fabricated using aluminum plate by a two-step anodic oxidation method [5].

Flexible PDMS substrates were prepared by spin coating PDMS monomer mixed with curing agent at a ratio of 10:1 at 300 rpm on AAO. PDMS film on the AAO surface was peeled off after the sample was placed in a baking oven at 75°C for 2 hours. The thickness of PDMS flexible film is about 1.04 mm. TiO_2 was prepared on PDMS via spray pyrolysis using a precursor solution of titanium diisopropoxide bis(acetylacetonate) in ethanol with a volume ratio of 1: 39 on heating stage with a temperature of 300°C. The thickness of TiO_2 was determined by the volume of precursor solution. Then TiO_2 films were in situ annealed on heating stage at 300°C for 60 min. We deposited Pd film with a thickness of 30 nm by DC magnetron sputtering at a deposition rate of 1.5 Å/s on the surface of TiO_2. Finally a couple of electrodes of 150 nm Ag was prepared on Pd film through E-beam evaporation. Figure 1 shows the schematic fabrication procedure of Pd/TiO_2 nanofilm hydrogen sensor. We chose the thickness of TiO_2 as 6, 10 and 12 nm, respectively.

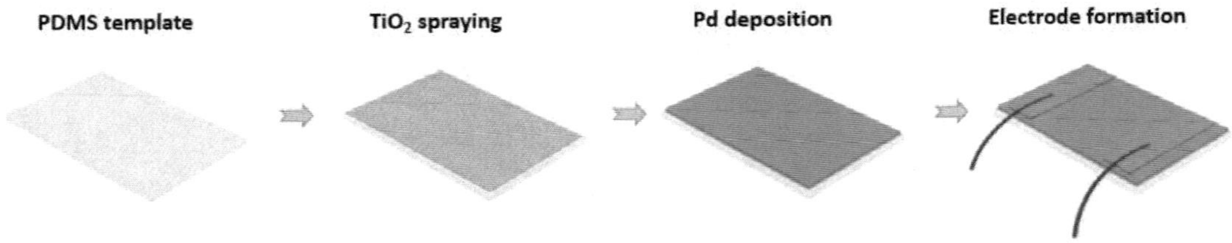

Figure 1. Schematic fabrication procedure of Pd/TiO_2 nanofilm hydrogen sensor.

3. Result and discussion

The samples were exposed to a wide range H_2 concentrations at room temperature from 0.2% to 1.8% in dry synthetic air for investigating sensing properties. Figure 2 shows the real-time resistance change of the samples at various H_2 concentrations. From figure 2, we can see the resistances of three samples decreased under H_2 atmosphere and reversed quickly under synthetic air atmosphere, which demonstrates that Pd/TiO_2 nanofilm in our work would be discontinuous and work on the "break junction" effect. Meanwhile, the baseline resistances of samples decreased with increasing the thickness of TiO_2 observed in figure 2. The decrease of resistance would be due to the enhanced thickness of Pd/TiO_2 nanofilm and the improved electrical conductivity.

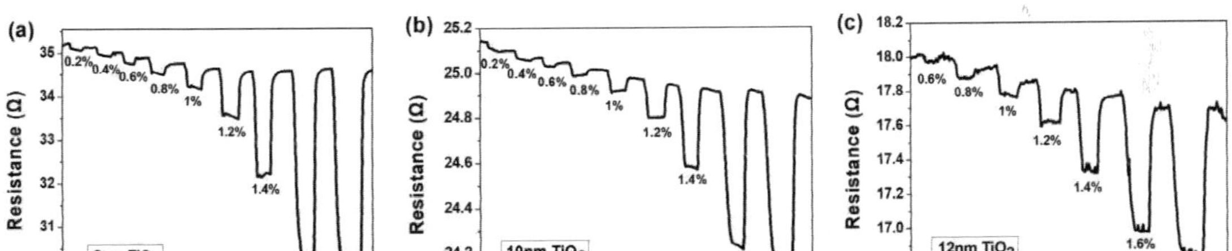

Figure 2. Sensing curve of three samples with different thickness of TiO_2 measured under different hydrogen concentrations. (a) 6nm TiO_2, (b) 10nm TiO_2, and (c) 12nm TiO_2.

In order to systematically evaluate the sensing performance, the steady-state response (R) of the sensors was defined as $R=R_0-R_{H2}/R_0$, where R_0 and R_{H2} represents the sensor resistance in pure air and hydrogen gas, respectively. Figure 3 shows the response of hydrogen sensor samples at different hydrogen concentration. From figure 3, the relationship between the response and the square root of H_2 partial pressure is not linear for three samples. As we all know, for most resistive Pd-based H_2 sensors, the relationship between the resistance change and the square root of H_2 partial pressure is approximately linear [13]. Therefore we can further conclude that the Pd/TiO_2 nanofilm on PDMS film works as the "break junctions". Meanwhile, the numerical value of sensitivity demonstrates the dominating role of Pd in the sensing response of Pd/TiO_2 nanofilm and the sensor with 6 nm TiO_2 has the optimal sensitivity compared with other samples.

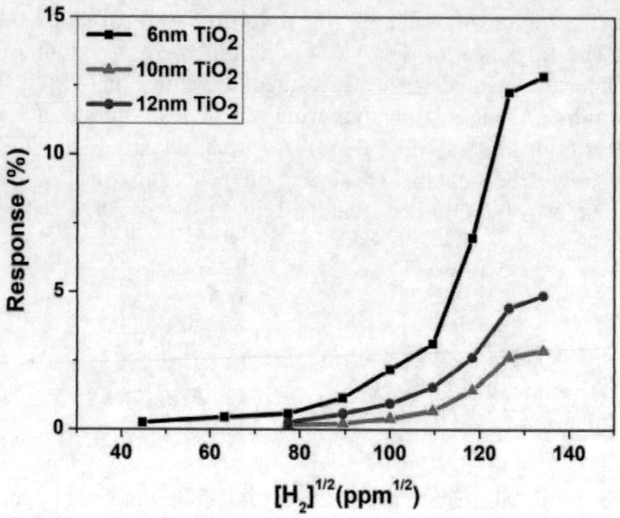

Figure 3. The steady-state response of hydrogen sensor samples at different hydrogen concentration.

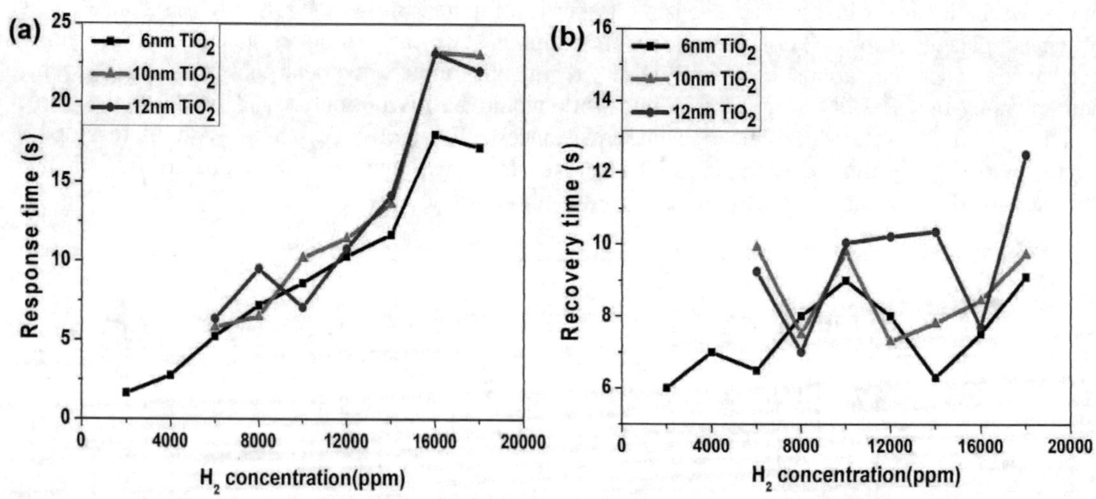

Figure 4. Response and recovery times of three samples with different thickness of TiO_2. (a) Response time, (b) Recovery time.

The response and recovery times for three samples were shown in figure 4. At room temperature, the sample with 6 nm TiO_2 has stable response and recovery at both low and high H_2 concentration. However, for the samples with 10 nm and 12 nm TiO_2, we could not get the response/recovery times at H_2 concentration of 0.2% and 0.4% for the unsteady bringing noise. The response times increased nearly in the linear range of H_2 detection at concentrations from 0.2% to 1.8%. Complete recovery was observed for three samples and recovery times were less than 13s at H_2 concentration range. Besides, we can see the hydrogen sensor with 6 nm TiO_2 shows fastest response in three samples. Especially, the response time and recovery time are respectively 1.6 s/6 s at H_2 concentration of 0.2% for the sample with 6 nm TiO_2, which are faster than those of the flexible hydrogen sensors based Pd reported before [11,14].

In order to investigate the enhanced performance of H_2 sensors with TiO_2 layer, the schematic H_2 sensing mechanism for Pd/TiO_2 nanofilm was described in figure 5. With the PDMS prepared by AAO as the substrate, Pd/TiO_2 nanofilm has large specific surface area and "break junctions" of nano-sized Pd particles were formed in nanofilm. Isolated nano-sized Pd particles adsorb hydrogen and form Pd hydride, which leads to a rapid volume expansion of Pd. When some particles touch in the process of expansion, new conducting pathways are formed and the electrical resistance of film is decreased. For TiO_2 particles, dissociatively formed H atoms in Pd will react with the adsorbed oxygen and inject electrons into the TiO_2 layer, which results in the resistance decrease further. With the exist of TiO_2, the hydrogen dissolution in Pd/TiO_2 nanofilm will be strengthened with the large concentration difference of H atoms at the surface and interior of Pd/TiO_2 nanofilm, shown in figure 5. With the increase of TiO_2 thickness, it would decrease the specific surface area of Pd/TiO_2 nanofilm and result in low hydrogen adsorption and solubility [15].

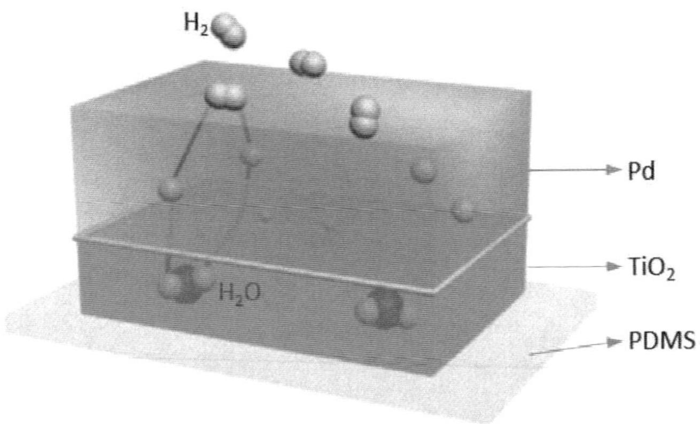

Figure 5. Schematic H_2 sensing mechanism for Pd/TiO_2 nanofilm.

4. Conclusion

In conclusion, flexible hydrogen sensors based on Pd/TiO_2 nanofilm with fast response were fabricated on PDMS by controllable thickness of TiO_2. We used AAO template to regulate the surface morphology of PDMS. We researched the sensor performance and the decreased resistance under hydrogen gas has demonstrated that Pd/TiO_2 nanofilm sensors work on "break junction" effect. The sensor with 6 nm thickness of TiO_2 achieves a fast response/recovery time of 1.6 s/6 s at 0.2% hydrogen concentration, respectively. The fast response of the sensor is mainly due to large specific surface area of Pd/TiO_2 nanofilm and enhanced dissolution of hydrogen into Pd with the help of TiO_2.

Acknowledgments

We acknowledge funding from the National Natural Science Foundation of China (Grant Nos. 61775026, 61505015, 61504020, 61405026), the Fundamental Research Funds for the Central Universities (Grant Nos. ZYGX2018J048, ZYGX2014J049), the National Key R & D Program (Grant No. 2014M552330) and China Postdoctoral Science Foundation (Grant No. 2014M552330).

References

[1] Momirlan M and Veziroglu T N 2005 *Int. J. Hydrogen Energy* **30** 795-802
[2] Hübert T, Boon-Brett L, Black G and Banach U 2011 *Sens. Actuators B: Chem.* **157** 329-52
[3] Ding D, Chen Z and Lu C 2006 *Sens. Actuators B: Chem.* **120** 182-6
[4] Ding D and Chen Z 2007 *Adv. Mater.* **19** 1996-9
[5] Hao M, Wu S, Zhou H, Ye W, Wei X and Wang X 2015 *J. Mater. Sci.* **51** 2420-6
[6] Wu S, Zhou H, Hao H, Wei X, Li S and Yu H 2016 *Appl. Surf. Sci.* **380** 47-51

[7] Mao S, Zhou H, Wu S, Yang J, Li Z, Wei X, Wang X, Wang Z and Li J 2018 *Int. J. Hydrogen Energy* **43** 22727-32
[8] Lim M, Kim D H, Park C, Lee Y, Han S, Li Z, Williams R and Park I 2012 *ACS Nano* **6** 598-608
[9] Krško O, Plecenik T, Roch T, Grančič B, Satrapinskyy Li, Truchly M, Durina P, Gregor M, Kús P and Plecenik A 2017 *Sens. Actuators B: Chem.* **240** 1058-65
[10] Chung M, Kim D, Seo D, Kim T, Im H and Lee H 2012 *Sens. Actuators B: Chem.* **169** 387-92
[11] Kim W, Jang B, Lee H and Lee W 2016 *Sens. Actuators B: Chem.* **224** 547-51
[12] Sun Y and Wang H 2007 *Appl. Phys. Lett.* **90** 213107
[13] Kirchheim R, Sommer F and Schluckebier G 1982 *Acta Metall.* **30** 1059-68
[14] Hu Y, Lei J, Wang Z, Yang S, Luo X, Zhang G, Chen W and Gu H 2016 *Int. J. Hydrogen Energy* **41** 10986-90
[15] Rumiche F, Wang H, Hu W, Indacochea J and Wang M 2008 *Sens. Actuators B: Chem.* **134** 869-77

IOP Conf. Series: Materials Science and Engineering **634** (2019) 012003 doi:10.1088/1757-899X/634/1/012003

Nano insulation materials exploiting the Knudsen effect

B P Jelle[1,2,5], S A Mofid[1], T Gao[1], M Grandcolas[3], M Sletnes[2] and E Sagvolden[4]

[1]Norwegian University of Science and Technology (NTNU), Department of Civil and Environmental Engineering, NO-7491 Trondheim, Norway
[2]SINTEF Building and Infrastructure, Department of Materials and Structures, NO-7465 Trondheim, Norway
[3]SINTEF Industry, Department of Materials and Nanotechnology, NO-0314 Oslo, Norway
[4]SINTEF Industry, Department of Sustainable Energy Technology, NO-0314 Oslo, Norway

[5]E-mail: bjorn.petter.jelle@ntnu.no

Abstract. As the world's focus is turned even stronger toward miscellaneous energy efficiency and saving aspects, the development of new high-performance thermal insulation materials for building applications will play an important role in this regard. The aim of the presented study is to develop an understanding for the governing thermal transport mechanisms and utilize the Knudsen effect in nanoporous insulation materials through theoretical concepts and experimental laboratory explorations, thus being able to synthesize nano insulation materials (NIM) with very low thermal conductivity values as a major goal. NIMs based on hollow silica nanospheres (HSNS) have been synthesized by a sacrificial template method, where the idea is that the heat transport by gas conductance and gas/solid state interactions decreases with decreasing pore diameters in the nano range as predicted by the Knudsen effect. HSNS with reduced thermal conductivity compared to their solid counterparts have been prepared where the hollow sphere cavities and voids between the spheres are filled with air at atmospheric pressure, i.e. eliminating the need for various measures like e.g. protective metallized foils to maintain a vacuum or expensive low-conducting gases in the cavities and voids. Hence, HSNS represent a promising stepping-stone toward the future high-performance thermal insulation materials.

1. Introduction

In the years ahead there will probably be a continued increased attention and focus on energy efficiency and saving aspects alongside utilization of renewable and non-polluting energy sources. Miscellaneous energy sources and harvesting opportunities are being explored like e.g. solar cells and building integrated photovoltaics (BIPV) [1-10]. However, the energy you are not using you do not have to produce. In this respect, the development of new high-performance thermal insulation materials for building applications will be important [11-13]. A crucial property of these materials is the thermal conductivity, where it is an expressed goal to reach as low thermal conductivity as possible.

Traditional thermal insulation materials have thermal conductivity values typically in the range 20 to 50 mW/(mK) with cellulose and cork being in the higher end having values between 40 to 50 mW/(mK), mineral wool like glass wool and rock wool being in the middle range with values between 30 to 40 mW/(mK), polystyrene products like expanded polystyrene (EPS) and extruded polystyrene (XPS) also having values in the middle range between 30 to 40 mW/(mK), and polyurethane (PUR) being in the lower end having values between 20 to 30 mW/(mK) [12]. In this respect, one should note that the thermal conductivity may increase to considerably higher values if the moisture content is increased in most of these materials. It should also be noted that even if PUR is safe in its intended use, it represents a potential serious health concern and hazard in case of a fire as PUR will when burning release poisonous gases of hydrogen cyanide (HCN) and isocyanates, where the HCN toxicity arises from the cyanide anion (CN^-) which prevents cellular respiration. These traditional thermal insulation materials may be cut and adjusted at the building site without any loss of thermal resistance, however, due to their relatively high thermal conductivities, walls, floors and roofs may become rather thick in order to achieve the desired thermal resistance of today's and the coming energy-efficient and zero emission buildings of the future.

Hence, there is a quest to make thermal insulation materials with substantially reduced thermal conductivity values.

State-of-the-art thermal insulation materials like vacuum insulation panels (VIP) [14-21], gas-filled panels (GFP) [22,23] and aerogels [24-29] have the potential of reaching substantially lower thermal conductivity values than the traditional thermal insulation materials. In addition, phase change materials (PCM) [30-32] may also become part of the thermal building envelope by absorbing and releasing energy when needed.

VIPs consist of an open-porous core surrounded by a laminate foil, where the laminate foil should have as low air and water vapour diffusion as possible in order to maintain a close-to-zero gas pressure, i.e. vacuum, inside the VIP core. Typically, VIPs with fumed silica cores have pristine (non-aged) thermal conductivity values between 3 to 4 mW/(mK), which with time will increase due to diffusion of air and moisture through the laminate foil and into the core (design values are often reported as 7 to 8 mW/(mK) taking into account diffusion/ageing effects and often also heat bridge effects caused by the enveloping laminate foil), and when perforated the thermal conductivity increases to about 20 mW/(mK). GFPs are in principle similar to VIPs, but with thermally low-conducting gases in their cores, where the pores in the GFPs are macroscopic pores and thus much larger than the VIP core pores which are in the micro and nano range. Although the potential of GFPs may seem high, they still have not come into relatively widespread use as e.g. VIPs, where currently VIPs may seem the better choice of these two. Note that both VIPs and GFPs can not be cut or adapted at the building site as that would cause a loss of vacuum and low-conducting gases, which none the less will also take place by diffusion in both VIPs and GFPs. Commercial aerogels as thermal insulation materials have thermal conductivity values typically in the range 12 to 20 mW/(mK), i.e. considerably higher values than for VIPs in their pristine condition and even than for typical VIP design values. Aerogels have an air content of 95 to 99 vol% and are thus mostly consisting of air by volume fraction, hence the aerogels are also very brittle. In addition, the aerogel products of today are very expensive. An advantage of aerogels is that they may be produced either as opaque, translucent or transparent materials, and they may also be cut and adjusted at the building site without any loss of thermal resistance.

Thus, there is a quest to make new thermal insulation materials as robust and flexible as the traditional thermal insulation materials and with as low thermal conductivity values as the state-of-the-art thermal insulation materials, more specifically as low as the VIPs in their pristine condition but with no increase of thermal conductivity during time.

The quest for making better thermal insulation materials is often concentrating the efforts around producing porous materials either filled with air, low-conducting gases (e.g. argon, krypton and xenon) or near-vacuum. A very interesting aspect in this regard is what happens when the pore diameters of the porous material are lowered into the nano range, i.e. that the thermal conductivity is substantially lowered due to the Knudsen effect, which basically is occurring when the pore diameter is reduced to a value smaller than the mean free path of the molecules in the pores [11-13]. For further details and elaborations around the Knudsen effect including the governing equations it is referred to the available literature within this research field.

In such a case and of immense significance is that very low values of the thermal conductivity can be reached even with air-filled pores at atmospheric pressure. Hence, by exploiting the Knudsen effect, there is a large potential for being able to manufacture high-performance thermal insulation materials in form of nano insulation materials (NIM) [11-13] which will be robust towards any perforations and may be cut as desired at the building site without loss of performance due to the air-filled nanopores. The term super insulation materials (SIM) has also come into common usage within this field, with thermal conductivity values below 26 mW/(mK) (stagnant air value), 20 mW/(mK) or 10 mW/(mK) as examples of common borders for SIMs.

The objective of the work presented herein is to investigate the pathway of creating nano insulation materials by making hollow silica nanospheres (HSNS) by a sacrificial template method. In this method, templates of polystyrene (PS) are first fabricated, thereafter the PS templates are coated with a silica layer, and finally the PS templates are removed by burning them away by a diffusion process through the silica layer.

2. Experimental

Hollow silica nanospheres (HSNS) have been made by a sacrificial template method. Polystyrene (PS) template spheres were first fabricated and then coated with a silica layer. The silica precursor was either tetraethyl orthosilicate (TEOS) or water glass (Na_2SiO_3). The samples were placed in a heat incubator and the PS template spheres were burnt away by a diffusion through the enveloping silica layer. The final result was then the HSNS. Basically, the HSNS synthesis follows the schematic procedure as illustrated in figure 1. Further details and elaborations concerning the experimental conditions may be found in our earlier studies and a forthcoming one [13,33-38].

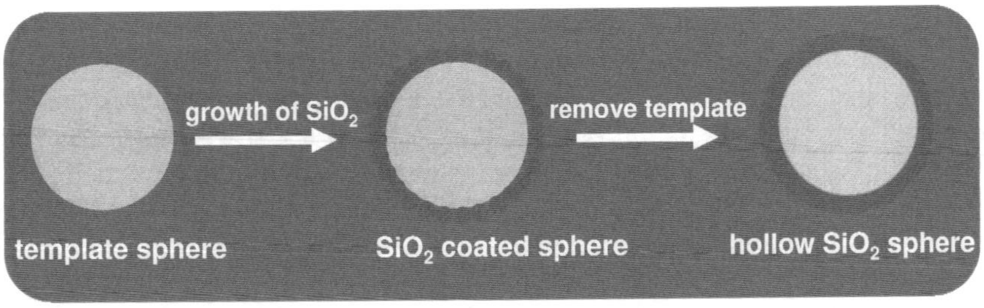

Figure 1. Illustration of the synthesis process when fabricating hollow silica nanospheres (HSNS) by the sacrificial template method.

The visual appearance of HSNS at nano level with their dimensional attributes were characterized by scanning electron microscopes (SEM) and transmission electron microscopes (TEM). Furthermore, the thermal conductivity λ of the HSNS was determined by calculation of the product $\lambda=\alpha\rho c_p$ where the diffusivity α was measured by a laser flash apparatus, the mass density ρ was calculated as $\rho=m/V$ from mass m and volume V measurements, and the specific heat capacity c_p was measured by differential scanning calometry (DSC). The details of these measurements and the corresponding HSNS syntheses will be published in a forthcoming and more extensive article [33]. In the following, a few excerpts and highlights will be given.

3. Results and discussion

The overall objective of this work is to gain insight into thermal transport mechanisms in nanoporous materials, thus enabling the opportunity of tailor-making materials with as low thermal conductivity as possible. That is, to fabricate nano insulation materials with the lowest possible thermal conductivity by utilizing the Knudsen effect.

The results presented herein are based on various syntheses of hollow silica nanospheres (HSNS) by the sacrificial template method [34-38], where polystyrene (PS) spheres have acted as sacrificial templates for these investigations. In general, the preparation of PS spheres with small dimensions and inner pore diameters well below 100 nm, which is needed in order to get the full effect of the Knudsen effect, has proven to be rather challenging in our experimental laboratory studies. Nevertheless, the further below 100 nm in inner sphere diameter one may reach, the lower thermal conductivity values are expected, although other parameters like e.g. sphere shell thickness, sphere packing density and mesoporosity, may also influence this. However, the actually obtained and henceforth measured values may sometimes be quite different from the anticipated ones.

Examples of synthesized PS spheres are shown in the SEM images given in figure 2, depicting several spheres together and measurement of the diameter of a single sphere. In this particular case, these PS template spheres have diameters of 85 nm (figure 2).

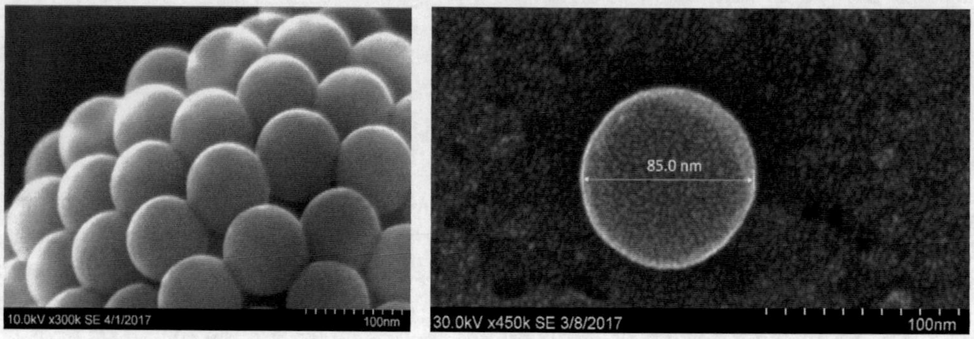

Figure 2. SEM images of PS template spheres with diameters of 85 nm.

After having prepared the PS template spheres, these templates were coated with a silica layer often consisting of small silica nanoparticles which give a relatively rough surface, whereas in some other cases the silica layer may look like a large, wrinkled silica sheet (not shown here), the wrinkled silica sheet surface being substantially smoother than the rough silica particle surface. These surface differences may greatly influence the interfacial interaction between the spheres when packed together, and hence also influence the thermal transport properties and the overall thermal conductivity.

Then, after the silica coating process, the PS template spheres were removed by heating the powder sample so that the PS could escape by diffusion through the silica shell, thereby leaving hollow silica nanospheres (HSNS) as the final result.

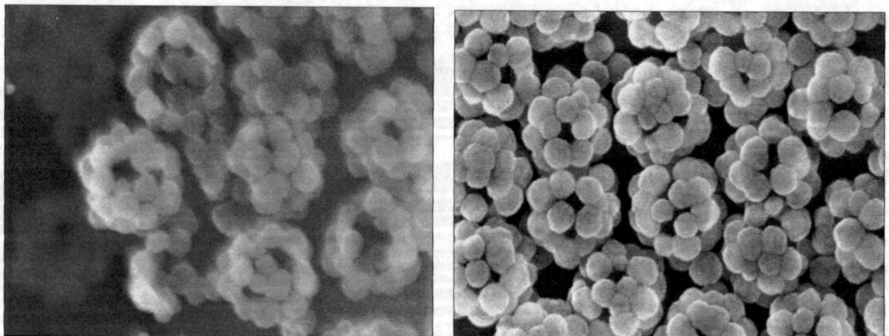

Figure 3. SEM images of HSNS with inner diameter 85 nm and two different shell thicknesses of 34 nm (left) and 67 nm (right).

Examples of the resulting HSNS are depicted in figures 3 and 4, the latter one also depicting the measurements of the silica nanoparticle diameters (shell thicknesses), where SEM images are given of HSNS with inner diameter 85 nm and two different shell thicknesses of 34 nm and 67 nm. As can be observed from the SEM images, the HSNS surfaces of these specific samples are rather rough and porous due to the relatively large silica particles constituting the HSNS shell.

Figure 4. SEM images of HSNS with inner diameter 85 nm and two different shell thicknesses of 34 nm (left) and 67 nm (right), also depicting the measurements of the silica particle diameters.

In the study reported herein several different synthesis parameters have been varied in order to study their influence on the HSNS appearance (e.g. morphology) and their thermal conductivity. A detailed study of these aspects will be published in a more comprehensive and in-depth article [33]. As a main result and as anticipated, the thermal conductivity of HSNS was found to decrease with decreasing PS template sphere diameter and thus the resulting HSNS inner diameter. Here, the inner diameters of the synthesized HSNS were ranging between 85 to 213 nm. Correspondingly, the thermal conductivities of these HSNS have been measured (calculated based on measured values) between 14.3 to 29.7 mW/(mK). That is, the lowest measured thermal conductivity was found for the HSNS with the smallest inner diameters.

Thus, although a rather challenging task, one should attempt to make PS spheres as small as possible, i.e. with diameters well below 100 nm, e.g. down to 40 nm or even below. Success with the PS template preparation will probably lead to success with synthesizing HSNS with much lower thermal conductivities. The goal of manufacturing super insulation materials (SIM) would in such a case have been achieved, except from the commercial aspects in these matters. However, there are a lot of issues to be addressed and investigated much more in-depth in the field of making SIMs. In addition to specific synthesis aspects, general viewpoint issues like robustness assessment [39], durability and accelerated climate ageing [40], as well as life cycle analysis [41], should also be considered.

Nevertheless, to have reached as low value as 14.3 mW/(mK) in thermal conductivity, without having utilized very small inner sphere diameters and hence without having fully exploited the Knudsen effect (i.e. far below 100 nm), is promising with respect to forthcoming and future optimization attempts in the quest for SIMs.

4. Conclusions

Nano insulation materials (NIM) have been made as hollow silica nanospheres (HSNS) by utilizing the Knudsen effect with pore sizes, i.e. inner diameters of the HSNS, in the nano range.

The HSNS have been manufactured by employing a sacrificial template method, where the polystyrene (PS) templates have been removed from the silica spheres by heating and thus a subsequent diffusion process through the silica shell, hence leaving hollow silica spheres with inner diameters in the nano range as the result from the synthesis.

In the specific syntheses and experiments reported within this study, HSNS with inner diameters ranging between 85 to 213 nm have been made, with corresponding thermal conductivity values calculated to be between 14.3 to 29.7 mW/(mK), respectively, based on measured values of diffusivity, specific heat capacity, mass and volume, the two latter ones giving the mass density.

Acknowledgments
This work is supported by the Research Council of Norway within the Nano2021 program through the SINTEF and NTNU research project "High-Performance Nano Insulation Materials" (Hi-Per NIM).

References

[1] Norton B, Eames P C, Mallick T K, Huang M J, McCormack S J, Mondol J D and Yohanis Y G 2011 Enhancing the performance of building integrated photovoltaics *Sol. Energy* **85** 1629-64
[2] Jelle B P, Breivik C and Røkenes H D 2012 Building integrated photovoltaic products: A state-of-the-art review and future research opportunities *Sol. Energy Mater. Sol. Cells* **100** 69-96
[3] Jelle B P 2013 The challenge of removing snow downfall on photovoltaic solar cell roofs in order to maximize solar energy efficiency - Research opportunities for the future *Energy Build.* **67** 334-51
[4] Breivik C, Jelle B P, Time B, Holmberget Ø, Nygård J, Bergheim E and Dalehaug A 2013 Large-scale experimental wind-driven rain exposure investigations of building integrated photovoltaics *Sol. Energy* **90** 179-87
[5] Fasana S and Nelva R 2013 Improvementt of the water resistancy in the integration of photovoltaic panels on traditional roofs *Constr. Build. Mater.* **48** 1081-91
[6] Jelle B P 2016 Building integrated photovoltaics: A concise description of the current state of the art and possible research pathways *Energies* **9** 1-30
[7] Jelle B P, Gao T, Mofid S A, Kolås T, Stenstad P M and Ng S 2016 Avoiding snow and ice formation on exterior solar cell surfaces - A review of research pathways and opportunities *Procedia Eng.* **145** 699-706
[8] Tripathy M, Sadhu P K and Panda S K 2016 A critical review on building integrated products and their applications *Renewable Sustainable Energy Rev.* **61** 451-65
[9] Andersson P-O, Jelle B P and Zhang Z 2017 Passive snow repulsion: A state-of-the-art review illuminating research gaps and possibilities *Energy Procedia* **132** 423-8
[10] Andenæs E, Jelle B P, Ramlo K, Kolås T, Selj J and Foss S E 2018 The influence of snow and ice coverage on the energy generation from photovoltaic solar cells *Sol. Energy* **159** 318-28
[11] Jelle B P, Gustavsen A and Baetens R 2010 The path to the high performance thermal building insulation materials and solutions of tomorrow *J. Build. Phys.* **34** 99-123
[12] Jelle B P 2011 Traditional, state-of-the-art and future thermal building insulation materials and solutions - Properties, requirements and possibilities *Energy Build.* **43** 2549-63
[13] Jelle B P, Gao T, Sandberg L I C, Tilset B G, Grandcolas M and Gustavsen A 2014 Thermal superinsulation for building applications - From concepts to experimental investigations *International Journal of Structural Analysis and Design* **1** 43-50
[14] Baetens R, Jelle B P, Thue J V, Tenpierik M J, Grynning S, Uvsløkk S and Gustavsen A 2010 Vacuum insulation panels for building applications: A review and beyond *Energy Build.* **42** 147-72
[15] Wegger E, Jelle B P, Sveipe E, Grynning S, Gustavsen A, Baetens R and Thue J V 2011 Aging effects on thermal properties and service life of vacuum insulation panels *J. Build. Phys.* **35** 128-67
[16] Sveipe E, Jelle B P, Wegger E, Uvsløkk S, Grynning S, Thue J V, Time B and Gustavsen A 2011 Improving thermal insulation of timber frame walls by retrofitting with vacuum insulation panels - Experimental and theoretical investigations *J. Build. Phys.* **35** 168-88
[17] Grynning S, Jelle B P, Uvsløkk S, Gustavsen A, Baetens R, Caps R and Meløysund V 2011 Hot box investigations and theoretical assessments of miscellaneous vacuum insulation panel configurations in building envelopes *J. Build. Phys.* **34** 297-324
[18] Alam M, Singh H and Limbachiya M C 2011 Vacuum insulation panels (VIPs) for building construction industry - A review of the contemporary developments and future directions *Appl. Energy* **88** 3592-602
[19] Haavi T, Jelle B P and Gustavsen A 2012 Vacuum insulation panels in wood frame wall constructions with different stud profiles *J. Build. Phys.* **36** 212-26
[20] Bouquerel M, Duforestel T, Baillis D and Rusaouen G 2012 Heat transfer modeling in vacuum insulation panels containing nanoporous silicas - A review *Energy Build.* **54** 320-36
[21] Kalnæs S E and Jelle B P 2014 Vacuum insulation panel products: A state-of-the-art review and

future research pathways *Appl. Energy* **116** 355-75

[22] Mills G L and Zeller C M 2008 The performance of gas filled multilayer insulation *Advances of Cryogenic Engineering: Transactions of the Cryogenic Engineering Conference* **53** 1475-82

[23] Baetens R, Jelle B P, Gustavsen A and Grynning S 2010 Gas-filled panels for building applications: A state-of-the-art review *Energy Build.* **42** 1969-75

[24] Baetens R, Jelle B P and Gustavsen A 2011 Aerogel insulation for building applications: A state-of-the-art review *Energy Build.* **43** 761-9

[25] Gao T, Jelle B P, Ihara T and Gustavsen A 2014 Insulating glazing units with silica aerogel granules: The impact of particle size *Appl. Energy* **128** 27-34

[26] Gao T, Jelle B P, Gustavsen A and He J 2014 Lightweight and thermally insulating aerogel glass materials *Applied Physics A: Materials Science & Processing* **117** 799-808

[27] Buratti C, Moretti E, Belloni E and Agosti F 2014 Development of innovative aerogel based plasters: Preliminary thermal and acoustic performance evaluation *Sustainability* **6** 5839-52

[28] Jelle B P, Baetens R and Gustavsen A 2015 Aerogel insulation for building applications *The Sol-Gel Handbook* Vol 3 D Levy and M Zayat eds (Weinheim, Germany: Wiley-VCH) pp 1385-412

[29] Sletnes M, Jelle B P and Risholt B 2017 Feasibility study of novel integrated aerogel solutions *Energy Procedia* **132** 327-32

[30] Demirbas M F 2006 Thermal energy storage and phase change materials: An overview *Energy Sources Part B* **1** 85-95

[31] Baetens R, Jelle B P and Gustavsen A 2010 Phase change materials for building applications: A state-of-the-art review *Energy Build.* **42** 1361-8

[32] Kalnæs S E and Jelle B P 2015 Phase change materials and products for building applications: A state-of-the-art review and future research opportunities *Energy Build.* **94** 150-76

[33] Mofid S A, Jelle B P, Zhao X, Gao T, Grandcolas M, Ng S and Yang R 2019 Utilization of size-tunable hollow silica nanospheres for building thermal insulation applications (To be published)

[34] Gao T, Jelle B P, Sandberg L I C and Gustavsen A 2013 Monodisperse hollow silica nanospheres for nano insulation materials: Synthesis, characterization, and life cycle assessment *ACS Appl. Mater. Interfaces* **5** 761-7

[35] Gao T, Jelle B P, Sandberg L I C and Gustavsen A 2015 Thermal conductivity of monodisperse silica nanospheres *Journal of Porous Media* **18** 941-7

[36] Gangåssæter H F, Jelle B P and Mofid S A 2017 Synthesis of silica-based nano insulation materials for potential application in low-energy or zero emission buildings *Energy Procedia* **122** 949-54

[37] Ng S, Jelle B P, Sandberg L I, Gao T and Mofid S A 2018 Hollow silica nanospheres as thermal insulation materials for construction: Impact of their morphologies as a function of synthesis pathways and starting materials *Constr. Build. Mater.* **166** 72-80

[38] Grandcolas M, Jasinski E, Gao T and Jelle B P 2019 Preparation of low density organosilica monoliths containing hollow silica nanospheres as thermal insulation materials *Mater. Lett.* **250** 151-4

[39] Jelle B P, Sveipe E, Wegger E, Gustavsen A, Grynning S, Thue J V, Time B and Lisø K R 2014 Robustness classification of materials, assemblies and buildings *J. Build. Phys.* **37** 213-45

[40] Jelle B P 2012 Accelerated climate ageing of building materials, components and structures in the laboratory *J. Mater. Sci.* **47** 6475-96

[41] Schlanbusch R D, Jelle B P, Sandberg L I C, Fufa S M and Gao T 2014 Integration of life cycle assessment in the design of hollow silica nanospheres for thermal insulation applications *Build. Environ.* **80** 115-24

IOP Conf. Series: Materials Science and Engineering **634** (2019) 012004 doi:10.1088/1757-899X/634/1/012004

Metastable states of graphene Dirac fermion in potential barrier

S Kuwata[1,3] and D Takehisa[2]

[1]Graduate School of Information Sciences, Hiroshima City University, Asaminami-ku, Hiroshima 731-3194, Japan
[2]Department of System Engineering, Hiroshima City University, Asaminami-ku, Hiroshima 731-3194, Japan

E-mail: kuwata@hiroshima-cu.ac.jp

Abstract. Weak quantization of the Dirac equation under a potential barrier (with height V_0) indicates that while the bound state of Dirac fermion (with energy $E_0 \in \mathbb{R}$ and mass m) is allowed for $|E_0| < m$, the Dirac fermion is metastable for $|E_0| > m$. Under the condition $V_0 \gg m$, which is well satisfied for typical graphene, it is found that the tunneling time Δt and the lifetime τ for the metastable state are related as $\Delta t \gtrsim 2\tau$. As m decreases, it follows that $\Delta t/\tau$ tends to increase and behaves as $\log(1/m)$ up to higher order logarithmic correction.

1. Introduction

Graphene, due to its novel electric and magnetic properties [1], has a significant application in nanotechnology [2]. Ideal graphene can be represented by the 2D massless Dirac equation. As contrasted with the Schrödinger equation, the Dirac equation has a somewhat mysterious property, called the Klein paradox [3] or the Klein tunneling [4] (if the width of a potential barrier is finite), where electrons can propagate the potential barrier without damping even if the potential height is large enough, compared with the incident energy. The Klein paradox can be resolved by interpreting antiparticles as particles with negative energy propagating backward in time [5], or more strictly, by second quantization [6-9].

However, from a practical point of view, it may be convenient to understand the Klein paradox without resorting to second quantization. As such a method, weak quantization [10,11] has been known, where the energy eigenvalues for the Dirac equation are treated as complex numbers. As in the case of confining potential [10,11], electrons in a potential barrier turn out to be metastable, with the density of states given by the exact Breit-Wigner spectrum.

The aim of this article is to show that, based on weak quantization, if the potential height is large enough compared with the effective mass for the corresponding 1D Dirac fermion, then the tunneling time is larger than twice the lifetime for the metastable state. This implies that $P \leq e^{-2}$ (≈ 0.135), where P represents the probability that electrons in the metastable state remain their state during the tunneling time. In Sec. 2, we deal with the 1D Dirac equation under a potential barrier, where the Weyl function is introduced to calculate the density of states. In Sec. 3, we calculate the density of states, from which the lifetime and tunneling time can be derived. Sec. 4 is devoted to conclusion.

2. Methods

We begin with the 1D Dirac equation under a potential barrier. In units of $\hbar = c = 1$, we have

$$\psi'(x) - [E - V(x)\, \mathrm{i}\sigma_y + m\sigma_x]\psi(x) = 0, \tag{1}$$

where σ_x and σ_y represent the Pauli matrices. For non-normal incidence of 2D massless Dirac fermion on the potential barrier, the problem can be reduced to that of the 1D massive Dirac fermion, with the effective mass m being proportional to the transverse momentum [12]. The potential barrier $V(x)$ (with height V_0 and width $2a$) is given by

$$V(x) = \begin{cases} V_0 & (|x| < a), \\ 0 & (|x| > a), \end{cases}$$

with $a, V_0 > 0$. Considering that ψ depends on the energy eigenvalue E, we can write ψ as ψ_E.

To obtain the density of states, it may be convenient to introduce two Weyl functions $W_+, W_-: \mathbb{C} \longrightarrow \mathbb{C}$ such that [10]

$$\operatorname{Im} W_\pm(E) = (\operatorname{Im} E) \int_{\Omega_\pm} |\psi_E(x)|^2\, \mathrm{d}x,$$

where $W_+ = [0, \infty)$, $W_- = (-\infty, 0]$, and E is chosen as a complex number (note that in Ref. [10], E was defined as $-E$). Then the density of states, ρ, of energy $E_0 \in \mathbb{R}$ can be given by

$$\rho(E_0) = \lim_{\nu \to 0_+} [W_+(E_0 + \mathrm{i}\nu) + W_-(E_0 + \mathrm{i}\nu)]. \tag{2}$$

It is fortunate that, due to the parity conservation for $\psi(x)$ by $V(-x) = V(x)$, W_+ and W_- are simply related as [10,11]

$$W_+(E)\, W_-(E) = -1. \tag{3}$$

It should be noticed that while W_+ and W_- are analytic for $\operatorname{Im} E > 0$ by their definition, they have singular points for $\operatorname{Im} E \le 0$. Suppose that all the singular points are contributed from an (infinite) number of poles of order one. Then $\rho(E_0)$ can be regarded as a Breit-Wigner spectrum, with the central energy $\langle E \rangle$ and the half width Γ given by the pole of $W_\pm(E)$ at $E = \langle E \rangle - \mathrm{i}\Gamma/2$ (see Appendix A).

3. Results

Following the procedure for calculating $W_+ = W_+(E)$, we obtain (see Appendix B)

$$W_+ = \frac{\alpha + \mathrm{i}\beta}{\gamma + \mathrm{i}\delta}, \qquad \begin{pmatrix} \alpha & \beta \\ \gamma & \delta \end{pmatrix} = \begin{pmatrix} -\lambda \cos\theta & \mu \sin\theta \\ \frac{\lambda}{\mu} \sin\theta & \cos\theta \end{pmatrix}, \tag{4}$$

where λ, μ and θ are given by

$$\lambda = \frac{\omega_{\mathrm{ex}}}{E + m}, \qquad \mu = \frac{\omega_{\mathrm{in}}}{E - V_0 + m}, \qquad \theta = a\omega_{\mathrm{in}},$$

with $\omega_{\mathrm{ex}} = \sqrt{E + m}\sqrt{E - m}$ and $\omega_{\mathrm{in}} = \sqrt{(V_0 - E) + m}\sqrt{(V_0 - E) - m}$. We address the problem for the cases of $|E_0| < m$ and $|E_0| > m$ separately.

3.1. $|E_0| < m$

In this case, $W_\pm(E_0)$ has one or more singular points (poles of order one). This implies that there is a possibility that $\rho(E_0)$ has a spectrum of delta function type through the well-known formula $\lim_{\nu \to 0_+} \frac{1}{x + \mathrm{i}\nu} = \mathrm{P}\frac{1}{x} - \mathrm{i}\pi\delta(x)$ (where P represents the principal value), indicating the existence of a bound state. The existence of the bound state, in itself, can be understood by the charge conjugate symmetry of equation (1), mentioned in Remark 2 below. Recalling that the charge conjugate state ψ_c has energy $(-E)$ under the potential $(-V)$, we find that the bound state can be realized for $-E_0 - m > -V_0$ (the subtraction by m comes from the elimination of the rest mass). Thus the bound state can be allowed for E_0 such that

$$-m < E_0 < \min(m, V_0 - m) \qquad (V_0 > 0).$$

Considering further that the peak positions of $\rho(E_0)$ are (almost) uniformly distributed (not accumulated), as in the case for $|E_0| > m$ below, we find that as m decreases, the number of the bound states tends to decrease. Hence in the limit of $m \to 0$, the number of the bound state turns out to be, at most, unity. In this sense, we can neglect the bound state for $m \ll 1$.

- **Remark 1** A more detailed analysis indicates that even for sufficiently small but non-vanishing m, there always exists a single bound state for $|E_0| < m$.

3.2. $|E_0| > m$

In this case, $W_\pm(E_0)$ has no singular point, so that $\lim_{\nu \to 0_+}$ and Im in equation (2) can be exchanged with each other. As a consequence, $\rho(E_0)$ can be calculated as

$$\begin{aligned}\rho(E_0) &= \left[-\begin{vmatrix}\alpha & \beta \\ \gamma & \delta\end{vmatrix}\left(\frac{1}{\gamma^2+\delta^2}+\frac{1}{\alpha^2+\beta^2}\right)\right]_{E\to E_0} \\ &= \left[\frac{\omega_{\rm ex}(E - V_0 + m)}{\omega_{\rm ex}^2 - V_0(E - m\cos 2a\omega_{\rm in})} + (m \to -m)\right]_{E\to E_0}\end{aligned} \qquad (5)$$

from which it is found that $\rho(E_0)$ is invariant under $m \to -m$, and also under $(E_0, V_0) \to (-E_0, -V_0)$.

- **Remark 2** The invariances of $\rho(E_0)$ under $m \to -m$ and $(E_0, V_0) \to (-E_0, -V_0)$ come from the mass inversion symmetry and charge conjugate symmetry of equation (1), respectively; $\psi_{\rm m} := \sigma_y \psi$ and $\psi_{\rm c} := \sigma_x \psi^*$ satisfy equation (1) with $m \to -m$ and $(E_0, V_0) \to (-E_0, -V_0)$, respectively.

Furthermore, ρ has the following limit:

$$\lim_{E_0 \searrow m} \rho(E_0) = 0 \qquad \left[m(1 \pm \cos 2a\omega_{\rm in}^{(0)}) \neq 0\right], \qquad (6)$$

where $\omega_{\rm in}^{(0)} = \omega_{\rm in}|_{E\to m}$. As an example of $\rho(E_0)$, see figure 1, in which the spectrum is well represented by the Breit-Wigner distribution. Notice that $\rho(E_0)$ satisfies equation (6) and that the peak positions are (almost) uniformly distributed, with the energy difference between neighboring peaks being around $\frac{\pi}{2}\frac{1}{a} \approx 1.5708 \times 10^{-2}$ (see the second column in table 1).

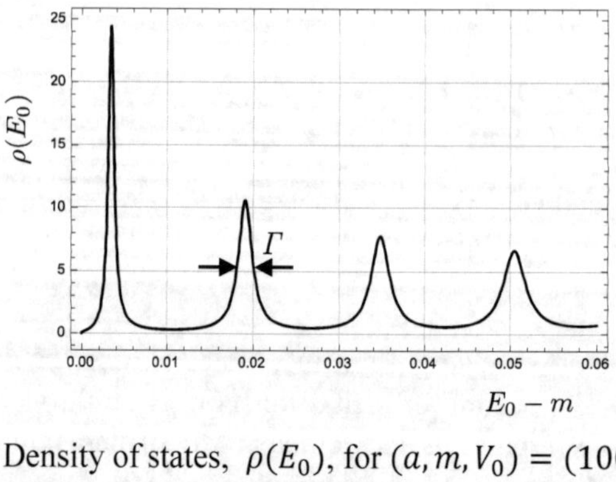

Figure 1. Density of states, $\rho(E_0)$, for $(a, m, V_0) = (100, 1, 100)$.

Recalling that the central energy $\langle E \rangle$ and the half width Γ are given by the pole (of order one) of $W_+(E)$ and $W_-(E)$ at $E = \langle E \rangle - \mathrm{i}\Gamma/2$, we obtain, together with equation (3), the following relations:

$$\begin{cases} +\dfrac{E-m}{\omega_{\text{ex}}}\sin\theta + \mathrm{i}\dfrac{E-V_0-m}{\omega_{\text{in}}}\cos\theta = 0, \\ -\dfrac{E-m}{\omega_{\text{ex}}}\cos\theta + \mathrm{i}\dfrac{E-V_0-m}{\omega_{\text{in}}}\sin\theta = 0, \end{cases} \quad (E = \langle E\rangle - \mathrm{i}\Gamma/2), \tag{7}$$

corresponding to the zero point of $\gamma + \mathrm{i}\delta$ and that of $\alpha + \mathrm{i}\beta$, respectively. For the values of $\langle E\rangle$ and Γ in figure 1, see table 1, where use has been made of equation (7) numerically.

Table 1. Values of $\langle E\rangle - m$ and Γ for the n-th metastable state for the same values of (a, m, V_0) as in figure 1. For later convenience, $\kappa = 2a\Gamma/v_{\text{g}}$, where $v_{\text{g}} = \frac{\sqrt{\langle E\rangle^2 - m^2}}{\langle E\rangle}$.

n	$\langle E\rangle - m$	Γ	κ
1	3.34021×10^{-3}	8.10331×10^{-4}	1.98782
2	1.90475×10^{-2}	1.92952×10^{-3}	2.00531
3	3.47548×10^{-2}	2.60249×10^{-3}	2.02532
4	5.04621×10^{-2}	3.13153×10^{-3}	2.04531

Before obtaining a general solution to equation (7), we consider a special solution such that $\Gamma = 0$. In this case, it follows that

$$(\Gamma = 0) \Rightarrow \begin{cases} \langle E\rangle - m = 0 = \cos a\omega_{\text{in}}^{(0)}, \\ \langle E\rangle - m = 0 = \sin a\omega_{\text{in}}^{(0)}, \end{cases} \tag{8}$$

satisfying $1 + \cos 2a\omega_{\text{in}}^{(0)} = 0$ and $1 - \cos 2a\omega_{\text{in}}^{(0)} = 0$, respectively. Recalling the condition in the bracket in equation (6), we find that $\lim_{E_0 \searrow m} \rho(E_0)$ is not vanishing in this case.

- **Remark 3** The relation of equation (8) is known as a super-critical condition [13], in which the Klein tunneling can be realized.

However, the relation of $\langle E\rangle = m$ implies that the incident momentum on the potential barrier is vanishing, so that such a state is difficult to observe. In this sense, we neglect the case of $\langle E\rangle = m$.

It seems that equation (7) is still somewhat more difficult than to solve in an analytical way. To simplify it, consider the condition such that

$$V_0 \gg m, \tag{9}$$

which holds for typical graphene, where $V_0 \sim 1$ eV. In this case, m in $(V_0 - E \pm m)$ may be neglected, so that ω_{in} can be approximated as

$$\omega_{\text{in}} = \omega_{\text{in}}^{(\text{r})} + \mathrm{i}\omega_{\text{in}}^{(\text{i})} \simeq (V_0 - \langle E\rangle + \mathrm{i}\Gamma/2) \qquad \left(\omega_{\text{in}}^{(\text{r})}, \omega_{\text{in}}^{(\text{i})} \in \mathbb{R}\right).$$

As a consequence, equation (7) is reduced to

$$\begin{cases} +(E-m)\sin\theta - \mathrm{i}\omega_{\text{ex}}\cos\theta \simeq 0, \\ -(E-m)\cos\theta - \mathrm{i}\omega_{\text{ex}}\sin\theta \simeq 0, \end{cases} \quad (E = \langle E\rangle - \mathrm{i}\Gamma/2). \tag{10}$$

Denoting the real and imaginary parts of ω_{ex} as $\omega_{\text{ex}} = e + \mathrm{i}f \ (e, f \in \mathbb{R})$, for the time being, we can rewrite equation (10) using a matrix form as

$$\begin{cases} \left(\sin\omega_{\text{in}}^{(\text{r})} \quad \cos\omega_{\text{in}}^{(\text{r})}\right)M_1 = (0 \quad 0), \\ \left(\cos\omega_{\text{in}}^{(\text{r})} \quad \sin\omega_{\text{in}}^{(\text{r})}\right)M_2 = (0 \quad 0), \end{cases}$$

where M_1 and M_2 are related as

$$\begin{cases} \frac{M_1+M_2}{2} = \begin{pmatrix} E_{\rm NR}\cosh a\hat{\Gamma} - e\sinh a\hat{\Gamma} & -\hat{\Gamma}\cosh a\hat{\Gamma} - f\sinh a\hat{\Gamma} \\ 0 & 0 \end{pmatrix}, \\ \frac{M_1-M_2}{2} = \begin{pmatrix} 0 & 0 \\ \hat{\Gamma}\sinh a\hat{\Gamma} + f\cosh a\hat{\Gamma} & +E_{\rm NR}\sinh a\hat{\Gamma} - e\cosh a\hat{\Gamma} \end{pmatrix}, \end{cases}$$

with $E_{\rm NR} := \langle E\rangle - m$ and $\hat{\Gamma} := \Gamma/2$, so that their determinants amount to

$$\begin{cases} \det M_1 = (f\hat{\Gamma} - eE_{\rm NR})\cosh 2a\hat{\Gamma} + \frac{1}{2}(e^2 + f^2 + E_{\rm NR}^2 + \hat{\Gamma}^2)\sinh 2a\hat{\Gamma}, \\ \det M_2 = -\det M_1. \end{cases}$$

Hence from $\det M_1 = 0$ or $\det M_2 = 0$ by $(\sin\omega_{\rm in}^{(\rm r)} \quad \cos\omega_{\rm in}^{(\rm r)}) \neq (0 \quad 0)$, we obtain the following relations:

$$\begin{aligned} \tanh 2a\hat{\Gamma} &= \frac{2(eE_{\rm NR} - f\hat{\Gamma})}{e^2+f^2+E_{\rm NR}^2+\hat{\Gamma}^2} \\ &= \frac{\sqrt{\Phi_m(\hat{\Gamma}, E_{\rm NR})}}{\sqrt{2}(E_{\rm NR}+m)}, \end{aligned} \tag{11}$$

where $\Phi_m(\hat{\Gamma}, E_{\rm NR}) = -\hat{\Gamma}^2 + 2mE_{\rm NR} + E_{\rm NR}^2 + \sqrt{(\hat{\Gamma}^2 + E_{\rm NR}^2)[\hat{\Gamma}^2 + (2m + E_{\rm NR})^2]}$. In the second line in equation (11), we have replaced e and f using $\hat{\Gamma}$ and $E_{\rm NR}$ through the relation of $e + if = \sqrt{E-m}\sqrt{E+m}$, with $E = (E_{\rm NR} + m) - i\hat{\Gamma}$. It may be convenient to derive the necessary condition for equation (11) as

$$(E_{\rm NR} + m)^2X^4 + (\hat{\Gamma}^2 - 2mE_{\rm NR} - E_{\rm NR}^2)X^2 - \hat{\Gamma}^2 = 0 \quad (X := \tanh 2a\hat{\Gamma}), \tag{12}$$

which has been obtained from squaring both sides of equation (11), shifting the terms $(-\hat{\Gamma}^2 + 2mE_{\rm NR} + E_{\rm NR}^2)$, and finally squaring both sides of the resultant relation.

In the rest of this section, we deal with the three following regimes: $2a\hat{\Gamma} \ll 1$, $2a\hat{\Gamma} \gg 1$, and $2a\hat{\Gamma} = \mathcal{O}(1)$, to claim that $2a\hat{\Gamma} \gtrsim v_{\rm g}$, where $v_{\rm g}$ represents the group velocity, mentioned in table 1. Notice that the relation $2a\hat{\Gamma} \gtrsim v_{\rm g}$ can be rewritten as $\Delta t \gtrsim 2\tau$, where $\Delta t\ (= 2a/v_{\rm g})$ and $\tau\ (= \Gamma^{-1})$ represent the tunneling time and the lifetime of the metastable state, respectively.

3.2.1. Case for $2a\hat{\Gamma} \ll 1$. In this limit, X can be expanded as

$$\tanh 2a\hat{\Gamma} = 2a\hat{\Gamma} + O\left((2a\hat{\Gamma})^3\right) \qquad (2a\hat{\Gamma} \to 0). \tag{13}$$

Neglecting the higher-order terms, we obtain from equations (12) and (13)

$$2a\hat{\Gamma} = \frac{1+(2a)^2(E_{\rm NR}^2+2mE_{\rm NR})}{1+(2a)^2(E_{\rm NR}+m)^2} \qquad (2a\hat{\Gamma} \ll 1). \tag{14}$$

Here we have eliminated the case of $2a\hat{\Gamma} = 0$, because $2a\hat{\Gamma} = 0$ leads to $E_{\rm NR} = 0$ by equation (11), super-critical solution. Then the ratio $\kappa := \frac{\Delta t}{\tau}$ can be rewritten as a function of $\langle E\rangle$ as

$$\begin{aligned} \kappa &= \frac{2a\Gamma}{v_{\rm g}} \\ &= 2\sqrt{g(\langle E\rangle)}, \quad g(\langle E\rangle) := \left(\frac{1+(2a)^2(\langle E\rangle^2 - m^2)}{1+(2a)^2\langle E\rangle^2}\right)\left(\frac{\langle E\rangle^2}{\langle E\rangle^2 - m^2}\right). \end{aligned}$$

Considering that $g(\langle E\rangle)$ is a monotonically decreasing function for $\langle E\rangle > m$, and that $g(\langle E\rangle) \to 1$ for $\langle E\rangle \to \infty$, we eventually find that

$$\kappa > 2.$$

- **Remark 4** By equation (14), the relation $2a\hat{\Gamma} \ll 1$ is realized for $m \gg a^{-1}, E_{NR}$. Thus, the limit of $m \to 0$ is incompatible with $2a\hat{\Gamma} \ll 1$. Actually, in the limit of $m \to 0$, the right-hand side of equation (14) amounts to unity, so that the expansion equation (13) is not a good approximation.

3.2.2. Case for $2a\hat{\Gamma} \gg 1$. In this limit, X^2 can be parameterized as

$$X^2 = 1 - \varepsilon \qquad (0 < \varepsilon \ll 1). \tag{15}$$

Substituting equation (15) into equation (12) and neglecting $O(\varepsilon^2)$, we obtain

$$\varepsilon = \frac{m^2}{\hat{\Gamma}^2 + E_{\mathrm{NR}}^2}.$$

Hence, the lower bound for κ can be written as

$$\begin{aligned}\kappa &\geq 2a\Gamma \\ &= 2a\tanh^{-1}\sqrt{1-\varepsilon} \\ &= 4\log(4/\varepsilon) + O(\varepsilon) \quad (0 < \varepsilon \ll 1)\end{aligned}$$

where the first inequality comes from $v_g \leq 1$. Hence we obtain $\kappa \gg 1$ for $2a\hat{\Gamma} \gg 1$.

3.2.3. Case for $2a\hat{\Gamma} = O(1)$. By $v_g \leq 1$, it follows from $\kappa \geq 2a\Gamma$ that

$$\kappa \gtrsim 2.$$

In either case, κ satisfies $\kappa \gtrsim 2$. Considering that $\tanh 2a\hat{\Gamma}$ is a monotonically increasing function of $2a\hat{\Gamma}$, we claim from equation (11) that $\kappa \gtrsim 2$ should hold for all values of $2a\hat{\Gamma}$.

4. Conclusion

We have calculated, based on weak quantization, the lifetime τ for the metastable 1D Dirac fermion under a potential barrier (with its height V_0), to reveal that τ and the tunneling time Δt are related as $\Delta t \gtrsim 2\tau$ in the limit of $V_0 \gg m$. The relation of $\Delta t \gtrsim 2\tau$ indicates that $P := \mathrm{e}^{-\Delta t/\tau} \lesssim \mathrm{e}^{-2}$ (≈ 0.135), where P represents the probability that the electrons in the metastable state remain their state during the tunneling time. As m decreases, P tends to be smaller and behaves as $P \sim (m/\log m)^2$ by $\mathrm{e}^{-\kappa} \sim \mathrm{e}^{-\log(4/\varepsilon)} \sim \varepsilon \sim \left(m/\hat{\Gamma}\right)^2 \sim (m/\log m)^2$. Thus in the regime of $m \to 0$, which corresponds to the normal incidence in graphene, it can be stated that almost all metastable electrons are decayed during the tunneling time. To obtain the decay mode, however, field theoretical consideration is necessary.

Acknowledgments
The authors are indebted to H Fujisaka for invaluable discussions.

Appendices

- **A** Breit-Wigner spectrum

Consider the density of states of a Breit-Wigner type as $\hat{\rho}(E_0 + \mathrm{i}\nu) := \left|\hat{\psi}(E_0 + \mathrm{i}\nu)\right|^2$, where $\hat{\psi}(E_0 + \mathrm{i}\nu) = \sum_n \frac{c_n}{E_0 - \langle E_n\rangle + \mathrm{i}(\nu + \Gamma_n/2)}$, with $\nu > 0, \Gamma_n \geq 0$, and $E_0, \langle E_n\rangle \in \mathbb{R}$; $c_n \in \mathbb{C}$. In this case, $\hat{\rho}$ can be rewritten as

$$\hat{\rho}(E_0 + \mathrm{i}\nu) = \mathrm{Im}\left[\widehat{W}(E_0 + \mathrm{i}\nu)\right],$$

where $\widehat{W}(E_0 + \mathrm{i}\nu) := \sum_n \gamma_n \frac{1}{E_0 - \langle E_n\rangle + \mathrm{i}(\nu + \Gamma_n/2)}$, with $\gamma_n = 2\mathrm{i}c_n \sum_m c_m^* \frac{1}{E_n - E_m - \mathrm{i}[2\nu + (\Gamma_n + \Gamma_m)/2]}$. Notice that $\widehat{W}(E)$ (for $E \in \mathbb{C}$) is analytic for $\mathrm{Im}\, E > 0$, and that it has a pole (of order one) at $E = \langle E_n\rangle - \mathrm{i}\Gamma_n/2$.

- **B** Calculation of $W_+(E)$

Let $\psi^{(i)}(x)$'s ($i = 1, 2$) be two solutions to equation (1) for $x \geq 0$ such that $\psi^{(1)}(0) = \binom{1}{0}$ and

IOP Conf. Series: Materials Science and Engineering **634** (2019) 012004 doi:10.1088/1757-899X/634/1/012004

$\psi^{(2)}(0) = \binom{0}{1}$. The function $W_+(E)$ is given by the boundary condition such that

$$\Psi(x) := \psi^{(1)}(x) + W_+(E)\,\psi^{(2)}(x) \qquad (x \geq 0)$$

satisfies $\lim_{x\to\infty} \Psi(x) = 0$ (for $\mathrm{Im}\, E > 0$). Taking account of the boundary condition of $\psi^{(i)}(x)$ $(i = 1, 2)$ at $x = a$, we obtain equation (4).

References

[1] Katsnelson M I 2012 *Graphene: Carbon in Two Dimensions* (Cambridge: Cambridge University Press)
[2] Castro Neto A H, Guinea F, Peres N M R, Novoselov K S and Geim A K 2009 *Rev. Mod. Phys.* **81** 109-62
[3] Klein O 1929 *Z. Phys.* **53** 157-65
[4] Calogeracos A and Dombey N 1999 *Int. J. Mod. Phys. A* **14** 631-43
[5] Holstein B R 1998 *Am. J. Phys.* **66** 507-12
[6] Hund F 1941 *Z. Phys.* **117** 1-17
[7] Schwinger J 1951 *Phys. Rev.* **82** 664-79
[8] Krekora P, Su Q and Grobe R 2004 *Phys. Rev. Lett.* **92** 040406
[9] Gavrilov S P and Gitman D M 2016 *Phys. Rev. D* **93** 045002
[10] Titchmarsh E C 1961 *Proc. London Math. Soc.* **S3-11** 159-68
[11] Giachetti R 2008 *Phys. Rev. Lett.* **101** 190401
[12] Stander N, Huard B and Goldhaber-Gordon D 2009 *Phys. Rev. Lett.* **102** 026807
[13] Dombey N, Kennedy P and Calogeracos A 2000 *Phys. Rev. Lett.* **85** 1787-90

IOP Conf. Series: Materials Science and Engineering **634** (2019) 012005 doi:10.1088/1757-899X/634/1/012005

Electrical conductivity of polyazomethine/reduced graphene oxide nanocomposites

S V Kostromin[1], A Podshivalov[2], M Asandulesa[3] and S Bronnikov[1,4]

[1]Russian Academy of Science, Institute of Macromolecular Compounds, Bolshoi Prospekt 31, 199004 St. Petersburg, Russian Federation
[2]ITMO University, Kronverkskiy Prospekt 49, 197101 Sankt Petersburg, Russian Federation
[3]Romanian Academy, "Petru Poni" Institute of Macromolecular Chemistry, Aleea Grigore Ghica Voda 41 A, 700487, Iaşi, Romania

E-mail: sergei_bronnikov@yahoo.com

Abstract. We produced nanocomposites comprised of polyazomethine and reduced graphene oxide by solution mixing. The nanofiller concentration ranged from 0.25 to 1.5 wt.%. Both neat polymer and nanocomposites samples were investigated using broadband dielectric spectrometry under the same conditions. The electrical conductivity σ of the samples was examined as a complex value being a combination of two parameters: *dc* conductivity (σ_{dc}) and *ac* conductivity (σ_{ac}). The temperature dependences of σ_{ac} were found to follow the Arrhenius equation, whereas the frequency dependences of σ_{ac} were shown to obey a power law. We proposed that the correlated barrier hopping could explain behaviour of σ_{ac} of the nanocomposites under study.

1. Introduction

Fabrication of polymer-based nanocomposites became an actual challenge of polymer science, which is determined by the engineering requirements for high performance materials. It is known that the presence of a small amount of nanofiller (a few percents or even less) in a polymer matrix can substantially improve performance characteristics of polymer-based materials [1,2].

Because of unique properties of carbon nanoparticles, they have been used extensively for the fabrication of polymer-based nanocomposites [3-5]. Graphene is known to exhibit superior physical and chemical properties. As a consequence, graphene and its derivatives seem to be effective carbon nanoparticles for creation of novel functional materials including stretchable electronic devices [6] as well as valuable nanofiller for polymer matrices [3,5,7].

The generally accepted method of graphene fabrication is graphene oxide (GO) reduction. Since GO initially contains significant amount of oxygen (in functional groups) on its surface and water residues within graphene layers, its intrinsic conductivity is quite low [5]. GO reduction decreases the amount of oxygen and removes water residues, thus enhancing its intrinsic conductivity. However, even reduced GO (rGO) often still contains oxygen-containing groups of GO, which could be favourable for chemical interaction with polymer matrix. Many conventional polymers have been used as matrices for production of nanocomposites comprising graphene or its derivatives. These nanocomposites demonstrate enhanced functional properties [5,7].

As a matrix, we chose polyazometine. Polyazomethines are polymers exhibiting high mechanical strength, thermal stability, film- or fiber-forming properties, and superior optical properties that specify their possible applications particularly in electronic and optical devices [8-10]. Polyazomethines are inherently insulators. Nevertheless, being filled with conductive nanoparticles, they exhibit enhanced conductivity [8].

When studying the polyazomethine/fullerene C_{60} nanocomposites using a broadband dielectric spectroscopy (BDS), we found out an appreciable enhancing of the their electrical conductivity above C_{60} percolation threshold (2 wt.%) [11,12]. The current study aims to fabrication of electroconductive polyazomethine/rGO nanocomposites varying in rGO loading and investigation of their electrical conductivity with BDS.

2. Materials and methods

2.1. Samples preparation

The chemical structure of the polymer which we used as a matrix for fabrication of electroconductive nanocomposites is presented in figure 1. It is polyazomethine (PAZ) with azobenzene side chains. Comprehensive information about its synthesis and principal characteristics can be found in [10,13]. GO was produced from graphite powder using a Hummers method [5]. Subsequently water dispersion of GO was used as a raw material for the rGO production.

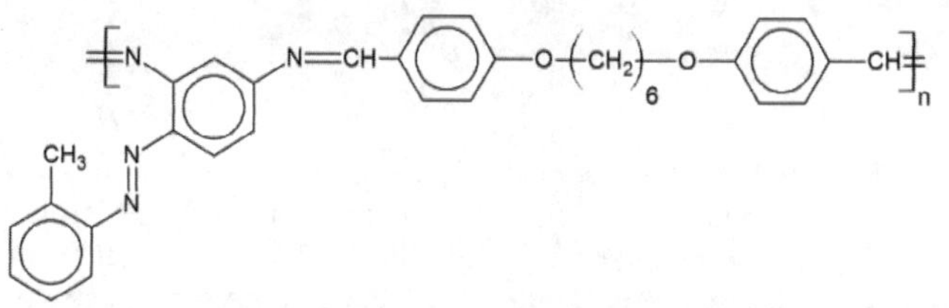

Figure 1. Chemical structure of PAZ.

According to [14], we obtained rGO through thermal annealing of dried GO at high temperature. To prepare nanocomposites, we used the solution mixing method. For every sample we ~~first~~ prepared a PAZ solution in dimethylforamide (DMF) and mixed it with specified volume of rGO (similarly dispersed in DMF) to provide necessary rGO loading in nanocomposite. For effective elimination of solvent, we subjected the mixture to distillation in vacuum. Moreover, a continuous vacuum drying of the nanocomposites samples at 60°C was applied. Following this way, a set of PAZ/rGO nanocomposites samples varied in rGO concentration (0.25, 0.5 and 1.5 wt%) were produced for further research.

2.2. Broadband dielectric spectroscopy

Complex dielectric permittivity ε^* of the nanocomposites was measured using a Novocontrol Alpha dielectric spectrometer (Novocontrol Technologies, Germany) in wide ranges of frequency ($10^0 - 10^6$ Hz) and temperature (-100 – 170ºC). The temperature control accuracy was provided by using a Quatro cryosystem (Novocontrol Technologies, Germany). The samples under study were placed in the measuring cell where they were subjected to alternate voltage with amplitude 1 V in an inert (nitrogen) atmosphere. We collected and analyzed the results performed at sequential cooling of the measuring cell with a sample from 170ºC to -150ºC. For BDS measurements we used the nanocomposite samples in the form of thin layers with a thickness of 50 μm.

The dielectric loss ε'' and electrical conductivity σ are known to be interrelated [15]:

$$\sigma(f,T) = 2\pi\varepsilon_0 f\varepsilon''(f,T) \tag{1}$$

where ε_0 is the permittivity of the free space and ε'' is the imaginary part of the complex dielectric permittivity $\varepsilon^* = \varepsilon' - i\varepsilon''$.

3. Results and discussion

3.1. General conductivity behavior

Figure 2 demonstrates the electrical conductivity of the PAZ/rGO nanocomposites at different rGO concentration (φ) depending on both temperature and frequency. At a low rGO concentration (0 – 0.5 wt.%), σ slightly increases with increasing φ. Nevertheless, higher rGO concentrations (φ = 1.5 wt.%) contribute to a more noticeable increase in σ. The nanocomposite with φ = 1.5 wt.% exhibits the most significant σ growth. Anyway, electrical conductivity of the nanocomposites tends to increase with increasing both temperature and frequency.

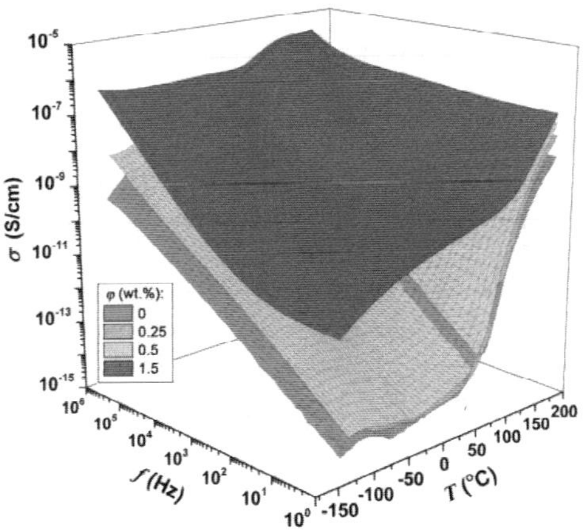

Figure 2. 3D representation of the electrical conductivity of the PAZ/rGO nanocomposites with different rGO concentration depending on both temperature and frequency.

According to [15], we presented the measured electrical conductivity σ as an additive parameter comprising *ac* (alternate current) conductivity (σ_{ac}) and *dc* (direct current) conductivity (σ_{dc}):

$$\sigma(f,T) = \sigma_{ac}(f,T) + \sigma_{dc}(T) \tag{2}$$

We should emphasize that σ_{ac} depends on both frequency f and temperature T, while σ_{dc} depends on temperature solely. Further the frequency and temperature dependences of conductivity and its components of the PAZ/rGO nanocomposites at different rGO concentration will be considered.

3.2. Electrical conductivity depending on frequency

Figure 3 depicts variation of the PAZ/rGO nanocomposites conductivity depending on frequency. It is clear that two components of electrical conductivity could clearly be recognized in accordance with equation (2): σ_{dc} and σ_{ac}.

Obviously we can estimate the *dc* conductivity for investigated nanocomposites using figure 3 only at rather high temperatures (above 100°C) except nanocomposite with φ = 1.5 wt.% which exhibits *dc* conductivity in the whole temperature range. Consequently figure 4 shows variation of σ_{dc} with temperature in the range from 100°C to 170°C. Hereby, *dc* conductivity of all the nanocomposites tends to increase with increasing temperature. Apparently, the presence of rGO in small quantities (0.25 – 0.5 wt.%) increases σ_{dc} slightly because these amounts of rGO nanoparticles are not enough to form an extensive conductive network through the whole sample [16]. As the content of rGO becomes higher, σ_{dc} increases substantially as far as a percolation network of rGO nanoparticles is formed.

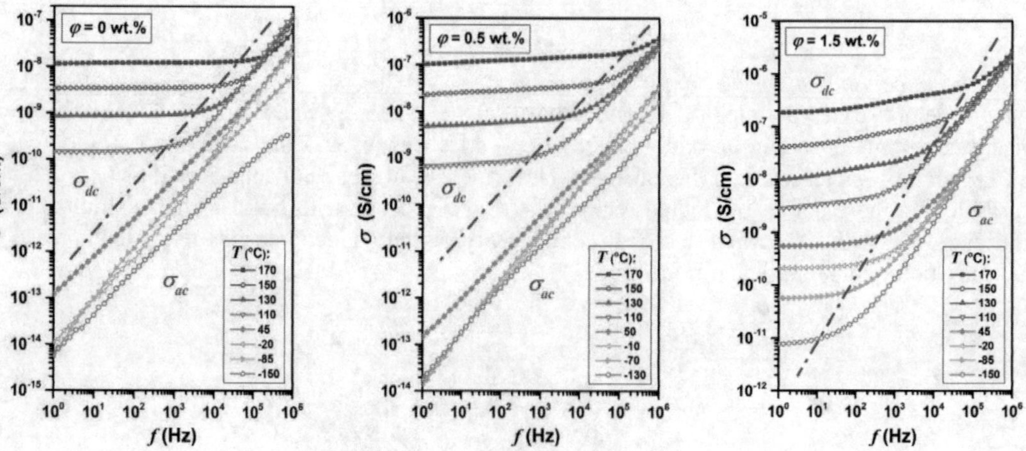

(a) (b) (c)

Figure 3. Conductivity of exemplary PAZ/rGO nanocomposites with different rGO content ((a) $\phi = 0$, (b) $\phi = 0.5$ wt.%, and (c) $\phi = 1.5$ wt.%) vs. frequency. Dash-dot lines depict the conditional boundaries between σ_{dc} and σ_{ac}.

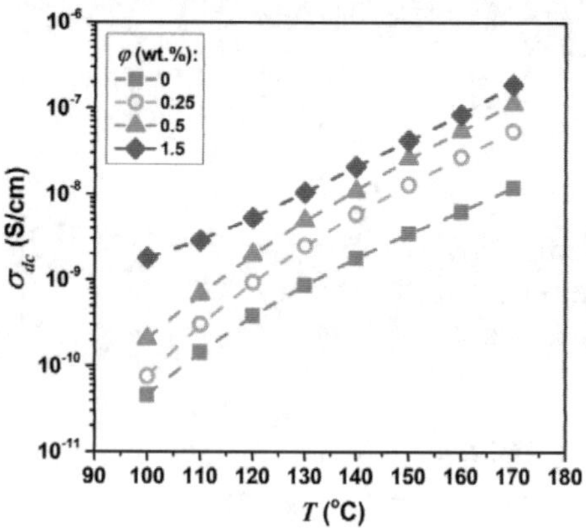

Figure 4. *dc* Conductivity of PAZ/rGO nanocomposites with different rGO loading vs. temperature.

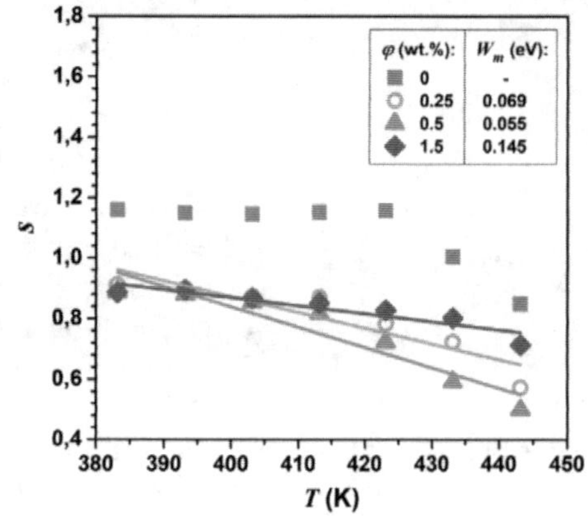

Figure 5. Power exponent s computed with equation (4) vs. absolute temperature.

Below we will analyze variation of σ_{ac} with frequency. As follows from figure 3, σ_{ac} increases dramatically with increasing frequency presumably following to a power law [17]:

$$\sigma_{ac} = Af^{S} \tag{3}$$

where A and s are characteristic constants of material. Further we will analyze parameter s as a function of temperature to clarify the mechanism of dielectric relaxation. In literature, two principal mechanisms of dielectric relaxation in solids are discussed. One mechanism is the quantum tunneling of either electrons or polarons through the barriers separating localized states. Another mechanism is electron hopping over the same barriers. Pike [18] modified the second mechanism and suggested the correlated barrier hopping (CBH) mechanism. Particularly he interpreted behaviour of *ac* conductivity of disordered solids. Following Pike's model, the parameter s in equation (3) could be presented as follows [19,20]:

$$s = 1 - \frac{6kT}{W_m} \tag{4}$$

where W_m is a height of the potential barrier between localized states to be overcome by the charge carrier. Reliability of equation (4) has been confirmed for polymers and polymer nanocomposites [19-21].

As follows from figure 5, the s parameter for all the samples (except neat polymer) goes down with increasing temperature rather in a linear manner, that is equation (4) is valid for the nanocomposites basically but neat polymer does not follow it. Therefore we can conclude that *ac* conductivity of the PAZ/rGO nanocomposites follows the CBH model. The energetic parameter of the CBH process, W_m, was calculated for all the nanocomposites. Its values are listed in the box within figure 5. It is clear that rGO concentration in nanocomposites does not influence the W_m value.

3.3. Electrical conductivity depending on temperature

Figure 6 represents conductivity of the PAZ/rGO nanocomposites depending on reciprocal temperature. In the region of low and medium temperatures, the conductivity becomes actually the *ac* conductivity ($\sigma = \sigma_{ac}$). The growth of frequency rather than the growth of the temperature causes a more significant increase in *ac* conductivity. However, in the region of glass transition σ_{ac} grows intensively and achieves a maximum value. Above the glass transition temperature, σ becomes in fact equal to σ_{dc} ($\sigma = \sigma_{dc}$). As follows from figure 6, σ_{dc} is a frequency independent parameter.

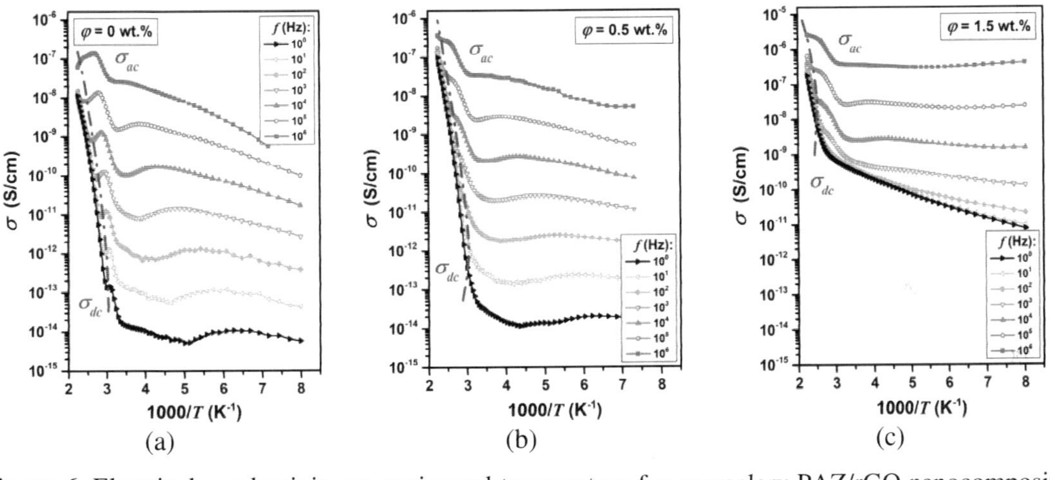

Figure 6. Electrical conductivity vs. reciprocal temperature for exemplary PAZ/rGO nanocomposites with different rGO concentration ((a) ϕ = 0, (b) ϕ = 0.5 wt.%, and (c) ϕ = 1.5 wt.%) at selected frequencies. Dash-dot lines depict the conditional boundaries between σ_{dc} and σ_{ac}.

Since *ac* conductivity is a thermally activated process, it should follow the Arrhenius equation:

$$\sigma_{ac} = \sigma_0 \exp\left(-\frac{E_a}{kT}\right) \tag{5}$$

where E_a is the activation energy of the charge carriers, k is the Boltzmann constant, and T is the absolute temperature.

Using figure 6 and equation (5), we computed the E_a values. The activation energy as a function of frequency is presented in figure 7. It is clear that, E_a decreases with increasing frequency. Such a tendency in activation energy indicates an increase in probability of the electron jumps between the atom sites [22]. Besides, E_a decreases with increasing rGO content. This finding explains an increase in σ_{ac} with increasing nanofiller concentration.

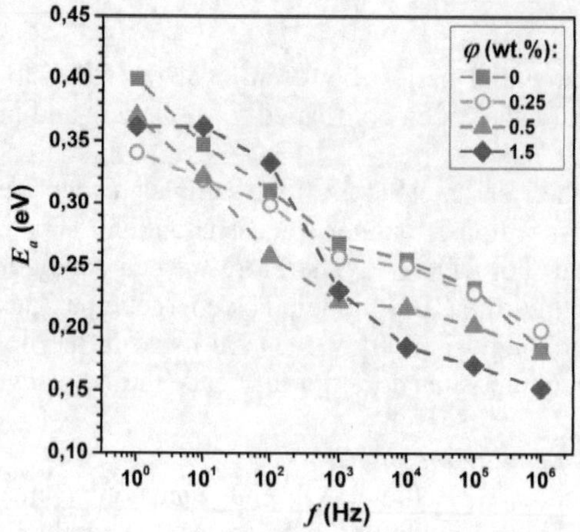

Figure 7. Activation energy of *ac* conductivity of PAZ/rGO nanocomposites with different rGO loading vs. frequency.

4. Conclusions

The PAZ/rGO nanocomposites with different rGO content were prepared. Their electrical conductivity was investigated with BDS. Two components of conductivity, σ_{dc} and σ_{ac}, were distinguished. We studied σ_{dc} depending on temperature exclusively, whereas σ_{ac} was analyzed depending on both temperature and frequency. It was shown that at small quantities (0.25 – 0.5 wt.%) of rGO, σ_{dc} increases slightly because these amounts of rGO nanoparticles are not enough to form an extensive conductive network [16]. When the content of rGO becomes higher (> 0.5 wt.%), σ_{dc} increases intensively due to percolation network formation. The percolation threshold of the PAZ/rGO nanocomposite ($\varphi \approx 1.5$ wt.%), was found to be lower as compared with that of PAZ/C_{60} nanocomposite ($\varphi \approx 2$ wt.%) [11,12]. Summing up, we concluded that two-dimensional anisotropic rGO nanoparticles are more suitable for enhancing the electrical conductivity of polymers as compared with isotropic nanoparticles of fullerene C_{60}. We successfully described the temperature and frequency dependences of σ_{ac} using the Arrhenius equation and a power function, respectively. The exponent s in the power function was shown to decrease with increasing temperature. These data allowed us to suggest that the correlated barrier hopping is an appropriate mechanism for explaining the σ_{ac} behaviour of the PAZ/rGO nanocomposites.

References

[1] Paul D R and Robeson L M 2008 *Polymer* **49** 3187-204
[2] Jancar J, Douglas J F, Starr F W Kumar S K, Cassagnau P, Lesser A J, Sternstein S S and Buehler M J 2010 *Polymer* **51** 3321-43
[3] Kuilla T, Bhadra S, Yao D, Kim N H, Bose S and Lee J H 2010 *Progr. Polym. Sci.* **35** 1350-75
[4] Wang C, Guo Z-X, Fu S, Wu W and Zhu D 2004 *Progr. Polym. Sci.* **29** 1079-141
[5] Phiri J, Gane P and Maloney T C 2017 *Mater. Sci. Eng. B*, **215** 9-28
[6] Paek J, Kim J, An B W, Park J, Ji S, Kim S-Y, Jang J, Lee Y, Park Y-G, Cho E, Jo S, Ju S, Cheong W H and Park J-U 2017 *FlatChem* **3** 71-91
[7] Potts J R, Dreyer D R, Bielawski C W and Ruoff R S 2011 *Polymer* **52** 5-25
[8] Iwan A 2015 *Renewable Sustainable Energy Rev.* **52** 65-79
[9] Hussein M A, Abdel-Rahman M A, Asiri A M, Alamry K A and Aly K I 2012 *Designed Monomers Polym.* **15** 431-63
[10] Bronnikov S, Kostromin S, Musteaţa V and Cozan V 2015 *Liq. Cryst.* **42** 1102-10

[11] Bronnikov S, Podshivalov A, Kostromin S, Asandulesa M and Cozan V 2017 *Phys. Lett. A* **381** 796-800
[12] Bronnikov S, Kostromin S, Musteaţa V and Cozan V 2016 *J. Polym. Res.* **23** 54
[13] Cozan V, Iftime M, Sava I and Bronnikov S 2015 *High Perform. Polym.* **27** 661-8
[14] Kostromin S, Saprykina N, Vlasova E, Timpu D, Cozan V and Bronnikov S 2017 *J. Polym. Res.* **24** 211
[15] Blythe T and Bloor D 2005 *Electrical Properties of Polymers* (Cambridge: Cambridge Univ. Press)
[16] Bronnikov S, Podshivalov A, Kostromin S, Asandulesa M and Cozan V 2016 *Eur. Polym. J.* **78** 213-25
[17] Jonscher A K, 1983 *Dielectric Relaxation in Solids* (London: Chelsea Dielectric Press)
[18] Pike G E 1972 *Phys. Rev. B* **6** 1572-80
[19] Singh R, Kumar J, Singh R K. Kaur A, Sinha R D P and Gupta N P 2006 *Polymer* **47** 5919-28
[20] Okutan M, Köysal O, San S E and Şentürk E 2009 *J. Non-Cryst. Solids* **355** 2674-7
[21] Hassen A, Hanafy T, El-Sayed S and Humanshu A 2011 *J. Appl. Phys.* **110** 114-9
[22] Okutan M, San S E, Basaran E and Yakuphanoglu F 2005 *Phys. Lett. A* **339** 461-5

IOP Conf. Series: Materials Science and Engineering **634** (2019) 012006 doi:10.1088/1757-899X/634/1/012006

Modeling the absorption of CO_2 in solvents enhanced by nanoparticle in polymeric membranes

N Ghasem[1,2]

[1]Department of Chemical and Petroleum Eng., UAE University, Al-Ain, UAE

E-mail: nayef@uaeu.ac.ae

Abstract. Lately nanomaterial received substantial consideration by several researchers. Solid nanoparticles prove to have extensive applications in composite materials and in the enhancement of mass and heat transfer. Adding nanoparticles to a base fluid improves the performance of mass and heat transfer. The effect of the addition of nanoparticles to base fluids, such as pure water and aqueous sodium hydroxide, show significant improvement in the percent removal of carbon dioxide from a gas composed of 15% CO_2 balance is CH_4 using gas-liquid hollow fiber membrane contactor. Results reveal that the rate of absorption of carbon dioxide increases with an increase in the concentration of nanoparticles up to certain extent beyond which insignificant removal rate is observed. Predicted results revealed that the percent removal of CO_2 in the existence of CNT and NaOH is 35% and 40%, respectively.

1. Introduction

The major source of the world energy supply is said to be natural gas. Methane is the key component of natural gas with trivial percentages of carbon dioxide (CO_2), hydrogen sulfide (H_2S), and nitrogen. There do exist water vapor, which is normally removed by dehydration process using ethylene glycol or its alternatives. The existence of acid gases in natural gas in addition of lowering the natural gas heating value, it can cause pipe corrosion [1-3]. Accordingly, prior to using of natural gas as an energy source in power generation and water desalination plants, the impurities should be removed. The conventional absorption methods used is primarily packed bed towers. In spite of the successful removal of acid gas using the traditional absorption columns, this method suffers from difficulties such as flooding at high inlet gas flow rate, channeling and foaming that can weaken the mass transfer between liquid and gas [1,4].

Several researchers thought that gas-liquid hollow fiber membrane contactor is an encouraging substitute gas absorption technology. Gas elimination process via membrane contactor is a non-dispersive process; gas is not dispersed in liquid. In membrane contactor absorption of pollutant component is taking place at the membrane liquid interface without dispersion of one phase into the other phase. Membrane contactor is supposed to overcome the drawbacks of conventional absorption columns [5]. Absorption of CO_2 using membrane contactor covers effect of liquid and gas flow rate, membrane partial wetting, membrane packing density, type of polymer used in the fabrication of hollow fiber membranes, variety of solvents used in the absorption process [6]. Various mathematical model were developed to describe the separation process using membrane contactor [7-10].

Solvent used in CO_2 absorption are mainly alkanolamine aqueous solutions, such as monoethanolamine (MEA), diethanolamine (DEA). In spite of the successful performance of these solvents in removing most of CO_2 from natural gas and flue gas, they have disadvantages of being corrosive to pipes and absorber walls and if used in membrane contactor they can cause membrane degradation [11]. Accordingly, seeking for user friendly absorbent took the attention of many researchers to replace the traditional amine family solvent. Water-based nanofluids is anticipated to be encouraging alternative solvents, various nanofluids were experimentally investigated for CO_2 absorption such Fe_3O_4, silica (SiO_2), alumina (Al_2O_3) and carbon nanotube (CNT) in distilled water and there effect on CO_2 absorption [12]. Absorption of carbon dioxide in hollow fiber membrane contactor were investigated using water based nanofluid containing solid nanoparticles (Al_2O_3, TiO_2, SiO_2). Results revealed there is an optimum concentration of the these nanoparticles on the percent removal of CO_2 [13]. Various mathematical models to describe the CO_2 absorption in the presence of water-based solid nanoparticles were developed. Models took into consideration Brownian motion and Grazing effect. These models focused on physical absorption [14-16]. In the present work a mathematical model will be used to simulate the chemical absorption of aqueous sodium hydroxide (NaOH)-based solid nanoparticles. The effect of liquid flow rate, gas flow rate on the percent removal will be investigated.

2. Model equations

The presence of solid nanoparticles in a nanofluid generates several mechanisms that can play a positive or negative impact on the rate of absorption of the pollutant gas. The adsorption of gas component by particles in the gas-liquid interfacial area and desorb in the bulk liquid increases mass transfer coefficient and is named Grazing effect. The random motion of particles dispersed in liquid phase is called Brownian motion. The particle motion produced momentum that enhanced the mass transfer coefficient due to intensifying liquid motion near the gas-liquid interface. There are negative effect of particles on mass transfer coefficient such as particle clustering, particle agglomeration, increase of liquid viscosity [17]. Accordingly, there is optimum nanoparticle concentration in the base fluid. The absorption mechanism in gas-liquid hollow fiber membrane contactor enhance by nanoparticles is illustrated in figure 1.

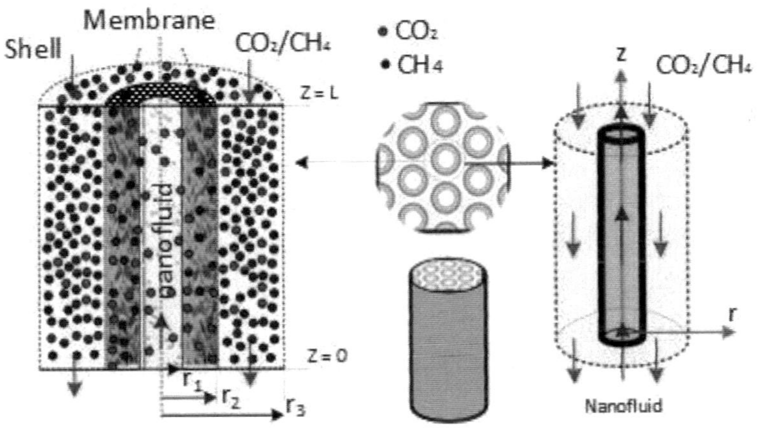

Figure 1. Schematic of segment of membrane contactor used in developed model.

The mathematical model is developed for the segment shown in figure 1 considering Happel's free surface and non-wetting model. The liquid nanofluid flow at z = 0 in the tube side, by contrast, the gas flow at z = L counter-currently. The model is developed considering isothermal steady state operation, ideal gas, incompressible liquid solvent, and negligible radial convection. The model consists of three segments: tube, membrane, and shell side. The system governing equations is as follows:

2.1. Tube side

The material balance equation represents CO_2 in the tube side is expressed as per equation (1):

$$D_{CO_2,t}\left(\frac{1}{r}\frac{\partial}{\partial r}\left(r\frac{\partial C_{CO_{2,t}}}{\partial r}\right)+\frac{\partial^2 C_{CO_{2,t}}}{\partial z^2}\right)=V_{z,t}\frac{\partial C_{CO_2},t}{\partial z}+\frac{k_pS_p}{1-\emptyset}\left(C_{CO_{2,t}}-C_s\right)-k_rC_{CO_{2,t}}C_{NaOH,t} \quad (1)$$

Where $D_{CO_2,t}$ is the diffusion coefficient of CO_2 in liquid phase in the tube side, $C_{CO_{2,t}}$ is concentration of carbon dioxide in the tube side, $V_{z,t}$ is the liquid velocity of the liquid inside the membrane tubes, k_p is the mass transfer coefficient between liquid and solid nanoparticles, S_p is the specific surface area of the nanoparticles, k_r is the reaction rate constant, C_s is the concentration of CO_2 at the liquid-solid interface [11]. The material balance equation represents NaOH in the tube side (equation (2)):

$$D_{NaOH,t}\left(\frac{1}{r}\frac{\partial}{\partial r}\left(r\frac{\partial C_{NaOH,t}}{\partial r}\right)+\frac{\partial^2 C_{NaOH,t}}{\partial z^2}\right)=V_{z,t}\frac{\partial C_{NaOH,t}}{\partial z}-2k_rC_{CO_{2,t}}C_{NaOH,t} \quad (2)$$

The fully developed velocity in the tube side is calculated as per equation (3):

$$V_{z,t}=2v_{z,avg}\left(1-\left(\frac{r}{r_1}\right)^2\right) \quad (3)$$

The adsorption of CO_2 in the solid particles is presented in equation (1) is calculated using the following auxiliary continuity equations of carbon dioxide adsorption in the solid nanoparticles (equation (4)).

$$\emptyset V_z\rho_p\frac{\partial q}{dz}=k_pa_p\left(C_{CO_{2,t}}-C_s\right) \quad (4)$$

Where q is the amount of CO_2 adsorbed by nanoparticle is determined by Langmuir isothermal adsorption mode [18] as expressed in equation (5):

$$q=q_m\frac{k_dC_s}{1+k_dC_s} \quad (5)$$

where q_m is the maximum adsorption capacity of nanoparticles, k_d is the Langmuir constant.

2.2. Membrane

The steady state diffusion mechanism of the gas components inside the membrane pores occurs by diffusion only (equation (6)):

$$D_{CO_2,m}\left(\frac{1}{r}\frac{\partial}{\partial r}\left(r\frac{\partial C_{CO_{2,m}}}{\partial r}\right)+\frac{\partial^2 C_{CO_{2,m}}}{\partial z^2}\right)=0 \quad (6)$$

2.3. Shell side

The steady state material balance equation of CO_2 in the shell side is represented by equation (7):

$$D_{CO_2,s}\left(\frac{1}{r}\frac{\partial}{\partial r}\left(r\frac{\partial C_{CO_{2,s}}}{\partial r}\right)+\frac{\partial^2 C_{CO_{2,s}}}{\partial z^2}\right)=V_{z,s}\frac{\partial C_{CO_{2,s}}}{\partial z} \quad (7)$$

The velocity in the shell side is given by equation (8) [19]:

$$V_{z,s}=2V_{z,av}\left[1-\left(\frac{r_2}{r_3}\right)^2\right]\left\{\frac{\left(\frac{r}{r_3}\right)^2-\left(\frac{r_2}{r_3}\right)^2+2ln\left(\frac{r_2}{r}\right)}{3+\left(\frac{r_2}{r_3}\right)^4-4\left(\frac{r_2}{r_3}\right)^2+4\ln\left(\frac{r_2}{r_3}\right)}\right\} \quad (8)$$

where $V_{z,av}$ is the average velocity of the gas stream in the shell side. The boundary conditions are summarized in table 1.

IOP Conf. Series: Materials Science and Engineering **634** (2019) 012006 doi:10.1088/1757-899X/634/1/012006

Table 1. Boundary conditions of mass transfer equations [20].

Boundary	Tube side	Membrane	Shell side
$z = 0$	$C_{CO_2,t} = 0$	Insulated	Convective flux
$z = L$	Convective flux	Insulated	$C_{CO_2,s} = C_{CO_2,o}$
$r = 0$	Symmetry	-	-
$r = r_1$	$C_{CO_2,t} = m * C_{CO_2,m}$	$C_{CO_2,m} = C_{CO_2,t}/m$	-
$r = r_2$	-	$C_{CO_2,m} = C_{CO_2,s}$	$C_{CO_2,s} = C_{CO_2,m}$
$r = r_3$	-	-	$\frac{\partial C_{CO_2,s}}{\partial r} = 0$

Where m is the solubility of CO_2 in the absorbent liquid. The nanofluid parameters used in the model is shown in table 2.

Table 2. The characterization of water-based carbon nanoparticles (CNT) [11].

Property	Value	Property of hollow fiber	Value
Morphology	Tubular	Fiber out diameter	1.10 mm
True density (g/cm^3)	2.2 g/cm^3	Inner diameter	0.42 mm
Particle capacity (q_m)	29.45 mol/kg	Number of fibers	11
Langmuir constant (k_d)	0.00049 m^3/mol	Module active length	0.21 m
Specific surface area (S_p)	500 m^2/g	Module diameter	0.08 m
Inner diameter	2.5 nm		

The governing model equations with the appropriate boundary conditions for coupling the tube, membrane, and shell sides are solved simultaneously using COMSOL Multiphysics version 5.4. The commercial software uses finite element method in the numerical solution of the set of differential equations.

3. Results and discussion

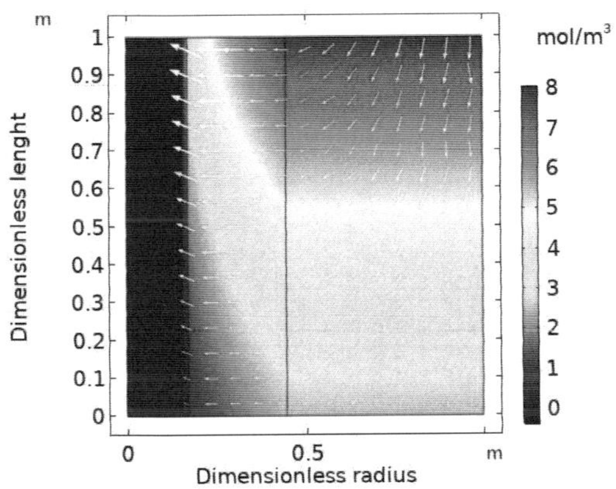

(a)

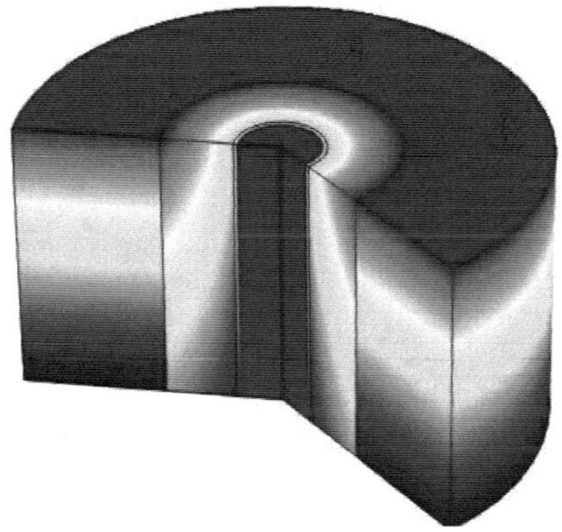

(b)

Figure 2. 2D Surface plot (a), 3D surface plot (b) for the CO_2 concentration profile, liquid flow rate 10 ml/min, gas flow rate 100 ml/min.

The 2D and 3D surface plot for the concentration of CO_2 throughout the membrane contactor is portrayed in figures 2(a) and 2(b), respectively. The CO_2 total flux and the track of the CO_2 diffusion pathway through the gas phase, membrane pores, and diffusion in water-based liquid nanofluid is categorized by white arrows. The carbon dioxide in the CO_2/CH_4 gas mixture enters at the upper side of the membrane module (i.e. z = L). The CO_2 is depleted from the gas mixture during the flow of the mixture in the gas phase. The CO_2 diffused through the membrane pores to the membrane-liquid interface based on concentration gradient where it is dissolved in the nanofluid.

The change in the concentration of CO_2 in the shell side along the length of the membrane is simulated in figure 3(a). It is clear that concentration of CO_2 decreased throughout the gases flow from z = L downward to z = 0 of the membrane modules, attributed to the continuous absorption of CO_2 into the solvent in the tube side. Figure 3(b) illustrated the CO_2 concentration outlined crosswise the membrane module at the top of module (z =1), middle of the module (z = 0.5) and close to the bottom of the module (z = 0.1) in the shell, membrane and tube side. The straight horizontal line (0.42 to 1) represents the CO_2 concentration in the shell side, the concentration is then start to drop, attributed to the diffusion of CO_2 in the membrane toward the solvent in the tube. The straight line between the dimensionless radiuses of 0 to 0.18 is the concentration of CO_2 in the solvent which is almost very small compared to the bulk concentration of sodium hydroxide solution. This is attributed to the continuous absorption of CO_2 during the gas flow inside the shell side of the membrane and its diffusion through the membrane micropores to the tube side where CO_2 dissolved in the liquid; being adsorbed in the solid nanoparticles, react with sodium hydroxide, and portion dissolved in water.

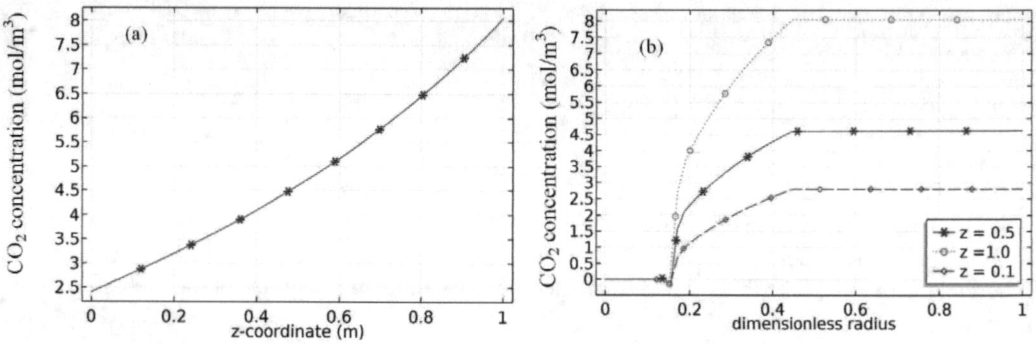

Figure 3. Concentration profile of CO_2 in the shell-membrane interface along the length of membrane (a), Concentration of CO_2 across various point (z = 1, z = 0, z = 0.1) of the module (b). Gas flow rate, 100 ml/min and liquid flow rate, 10 ml/min.

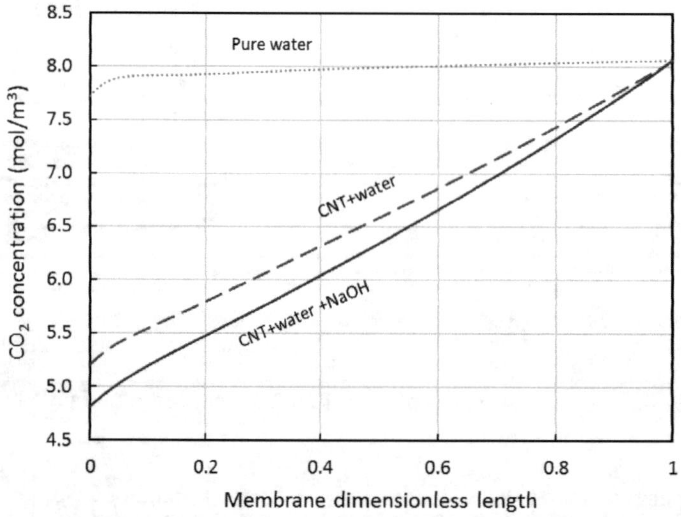

Figure 4. Effect of solvent type on CO_2 concentration along the membrane-gas interface, liquid flow rate, 10 m/min and gas flow rate of 100 ml/min.

The effect of solvent type is exposed in figure 4. The figure revealed that the drop in CO_2 concentration in pure water is insignificant. The presence of CNT in pure water causes a significant drop in the CO_2 concentration sideways the membrane-gas interface, this is attributed to the fact that the dynamic motion of the solid nanoparticles in the water-based nanofluid and the accompanying grazing and Brownian motion enhanced the CO_2 removal rate meaningfully. The percent removal of CO_2 in this case is around 36%. The addition of aqueous sodium hydroxide to water-based CNT nanoparticle fluid improves the removal rate of CO_2 as shown from the drop of the CO_2 concentration during gas flow in the shell side. The percent removal of carbon dioxide in the presence of 0.1 M NaOH is approximately 40% due to the chemical reaction between CO_2 and NaOH.

4. Conclusion

The CO_2 absorption using nanofluid composed of water, CNT, and sodium hydroxide in gas-liquid hollow fiber membrane contactor was modelled and simulated using computational fluid dynamics method. The numerical model composed of the material balance equations represent the component mass transport in the tube, membrane and shell side. The coupling equations were solved using COMSOL software version 5.4. The predicted results revealed that the presence of CNT in pure water increased the percent removal of CO_2 significantly compared pure water as absorbent without the addition of CNT nanoparticles. The addition of NaOH to the water-based CNT nanofluid improved the CO_2 removal rate. The percent removal of CO_2 in the water-CNT-NaOH nanofluid is estimated to be around 40% and can be increased by increasing liquid inlet flow rate.

References

[1] Ghobadi J, Ramirez D, Jerman R, Crane M and Khoramfar S 2018 CO_2 separation performance of different diameter polytetrafluoroethylene hollow fiber membranes using gas-liquid membrane contacting system *J. Memb. Sci.* **549** 75-83

[2] Rahim N A, Ghasem N and Al-Marzouqi M 2015 Absorption of CO_2 from natural gas using different amino acid salt solutions and regeneration using hollow fiber membrane contactors *J. Nat. Gas Sci. Eng.* **26** 108-17

[3] Rahim N A, Ghasem N and Al-Marzouqi M 2014 Stripping of CO_2 from different aqueous solvents using PVDF hollow fiber membrane contacting process *J. Nat. Gas Sci. Eng.* **21** 886-93

[4] Ghobadi J, Ramirez D, Khoramfar S, Jerman R, Crane M and Hobbs K 2018 Simultaneous absorption of carbon dioxide and nitrogen dioxide from simulated flue gas stream using gas-liquid membrane contacting system *Int. J. Greenh. Gas Control.* **77** 37-45

[5] Zhang H-Y, Wang R, Liang D T and Tay J H 2008 Theoretical and experimental studies of membrane wetting in the membrane gas–liquid contacting process for CO_2 absorption *J. Memb. Sci.* **308** 162-70

[6] Ghasem N, Al-Marzouqi M, Al-Marzouqi R, Dowaidar A and Vialatte M 2010 Removal of CO_2 from gas mixture using hollow fiber membrane contactors fabricated from PVDF/Triacetin/Glycerol cast solution *Eur. Conf. Chem. Eng. ECCE'10* (Tenerife, Spain), pages 136-141

[7] Ghasem N, Al-Marzouqi M and Abdul Rahim N 2013 Modeling of CO_2 absorption in a membrane contactor considering solvent evaporation *Sep. Purif. Technol.* **110** 1-10

[8] Khan M J H, Hussain M A, Mansourpour Z, Mostoufi N, Ghasem N M and Abdullah E C 2014 CFD simulation of fluidized bed reactors for polyolefin production - A review *J. Ind. Eng. Chem.* **20** 3919-46

[9] Ghasem N and Al-Marzouqi M 2017 Modeling and experimental study of carbon dioxide absorption in a flat sheet membrane contactor *J. Membr. Sci. Res.* **3** 57-63

[10] Rosli A, Shoparwe N F, Ahmad A L, Low S and Lim J K 2019 Dynamic modelling and experimental validation of CO_2 removal using hydrophobic membrane contactor with different types of absorbent *Sep. Purif. Technol.* **219** 230-40

[11] Hajilary N and Rezakazemi M 2018 CFD modeling of CO_2 capture by water-based nanofluids using hollow fiber membrane contactor *Int. J. Greenh. Gas Control.* **77** 88-95
[12] Peyravi A, Keshavarz P and Mowla D 2015 Experimental investigation on the absorption enhancement of CO_2 by various nanofluids in hollow fiber membrane contactors *Energy and Fuels.* **29** 8135-42
[13] Mohammaddoost H, Azari A, Ansarpour M and Osfouri S 2018 Experimental investigation of CO_2 removal from N_2 by metal oxide nanofluids in a hollow fiber membrane contactor *Int. J. Greenh. Gas Control.* **69** 60-71
[14] Rezakazemi M, Darabi M, Soroush E and Mesbah M 2019 CO_2 absorption enhancement by water-based nanofluids of CNT and SiO_2 using hollow-fiber membrane contactor *Sep. Purif. Technol.* **210** 920-6
[15] Darabi M, Rahimi M and Molaei Dehkordi A 2017 Gas absorption enhancement in hollow fiber membrane contactors using nanofluids: Modeling and simulation *Chem. Eng. Process. Process Intensif.* **119** 7-15
[16] Dai Z *et al* 2019 CO_2 absorption enhancement by water-based nanofluids of CNT and SiO_2 using hollow-fiber membrane contactor *Sep. Purif. Technol.* **210** 920-6
[17] Villeneuve K, Albarracin Zaidiza D, Roizard D, Rode S, Nagy E, FeczkóT and Koroknai B 2007 Enhancement of oxygen mass transfer rate in the presence of nanosized particles *Chem. Eng. Sci.* **62** 7391-8
[18] Lu S, Xing M, Sun Y and Dong X 2013 Experimental and theoretical studies of CO_2 absorption enhancement by nano-Al_2O_3 and carbon nanotube particles *Chinese J. Chem. Eng.* 21 983-90
[19] Happel J 1959 Viscous flow relative to arrays of cylinders *AIChE J.* **5** 174-7
[20] Fasihi M, Shirazian S, Marjani A and Rezakazemi M 2012 Computational fluid dynamics simulation of transport phenomena in ceramic membranes for SO_2 separation *Math. Comput. Model.* **56** 278-86

The influence of the powder additive upon selected mechanical properties of a composite

R Szczepaniak[1,3], K Rolecki[2] and A Krzyzak[1]

[1]Department of Airframe and Engine, Faculties of Aeronautics, Polish Air Force Academy, Dywizjonu 303 no. 35 street, 08-521 Deblin, Poland
[2]8 Transport Aviation Base, Cracow, Poland

E-mail: r.szczepaniak@law.mil.pl

Abstract. Five series of samples were prepared, each batch consisting of 20 epoxy-resin items, CES R70 with CES H72 hardener and 4 types of powder additives: montmorillonite (MMT), Expancell d25 Microsphere (microsphere), nanocarbon (NanoC) and silicon carbide (SiC). All the additives were added in quantities corresponding to 15% of the total volume share of the material. The test samples for tensile test and three-point transverse bend test were made in accordance with the standards: PN-EN ISO 527-1 and PN-EN ISO 178. The addition of powder materials into the resin affected the change of the examined parameters when investigating the mechanical properties. The static tensile test proved the supposition about changing the strength of the material. It also confirmed the change in fragility of the prepared samples in relation to the base material. Similar conclusions may be drawn from the development of the research findings of the materials with regard to bending and hardness of the produced samples.

1. Introduction

A combination of high material strength and stiffness with low weight is the main advantage of fibre-reinforced composites, which are being increasingly used in various industries, including aviation (lifting surface of the aircraft and helicopter), motoring (carbon-fibre and honeycomb composite structure in Formula One), recreation (snowboards, fishing rods, boats, etc.) and many other (such as wind energy) [1,2]. The properties of glass-reinforced and carbon-reinforced fibres are commonly known [3]. One of the most important investigations aimed at determining the usefulness of a given material for air industry applications is to specify its strength characteristics. Moreover, research is being conducted to determine the strength properties of resins used in aerospace technology [4,5], bearing in mind high fragility of the matrix and low binding between the matrix and the base-material [6,7]. In order to increase the mechanical and thermal strength of laminates even further, there are additional reinforcements in the form of particles [8,9,10]. In the available literature, numerous studies address the issue of the material parameters of modified composites, both for fibre reinforcement [11] and powder reinforcement, especially when using carbon nanotubes [11,12,13,14,15], SiC [16,17] and MMT applications [18,19]. The study measures the strength properties of powder composites (including dispersion-strengthened composites) with a 15% volume share of the additive in order to determine the effect of particular additives upon the change of composite ultimate tensile strength and bending strength. The article presents partial findings of research into mechanical properties of composites which exert an influence on the wear process of composites in friction couples, due to the fact that in aviation it is essential to keep the highest level of safety, also including material strength, combined with other properties which may be mutually exclusive, e.g. abrasive resistance [example 16] or ablative properties of materials [18].

2. Experimental

Tests were carried out on the strength of composite powder in accordance with the PN-EN ISO 527-1 and PN-EN ISO 178 norms in order to determine the strength properties during tensile and bending tests of the examined composite, depending on the type of the used powder additive. The investigations were conducted on the Zwick Roell Z5.0 machine used for static tests. Furthermore, imaging of the fracture structure of the damaged material was performed on a Hitachi electron microscope TM3030Plus, also measuring the toughness of the examined materials by means of the durometer Shore/IRHD DIGI Test II manufactured by Bareiss.

2.1. Test materials

The strength tests were conducted on a powder composite. Composite samples were prepared on the basis of the CES R70 matrix resin, cured with CES H72 hardener, in proportions 100:54 wt. (recommended by the manufacturer). The powder fillers were: montmorillonite (MMT), Expancell d25 Microsphere (microsphere), nanocarbon (NanoC) and silicon carbide (SiC). The amount of a powder additive in each sample equalled 15 % of the composite volume. The curing occurred at ambient temperature. The author prepared 20-sample batches for research with every additive as well as base samples without any additive, both for static tensile and bending tests. The samples for experiments were created by casting resin with an additive into special silicone moulds, whose dimensions are specified in figures 1 and 2.

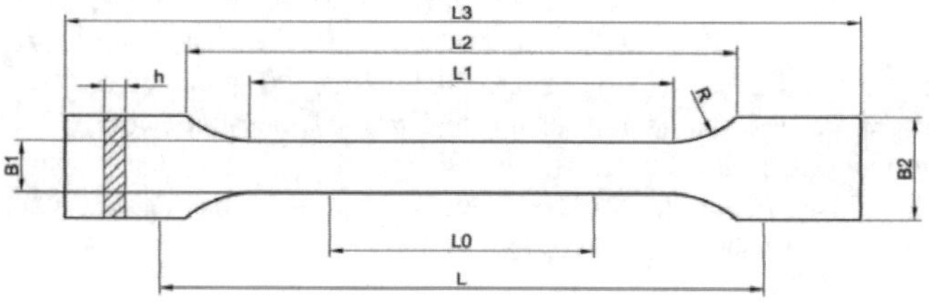

Figure 1. Universal static moulding - tensile test.

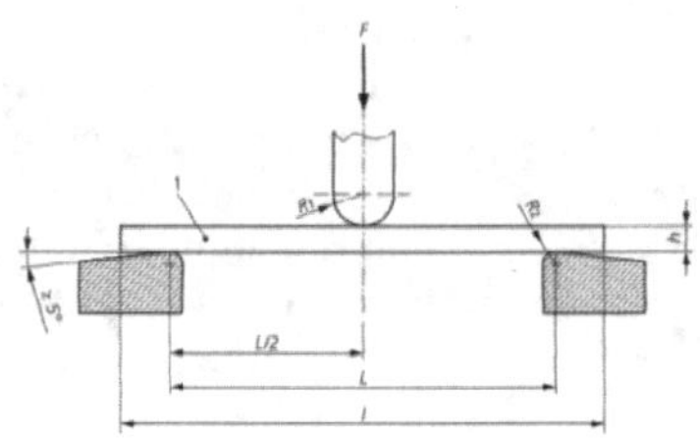

Figure 2. Universal static moulding - bending test.

2.2. Geometry of samples used to research

The samples (mouldings) were prepared as flat samples with necking (figure 1). The following dimensions of samples were adopted: thickness 4.0±0.2 mm, width of the measurement part 10±0.2 mm and overall length above 150 mm. In the case of mouldings made with resin and an additive, the measuring part equalled 80 ± 2 mm.

Next, the author prepared samples for testing in a shape illustrated in figure 2. The dimensions of samples prepared for the examination of bending, rectangular-shaped, have been listed in table 1.

Table 1. Dimensions of a moulding for a static bending test.

Mould sizes	
l – specimen length	80 mm
b – specimen width	10 mm
h – specimen thickness	60 mm
L – distance between supports	64 mm
R1 – punch radius	5 mm
R2 – support radius	5 mm

After conducting mechanical investigations (tensile and bending tests), selected samples from each batch underwent testing on a durometer in order to determine the hardness by Shore (D-scale) method. Also the morphology of the surface of the fracture of the examined materials was featured with an electron microscopy.

In order to create more transparent graphs, the author adopted labels of the examined samples in accordance with those listed in table 2.

Table 2. Nomenclature of samples for examinations with powder additives.

Composite	Tensile test	Bending test
CES R70 + CES H72	R1	Z1
CES R70 + CES H72 + microsphere	R2	Z2
CES R70 + CES H72 + MMT	R3	Z3
CES R70 + CES H72 + SiC	R4	Z4
CES R70 + CES H72 + nanocarbon	R5	Z5

3. Results and discussion

3.1. Tensile test

The first type of the experimental studies of the powder-strengthened composites under scrutiny was a static tensile test. Figures 3 and 4 show an exemplary line chart of the obtained graphs with one exemplary course of elongation. The composites with the addition of MMT and SiC have the properties of brittle materials (destruction occurs approximately after 1.1 – 1.2% of stretching), whereas the resin and composites with the addition of nanocarbon and microspheres have the qualities of plastic materials.

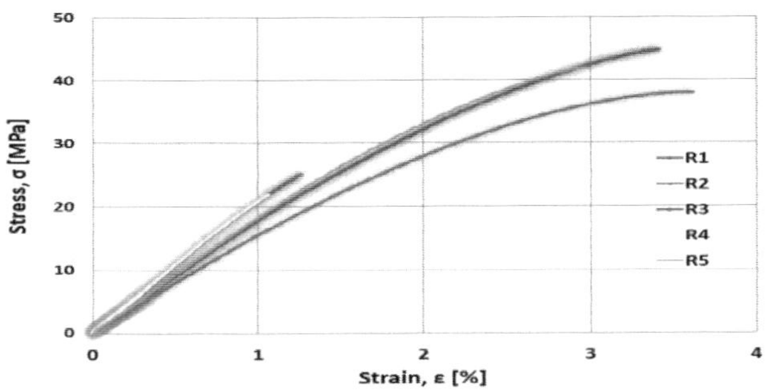

Figure 3. Exemplary characteristics of composite stretching.

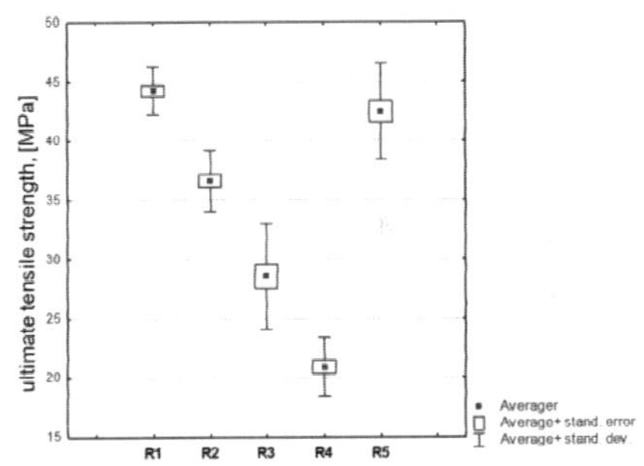

Figure 4. Distribution of tensile strength of particular composites.

Table 3. Results of a statistical analysis of failure strain in samples during elongation [MPa].

	average	variance	standard deviation	standard error	skewness	kurtosis
R1	44.230	44.650	4.140	2.034	0.455	-0.200
R2	36.600	36.900	6.800	2.607	0.583	-0.739
R3	28.570	29.700	20.270	4.503	1.007	-0.291
R4	20.895	20.850	6.230	2.495	0.558	0.528
R5	42.456	43.700	16.730	4.091	0.915	-1.169

It is possible to observe a significant drop in strength of a composite material after the addition of powder, apart from carbon nanotubes, whose value is approximately equal to the value obtained for a

IOP Conf. Series: Materials Science and Engineering **634** (2019) 012007 doi:10.1088/1757-899X/634/1/012007

composite without the additive. The basic values resulting from a statistical analysis have been presented in table 3.

3.2. Bending test

The second type of the experimental studies of the powder-strengthened composites under scrutiny was a bending test. In the figures below (figures 5 and 6), there is an exemplary graph of the obtained courses for bended composites. The composite with the addition of SiC became damaged after the bending of approximately 1.8 mm, which corresponds to its brittle properties. On the other hand, the composites with the addition of microspheres were damaged after the bending of 13 mm on average (figure 5).

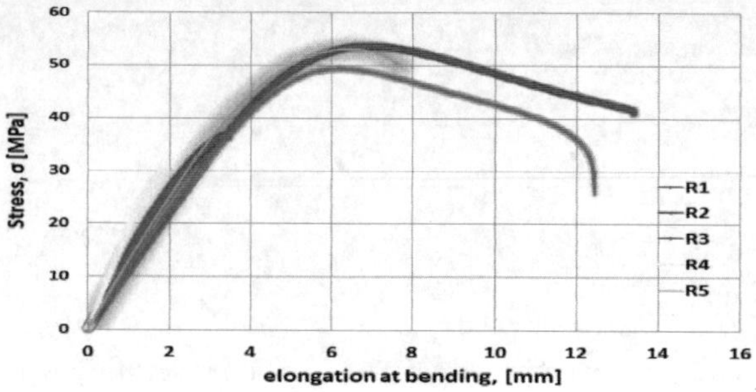

Figure 5. Exemplary characteristics of composite bending.

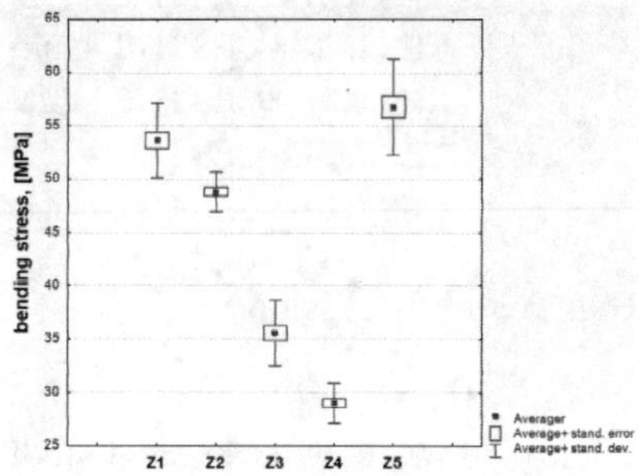

Figure 6. Distribution of bending strength of particular composites.

Similarly, as in earlier research, there is a statistical analysis for all examined materials with regard to the bending strength parameter, with the results presented in table 4. It is possible to observe a significant drop in strength of the composite material after the addition of powder, apart from carbon nanotubes, whose value is even higher than that of the epoxy resin without the additive.

Table 4. Results of a statistical analysis of failure strain in samples during bending [MPa].

	average	variance	standard deviation	standard error	skewness	kurtosis
Z1	53.635	53.65	12.05	3.472	0.776	-0.047
Z2	48.830	49.20	3.50	1.870	0.418	-0.587
Z3	35.555	35,65	9.48	3.079	0.689	-0.283
Z4	29.010	29.05	3.41	1.846	0.413	0.421
Z5	56.776	58.75	20.12	4.486	1.003	-0.403

3.3. Measurement of hardness

After conducting mechanical research, the author measured hardness of a composite material. The results are listed in table 5, in which there are hardness values and the most important findings of a simple statistical analysis of the obtained results. It is possible to observe very similar values of hardness for Z1, Z3 and Z5 composites. It undoubtedly influenced a uniform distribution of the additive in the resin with MMT additives and nanocarbons, which are characterized with small dimensions.Moreover, Z4 (composite with the addition of SiC) achieved much higher hardness values due to high SiC hardness (when it exceeds 9 points on Mohs scale, it is used as an abrasive material). On the other hand, Z2 (the composite with the addition of a microsphere) has low hardness and low density, which resulted in a considerable drop in the value of hardness of the whole composite.

In order to observe the structure of the fracture of the examined material, the author made an imagery of the surface by means of electron microscopy and the results are presented below for a selected sample from each measurement on completion of fracture testing.

Table 5. Results of a statistical analysis on sample hardness after strength tests [-].

	average	variance	standard deviation	standard error	skewness	kurtosis
Z1	77.66	0.955	0.977	0.219	-1.041	-0.148
Z2	54.44	1.070	1.034	0.231	0.172	0.306
Z3	78.33	2.590	1.609	0.360	-0.623	-1.266
Z4	86.04	1.140	1.067	0.239	-0.141	-0.787
Z5	77.70	0.490	0.700	0.157	-0.248	-1.285

All the obtained research findings do not follow normal distribution (proved by skewness and kurtosis results, see tables 3-5). Therefore, the significance of the influence of the fillers upon the examined properties were conducted by means of a non-parametric Mann-Whitney U test to compare outcomes between two independent samples. The Mann-Whitney U test is based on calculating U statistics with the total number of observations $n \leq 20$. The hypothesis of changing the significance is fulfilled only when U statistics is smaller than the critical Ukr value read from tables, adequate to the assumed confidence level α ($\alpha < 0.05$) and the number of groups n1 and n2. Table 6 lists the findings of the confidence test with regard to the impact of the filler on the composite properties against pure resin.

Table 6. U Mann-Whitney test results significant impact of filler if p<0.05).

Property	Microsphere		MMT		SiC		nanocarbon	
	U	p	U	p	U	p	U	p
hardness	0.00000	0.000000	130.0000	0.060112	0.00	0.000000	188.5000	0.766046
Tensile strenght	2.00000	0.000000	0.0000	0.000000	0.00	0.000000	155.0000	0.228695
Bending strength	44.00000	0.000026	0.0000	0.000000	0.00	0.000000	124.0000	0.041124

Due to the use of an Hitachi electrone microscope, it was possible to capture the image of structures and surfaces of examined conducting and non-conducting specimens, without the necessity to make initial processing and applying additional elements. The article presents only representative images obtained during the examinations of composite materials with additives. In the conducted examinations, the author used the magnification of 50 to 200 (depending on the used additive in photos at applied magnifications). In order to illustrate a composite, the following observation mode was used: secondary electron (SE) to depict the surface morphology (figures 7, 8, 9) and backscattered electron (BSE) to show the compositional information in the sample (figures 10 and 11), in which the used additives are visible.

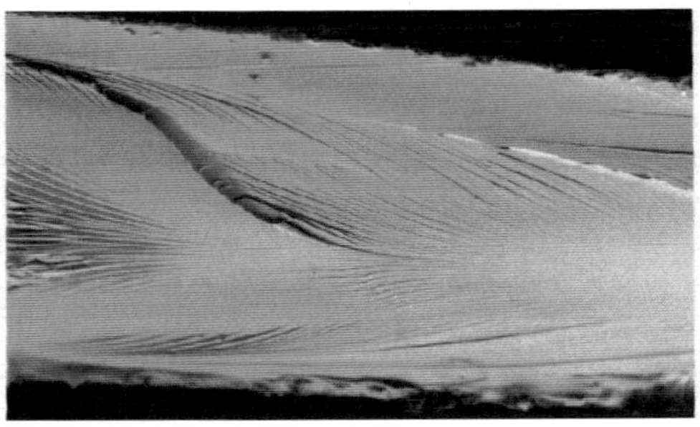

Figure 7. CES R70 +CES H72 magnification 50x SE.

Figure 8. CES R70 +CES H72 + microsphere magnification 50x SE.

IOP Conf. Series: Materials Science and Engineering **634** (2019) 012007 doi:10.1088/1757-899X/634/1/012007

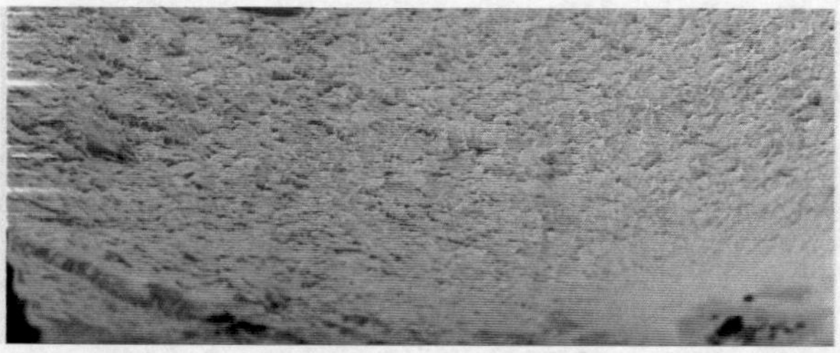

Figure 9. CES R70 +CES H72 + NanoC magnification 200x SE.

Figure 10. CES R70 +CES H72 + SiC magnification 100x BSE.

Figure 11. CES R70 +CES H72 + MMT magnification 200x BSE.

4. Conclusions

When analysing the above results of strength tests of powder composites and observations on an electron microscope, it is possible to formulate the following conclusions:

- The addition of powders to resin significantly changes the strength of the composite both during tensile testing and bending. Only the composite with the addition of nanocarbon showed increased strength to bending, approximately equal to 56.8 MPa, i.e. 5.8% than resin. The remaining composites show a drop in strength to bending even by 45.9%.
- The composites with the addition of SiC and MMT (figure 3) are characterized by a higher Young's modulus, at the level of 2.1 and 2.3.GPa. The lower Young's modulus can be observed in the composite with nanospheres, at the level of 1.6 GPa, i.e. 6% less than resin.
- In case of SiC nanofillers and the microspheres, the influence of the fillers' hardness on composite hardness was observed. In relation to resin hardness, the hardness of the SiC composite grew by approximately 10.8%. In case of the composites with nanospheres, the hardness fell by approximately 29.9%.
- Only the addition of carbon nanotubes did not diminish tensile strength or bending strength. Besides, it made the materials more brittle (apart from the composite containing microspheres). The changes are statistically significant.
- The use of various types of additives changes the internal structure of a composite, as observed on an electron microscope. It clearly affects the strength properties of materials with the powder additives. Nanocarbon, as an additive, significantly influences the change in the mechanical properties only in the three point bending test.

Acknowledgments

This study was supported by Ministry of National Defence Republic of Poland, Research Project No GB/5/2018/209/2018/DA and by Ministry of Science and Higher Education Research Project No. 30/2018.

References

[1] Mouritz A P, Gellert E, Burchill P and Challis K 2001 Review of advanced composite structures for naval ships and submarines *Compos. Struct.* **53** 21-41
[2] Soutis C 2005 Carbon fiber reinforced plastics in aircraft construction *Mater. Sci. Eng. A.* **412** 171-6
[3] Yasmin A and Daniel I M 2004 Mechanical and thermal properties of graphite platelet/epoxy composites *Polymer* **45** 8211-9
[4] Callister W D 2003 *Materials Science and Engineering: An Introduction* (New York: Wiley) p 505, 550
[5] Hameed N, Sreekumar P A, Francis B, Yang W and Thomas S 2007 Morphology, dynamic mechanical and thermal studies on poly(styrene-co-acrylonitrile) modified epoxy resin/Glass fibre composites *Compos Part A Appl S* **38** 2422-32
[6] Sridharan S 2008 *Delamination Behaviour of Composites* (New York: CRC Press)
[7] Ogihara S and Takeda N 1995 Interaction between transverse cracks and delamination during damage process in CFRP cross-ply laminates *Compos. Sci. Technol.* **54** 395-404
[8] Qi C, Li C and Huang Y 2006 *AIAA 57th Int Astronautical Congress* **8** 5237-40
[9] Srivastava V K and Wahne S 2007 Wear and friction behaviour of soft particles filled random direction short GFRP composites *Mater. Sci. Eng. A.* **458** 25-33
[10] Yusriah L, Mariatti M and Abu Bakar A 2010 Mechanical properties of particulate filler/woven glass fabric filled vinyl ester composites *J Vinyl Addit Techn* **16** 98-104
[11] Sonsakul K and Boongsood W 2017 Effects of glass scraps powder and glass fiber on mechanical properties of polyester composites *IOP Conf. Series: Materials Science and Engineering* **273** 012006
[12] Ku H, Wong P, Huang J, Fung H and Mohan T 2011 Tensile tests of glass powder reinforced epoxy composites: pilot study *ICKEM 2011 1st Int Conf on Key Engineering Materials* (Sanya, China) pp 25-7
[13] Devendra K and Rangaswamy T 2013 Strength characterization of E-glass fiber reinforced epoxy composites with filler materials *J. Miner. Mater. Char. Eng.* **1** 353-7
[14] Li Y, Zhang H, Peijs T and Bilotti E 2016 Graphene delivery systems for hierarchical fiber reinforced composites *MRS Advances* **1** 1339-44
[15] Zhang H, Kuwata M, Bilotti E and Peijs T 2015 Integrated damage sensing in fibre-reinforced composites with extremely low carbon nanotube loadings *J. Nanomater.* **2015** 785834
[16] Rathod V T, Kumar J S ans Jain A 2017 Polymer and Ceramic Nanocomposites for Aerospace Applications *Appl Nanosci* **7** 519-48
[17] Klasik A, Sobczak J, Pietrzak K, Makowska K and Wojciechowski A 2011 Praktyczna weryfikacja metody recyklingu kompozytowych tarcz hamulcowych *Transp. Samoch.* **4** 95-108
[18] Szczepaniak R, Czubak K, Przybylek P and Stabryn S 2017 The analysis of the properties of ablative composites with carbon and MMT nanofiller reinforcement *J. KONES* **24** 305-12

IOP Conf. Series: Materials Science and Engineering **634** (2019) 012008 doi:10.1088/1757-899X/634/1/012008

Novel acid soluble consolidating material to overcome lost circulation problems in reservoir intervals

J Feng[1,2], L H Zheng[1], L Li[2,5], H J Hao[2] and C Zhang[3] and Z Zhang[4]

[1]College of Petroleum Engineering, China University of Petroleum (Beijing), Beijing 102249, China
[2]CNPC Engineering Technology R & D Company Limited, Beijing 102206, China
[3]PetroChina Qinghai Oilfield Company, Dunhuang 736202, China
[4]Petrochina Tarim Oilfield Company, Korla 841000, China

E-mail: lilongdr@cnpc.com.cn

Abstract. Based on lost circulation problem, a new kind of acid soluble consolidating material was developed, and the sealing characteristics were evaluated. This acid soluble consolidating material had the characteristics of rapid water loss, high filter-cake consistency and high acid dissolution. The results indicated that the obtained curing slug had good stability at high temperature, and the sealing strength of curing slug was as high as 10 MPa. High temperature and high pressure (HTHP) sealing analog experiment indicated that the maximum pressure-bearing capacity of plugged zone was up to 18.5 MPa. Using this consolidating material to seal the shale is a very powerful and economical approach to address lost circulation problem in troublesome thief zones.

1. Introduction

During the drilling operation of abnormal high pressure formation, the fracturing induced leakage is often encountered [1-3]. Due to great differences among leakage points, it is very difficult to locate the leakage zone of these induced leakages, so it will bring great challenge to plugging technology [4-7].

Due to the particular characteristics of multiple leakage points, it is very difficult to determine the accurate leakage zone [8-10]. The conventional plugging technology, including cement plugging technology, bridge plugging technology, gel polymer plugging technology, and so on, has the characteristics of easily repeated leakage, long plugging operation time and poor plugging effect, so the drilling cycle and drilling cost should be increased greatly [11-13]. In recent years, the expansion pipe plugging technology has provided an effective means for the complex lost circulation, but there are disadvantages such as high cost and complicated application technology.

A new type of acid soluble consolidating material (ASCM) is developed by one-pot method, and the properties of ASCM are evaluated by HTHP water loss at high temperature, consolidating ability and pressure bearing capacity. The results indicate that ASCM has the characteristics of high water loss, high acid dissolution and high bearing pressure. In the process of plugging operation, due to low viscosity and small particle size, ASCM could enter the thief zones and exhibit rapid water losses, so the leakage channel could be plugged by obtained filter cake. Using ASCM to control lost circulation is a very powerful and economical approach to address lost circulation problems in multiple leakage zones. In the future, ASCM might hold great promise to resolve instability problems in reservoir intervals.

IOP Conf. Series: Materials Science and Engineering **634** (2019) 012008 doi:10.1088/1757-899X/634/1/012008

2. Preparation of acid soluble consolidating material

Given the nature of plugging material, the responsive consolidating material is composed by main material, suspending agent, curing agent and filter aid. All things considered, ASCM should have the characteristics of rapid water loss, good stability, high filter-cake consistency and high acid dissolution.

2.1. Host materials

The primary feature of ASCM is high water loss. The host materials (HM) should have good porous structure and compressibility, and they could play an important role in filling and sealing the leakage channel. As shown in table 1, four kinds of porous structural materials are evaluated.

Table 1. The evaluation of properties of host materials.

Number	Porous materials	Filtration time/s	Filter loss/mL	Cake thickness /mm	Acid solubility /%	Consisto meter
1	HM-1	38	145	9	75	hard
2	HM-2	35	150	9	65	hard
3	HM-3	34	130	10	69	hard
4	HM-4	23	120	8	99	medium

The experimental results indicate that HM-2 has big water loss, low acid dissolution, and high filter-cake consistency. HM-4 has big water loss, excellent acid dissolution, and low filter-cake consistency. So both HM-2 and HM-4 are chosen as composite host materials. Based on the requirement for acid solubility, the host materials could be adjusted by the proportion of HM-2 and HM-4.

2.2. Suspending agent

The suspending agent is used to suspend the host materials and other materials, and it could improve the suspension stability of plugging slurry. The suspending agent (SA) could play a role in bridging in the lost circulation channel, and improve the filter-cake consistency and toughness. The properties of suspending agents are evaluated, as shown in table 2.

Table 2. The evaluation of the properties of suspending agents.

Number	Suspending agent	Stability/%	Filter loss/mL	Cake thickness/mm
1	SA-1	85	160	8
2	SA-2	95	170	9
3	SA-3	90	175	10

The experimental results demonstrate that SA-2 has good suspension effect. SA-2 has good water dispersibility, so it could improve the suspension stability of plugging slurry greatly. In addition, SA-2 has good compatibility and bonding strength, so it could play a role in toughening modification of solidified filter cake.

2.3. Curing agent

The plugging material could form filter cake easily in the leakage channel, but the pressure-bearing capacity of filter cake is low. To improve the strength of filter cake, the curing agent (CA) should be used, so the filter cake could be solidified. The properties of curing agents are tested in table 3.

The results indicate that three kinds of curing agents have little influence on the strength of filter cake. The filter cake formed by CS-3 has the highest strength, so CS-3 has the best curing effect. The curing effect of CS-3 is being improved with the increase of dosage of CS-3. But when the dosage of CS-3 is more than 10%, the gelation of plugging slurry could be increased greatly. The results indicate that CS-3 has enough curing effect when the dosage of CS-3 is 5~10%.

IOP Conf. Series: Materials Science and Engineering **634** (2019) 012008 doi:10.1088/1757-899X/634/1/012008

Table 3. The evaluation of properties of curing agents.

Number	Curing agent	Curing strength /KN	Cake thickness /mm	Filtration time /s	Filter loss /mL
1	CS-1	16	11	25	190
2	CS-2	12	9	23	185
3	CS-3	20	10	23	180

2.4. Filter aid

The filter aid is used to further improve the filtration performance of plugging agent, and accelerate the filtration rate of plugging slurry. The effect of the dosage of filter aid (FA) on the properties of plugging slurry is shown in table 4.

Table 4. The influence of the dosage of FA on the properties of plugging slurry.

Number	Dosage /%	Filtration time/s	Filter loss /mL	Cake thickness /mm	Stability /%	Curing strength /KN
1	4	30	180	10	95	19
2	6	25	180	9	95	20
3	8	18	185	10	90	20
4	10	16	190	10	85	20
5	12	16	190	10	85	21
6	15	15	190	11	80	21

The experimental results indicate that this filter aid could accelerate the filtration rate, increase the filter loss, and improve the curing strength of filter cake. The filtration rate of plugging agent could be improved with increase of the dosage of filter aid. But when the dosage of filter aid is more than 12%, the stability of plugging agent could be weakened. All things considered, the filter aid has enough filtration rate when the dosage is 8~12%.

2.5. Acid soluble consolidating material

Taking into consideration of various components, the proportion of main material, suspending agent, curing agent and filter aid is optimized, and a new kind of acid soluble consolidating material is developed.

Formula of ASCM: 40%HM-2+30%HM-4+12%SA-2+8%CS-3+10%FA

After ACSM and water are mixed and stirred fully, a type of plugging slurry with good suspension and strong fluidity is prepared. The filter cake formed at high pressure could be used to seal the formations with porous leakage or fractured leakage.

3. Results and discussion

3.1. The evaluation of HTHP water loss at different temperatures

For a consolidating material, rapid water loss and big water loss at high temperature and high pressure (HTHP) conditions is very important. The high-temperature and high-pressure filter press is used to test the filtration properties of ASCM. The water loss and filtration time of plugging slurry at high temperature are shown in table 5.

The experimental results indicate that this plugging slurry containing ASCM has high water loss in 60-75 second. In addition, the water loss could be reduced with the increase of temperature. The water loss remains unchanged when the temperature is above 150°C. So ASCM has good filtration properties at high temperature.

Table 5. The test of HTHP water loss of plugging slurry containing ASCM.

Temperature	80°C	100°C	120°C	150°C	180°C
Water loss/mL	325	320	310	310	305
Filtration time/s	75	68	67	60	61

[a] Formula for plugging slurry: fresh water + 20% ASCM.

3.2. The consolidating ability of ASCM

The high-temperature stability of solidified filter cake is of great importance for the completion operation. The maintenance intensity of obtained curing slug at 180°C is shown in figure 1. The curing strength could be enforced with the increase of curing time, but the curing strength is reduced when the curing time is more than 15 d. Therefore, the curing slug has good stability at high temperature when the curing time is less than 15 d. In addition, the sealing strength could be up to 10 MPa, so it could meet the requirement of completion operation.

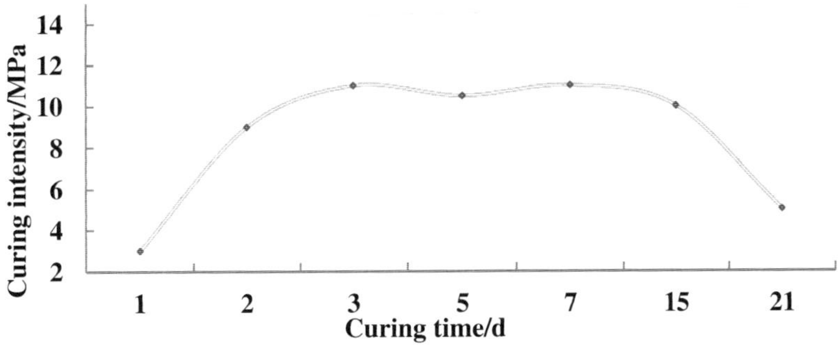

Figure 1. The relation curve between the curing intensity of curing slug and curing time.

3.3. The test of acid dissolution of ASCM

For the leakage zone in reservoir intervals, the plugging material should have good solubility in acid. Different quantity of hydrochloric acid with the concentration of 15% is added to the same amount of ASCM, and ASCM samples are dissolved rapidly in hydrochloric acid. After soaking for 24 h, filtration and desiccation, the sample weight before and after acidification is weighed and measured, and the acid dissolution is calculated as bellow,

$$\eta=[M_1-M_2/M_1]\times 100\% \quad (1)$$

Where η refers to acid dissolution, %; M_1 refers to ample weight before acidification, g; M_2 refers to sample weight after acidification, g.

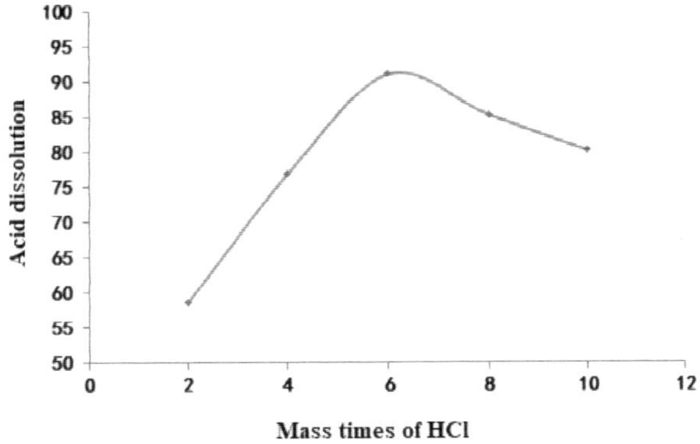

Figure 2. The curve of acid dissolution of ASCM.

The curve of acid dissolution of ASCM samples is shown in figure 2. The experimental results indicate that the acid dissolution of samples is raised with the increase of the dosage of hydrochloric acid. The maximum acid dissolution of ASCM is 91%. So ASCM has good acid dissolution.

The curing slug is obtained by solidified filter cake from ASCM. The acid dissolution of curing slug is evaluated by immersion in hydrochloric acid. The curve of acid dissolution of curing slug is shown in figure 3. The end-face immersion rolling test indicates that the acid dissolution of curing slug is above 85% when the immersion time is above 16 h.

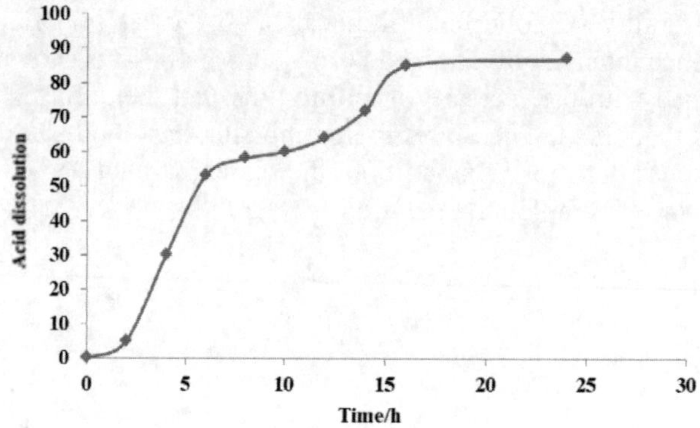

Figure 3. The curve of acid dissolution of curing slug from ASCM.

3.4. The evaluation of plugging performance of ASCM

The plugging slurry containing ASCM is prepared, and the plugging performance of ASCM is evaluated by HTHP sealing analog device. Using 10 mm slotted templet to simulate the leakage zone, the sealing strength of plugged zone produced from plugging slurry is shown in figure 4.

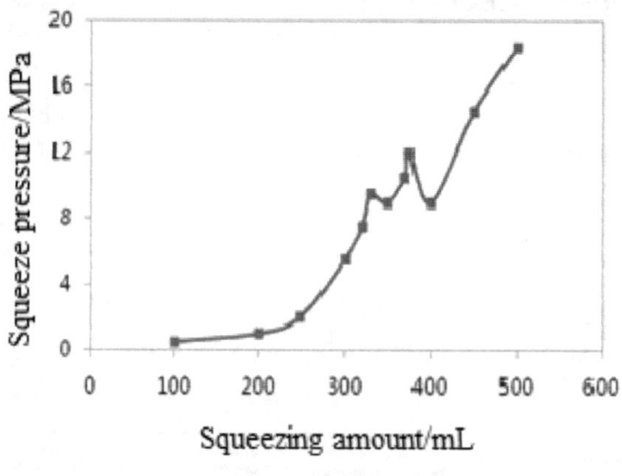

Figure 4. The squeeze pressure curve of plugging slurry containing ASCM.

The results indicate that the plugging slurry could reside and pile up in the simulated leakage zone. When the squeezing amount is more than 250 mL, the complete filtration of plugging slurry is encountered. The squeeze pressure rises gradually with the increase of squeezing amount. The plugging slurry containing ASCM is compacted gradually in the leakage zone, and the highest pressure-bearing capacity is as high as 18.5 MPa.

4. Conclusions

- A new type of acid soluble consolidating material is prepared. ASCM has the characteristics of rapid water loss, big water loss and high filter-cake consistency.

- The curing slug has good stability even at 180°C, the sealing strength of curing slug is as high as 10 MPa in 15 d, so ASCM could meet the requirement of completion operation.
- The acid dissolution of ASCM could reach up to 91%. The highest pressure-bearing capacity of plugged zone is up to 18.5 MPa. ASCM could seal the reservoir with porous leakage or fractured leakage effectively, and thus improve wellbore strength greatly.

Acknowledgments
This study is supported by the National Science and Technology Major Project (No. 2016ZX05020-004, 2016ZX05051, 2017ZX05030-004) and CNPC Project (No. 2018E-18, 2018D-5009-05, 2017D5008-03, 2016E-0109, 2016E-0608).

References

[1] Mansour A, Ezeakacha C, Taleghani A D, Li G and Salehi S 2017 Smart lost circulation materials for productive zones *Soc. Pet. Eng.* **1** 1-17
[2] Li L, Xu X G, Sun J S, Yuan X B and Li Y M 2012 Vital role of nanomaterials in drilling fluid and reservoir protection applications *Soc. Pet. Eng.* **1** 1-8
[3] Li L, Li Y M, Yang Y P and Ma C 2012 Application of nanomaterials in reservoir protection *Appl. Mech. Mater.* **204** 699-702
[4] Yin D, Li L, Xu X G and Wang J H 2016 Application of high density non-aqueous fluid technology in the efficient development and production of super-deep complicated formations in the Tian Mountain Front Block *Soc. Pet. Eng.* **1** 1-13
[5] Li L, Yuan X B and Sun J S 2013 Vital role of nanotechnology and nanomaterials in the field of oilfield chemistry *Soc. Pet. Eng.* **1** 1-7
[6] Li L, Sun J S and Xu X G 2012 Study and application of nanomaterials in drilling fluids *Adv. Mater. Res.* **535** 323-8
[7] Kaageson-Loe J N, Sanders M W and Growcock F 2008 Particulate-based loss prevention material? The secrets of fracture sealing revealed *Soc. Pet. Eng.* **1** 1-10
[8] Ivan C D, Bruton J R and Thiercelin M 2002 Making a case for re-thinking lost circulation treatments in induced fractures *Soc. Pet. Eng.* **1** 1-8
[9] Li L, Xu X G, Zhu J Z and Wang J H 2016 Application of innovative high-temperature high-density oil-based drilling fluid technology in the efficient exploration and development of ultra-deep natural gas resources in West China *Soc. Pet. Eng.* **1** 1-12
[10] Li L, Yin D and Li L 2018 Application of innovative high density high-performance water-based drilling fluid technology in the efficient development and production of ultra-deep complicated formations in the Tian Mountain Front Block *Soc. Pet. Eng.* **1** 1-12
[11] Mata F and Veiga M 2004 Crosslinked cements solve lost circulation problems *Soc. Pet. Eng.* **1** 1-5
[12] El-Sayed M, Ezz A, Aziz M and Waheed S 2007 Successes in curing massive lost-circulation problems with a new expansive LCM *Soc. Pet. Eng.* **1** 1-5
[13] Zhu J Z, Li L, Li L and Li J X 2015 Application of UDM-2 drilling fluid technology in the development of upper-deep oil and gas resources in Tarim Basin *Soc. Pet. Eng.* **1** 1-15

IOP Conf. Series: Materials Science and Engineering **634** (2019) 012009 doi:10.1088/1757-899X/634/1/012009

Novel hybrid composite based on bio-PET with basalt/carbon fibre

S Kuciel[1,2] and K Mazur[1]

[1]Tadeusz Kościuszko Cracow University of Technology, Institute of Materials Engineering, ul. Warszawska 24, 31-155 Krakow, Poland

E-mail: stask@mech.pk.edu.pl

Abstract. The aim of this work was to evaluate the possibility of manufacturing hybrid composites based on one of the latest generations of bio derivative polyethylene terephthalate with equal amounts of basalt and carbon fibres. This combination of two types of fibres gives the opportunity to obtain synergistic strength and performance properties. Samples obtained by injection moulding were characterized by morphological, mechanical (tensile, flexural, impact) and processing (softening temperature, thermal expansion and shrinkage) properties. Light-weight and bio-based hybrid composites based on bio-PET matrix with already 10 wt% basalt fibres and 10 wt% carbon fibres are characterized by more than 90 % increase in tensile strength and they have almost four times increase in the modulus of elasticity at a wide range of temperatures. Due to the good mechanical properties of hybrid composites, they can be successfully used in various industry sectors, mainly for consumer goods, household appliances and in automotive industry. Products made of bio-based sources pose a significantly lower threat to the environment than those from petrochemical sources.

1. Introduction

Exhausting sources of crude oil and an increase in public awareness about environmental protection have significantly influenced the interest of academic and industrial society in bio-based materials. However, still composites based on polyesters, polyamides and other types of polymers, which are derived from petroleum oils are often used in various industries. The production capacity of bio-based polymers represented only a 2% share of the total polymer production in 2013 and will increase to 4% by 2020 [1]. PET is the most commonly used semi-aromatic polyester. It is the fourth-most-produced plastic [2], with a global supply of more than 19.8 million tons in 2012 [3]. PET has been widely used as beverage bottles, food containers, fibres and fabrics, packing films, photographic and recording tapes, engineering resins, and so on [4]. However, alarming information about exhausting sources of crude oil has become an impulse to replace PET with a bio-based polymer. A competitor for PET is its counterpart produced from renewable sources- bioPET. BioPET accounts for 26% of the total production of bio-based composites. The biggest producer is Coca-Cola®, which uses for production of their bottles bioPET based on ethylene glycol (EG) and terephthalic acid (TA), which are produced from vegetable sugars [5]. Although bioPET has many advantages, its biggest disadvantage is the price of 2.25 €/kg, which is about 30% higher than traditional petrochemical polymers [6].

In light of this context, polymer composites are modified more and more often by the addition of fibres. BioPET modifications are not often described in the literature and most often the reinforcement composites are by introducing nanoparticles or lignocellulose fibres. In the work of Jang *et al* [7] bioPET (ECOZEN T95 with 9% mean biomass content) with the addition of nano-clay in the content of 0 up to 5 wt% has been tested. Studies have shown that with the increase in nano-clay content, the thermal and mechanical properties of the tested composites have decreased. After the additional introduction of supercritical carbon dioxide ($scCO_2$), the highest composites properties were obtained, exceeding the values obtained for neat bioPET. It was concluded that the addition of $scCO_2$ proved to be an effective exfoliation agent. Another work based on bioPET was the work of Montava-Jorda *et al* [8] who added cotton textile waste (CTW) from 1 to 10 wt% to bioPET provided by NaturePlast. The highest properties were obtained for composites with an addition of 3-5 wt% of CTW. The addition of CTW in each case increased the rigidity and hardness of composites, however, with increasing fibre content, ductility and toughness decreased. Improving bioPET properties as shown by research is not an easy task. Reinforcing polymer composites with one type of filler does not always give such results; therefore, more and more attention is focused on the production of hybrid composites. The purpose of combine two or even more fibre types in a single composite is to maintain the advantages of both fibres and improve the disadvantages. Therefore, the aim of this research work was to develop, for the first time, hybrid composites based on bioPET reinforced with basalt/carbon fibre. The main goal was to increase the mechanical and processing properties. The samples made using injection moulding were subjected to the characteristics of basic physical and mechanical properties, and additionally Vicat softening point, shrinkage and thermal expansion were determined.

2. Methodology

2.1. Materials

The materials used in this investigation were bio-based PET (ECOZEN® BS 400 – SK chemicals, Korea) carbon fibre (Zoltek PX35, Toray Group, USA) with nominal diameter of 7.2 μm, length of 4 mm and basalt fibre (Basaltex BCS17-6.4-KV16, Basaltex, Belgium) with nominal diameter of 17 μm, length of 6.35 mm. Standard dumbbell samples were injection moulded using the Engel ES 200/40 (Engel Austria, Schwertberg) machine after compounding fibre and bioPET by means of a twin screw extruder. The temperatures in the subsequent zones ranged from 230°C to 260°C and the injection speed ranged from 60 to 90 mm/s.

2.2. Method of testing

2.2.1. Density. The density of the produced composites was measured using the hydrostatic method at room temperature on a scale type RADWAG WAS 22W, (Radwag, Radom, Poland) according to EN ISO 1183. The samples were measured in ethanol medium.

2.2.2. Water absorption. Water absorption was carried out according to ASTM D570-98 standard. The specimens of bioPET and composites were immersed in distilled water at room temperature and weighed after 1 and 30 days using electronic weighing balance (RADWAG WAS 22W). Samples were removed from water bath and their weight was measured after surface drying using tissue paper. Water absorption was calculated using equation:

$$\%W = \frac{W_n - W_0}{W_0} \times 100$$

where W_0 is the initial weight of the sample, W_n is the weight of the saturated sample, %W gives the percentage increase in weight

Additionally, to assess the influence of absorbed water all compositions were tested in a conditioned state and after 30 days of soaking in water.

2.2.3. Mechanical properties. Specimens of mechanical testing were conditioned at 22°C/50% relative humidity for at least 80 h according to ISO 291 for test room conditions. Tensile tests were carried out according to PN-EN ISO 527-1:2012. MTS Criterion Model 43 testing machine (MTS Systems Corp.,

IOP Conf. Series: Materials Science and Engineering **634** (2019) 012009 doi:10.1088/1757-899X/634/1/012009

Eden Prairie, MN, USA) with maximum load range of up to 30 kN and with a traverse speed of 5 mm/min was used. For the accurate measurement of displacement allowing the determination of the tensile modulus, an MTS 634.31F axial extensometer was used. In addition to the tensile tests and three-point flexural tests were also conducted for the compositions according to PN-EN ISO 178:2011. Machine control was carried out from the level of an MTS TestSuite TW software 1.0 (MTS Systems Corp., Eden Prairie, MN, USA). Charpy impact test was carried out in accordance with PN-EN ISO 179-2 standard using Zwick/Roell MTS-SP (Zwick Roell Group, Ulm, Germany) testing machine. The average un-notched impact strength was calculated. The mechanical tests were performed at various temperatures of potential application. Tests were carried out at −24, 22 and at 80°C using the temperature chamber (Instron, Norwood, USA).

2.2.4. Processing properties. The softening point was determined by the Vicat method (PN-EN ISO 306:2014-02) by Ceast 6510 (Pianezza, Italy) machine using method A50 with a force of 10 N and a heating rate of 50°C/h. The percentage linear shrinkage (LS) of the specimen was calculated as:

$$LS = \frac{L_1 - L_2}{L_1} x100\%$$

where, LS is linear shrinkage, L_1 is selected dimension in the mould [μm] and L_2 is the same dimension measured on the mould at a certain temperature and pressure [μm]

2.2.5. Composite morphology. SEM structure images were taken on tensile-test fracture surfaces of specimens using scanning electron microscope JEOL JSN5510LV (JEOL Ltd., Tokyo, Japan). The accelerating voltage was 20 kV. All samples were coated with gold using Cressington 108 auto sputter coater (Cressington Scientific Instruments, Watford, UK) before observation.

2.2.6. Thermal expansion. The thermal properties of composites were performed on the NETZSCH 402 F1 Hyperion device (NERZSCH Group, Selb, Germany), where the samples were placed vertically. The dilatometric analysis consists in measuring the length of the sample (L) as a function of temperature (T). The length and temperature data were recorded and analysed with Proteus software. The samples were cooled from 30°C temperature to -60°C, heated to 140°C and finally cooled to -60°C temperature. The heating and cooling rates were kept constant at 10°C/min. The coefficient of linear thermal expansion (αL) was calculated as:

$$\propto_L = \frac{1}{L}\frac{dL}{dT}, \ 1/K$$

where L is the linear dimension of the test sample and dL/dT is the rate of change in the linear dimension per unit temperature.

3. Results

The composition of bioPET produced with a mixture of basalt fibres (BF) and carbon fibres (CF) and the results of basic physical and mechanical properties are presented in table 1. The addition of an equilibrium fibre mixture increases the density of the composition slightly up to 10% at the same time improving processing properties. As can be observed, as the fibre content increased, the thermal resistance of the composites increased, while the percentage of shrinkage decreased. The highest value improvement was observed for bioPET10.10 composites - 10% and over 60% for Vicat softening point (VSP) and shrinkage, respectively. Combination of basalt and carbon fibres creates a synergistic effect on the improvement of the thermal properties of the composite. As the fibre content increases, VST increases, which is due to the restriction of the movement of PET molecular chains by fibres, the increase in the Young's modulus at elevated temperatures and the increase in thermal resistance. Modular CFs are responsible for a significant increase in stiffness, while an increase in the thermal stability rigid BFs. The reduction of shrinkage is caused by particularly long basalt fibres, which show a much smaller than the matrix shrinkage, additionally they connect the distal areas of the composite, stabilizing its dimensions. Improvement of processing properties with a simultaneous small increase in mass significantly widens the scope of application of manufactured composites.

Table 1. Composition and basic physic and processing properties of the materials.

Index	Composition	Density [g/cm^3]	Vicat softening temperature [°C]	Shrinkage [%]
bioPET	Neat bioPET ECOZEN® BS 400	$1.27^{\pm0.02}$	$190.6^{\pm7}$	$1.2^{\pm0}$
bioPET5.5	ECOZEN® BS 400 + 5wt% BF + 5wt% CF	$1.32^{\pm0.04}$	$202.4^{\pm10}$	$0.65^{\pm0.01}$
bioPET7.7	ECOZEN® BS 400 + 7.5wt% BF + 7.5 wt% CF	$1.37^{\pm0.04}$	$205.7^{\pm9}$	$0.54^{\pm0.02}$
bioPET10.10	ECOZEN® BS 400 +10 wt% BF+ 10 wt% CF	$1.41^{\pm0.03}$	$209.8^{\pm12}$	$0.46^{\pm0.01}$

Table 2. Comparison of basic strength results.

Index	Tensile strength, [MPa]	Tensile modulus [MPa]	Energy to break [J]	Flexural strength [MPa]	Flexural modulus [MPa]	Impact Strength, [kJ/m^2]
bioPET	$52.3^{\pm0.7}$	$2\,081^{\pm75}$	$7.9^{\pm0.2}$	$86.4^{\pm0.9}$	$2\,128^{\pm95}$	$19.9^{\pm0.9}$
bioPET5.5	$76.3^{\pm1.1}$	$5\,864^{\pm105}$	$3.8^{\pm0.1}$	$126.3^{\pm1.0}$	$4\,860^{\pm121}$	$22.8^{\pm1.3}$
bioPET7.7	$95.1^{\pm0.9}$	$7\,687^{\pm103}$	$4.1^{\pm0.09}$	$149.2^{\pm0.8}$	$7\,418^{\pm89}$	$24.9^{\pm0.9}$
bioPET10.10	$97.1^{\pm1.2}$	$9\,603^{\pm121}$	$4.5^{\pm0.1}$	$159.2^{\pm1.4}$	$8\,336^{\pm152}$	$29.9^{\pm1.6}$

To determine the effect of hybridization on the mechanical properties of fibre mixtures (CF and BF) of composites based on bioPET, tensile, bending and Charpy impact strength tests were carried out (table 2). As the fibre content increased, there was a significant linear improvement in mechanical properties, this was mainly observed in the Young's modulus and bending modulus. The bioPET 10.10 composites were characterized by an almost twofold increase in tensile strength (52.3 MPa – bioPET and 97.1 MPa – bioPET10.10) and almost fivefold increase in the Young's modulus (~2 GPa – bioPET and 10 GPa – bioPET10.10). In addition, an increase in the impact strength was observed, for bioPET10.10 composites, where the growth was as higher as 50%. Mainly the increase in impact strength was caused by the addition of basalt fibres which are characterized by high values of impact resistance [9].

Due to the fact that bioPET reinforced composites can be used in various industrial sectors, where temperatures ranging from -50 (outdoor parking lots) up to 125°C (under the hood), mechanical tests have been carried out at temperatures: -24ºC and +80ºC [10]. In figure 1, the values obtained during the tensile test were presented: tensile strength (left) and Young modulus (right) at three temperatures. The highest properties were observed at reduced temperatures, while the lowest at +80°C. The decrease in strength properties is slightly different between different test temperatures, which proves bioPET stability at extreme temperatures, which may indicate a higher degree of crystallinity of the tested composites. Polymers with high crystallinity have a higher glass transition temperature T_g (T_g is 67°C for amorphous PET and 81°C for crystalline PET) and have higher modulus, toughness, stiffness, tensile strength, hardness and more resistance to solvents, but less impact strength [11].

Table 3 shows the values obtained during the three-point bending and Charpy impact strength without notch at three temperatures. Similar relationship as in the case of tensile strength was obtained during the bending test. The highest values were achieved at reduced temperatures for the highest fibre content. Along with the increase of fibre content, the bending strength and the module increase by about 2 times and by 3.5 times, respectively for bioPET10.10 composites. In the case of the impact strength value, the highest values were obtained at elevated temperatures, this is related to the increase in the plasticity of the materials. At elevated temperatures, along with the increase in fibre content, the values of impact strength decreased, due to higher stiffness.

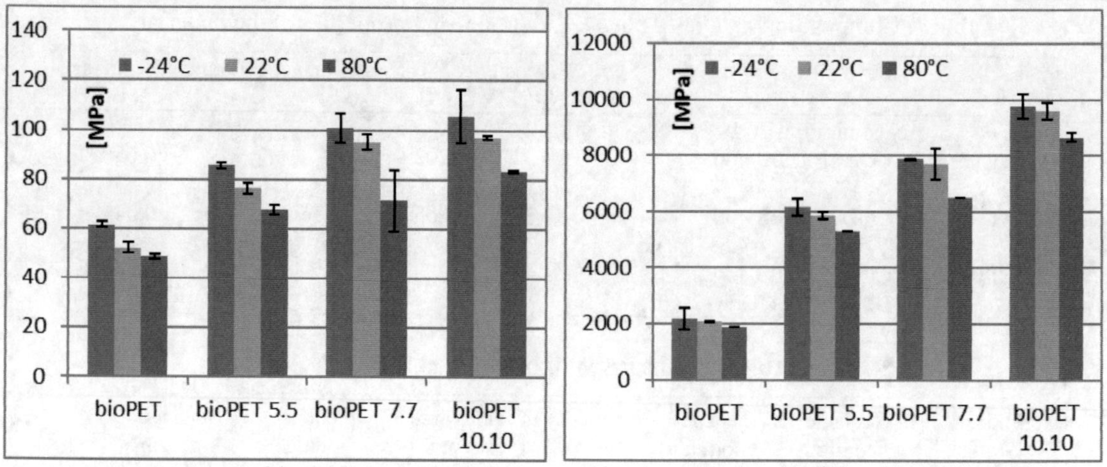

(a) (b)

Figure 1. Comparison of tensile properties: tensile strength (a) and Young modulus (b) at three temperatures.

Table 3. Results of bending and impact test at three temperatures.

Index		Flexural strength [MPa]	Flexural modulus [MPa]	Impact Strength [kJ/m^2]
bioPET	-24ºC	$93.8^{\pm0.8}$	$2\,222^{\pm195}$	$13.1^{\pm0.5}$
	22ºC	$86.4^{\pm1.0}$	$2\,128^{\pm121}$	$19.9^{\pm1.2}$
	80ºC	$81.9^{\pm1.3}$	$1\,861^{\pm129}$	$25.5^{\pm1.4}$
bioPET 5.5	-24ºC	$141.7^{\pm1.1}$	$5\,255^{\pm65}$	$16.0^{\pm1.4}$
	22ºC	$126.3^{\pm1.0}$	$4\,860^{\pm121}$	$22.8^{\pm1.2}$
	80ºC	$109.5^{\pm0.9}$	$4\,061^{\pm189}$	$26.6^{\pm2.1}$
bioPET 7.7	-24ºC	$162.2^{\pm1.5}$	$6\,731^{\pm175}$	$16.1^{\pm1.2}$
	22ºC	$149.2^{\pm1.0}$	$7\,418^{\pm121}$	$24.9^{\pm1.7}$
	80ºC	$110.6^{\pm1.6}$	$5\,122^{\pm139}$	$25.1^{\pm1.8}$
bioPET 10.10	-24ºC	$179.5^{\pm1.8}$	$8\,344^{\pm205}$	$13.1^{\pm1.4}$
	22ºC	$159.2^{\pm1.0}$	$8\,336^{\pm121}$	$19.9^{\pm1.1}$
	80ºC	$128.9^{\pm2.1}$	$7\,724^{\pm129}$	$20.1^{\pm2.9}$

Table 4. Water absorption of the tested composites.

Time	bioPET	bioPET 5.5	bioPET 7.7	bioPET 10.10
1 day [%]	0.07	0.065	0.063	0.06
30 days [%]	0.26	0.24	0.22	0.20

Table 4 presents the results of water sorption for the tested composites after 1 day and 30 days of soaking. As can be seen, the addition of fibres caused a decrease in water sorption. As the fibre content increases, the amount of water taken up by the material decreases. For bioPET10.10 (0.20%) the difference between bioPET neat (0.26%) after 30 days was 0.06% (table 4). Additionally, the influence of water sorption after 30 days on strength properties was examined (figure 2). After 30 days, it can be observed a decrease in strength properties. The reduction of mechanical properties after 30 days of hydrothermal aging is the effect of water sorption at the fibre boundaries and stress relaxation between the fibres and the matrix, and disintegration of chains due to the reduction of the molecular weight of the polymer matrix. Under the influence of water macromolecular chains break with gradual decrease of molecular weight, up to monomers. However, it is worth noting that the decrease in properties was slight to a maximum of 5%.

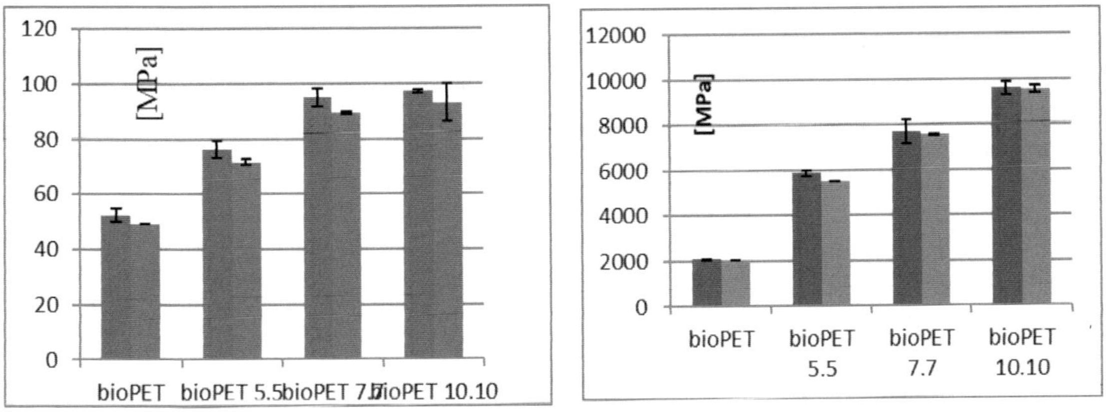

Figure 2. Comparison of tensile test result immediately after injection (red) and after 30 days of soaking in the water (blue).

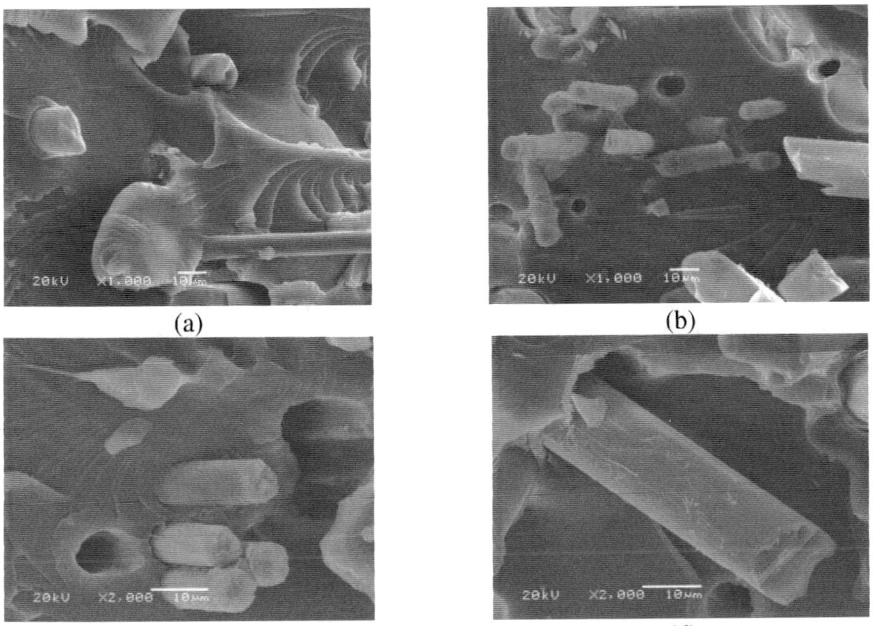

(a) (b) (c) (d)

Figure 3. SEM images of the bioPET composites with basalt and carbon fibres (the fracture after elongation test): (a) bioPET 5.5, (b) bioPET 10.10, (c) bioPET 5.5 and (d) bioPET 10.10.

SEM was carried out to evaluate the state of dispersion and adhesion of BF and CF in bioPET matrix. Micrographs of the fractured surfaces of tensile specimens are presented in figure 3(a) bioPET 5.5 at 1000x magnification –area of developed fracture of bioPET with broken carbon fibre, figure 3(b) bioPET 10.10 at 1000x magnification – area of mixing basalt and carbon fibres -one broken basalt fibre and "whole" after pulling out of any carbon fibre on ductile fracture of bioPET figure 3(c) bioPET 5.5 at 2000x magnification a group of parallel carbon fibres, figure 3(d) bioPET 10.10 at 2000x magnification - a single basalt fibre in the bioPET matrix Fractured specimens after elongation test are characterized by a combined characteristic from a brittle fracture (for higher addition of fibres) to a ductile fracture (for bioPET 5.5 with 'pull out' effect for parts of BF). Various cross sections of BFs were observed with measured diameter of basalt fibre: 16-23 μm. SEM images indicate better adhesion to CF than BF. On CF, it can be seen the residual of the matrix and much lower in the pull out phenomenon than in the case of BF. Additionally, good fibre adhesion is confirmed by durability tests.

The values of linear coefficient of thermal expansion of tested materials have been presented in table 5. Neat bioPET had the highest value. This is due to the fact that thermoplastic polymers have high coefficients of linear thermal expansion due to weak secondary bonds between the chains. While the filler content added to the polymer matrix increased, the linear coefficient of thermal expansion decreased. It decreased almost three times. This is a positive phenomenon due to the dimensional stability of elements made of such composites, better possibility of their cooperation and accuracy of fixing.

Table 5. Results of dilatometry test.

Index	Linear coefficient of thermal expansion $\alpha \times 10^{-6}$, 1/K -20°C to 100°C	Softening temperature [°C]	Maximum of extension [%]	The shrinkage start temperature [°C]
bioPET	8.88	104.5	0.66	128.3
bioPET 5.5	3.14	107.3	0.25	150.8
bioPET 7.7	2.70	108.1	0.24	151.1
bioPET 10.10	2.73	108.1	0.21	154.8

4. Conclusion

The conducted research confirmed the possibility of producing an effective reinforcement for bioPET matrix by carbon and basalt fibres added simultaneously. The strength increase for the bioPET 7.7 composites was up to 85% and the tensile modulus increased almost four times. The impact strength remained constant, thermal resistance increased and the technological shrinkage decreased twice. Such composites become an attractive materials for technical products operating at elevated temperatures. To sum up, polymers from renewable sources are an attractive structural material that is not inferior to traditional polymers the from crude oil or gas. Due to the environmental protection requirements and the objective necessity of using renewable raw materials, research on new types of polymers available on the market due to the modern achievements of material engineering should be developed.

References

[1] Aeschelmann F and Carus M 2015 Biobased building blocks and polymers in the world: Capacities, production, and applications–status quo and trends towards 2020 *Ind Biotechnol* **5** 1-500

[2] Biron M 2007 Elsiever Science *Therm and therm comp.* chap. **2** *Thermoplastics and Thermoplastic Composites* 33-153

[3] Tsiropoulos I, Faaij A P C, Lundquist L, Schenker U, Briois J F and Patel M K 2015 Life cycle impact assessment of bio-based plastics from sugarcane ethanol *J Clean Prod.* **90** 114-127

[4] Jiang Y and Loos K 2016 Enzymatic synthesis of biobased polyesters and polyamides *Polymers* Polymers **8** 243-251

[5] Shen L, Worrell E and Patel M K 2012 Comparing life cycle energy and GHG emissions of bio-based PET, recycled PET, PLA, and man-made cellulosics *Biofuels, Bioprod Biorefining* **6** 625-399

[6] Carus M, Eder A and Beckmann J 2014 Green premium prices along the value chain of biobased products *Ind Biotechnol* **10** 83-8

[7] Jang K, Lee J W, Hong I K and Lee S 2013 Effect of supercritical carbon dioxide as an exfoliation aid on bio-based polyethylene terephthalate glycol-modified/clay nanocomposites

Korea Aust Rheol J **25** 145-52

[8] Montava-Jordà S, Torres-Giner S, Ferrandiz-Bou S, Quiles-Carrillo L and Montanes N 2019 Development of sustainable and cost-competitive injection-molded pieces of partially bio-based polyethylene terephthalate through the valorization of cotton textile waste *Int J Mol Sci.* **20** 1378

[9] Vara Prasad V and Talupula S 2018 A review on reinforcement of basalt and aramid (Kevlar 129) fibers *Mater Today Proc* **5** 5993-6008

[10] Bouzouita A 2016 Elaboration of polylactide-based materials for automotive application: Study of structure-process-properties interactions Amani Bouzouita Mechanics of materials [physics.class-ph]. *Université de Valenciennes et du Hainaut-Cambresis*; Université de Mons English. NNT: 2016VALE0027

[11] Yaraş A, Elçiçek H, Demirel B, Yaraş A and Elçiçek H 2011 Crystallization behavior of PET materials BAÜ Fen Bil. Enst. Dergisi Cilt **13** 26-35

IOP Conf. Series: Materials Science and Engineering **634** (2019) 012010 doi:10.1088/1757-899X/634/1/012010

Effect of wood/basalt hybridization on crystallization and mechanical properties of PLA

K Mazur[1,2] and S Kuciel[1]

[1]Tadeusz Kościuszko Cracow University of Technology, Institute of Materials Engineering, ul. Warszawska 24, 31-155 Krakow, Poland

E-mail: mazur117@o2.pl

Abstract. In this study hybrid composites based on polylactide were reinforced with wood and basalt fibres. Composites have undergone physical and mechanical tests and the water absorption after 1 and 14 days as well as the effect of biodegradation on strength properties have been determined. Additionally the effects of annealing and temperature of tests on mechanical properties were investigated. The addition of fibres considerably increased the susceptibility of the tested composites and for composites with addition of 7.5 wt% of each wood and basalt fibres increased by more than 35% at +23°C and was by 12 times higher at 80 °C. Heat treatment of composites positively influenced the results, mainly at elevated test temperatures. For pure PLA, the increase in Young's modulus was over 22 times (from 157 MPa to 3 473 MPa). The addition of fibres resulted in a reduction of the differences between the materials before and after annealing. Impact strength was more than double for composites after heat treatment at room temperature.

1. Introduction

In recent years, biodegradable polymers have become more and more popular especially in the scientific research, due to their excellent mechanical properties, comparable to petrochemical polymers. An additional advantage of biodegradable polymers is their source of origin (renewable resources and nontoxic to the human body) and the fact that after the end of their life cycle they decompose into products naturally occurring in nature. Polylactide (PLA) is the most commercially used polymer in this group. PLA has been used in many fields, as a drug delivery, tissue engineering, and biodegradable fibres. However, due to the good mechanical properties, it is also increasingly used in other industries, such as automotive or aerospace. In these areas of application, fibres are often used to improve mechanical properties. Mainly, in the case of biodegradable composites the fibres come from natural sources, which not only improve their properties, but additionally leave the material completely biodegradable. The most common fibres reinforcing biodegradable polymers are fibres consisting of lignocellulose and hemicelluloses. In literature, the most common lignocellulosic fibres which reinforce the polymers are flax, hemp, jute, kenaf, and abaca [1,2]. Unfortunately, fibres containing lignocelluloses have some disadvantages, such as heat or water sensitivity, as well as poor adhesion due to their high hygroscopicity compared to hydrophobic polyolefin matrices. Moreover, the addition of natural fillers makes them difficult to manufacture due to the fact that natural fillers, mainly those containing lignocellulose, are heat-sensitive components [3]. An alternative to lignocellulosic fibres are mineral fibres, naturally occurring in nature, which after incorporation into the polymer do not impair its biodegradability. The new natural reinforcement of polymer composites is basalt. Basalt fibres (BF) are characterized by excellent mechanical properties and high thermal resistance. They form a bridge between the most commonly used synthetic fibres, i.e. glass and carbon fibres. In comparison with glass fibres (GF), they have higher mechanical properties and higher thermal and chemical resistance, and compared to carbon fibres (CF), although they have lower strength and stiffness, they are cheaper and have better thermal and chemical resistance.

To accelerate the crystallization and improve crystallinity of PLA in order to increase the properties, numerous strategies have been employed, for instance, by blending, copolymerization, and manipulation of the processing. One of the most commonly described in the literature procedure to increase the crystalline phase in PLA is the addition of a nucleating agent, however, in the case of medical applications of PLA composites, this process is ruled out because often the nucleating agents are not biocompatible. Therefore, the appropriate selection of parameters during injection becomes an alternative and, as results from earlier tests, an effective attempt to increase the degree of crystallinity. Therefore, tailoring PLA's crystallization behaviour has been identified as an important step in manufacturing in order to improve the properties [4]. The improvement of PLA properties by means of heat treatment is the subject of research by scientists. All works show that this process will significantly improve the strength properties of polymers. Srithep *et al* [5] investigated the effect of annealing time and temperature on the crystallinity and heat resistance of injection moulded PLA. Their study showed that after annealing at 80°C for 30 min improved mechanical properties- 17% of tensile strength.

Not only the addition of one type of fibre or heat treatment of PLA composites can significantly improve their properties. In recent years, the method of hybridizing composites has become an increasingly popular method of reinforcing polymer composites. This method consists in introducing at least two different components into the matrix In the case of biodegradable polymers, the production of reinforcing fibres is important. Therefore, the simultaneous addition of two natural fibres, which are wood fibre and basalt fibre, causes not only the attribution of properties, but also leaves the material environmentally friendly. Therefore, this work concerns hybrid WF/BF composites. WF will be responsible for the increase of stiffness and density reduction of composites, while BF for the increase of mechanical properties. In addition, the influence of cold crystallization on the mechanical properties of the hybrid composites was determined. So far, the effect of hybridization of lignocellulosic and mineral properties on PLA has not been investigated.

2. Methodology

2.1. Materials

Polylactide grade PLI003 manufactured by Nature Plast (Isf, France). This polymer has the following properties: density — 1.24 g/cm^3; fluidity index (190°C/2.16 kg) — 35 g/10 min; tensile modulus — 3500 MPa; un-notched Charpy impact strength — 16 kJ/m^2; and thermal resistance (HDT B) — 55°C.

Wood fibres (Lignocel C120) manufactured by J. Rettenmaier&Söhne GmbH, (Rosenberg, Germany). The particles size is 70- 150 μm.

Chopped basalt fibres (BCS17-6.4-KV16) provided by Basaltex Inc. (Wevelgem, Belgium) whose diameter amounted to 17 μm, length to 6.35 mm, density to 2.67 g/cm^3, tensile modulus to 84 GPa, and tensile strength to 1 180 MPa;

Biocomposites:

PLA biocomposites were obtained using an injection moulding equipment Engel ES 200/40 HSL (Schwertberg, Austria). Injection moulding was carried out in the laboratory of plastics technology operating under GrupaAzoty SA (Tarnów, Poland). The size of the samples was 10x4x150 mm (dumb-bell shape).

Additionally, some of the injection moulded samples underwent isothermal heat treatment at 80°C for 1 h. Samples were placed to hot oven and they were hold between two glass sheets to prevent their deformation.

2.2. Method of testing

2.2.1. Density. The density of the produced composites was measured using the hydrostatic method at room temperature on a scale type RADWAG WAS 22 W, (Radwag, Radom, Poland) according to EN ISO 1183. The samples were measured in ethanol medium.

2.2.2. Morphology of fracture surfaces. Scanning electron microscope (SEM) analysis was conducted to describe the nature of failure and explain damage morphology employing low vacuum microscope JEOL 5510LV. The examined samples fractured during the tensile test were coated with a thin gold layer using vacuum coater (Carington) before observation.

2.2.3. Impact strength. The un-notched Charpy impact strength (a_{cU}) was measured using Zwick/Roell MTS-SP (Ulm, Germany) testing machine with a 1.5 J hammer.

2.2.4. Mechanical properties. The tensile test was performed by virtue of universal MTS Criterion Model 43 testing machine. The tensile test was performed with a constant cross-head speed of 5 mm/min, and the elongation was determined using the MTS axial extensometer and crosshead displacement. Furthermore, the test was run at room temperature (+23±2°C) and at decreased (-23°C) and elevated (+80°C) temperature range, which reflected the lowest and highest temperatures at which PLA composites can find practical application. Before the test, the samples had been conditioned and at the test temperature for 60 minutes.

The reported average data were based on the results of the tests concerning at least five standardized samples, along with calculated standard deviations to evaluate test reproducibility.

2.2.5. Hydrolitic degradation. The standardized samples were subjected to hydrothermal aging. They were immersed in a saline solution (2 wt. % salt) at 40 °C, to simulate human body environment. The samples were removed from the tank after 1 day and 14 days. The water remaining on the surface was wiped with a tissue paper and weighed. Finally, the water uptake was calculated as the mass difference, expressed as a percentage with the following formula:

$$M_t = \left[\frac{W_t - W_0}{W_0}\right] \times 100 \qquad (1)$$

where M_t - percentage of water content, W_t -instantaneous weight of the sample, and W_0 - initial weight of the sample.

Additionally, the tensile test was performed after 7 and 14 days of hydrothermal aging.

3. Results

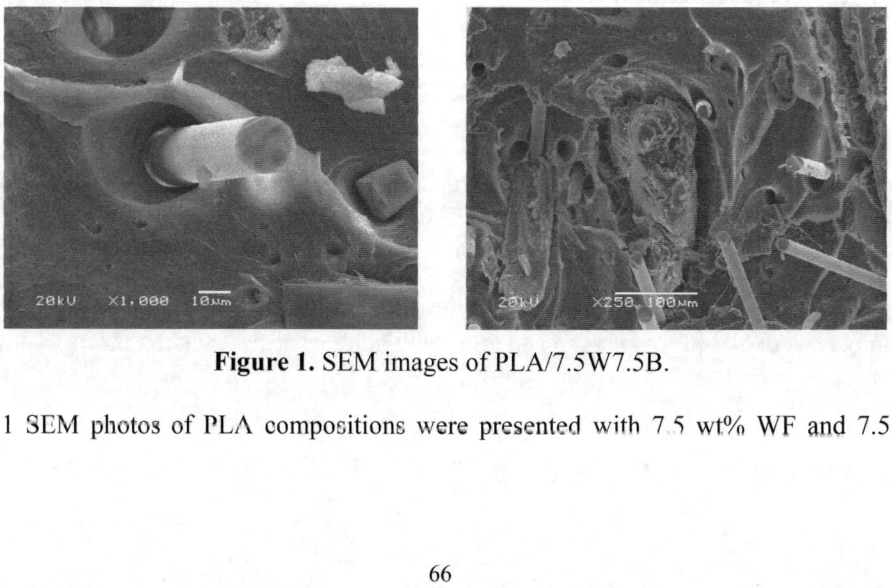

Figure 1. SEM images of PLA/7.5W7.5B.

In figure 1 SEM photos of PLA compositions were presented with 7.5 wt% WF and 7.5 wt% BF without annealing at the breakthroughs. Plastic breakthrough and evenly distributed fibres are observed. There are no visible agglomerations, and basalt and wood fibres mutually intertwine. There is a clear difference between the construction of WF, which are about 150 μm and made of several microfibrils, and thin, long BF. In the case of BF, the pull out phenomenon is more visible than in the case of WF. The remains of the matrix are visible on the BF, which may indicate good fibre-matrix adhesion, which has been confirmed in mechanical tests.

Table 1 shows the density values, impact strengths determined during the Charpy test without notch and the Vicat softening point (VSP). As can be seen, there was an increase in the density of the tested composites by about 5%, however, this increase is insignificant, because the composites with only 15 wt% of BF based on the PLA matrix have a density of 1.36 g/cm^3 [6]. Addition of only WF would result in lowering density, while reducing strength values, so the creation of hybrid composites allows to create a material with relatively good strength properties compared to density. The addition of fibres will not significantly affect impact values, which is a beneficial phenomenon, because the decreased in impact strength of the lignocellulosic fibres described in the literature has been compensated by the addition of BF, which are characterized by high impact strength. Whereas, materials with an increased degree of crystallization were characterized by an over-two-fold increase in the impact strength- PLA/7.5W7.5B from 21.5±0.4 to 47.7±3.2 (kJ/m^2). With the increasing fibre content VST increased. It is due to the increased stiffness of composites and high thermal stability of rigid BFs. The same situations like for impact strength occurred for VSP after heat treatment. The results are more than twice higher than for untreated composites.

Table 1. Density values, Charpy impact strength and Vicat softening point depending on the state.

	Samples	Density (g/cm^3)	State	(a_{cU}) (kJ/m^2)	Vicat softening point (°C)
PLA	Polylactide	1.25±0.01	Semi-amorphous	20.4±0.5	60.8±1.9
cPLA	Nature Plast		Crystallized	40.6±1.2	161.2±2.5
PLA/3.5W3.5B	Polylactide + 3.5/3.5	1.27±0.01	Semi-amorphous	19.9±0.3	71.4±1.9
cPLA/3.5W3.5B	wt.% wood/basalt		Crystallized	42.9±1.8	161.5±3.1
PLA/7.5W7.5B	Polylactide + 7.5/7.5	1.32±0.02	Semi-amorphous	21.5±0.4	77.4±2.1
cPLA/7.5W7.5B	wt.% wood/basalt		Crystallized	47.7±3.2	164.9±2.7

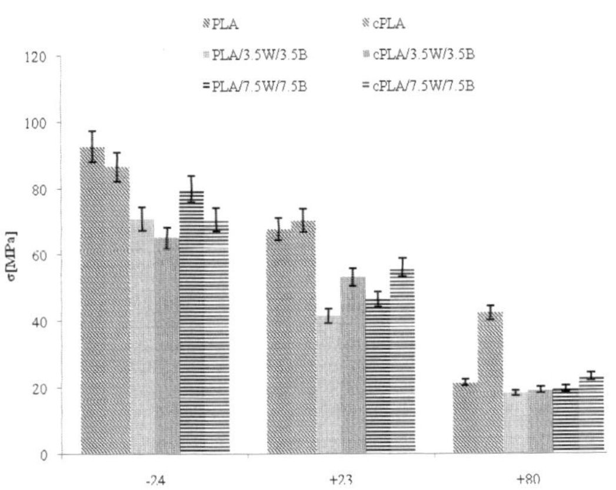

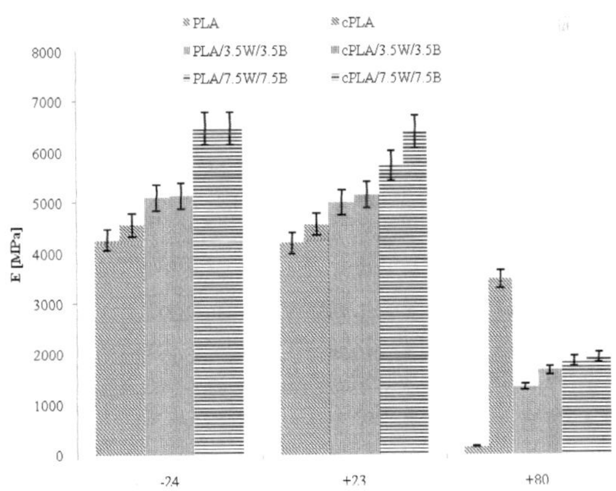

Figure 2. The values of tensile strength (left) and elastic modulus (right) depend on temperature and state.

In order to determine the mechanical properties of the produced materials, a tensile test was carried out. In figure 2. the values of tensile strength and Young's modulus have been presented in two states at different temperatures (-24°C, 23°C and +80°C). The addition of fibres resulted in a decrease in the tensile strength, however, as the fillers increase, the results are improved. This decrease is related to the addition of lignocellulosic fibres, the addition of which reduces the tensile strength [7]. The simultaneous addition of BF will not significantly increase the tensile strength in small amounts - 3.5 wt%. On the other hand, the increase in the content of BF, which have high strength properties, equalized values in relation to PLA at extreme temperatures. The addition of fibres improved the Young's modulus values at all tested temperatures. As the percentage of fibres increases, the value of the elastic modulus increases. As research shows, both WF and BF added separately significantly increase Young's modulus [8]. The highest increase in stiffness caused by the addition of fibres occurred at high temperatures, where the difference was 88.2% and 91.7% for PLA/3.5W3.5B and PLA/7.5W7.5B, respectively. These differences are due to the glass transition temperature PLA, which is within 60°C, which makes the unmodified polymer go from a brittle to a plastic state, which increases its deformation at the expense of strength properties. The addition of fibres has increased this temperature and thus increased stiffness. It is worth noting that the highest values for both tensile strength and Young's modulus were obtained at reduced temperatures.

The contribution of the crystalline structure, molecular mass and structure of the polymer chain is responsible for the strength properties of injected polymer composites. Therefore, in the tests, the composites were heat treated to obtain the crystalline phase. Thermal treatment of materials influenced the increase of PLA crystallinity, thus improving the tensile strength values at 23 and +80°C and slightly decreased at -24°C. The higher the temperature of the tests, the higher positive effect of crystallization was observed. The effect of crystallization on the increase in the properties of PLA composites is widely described in the literature [9]. The crystalline phase produced in PLA is stronger and stiffer than the amorphous phase. The highest improvement in properties was obtained for PLA at 80°C, increase by 50%. The increase in the crystalline phase in the composites improved the Young's modulus. In particular, this is observed in the case of elevated temperatures, where for PLA without crystallization, the decrease was over 95%, with the increase in the proportion of fibres the differences decrease. The addition of fibres made the reinforcement effect due to crystallization lower than in the case of neat PLA. Similar results were obtained by Wang *et al* [10], who investigated the effect of the crystallization time on the properties of PLA composites with GF (0wt%, 5wt%, 10wt%, 15wt% and 20wt%). The crystallization performed (80°C for 90 min) positively influenced the strength values of the composites. There was an increase in tensile strength and Young's modulus for pure PLA-60.4% and 34.3%, and for composites with 20wt% GF-20.6% and 35.8%, respectively.

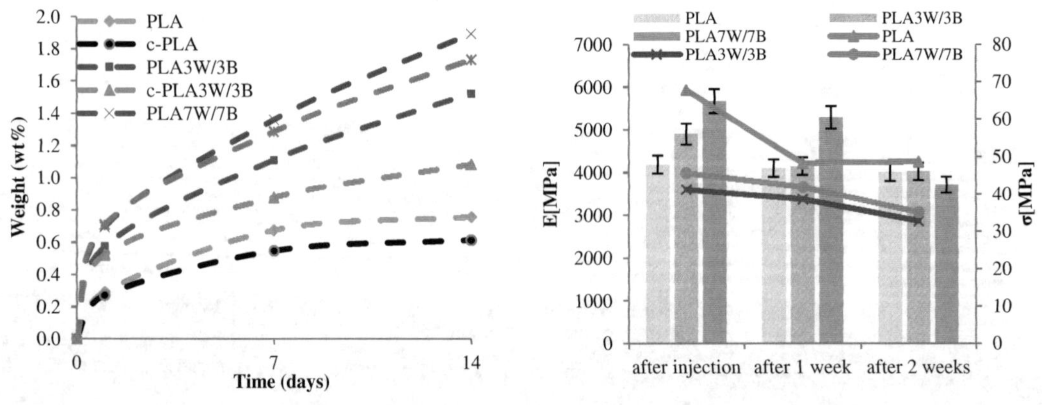

Figure 3. Weight gain for PLA composites, tensile strength and Young's modulus after incubation in water.

Biodegradable composites absorb water more quickly than traditional petrochemical composites, therefore, in order to comprehensively characterize biocomposites, it is important to determine water absorption and its impact on mechanical properties. In figure 3 were shown weight gain (left) and tensile properties after injection, 1 week and 2 weeks of immersion in water. The combination of hydrophilic ligocellulose fibres with non-absorbing mineral fibres allows to optimize some parameters. In the previous work, where the addition of only one type of fibre 7.5wt% WF, water absorption after 14 days was 2.15%, and 15wt% WF 3.51% [6]. Comparing these results with the results of hybrid composites, it can be noticed that the introduction of 7.5wt% WF and 7.5wt% BF resulted in a reduction of water absorption up to 1.8% and PLA/3.5W3.5B- 1.5%. The increase in the degree of crystallization will positively affect the water absorbency results for all materials tested. Similar results were obtained in the study of Kuciel *et al* [11]. Where the hybrid composites on the PLA matrix were tested with CF and BF (7.5/7.5 and 12.5/12.5wt%). It has been confirmed that, both the introduction of carbon and basalt fibres as well as the increase in crystallinity reduced the possibility of water sorption. After 7 days of incubation in water for PLA12.5/12.5 after heat treatment, the water absorption was 20% lower than for the neat heat treated polylactide and as much as 50% lower than for the neat and non-heat-treated polylactide.

4. Conclusion

In the work, hybrid composites based on PLA reinforced with wood and basalt fibres with improved mechanical properties have been created. Reinforcement composites with natural fibres not only creates a material with high strength properties, but also leaves it harmless to the environment after the biodegradation process. The content of lignocellulosic fibres, which negatively influence the strength properties of composites, was compensated by the addition of basalt fibres, leaving the material still low in density. Furthermore, after the addition of fibres, water absorption decreased, and the strength properties after 14 days of soaking in water only slightly decreased. In order to increase the degree of crystallinity, heat treatment was carried out. Increased crystallization after heat treatment leads to a significant improvement in the mechanical properties of pure PLA and PLA/WB composites. Both the addition of fibres and heat treatment significantly improved processing properties (VSP). As the presented research shows, the created hybrid composites, after optimizing the production process, taking into account mainly the additional heat cycle, can be successfully used in various industry sectors. They have high mechanical properties, even after long use in water environments, which is often a problem for biodegradable composites. In addition, their advantage is the low harm to nature after the end of the life cycle, because they break down into products naturally occurring in nature.

Acknowledgments

The authors acknowledge the support of the PROM programme no. PPI/PRO/2018/1/00013/U/001 which is co-financed by the European Social Fund under the Knowledge Education Development Operational Programme.

References

[1] Huda M S, Drzal L T, Misra M and Mohanty A K 2006 Wood-fiber-reinforced poly (lactic acid) composites: Evaluation of the physicomechanical and morphological properties *J Appl Polym Sci.* **102** 4856-69
[2] Bledzki A K and Jaszkiewicz A 2010 Mechanical performance of biocomposites based on PLA and PHBV reinforced with natural fibres - A comparative study to PP *Compos Sci Technol* **70** 1687-96
[3] Espinach F X, Delgado-Aguilar M, Puig J, Julian F, Boufi S and Mutjé P 2015 Flexural properties of fully biodegradable alpha-grass fibers reinforced starch-based thermoplastics *Compos Part B Eng.* **81** 98-106
[4] Mekonnen T, Mussone P, Khalil H and Bressler D 2013 Progress in bio-based plastics and plasticizing modifications *J Mater Chem A.* **1** 13379
[5] Srithep Y, Nealey P and Turng L S 2013 Effects of annealing time and temperature on the crystallinity and heat resistance behavior of injection-molded poly (lactic acid) *Polym Eng*

Sci. **53** 580-8

[6] Mazur K and Kuciel S 2018 Biodegradable composites based on polylactide reinforced with basalt and wood fibres *1st Conf. Materilas Design and Applications* Porto (Portugal)

[7] Bledzki A K and Faruk O 2006 Injection moulded microcellular wood fibre-polypropylene composites *Compos Part A Appl Sci Manuf.* **37** 1358-67

[8] Awal A, Rana M and Sain M 2015 Thermorheological and mechanical properties of cellulose reinforced PLA bio-composites *Mech Mater.* **80** 87-95

[9] Chen X, Ren J, Zhang N, Gu S and Li J 2015 Effects of heat treatment on the thermal and mechanical properties of ramie fabric-reinforced poly (lactic acid) biocomposites *J Reinf Plast Compos.* **34** 28-36

[10] Wang G, Zhang D, Li B, Wan G, Zhao G and Zhang A 2019 Strong and thermal-resistance glass fiber-reinforced polylactic acid (PLA) composites enabled by heat treatment *Int J Biol Macromol.* **129** 448-59

[11] Kuciel S and Romańska P 2018 Hybrid composites of polylactide with basalt and carbon fibers and their thermal treatment *Materials (Basel).* **12** 95-105

IOP Conf. Series: Materials Science and Engineering **634** (2019) 012011 doi:10.1088/1757-899X/634/1/012011

Mechanical properties and freeze-thaw durability of recycled aggregate pervious concrete

H B Liu[1], W J Li[1], H Yu[1], G B Luo[1] and H B Wei[1,2]

[1]College of Transportation, Jilin University, Changchun 130025, China

E-mail: weihb@jlu.edu.cn

Abstract. Recycled aggregate (RA) was obtained from waste cement concrete brick with the method of crushing and sieving. In order to research the effect of RA content on the properties of pervious concrete, RA was added to replace natural aggregate with equivalent quality method at different levels (25%, 50%, 75% and 100%). The natural aggregate pervious concrete and RA pervious concrete were prepared in the lab. The porosity, permeability, compressive strength, flexural strength and freeze-thaw (F-T) durability of RA pervious concrete were tested. The results indicated that the compressive strength and flexural strength of RA pervious concrete decreased with the increase of RA content. The rapid decreases for compressive strength and flexural strength occurred when the RA content was 25%-75% and 0%-25%, respectively. The porosity and permeability decreased slowly with the increasing RA content. F-T durability of RA pervious concrete was poorer than the natural aggregate pervious concrete. The incorporation levels had negative effect on the F-T durability, but the compressive strength loss of all groups after 25 F-T cycles met the requirement of the national standard. Although the mechanical properties and F-T durability of RA pervious concrete are not satisfied, it can still be used for sidewalks, parks and other light traffic roads.

1. Introduction

In recent years, with the quick economic development and the rise of city construction, the massive amount of the construction waste, if not treated in time, will cause serious environmental pollution [1]. Pervious concrete is considered as an environmentally friendly concrete due to its porous structure and high permeability. It has many environmental benefits, such as absorbing automobile exhaust, restoring groundwater supply, improving water quality and reducing water and soil pollution [2,3]. Moreover, due to its substantial pores, pervious concrete can not only improve the visibility and slip resistance of the road surface, but also abate the traffic noise and mitigate the urban heat island [4].

The recycle and reuse of non-degradable construction concrete waste has great social and economic potential [5,6]. RA is used in pervious concrete can effectively solve the environmental pollution caused by concrete waste. And it is an economic and environmental solution for the sustainable development [7]. In general, RA obtained from construction concrete waste is used to partially or completely replace the natural aggregate in the pervious concrete, which can achieve the purpose of cyclic utilization [8].

Zaetang *et al* [9] researched the mechanical properties of pervious concrete. The natural aggregate (NA) was replaced by the recycled concrete block aggregate (RBA) and recycled concrete aggregate (RCA) respectively with equivalent quality method at different incorporation levels (20%, 40%, 60%, 80% and 100%). The results showed that the compressive strengths of RBA and RCA pervious concrete increased first and then decreased with the increase of RA content. Both RBA and RCA improved the compressive strength of pervious concrete. However, the RA content had little impact on the flexural strength of previous concrete. The experiment conducted by Güneyisi *et al* [10] indicated that the compressive strength of RA pervious concrete decreased with the increase of RA content. Zhu *et al* [11] obtained the same conclusions that the compressive strength and flexural strength of RA pervious concrete decreased with the increasing RA content.

The increase of RA content leaded to the loose structure and low density of pervious concrete, which resulted in the increasing porosity [12]. The permeability of RA pervious concrete, reported by Yap *et al* [13], increased with the increasing RA content due to the large porous network in RA mixture. Zhu [14] concluded that the RA content had little impact on the effective porosity and permeability of pervious concrete.

The F-T durability of pervious concrete is very important due to the porous structure [15]. Generally, the relative dynamic modulus of natural aggregate previous concrete without admixtures decreased to 60% of the initial value after 50 F-T cycles. Wang [16] conducted a comparative research on the F-T resistance of RA previous concrete (RPC), natural limestone aggregate previous concrete and ordinary concrete. The aggregate particle size, water-cement ratio and porosity were 5-20mm, 0.25 and 15%, respectively. The result indicated that the compressive strength loss and the mass loss of RPC were 5.6% and 2.3% respectively after 25 F-T cycles.

In addition, Sriravindrarajah *et al* [17], Berry *et al* [18], Chen *et al* [19] also conducted experimental research on the property of RA pervious concrete. Although many researchers have studied the basic mechanical properties and permeability of pervious concrete under different RA content, the research methods adopted were different and the conclusions obtained were not consistent. At the same time, there was little research on the relationship between the RA content and the F-T durability of pervious concrete. Therefore, in this paper, the RA obtained from waste concrete brick was used to research the influence of RA content on the mechanical properties and F-T durability of pervious concrete.

2. Materials

2.1. Materials

Portland cement with strength grade of P.O 42.5 was selected in this study. Single gradation of coarse aggregate was adopted to meet the requirement of permeability for pervious concrete. Two different aggregates, natural aggregate and recycled concrete aggregate showed in figure 1, were used as coarse aggregate. The natural granite coarse aggregate was obtained from a concrete mixing station in Jilin Province, China. The RA, produced by waste concrete brick, was obtained from Jilin Xusheng Environmental Protection Building Material Co., Ltd. (Jilin, China). The particle size of the two aggregates was 4.75-9.5 mm and the basic properties were shown in table 1. The water reducer was applied to improve the workability of pervious concrete.

(a)

(b)

Figure 1. The natural aggregate (NA) and recycled aggregate (RA). (a) Natural aggregate and (b) Recycled aggregate.

Table 1. Basic properties of coarse aggregate.

Physical index	Natural aggregate (NA)	Recycled aggregate (RA)
Particle size (mm)	4.75-9.5	4.75-9.5
Apparent density (kg/m^3)	2786	2673
Bulk density (kg/m^3)	1534	1450
Bulk porosity (%)	44.9	45.8
Crushed value (%)	9.7	11.1
Needle-like particle content (%)	7.1	1.6
Water absorption rate (%)	1.63	5.08

2.2. Mixture proportion

The volumetric method was used for mixture design according to the Chinese national standard CJJ/T 135-2009. According to thc previous researches conducted by our group [20,21], the water-binder ratio and porosity were selected as 0.30 and 15%. The amount of water reducer was 0.5% of the mass of cement. The RA was added with equivalent quality method and the incorporation levels were 25%, 50%, 75%, and 100%. The mixture proportion is shown in table 2.

Table 2. Mixture proportion (in kg/m^3).

Mix ID	RA content	Coarse aggregate		Cement	Water	Superplasticizer
		NA	RA			
NA	0%	1503	0	483	145	2.41
RA25	25%	1127	376	483	145	2.41
RA50	50%	752	752	483	145	2.41
RA75	75%	376	1127	483	145	2.41
RA100	100%	0	1503	483	145	2.41

2.3. Specimen preparation

The cement paste encapsulating aggregate method and rodding method were used to prepare specimens. The 100 × 100 × 100 mm cubic specimen and 100 × 100 × 400 mm prismatic specimen were casted in the lab. The cubic specimen was used for porosity, permeability, compressive strength and F-T test and the prismatic specimen was used for flexural strength test. A total of 75 cubic specimens and 15 prismatic specimens were prepared. All specimens were demoulded after 24h and placed in standard-cured room with a temperature of 20 ± 2°C and a relative humidity of 95%.

2.4. Methods

2.4.1. Porosity. The pore structure of pervious concrete can be divided into three types: connected pore, semi-connected pore and closed pore. The connected pore and semi-connected pore are called effective pore. The effective porosity of the specimen can be calculated by following formula:

$$P = [1 - \frac{m_2 - m_1}{\rho V_0}] \times 100\% \quad (1)$$

where P is effective porosity (%); m_1 is the quality of the specimen submerged in water (g); m_2 is the saturated surface-dry quality of the specimen (g); ρ is the water density (g/cm^3), and V_0 is the volume of the specimen (cm^3).

2.4.2. Permeability. The constant head permeability method was adopted in this paper. The basic principle is that the discharged amount of water is directly related to the time when the water head height is constant. The permeability coefficient of pervious concrete is calculated by using equation (2).

$$k = \frac{QL}{AHt} \tag{2}$$

where k is the permeability coefficient of specimen (mm/s); Q is the discharged amount of water in t time (mm^3); L is the height of the specimen (mm); A is the upper surface area of the specimen (mm^2); H is the height of water head (mm), H = 150 mm; t is the time (s), t = 300 s.

2.4.3. Compressive strength and flexural strength. The compressive strength and flexural strength of pervious concrete are tested according to the Chinese national standard GB/T50081-2002. The flexural strength is tested by three-point bending test and the distance between two supporting points is 300 mm.

2.4.4. F-T durability. Rapid F-T method is used to evaluate the F-T durability of pervious concrete according to the Chinese national standard GB/T50082-2009. The crack and damage of the specimen were observed, and the compressive strength of the specimen was tested every 25 F-T cycles.

3. Results and analysis

3.1. Porosity and permeability

The experiment results with standard deviations of effective porosity and permeability coefficient are shown in table 3. The effect of RA content on the effective porosity and permeability coefficient is shown in figure 2. The relationship between effective porosity and permeability coefficient of pervious concrete is shown in figure 3.

Table 3. Effective porosity and permeability coefficient of pervious concrete.

Mix ID	Effective porosity (%)		Permeability coefficient (mm/s)	
	Mean	Standard Deviation	Mean	Standard Deviation
NA	14.4	0.13	3.91	0.08
RA25	14.2	0.16	3.88	0.09
RA50	13.7	0.27	3.73	0.1
RA75	13.4	0.19	3.61	0.07
RA100	13.1	0.19	3.56	0.11

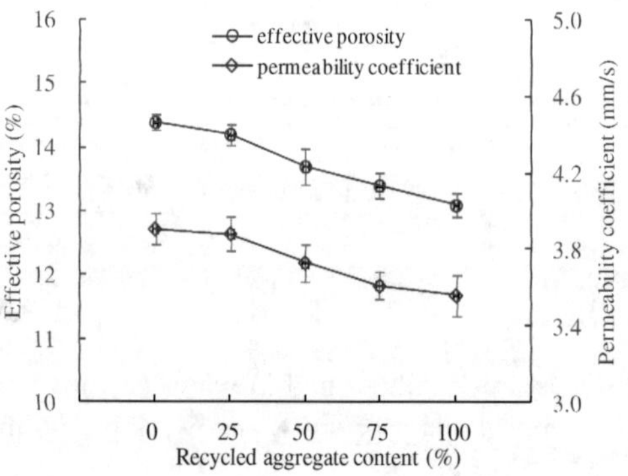

Figure 2. Effective porosity and permeability coefficient versus RA content.

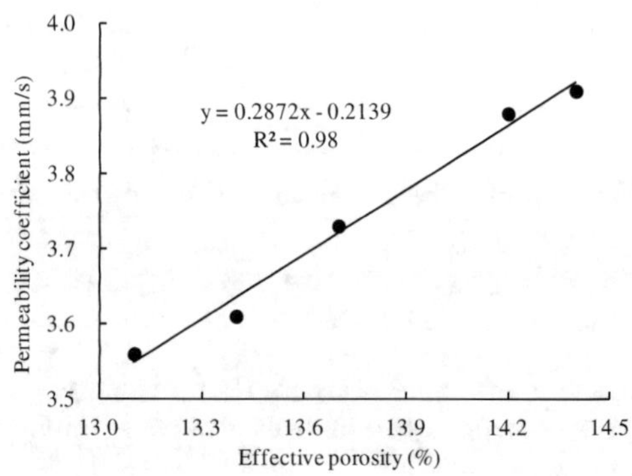

Figure 3. The relationship between permeability coefficient and effective porosity.

The effective porosity has great impact on the permeability of pervious concrete. Figure 2 indicated that the effective porosity and permeability of pervious concrete decreased with the increasing RA content. When the RA content increased from 0% to 100%, the effective porosity decreased from 14.4% to 13.1% and the permeability coefficient decreased from 3.91 mm/s to 3.56 mm/s. The reason was that the density of RA was less than natural aggregate. The coarse aggregate in pervious concrete was added with equivalent quality method, which resulted in the increase of coarse aggregate volume and the decrease of the pore number. Thus, the porosity and permeability of pervious concrete were decreased. Figure 3 indicate that the permeability coefficient of pervious concrete increased with the increasing effective porosity. The permeability coefficient was positively correlated with effective porosity, which also explained the reason why permeability coefficient showed a downward trend with the increasing RA content in figure 2.

3.2. Compressive strength and flexural strength

Table 4 shows the compressive strength and flexural strength of pervious concrete. The influence of RA content on the compressive strength of pervious concrete is shown in figure 4. The influence of RA content on the flexural strength of pervious concrete is shown in figure 5.

Table 4. Compressive strength and flexural strength of pervious concrete.

Mix ID	Compressive strength (MPa)		Flexural strength (MPa)	
	Mean	Standard Deviation	Mean	Standard Deviation
NA	22.2	0.42	4.84	0.11
RA25	21.7	0.37	4.20	0.05
RA50	19.7	0.59	3.97	0.09
RA75	16.7	0.31	3.75	0.07
RA100	15.9	0.35	3.73	0.08

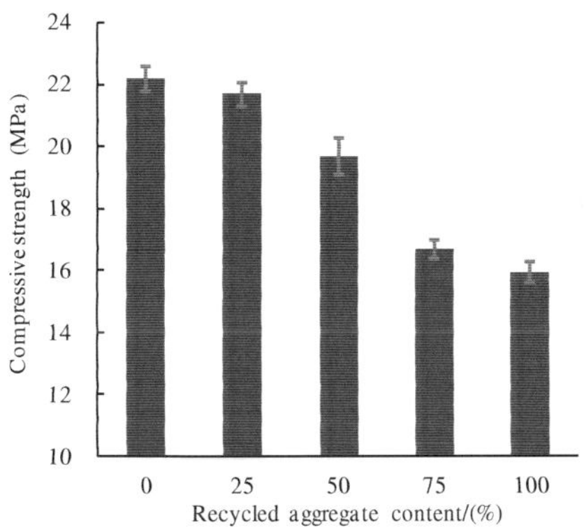

Figure 4. Compressive strength of pervious concrete versus RA content.

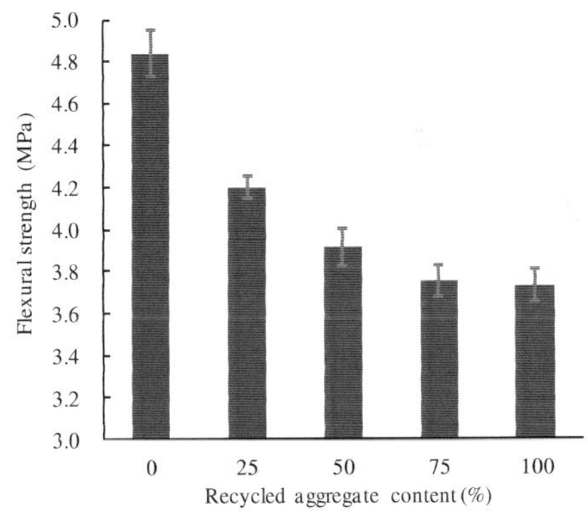

Figure 5. Flexural strength of pervious concrete versus RA content.

Figure 4 showed that, with the increasing RA content, the compressive strength of pervious concrete presented a downward trend. The compressive strength of pervious concrete decreased from 22.2 MPa to 15.9 MPa with the loss of 28.3%. In addition, the decrease of compressive strength was strongly affected by the incorporation level of RA content. When the RA content was less than 25%, the compressive strength decreased slowly, when the RA content was more than 25%, the compressive strength decreased rapidly. The reasons for the decrease of compressive strength were as follows: (a) the RA was obtained by crushing the waste cement concrete brick, the internal damage of RA was produced during the crushing process, which resulted in the lower strength of RA than natural aggregate; (b) due to the existence of hardened cement paste on the surface of RA, the adhesive property between RA and cement paste was weaker than that of natural aggregate; (c) the water imbibition of RA was larger than that of natural aggregate, part of the water in the mixture was absorbed by the RA, which resulted in incomplete hydration of cement. All the above decreased the compressive strength of RA pervious concrete.

Figure 5 indicated that the flexural strength of pervious concrete decreased with the increasing RA content, which was consistent with the effect of RA content on the compressive strength. The flexural strength decreased from 4.84 MPa to 3.73 MPa with the loss of 22.9%. The decrease degree of flexural strength was a little different from that of compressive strength. Compared with the decrease trend of compressive strength, when the RA content was less than 50%, the flexural strength decreased rapidly, when the RA content was more than 50%, the flexural strength decreased slowly. The decrease of flexural strength of RA pervious concrete also attributed to the low strength of RA, weak adhesive property and large water absorption of RA.

3.3. F-T durability

The compressive strength and its loss of pervious concrete after 25, 50, 75 and 100 F-T cycles are shown in table 5. The effect of F-T cycles on the compressive strength of RA pervious concrete is shown in figure 6. The compressive strength loss of RA pervious concrete under F-T cycles are shown in figure 7.

Table 5. Compressive strength and its loss of pervious concrete after F-T cycles.

Mix ID	Compressive strength and Its Loss After F-T Cycles							
	25 Cycles		50 Cycles		75 Cycles		100 Cycles	
	Compressive strength (MPa)	Loss (%)	Compressive strength (MPa)	Loss (%)	Compressive strength (MPa)	Loss (%)	Compressive strength (MPa)	Loss (%)
NA	21.2	4.5	18.1	18.5	16.7	24.8	14.2	36.0
RA25	20.5	5.5	17.3	20.3	15.9	26.7	12.9	40.6
RA50	18.4	6.6	14.2	22.8	12.5	36.5	11.2	43.1
RA75	14.7	12.0	11.9	28.7	10.4	37.7	8.4	49.7
RA100	13.5	15.1	10.9	31.4	9.6	39.6	7.1	55.3

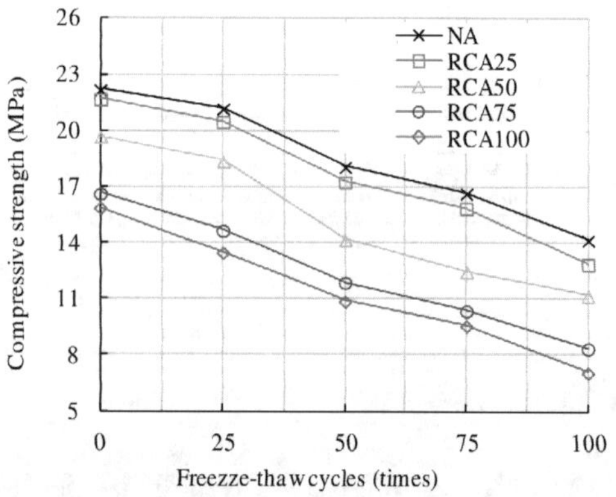

Figure 6. Compressive strength versus F-T cycles.

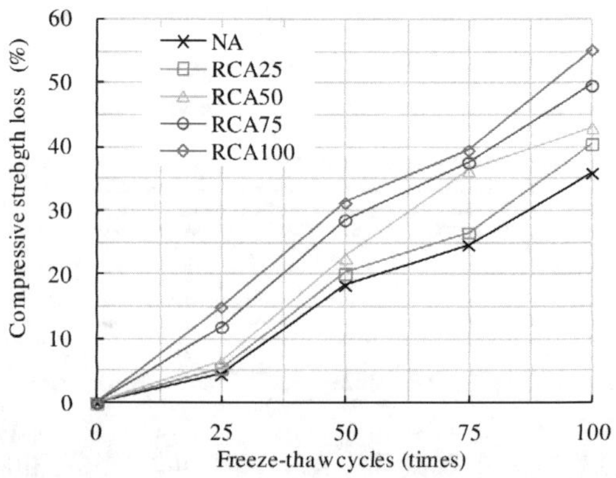

Figure 7. Compressive strength loss versus F-T cycles.

According to the figure 6, the conclusion can be drawn that, for all pervious concrete mixtures, when the incorporation levels of RA content were the same, the compressive strength of pervious concrete decreased with the increase of F-T cycles, which showed that the F-T resistance of pervious concrete gradually deteriorated under F-T cycles. When the number of F-T cycles was the same, the compressive strength of RA pervious concrete decreased with the increase of RA content. Compared with the compressive strength of 14.2 MPa for natural aggregate pervious concrete under 100 F-T cycles, the compressive strengths of 75% and 100% RA pervious concrete were only 8.4 MPa and 7.1 MPa. The F-T resistance was greatly reduced.

Figure 7 showed that the compressive strength loss of pervious concrete gradually increased with the increasing F-T cycles, and the compressive strength loss increased with the increasing RA content when the F-T cycles were the same. Compared with the natural aggregate pervious concrete, the RA pervious concrete had higher compressive strength loss. And the higher RA content, the higher compressive strength loss. For all pervious concrete mixtures, when the F-T cycles were 25, the compressive strength loss was 4.5% to 15.1%, which met the requirement of 20% for pervious concrete in the Chinese national standard CJJ/T 135-2009 and indicated the application of RA pervious concrete in F-T regions was feasible.

4. Conclusions

In this study, the RA pervious concrete was prepared and investigated. According to the experiment results, several conclusions can be concluded.

- Due to the adoption of equivalent quality replacement method, the effective porosity and permeability of pervious concrete gradually decreased with the increasing RA content.
- The addition of RA decreased the compressive strength and flexural strength of pervious concrete, but the pervious concrete contained different content of RA presented acceptable compressive strength and flexural strength.
- The F-T durability of RA pervious concrete was not equal to that of natural aggregate pervious concrete and the F-T durability of pervious concrete decreased with the increase of RA content.
- Although the addition of RA was negative to the performance of pervious concrete, it was still practicable to use it in sidewalks, parks and light traffic roads.

Acknowledgments

The authors express their appreciation for the financial support of Science Technology Development Program of Jilin Province (grant number 20180201026SF) and the Transportation Science and Technology Program of Jilin Province (grant number 2018-1-9).

References

[1] Bravo M, de Brito J, Pontes J and Evangelista L 2015 Mechanical performance of concrete made with aggregates from construction and demolition waste recycling plants *J Clean Prod.* **99** 59-74

[2] Yahia A and Kabagire K D 2014 New approach to proportion pervious concrete *Constr Build Mater.* **62** 38-46

[3] Zhang S H and Guo Y P 2015 Analytical equation for estimating the stormwater capture efficiency of permeable pavement systems *J Irrig Drain Eng.* **141** 06014004

[4] McCain G N and Dewoolkar M M 2010 Porous concrete pavements: mechanical and hydraulic properties *Transport Res Rec.* **2164** 66-75

[5] Ibrahim H A and Razak H A 2016 Effect of palm oil clinker incorporation on properties of pervious concrete *Constr Build Mater.* **115** 70-7

[6] Tam V W Y, Butera A and Le K N 2016 Carbon-conditioned recycled aggregate in concrete production *J Clean Prod.* **133** 672-80

[7] Su H, Yang J, Ling T C, Ghataora G S and Dirar S 2015 Properties of concrete prepared with waste tyre rubber particles of uniform and varying sizes *J Clean Prod.* **91** 288-96

[8] Sagoe-Crentsil K K, Brown T and Taylor A H 2001 Performance of concrete made with commercially produced coarse recycled concrete aggregate *Cement Concrete Res.* **31** 707-12

[9] Zaetang Y, Sata V, Wongsa A and Chindaprasirt P 2016 Properties of pervious concrete

containing recycled concrete block aggregate and recycled concrete aggregate *Constr Build Mater.* **111** 15-21

[10] Güneyisi E, Gesoğlu M, Kareem Q and Ipek S 2016 Effect of different substitution of natural aggregate by recycled aggregate on performance characteristics of pervious concrete *Mater Struct.* **49** 521-36

[11] Zhu X Y, Chen X D, Shen N, Tian H X, Fan X Q and Lu J 2018Mechanical properties of pervious concrete with recycled aggregate *Comput Concrete* **21** 180-93

[12] Wang Y M, Deng Z H, Tan Y H and Yang H F 2015 Permeability and strength of porous concrete made with recycled aggregate *Concrete* **7** 26-30

[13] Yap S P, Chen P Z C, Goh Y X and Ibrahim H A 2018 Characterization of pervious concrete with blended natural aggregate and recycled concrete aggregates *J Clean Prod.* **181** 155-65

[14] Zhu J C 2014 Study on the content of recycled aggregate in recycled pervious concrete *China Concrete and Cement Products* **10** 6-11

[15] Lund M S M, Kevern J T, Schaefer V R and Hansen K K 2017 Mix design for improved strength and freeze-thaw durability of pervious concrete fill in Pearl-Chain Bridges *Mater Struct.* **50** 1-15

[16] Wang J Q 2016 Test research on the shrinkage and frost-resistance of recycled aggregate porous concrete *Industrial Construction* **46** 103-6

[17] Sriravindrarajah R, Wang N D H and Ervin L J W 2012 Mix design for pervious recycled aggregate concrete *Int J Concr Struct M.* **6** 239-46

[18] Berry B M, Suozzo J M, Anderson I A and Dewoolkar M M 2012 Properties of pervious concrete incorporating recycled concrete aggregate *Transportation Research Board Annual Meeting* (Washington, D.C)

[19] Chen S K 2018 Influence of recycled aggregate proportion performance of recycled pervious concrete performance *Acta Mater Compos Sin.* **35** 1590-8

[20] Liu H B, Luo G B, Wei H B and Yu H 2018 Strength, permeability, and freeze-thaw durability of pervious concrete with different aggregate sizes, porosities, and water-binder ratios *Appl Sci.* **8** 1217

[21] Liu H B, Luo G B, Gong Y F and Wei H B 2018 Mechanical properties, permeability, and freeze–thaw resistance of pervious concrete modified by waste crumb rubbers *Appl Sci.* **8** 1843

Effect of halloysite nanotube loading on structure, mechanical and thermal properties of poly (l-lactic acid) / poly-(butylene succinate) blend

H Y Liu[1,2], L Chen[1], W Li[1] and K W Wang[1]

[1]School of Mechanical Engineering, Dongguan University of Technology, Dongguan 523808, China

E-mail: liuhuanyu@dgut.edu.cn

Abstract. Poly (l-lactic acid) (PLA)/poly-(butylene succinate) (PBS)/ Halloysite Nanotubes (HNT) composites were prepared via melting compounding in a twin-screw extruder. The effects of HNT addition on mechanical properties, thermal properties of PLA/PBS blends were investigated. The structure of ternary composites was studied by Transmission Electron Microscopy (TEM) and the results revealed that a small incorporation of HNT dispersed well in the two polymers. Ternary composites presented improved tensile strength and elongation at break at low loading levels of HNT. Dynamic Mechanical Analysis (DMA) results showed that the addition of HNT significantly increased storage modulus of PLA/PBS composites. The differential scanning calorimetry analysis indicated that HNT can sever as nucleating agent for PLA in ternary composites, which lead to decreased cold crystallization temperature and increased crystallinity of PLA.

1. Introduction

Due to increasing environmental concerns, biodegradable polyesters are receiving considerable attention. In competition of biodegradable polyester, the polymers (PLA and PBS) originated from renewable resource have been extensively investigated due to environmental protection [1]. In particular, PLA has attractive properties, such as good strength, good biocompatibility and barrier properties [2]. For these reasons, PLA is expected to be an interesting candidate for biomaterials and construction materials in near future [3]. However, PLA's poor flexibility, high price and low heat resistance are major obstacles for the possibility of commercial applications, particularly for packaging industries [4-6]. In this respect, Adding polycaprolactone (PCL) [7] and poly (butylene succinate) PBS [8-10], which are more tough and still biodegradable polyesters, into PLA could overcome the brittleness of PLA. Specially, researchers found much valuable information in PLA/PBS blends. Zhao *et al* [11] reported that elongation at break in PLA/PBS blends was considerably higher than that of pure PLA, but with an acceptable loss of strength. Furthermore, the crystallization of PLA was enhanced after the addition of PBS even though PBS was in a molten state. Takagi *et al* [12] and Shibata *et al* [13] found that the cold-crystallization of PLA was enhanced after adding PBS.

However, some studies [12-14] concluded that phase separation led to probable deterioration in some properties, since PLA was immiscible with PBS. Adding nanoparticles into PLA/PBS blends was an effective way of improving the compatibility of PLA and PBS [15-17]. In order to improve the interfacial bonding between PLA and PBS, numerous modification methods were introduced. However, the procedures were neither easy to handle nor environmental-friendly. Because it was toxic in modification nanoparticles such as Carbon Nanotube (CNT) [18], these nanocomposites could not be directly utilized in biomedical area.

The use of Halloysite nanotube (HNT) does not add risk to the environment because it is a green filler and a natural product. As demonstrated elsewhere [19], the composition of HNT and kaolin were resemble. HNT had a predominantly hollow tubular structure with nanoscale lumen. Its structure was relatively high length-to-diameter ratio, and low hydroxyl group density on their surface. Unlike other nanostructured clays that must be exfoliated, HNT did not require exfoliation. Furthermore, HNT provided strength increase without the loss of ductility [20]. In addition, its use did not cause any potential risks to environment as other nanofillers potentially did [18]. Compared to the CNT, HNT was easily available, abundant, rather cheap and eco-friendly. Given to these characteristics, HNT has been regarded as a promising competitor and alternative to both CNT and OMMT. Many HNT nanocomposites were prepared and studied by researchers [21-23]. Murariu [21] reported that the addition of HNT can result in an increase of thermal and mechanical properties at a small loading. Excellent tensile properties and good impact properties were observed in PLA /HNT nanocomposites with a content of 3–12 wt.% nanofiller [21]. However, from recent literatures, the application of HNT in PLA/PBS and how HNT affects the properties of PLA/PBS composites have not been reported. Further research in the effect of HNT on PLA/PBS properties will help in the development of several properties of PLA/PBS composites.

In this article, our major work is on the effect of HNT addition on tensile properties, dynamic mechanical properties, and thermal properties of PLA/PBS blends. Previous work [23] revealed that excellent properties were obtained when PLA/PBS weight ratios was 70/30 wt.%. Therefore, PLA/PBS/HNT composites having fixed composition of PLA and PBS of 70% and 30% (weight ratio) with various HNT loadings are prepared.

2. Experiment

2.1. Materials

PLA with a comercial grade 4032D used in the study has a specific gravity 1.25 g/cm^3, which was supplied by Natural Works LLC (USA). The PLA has a glass transition temperature about 60°C. It has a melting temperature about 163°C. Additonally, PBS (BIONOLLE #1903MD) was supplied by Showa High Polymer (Japan). It has density=1.26 g/cm^{-3} and a melting temperature of ~115°C. Halloysite nanotube (HNT) was obtained by Aldrich. It has nanotubes length=1~3 um, pore volume=1.26-1.34 ml/g. Before processing, HNT was dried at 80°C for 10 h in vacuum, whereas polymers were dried for 8 h.

2.2. Preparation of polymer blend and blend-clay composites

After dried, PLA/PBS (70 wt.%/30 wt.%) pellets were premixed. The ratio was determined because it resulted in a good balance of properties. Importantly, PLA was our target polymer, whose properties need to be improved. Different kinds of HNT (1, 3, 6, 9 wt.%) were blended with premixed polymer in a plastic bag. A BRABENDER counter-rotating twin-screw extruder was used to prepare composites by means of melt blending. The extruder is from Germany with a screw diameter of 25 mm and a length/diameter ratio of 20:1. The temperatures from feeding port to die were 160, 170, 195, 195, 195, 195, 175, and 175°C. The screw speed was set at 100 rpm. A water bath was introduced to cool the composites extrudate from the die. Then, the extruded blends into pellets was cuted by a pelletizer. Next, more than 4 h at 80°C was needed to dried these extruded pellets. After then, the pellets were made into sheets of 150 mm x 80 mm x 2 mm by compression molding mold at 190°C. Finally, the sheets were prepared into the samples for TEM, tensile properties, dynamic mechanical properties and dynamic rheological test. The dried pellets from the fresh extruder at the exit of die were prepared for differential scanning calorimetry analysis specimens. It is noted that B0, B1, B3, B6 and B9 represented the blends with 0, 1, 3, 6, 9 wt.% HNT contents, respectively.

2.3. Characterization

2.3.1. TEM. he dispersion of nanoparticles in composites was evaluated by Transmission Electron Microscopy (TEM) at an acceleration voltage of 200 kV. TEM is a JEOL field emission transmission electron microscopy named JEM-100XII. Before testing, a Leica EMUC6/FC6 microtome was used to make circular sheet into ultrathin sections with a thickness of ~100 nm.

2.3.2. Tensile test. An INSTRON universal machine (model 5566, USA) was used to evaluate tensile properties of ternary composites. All samples were tested at room temperature in accordance with GB/T 1040.2 (2006), and had the same tension mode with a single-strain rate of 5 mm•min^{-1}. Final data were determined in an average of five independent test.

2.3.3. Dynamic mechanical analysis (DMA). NETZSCH DMA labelled 204C was introduced to evaluated dynamic mechanical properties of neat PLA/PBS and PLA/PBS/HNT blends. All samples were tested in tensile mode at temperature ranged from − 65°C to +150°C with a heating rate of 3°C/min, and then with a frequency of 1 Hz and oscillation amplitude of 0.04 mm. The measurements was fixed at the same condition. The storage modulus and the loss modulus were obtained from DMA test. Specimen dimensions in test were approximately 1×5×32 mm (thick-width-length).

2.3.4. Differential scanning calorimetric (DSC). A NETZSCH DSC 204 F1 analyzer from Germany was used to evaluate thermal properties. All samples were tested at a weight of 6-10 mg with DSC protocol. In DSC protocol, samples were heated from room temperature to 210°C in a nitrogen atmosphere at a heating speed of 10°C/minm, and the temperature of 210°C was maintained for 5 min, then declined to room temperature at a cooling rate of 10°C/min. A second heating scan monitored was determined between room temperature to 210°C at a rate of 10°C/min. Melting temperature (T_m), melting enthalpy (ΔH_m), glass transition temperature (T_g) and cold crystallization temperature (T_{cc}) were obtained from the second heating mode. The fractional crystallinity (χ_c) was determined by the following equation:

$$\chi_c(\%) = \frac{\Delta H_m - \Delta H_c}{\Delta H_0 \times wt.\%} \times 100\% \tag{1}$$

Where χ_c was the crystallinity of polymer (PLA or PBS) in the composite, ΔH_c represented the cold crystallization enthalpy, ΔH_m represented the melting enthalpy, ΔH_0 represented the theoretical enthalpy for a fully crystalline, and wt.% represented the polymer weight fraction in HNT composites. According to previous work in our team, ΔH_0 for PBS [24] and PLA [25] had been assigned to be 200 J/kg and 93 J/kg, respectively.

3. Results and discussion

3.1. Morphology of PLA/PBS/HNT ternary composites

Obviously, polymer properties are evidently influenced by the dispersion of nanoparticles. A well dispersion of HNT nanoparticles is the desire thing, which will significantly influence thermal, tensile properties of PLA/PBS/HNT ternary composites.

The morphology of ternary composites was mainly evaluated by TEM analyses. Figures 1(a)-1(d) represents typical TEM pictures of PLA/PBS/HNT ternary composites. From the picture, HNT well dispersed in samples with 3 wt.% and 6 wt.% HNT loadings. When increasing the loading of HNT to 9 wt.%, the individually separated HNT and aggregated HNT coexisted in the sample. The results revealed that HNT well dispersed in some samples without complicated surface-treatment procedure.

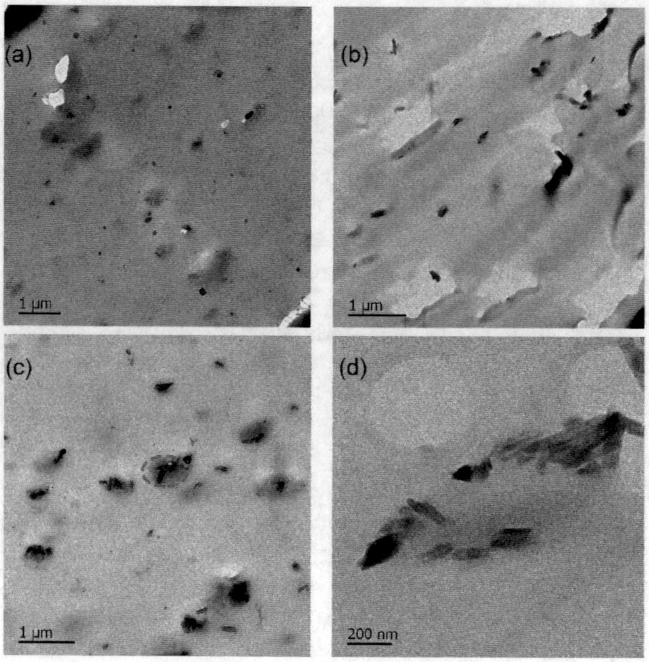

Figure 1. TEM images of PLA/PBS/HNT composites: (a) B3, (b) B6, (c) B9. (d) B9 in 200 nm scale. The red circle in (c) represents HNT aggregates.

3.2. Tensile properties and dynamic mechanical properties

Tensile strength and the elongation at break of PLA/PBS/HNT composites were showed in figure 2. In figure 2, tensile strength of ternary composites initially increased before 6 wt.%, and then monotonically declined with the introduction of more HNT. Specially, tensile strength was improved by almost 130% separately at 6 wt.% HNT loading than that of blank PLA/PBS blend (31.48 MPa). However, at high HNT loading (9 wt.%), tensile strength significantly decreased to 33.64 MPa while it was higher than that of pristine blend without HNT. Regarding to the elongation at break of ternary composites, the elongation at break climbed sharply to 17.86 wt.% at 6 wt.%, which was nearly 2 times better than that of blank PLA/PBS (9.78 wt.%), whereas a dramatic reduction occurred in ternary composite filled with 9 wt.% HNT.

There are some works reporting reinforcing effect for PLA/PBS blend when adding different nanoclay. For example, PLA/PBS with four organically modified pristine montmorillonites were prepared and researched [26]. Blends with pristine montmorillonite and nanoclay (C25A) showed poor mechanical properties. However, the maximum of the elongation at break and yield strength of PLA/PBS with organically modified clay (C20A) were obtained at organically modified clay content of 2 wt.%, which were 18%~5% increment compared with blank PLA/PBSA respectively [26]. For example, both the elongation at break and tensile strength of the PLLA/PBS blend decreased precipitously as a result of nanoclay (C25A) incorporation [15]. These results revealed that the structural characteristics of nanoparticles, dispersion state as well as interfacial interactions caused the different reforcing effects in clay composites. Comparing with nanoclay, HNT without modification methods showed good reinforcing properties for PLA/PBS. The excellent reinforcing of HNT for PLA/PBS could owe to the uniform distribution of HNT and the interfacial interaction between HNT and polymers [27]. However, in the case of PLA/PBS with 9 wt.% HNT loading, tensile strength and elongation at break started to decrease, whereas they were higher than those of blank PLA/PBS. This decline was ascribed to aggregated HNT (as shown in TEM test), which could be the stress-centralized points during loading process.

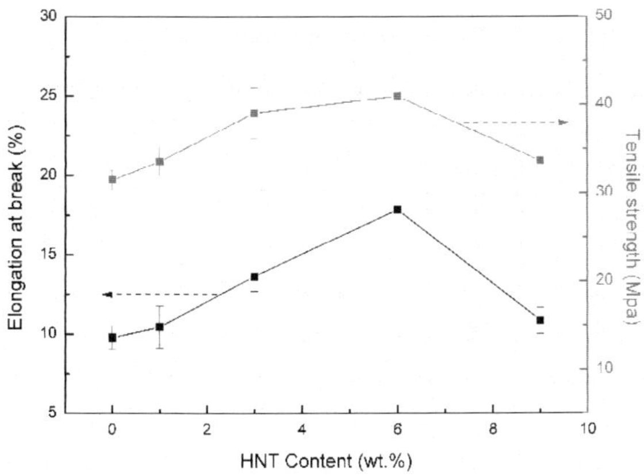

Figure 2. Tensile strength and the elongation at break of ternary blends with different HNT loadings.

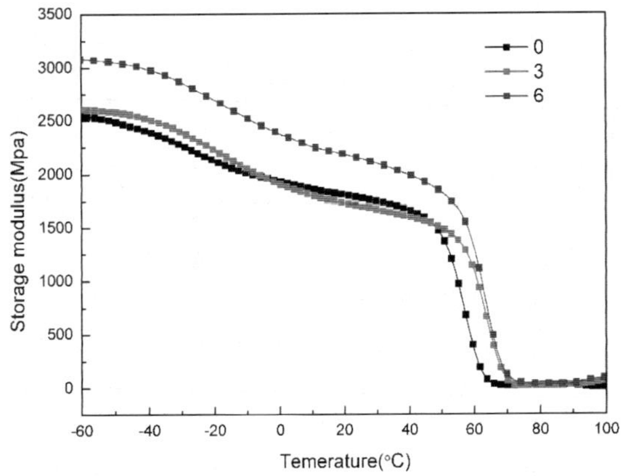

Figure 3. Temperature dependence of storage modulus for polymer ternary blends (Samples B0, B3, B6, each having 0, 3, 6 wt.% HNT respectively).

Figure 3 represents the temperature dependencies of storage modulus for pure PLA/PBS and the ternary blends obtained by the addition of 3 wt.%, 6 wt.% HNT. All investigated samples for storage modulus presented similar behaviors with the increase of temperature. Specially, storage modulus decreased gradually before 40°C, and then followed by a dramatic decline in the ranges of 40-60°C. The storage modulus of PLA/PBS composites was significantly enhanced by the incorporation of HNT. The interactions between polymer matrix lead to the rise. Thus, HNT could restrict the mobility of polymer chains adsorbed on HNT's surface, resulting in the formation of a stiffened interphase [27].

3.3. Thermal properties

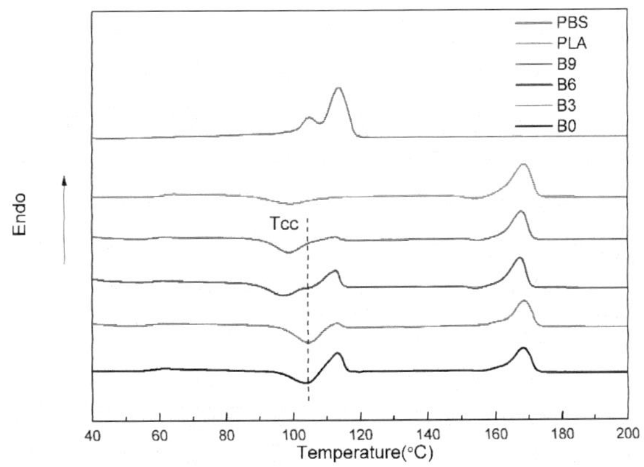

Figure 4. DSC thermograms of neat polymer, ternary composites.

Figure 4 reveals DSC thermograms of pure PLA/PBS and HNT composites during a second heating scan. As seen from figure 4, two melting endotherm peaks were discernible for neat PBS while a single one was found for ternary composites. The corresponding data are collected in table 1. Firstly, from these DSC results, it comes out that PLA in blank PLA/PBS composite showed a broad cold crystallization exothermic peak at about 103.6°C. Here, PLA' crystal is not perfect in cold crystallization because of the lower crystal ability of PLA chain, and the imperfect crystals form in cold crystallization would reorganize into more orderly crystal structures [28]. Furthermore, T_{cc} of

PLA shifted to a lower temperature after the introduction of HNT. On introduction of 6 wt.% HNT into the blend, the T_{cc} of sample B6 decreased to a lower temperature of nearly 97.2°C, which was 6.4°C lower than that of sample B0. This is because of the nucleating effect of HNT in PLA cooling crystallization process. The heterogeneous nucleation increases the crystallinity of PLA. Similar heterogeneous nucleation of nanosized particles for PLA were found [29-32].

Table 1. DSC data of PLA/PBS/HNT blends.

Blend samples	PLA component					PBS component		
(PLA/PBS/HNT)	T_{m1}(°C)	T_{cc}(°C)	ΔH_c(J/g)	ΔH_{m1}	χ_c(%)	T_{m2}(°C)	ΔH_{m2}(J/g)	χ_c(%)
Pure PLA	168.5	99	18.46	38.71	21.77			
B0	163.8	103.6	15.92	20.01	6.28	113.4	6.008	10.01
B3	168.2	104.3	16.41	22.67	9.62	113.4	13.78	22.97
B6	167.5	97.2	10.31	27.12	25.82	113.3	8.447	14.08
B9	167.4	98.7	13.31	28.02	22.60	113.2	9.774	16.29

As seen from DSC data, however, for the PBS component, T_{m2} of ternary blends were same with T_{m2} of blank PLA/PBS, indicating that HNT had little influence on T_{m2} of the ternary blends. The melt enthalpy of PBS component in ternary blends grew with the addition of HNT, though the increasing trend was not consistent.

4. Conclusions

PLA/PBS/HNT ternary blends were successfully obtained by direct melt mixing via a twin-screw extruder, and their structure, mechanical properties, thermal properties were studied. Results revealed that a small incorporation of HNT dispersed well in the two polymers. Both tensile strength and elongation at break for ternary composites were significantly improved due to the HNT incorporation. DMA results revealed HNT significantly enhanced the storage modulus of PLA/PBS composites. DSC analysis illustrates that HNT can act as nucleating agent in process of PLA crystallization of ternary composites, leading to the decrease of cold crystallization temperature and the increase of PLA crystallinity.

Acknowledgments

Authors wish to acknowledge the Research Start-up Funds of DGUT (Grant No. GC300502-48), Research Start-up Funds of DGUT (Grant No. GC300502-49), Innovated Service Foundation for Technology Industry(Grant No. 2001104-39), the Major Scientific Research Project of University in Guangdong Province (Grant No. 2017GXJK172), the Research and Reform Project of Higher Education in Guangdong Province(Grant No. E1201601).

References

[1] Hao Y, Yang H, Pan H, Zhang H and Ran X 2018 *Polym-Plast. Technol.* **57** 1-11
[2] Ray S S 2012 *Accounts. Chem. Res.* **45** 1710-20
[3] Park J W and Im S S 2002 *J. Appl. Polym. Sci.* **86** 647-55
[4] Kim J K, Park D, Lee M and Ihn K J 2001 *Polymer* **42** 7429-41
[5] Martin O and Avérous L 2001 *Polymer* **42** 6209-19
[6] Shibata M, Inoue Y and Miyoshi M 2006 *Polymer* **47** 3557-64
[7] Cohn D and Salomon A H 2005 *Biomaterials* **26** 2297-305
[8] Yokohara T and Yamaguchi M 2008 *Eur. Polym. J.* **44** 677-85
[9] Wang R, Wang S, Zhang Y, Wan C and Ma P 2009 *Polym. Eng. Sci.* **49** 26-33
[10] Ojijo V, Sinha R S and Sadiku R 2013 *ACS. Appl. Mater. Inter.* **10** 4266-76
[11] Zhao P, Liu W, Wu Q and Ren J 2010 *J. Nanomater.* **2010** 1-5
[12] Takagi J, Nemoto T, Takahashi T, Taniguchi T and Koyama K 2003 Seikei-Kakou **15** 581-7
[13] Shibata M, Inoue Y and Miyoshi M 2006 *Polymer* **47** 3557-64

[14] Hirotsu T, Nakayama K, Tagaki C and Watanabe T 2004 *J. Photopolym. Sci. Technol.* **17** 179-94
[15] Chen G, Kim H, Kim E and Yoon J 2005 *Polymer* **46** 11829-36
[16] Ojijo V, Cele H and Ray S S 2012 *Macromol. Mater. Eng.* **296** 865-77
[17] Wang X, Zhuang Y and Dong L 2012 *J. Appl. Polym. Sci* **126** 1876-84
[18] Cavallaro G, Lazzara G and Milioto S 2011 *Langmuir* **27** 1158-67
[19] Liu M, Li W, Rong J and Zhou C 2012 *Colloid. Polym. Sci.* **290** 895-905
[20] Wu W, Cao X, Zhang Y and He G J 2013 *J. Appl. Polym. Sci.* **130** 443-52
[21] Murariu M, Dechief A, Paint Y, Peeterbroeck S, Bonnaud L and Dubois P 2012 *J. Polym. Environ.* **20** 932-43
[22] Prashantha K, Lecouvet B, Sclavons M, Lacrampe M F and Krawczak P 2013 *J. Appl. Polym. Sci.* **128** 1895-903
[23] Chen R Y, Zou W, Wu C R, Jia S K, Huang Z, Zhang G Z and Qu J P 2014 *Polym. Test.* **34** 3
[24] Wu W, Cao X, Luo J, He G and Zhang Y 2014 *Polym. Composite* **35** 847-55
[25] Zhang H Z, Huang J T, Yang L, Chen R Y, Zou W, Lin X and Qu J P 2015 *Rsc. Adv.* **5** 4639-47
[26] Ojijo V, Sinha R S and Sadiku R 2012 *Acs. Appl. Mater. Inter.* **4** 2395-405
[27] Liu M, Zhang Y and Zhou C 2013 *Appl. Clay. Sci.* **35** 75-6
[28] Notta-Cuvier D, Murariu M, Odent J, Delille R, Bouzouita A, Raquez J, Lauro F and Dubois P 2015 *Macromol. Mater. Eng.* **300** 684-698
[29] Luo B, Hsu C, Li J, Zhao L, Liu M, Wang X and Zhou C 2013 *J. Biomed. Nanotechnol.* **9** 649
[30] Wu W, Cao X, Zhang Y and He G 2013 *J. Appl. Polym. Sci.* **130** 443-52
[31] Prashantha K, Lecouvet B, Sclavons M, Lacrampe M F and Krawczak P 2013 *J. Appl. Polym. Sci.* **128** 1895-903
[32] Xu W, Luo B, Wen W, Xie W, Wang X, Liu M and Zhou C 2015 *J. Appl. Polym. Sci.* **132** 41451-60

IOP Conf. Series: Materials Science and Engineering **634** (2019) 012013 doi:10.1088/1757-899X/634/1/012013

Study on performance of micro heat dissipation system base on liquid metal

H R Yun[1,2] and X X Zhang[1,2,3]
[1]Tianjin Municipal Key Lab of Advanced Fiber and Energy Storage Technology, Tianjin 300387, China
[2]School of Material Science and Engineering, Tianjin Polytechnic University, Tianjin 300387, China

[3]E-mail: zhangpolyu@aliyun.com

Abstract. This paper introduces an advanced liquid metal microcirculation system that uses a liquid metal cycle driven by an electromagnetic pump to achieve heat dissipation. The overall thickness of the structure can be as flimsy as 0.9mm, and the heat dissipation capacity is more than twice that of the corresponding ultra-thin heat pipe. This paper analyzes the performance of the complex by experimenting with the sample and comparing it with the software simulation results, and demonstrates the theoretical analysis method of the liquid metal microcirculation system.

1. Introduction

With the rapid development of electronic technology, electronic devices have entered miniaturization and the cooling system requires less space and higher heat. Microchannel heat dissipation is an advanced cooling method that solves the problem of heat dissipation in extremely high heat flux electronic chips. Currently, microchannel research includes theoretical mechanisms, numerical simulations, structural optimization, processing techniques, industrial applications and two-phase flow [1,2]. Typical microchannel driven pumps include impeller pumps, piezoelectric pumps, electromagnetic pumps, electroosmotic pumps, etc. Cooling media mainly include air, freon, water, alcohol and ethylene glycol [3-5], including two-phase flow and nanofluid [6-8]. Liquid metal (gallium-based low melting point metal) has a series of excellent characteristics, such as low melting point, high boiling point, high thermal conductivity, high heat capacity, stable nature and non-toxicity [9]. At present, research on the basic properties and application fields of liquid metals is gradually deepening [10-15]. High-performance microchannel devices for liquid metals were first proposed, and micro-heat dissipation systems based on liquid metal research are becoming an important research direction in the field of heat dissipation.

In this paper, a new micro-heat dissipation structure is studied, which uses a low-melting-point metal as a medium to circulate in a micro-flow channel with an equivalent diameter of less than 1 mm, and then dissipates heat through a heat-dissipating fin and a cooling fan. The overall thickness of the micro system can reach 0.9 mm, the thickness of the miniature electromagnetic pump, the heat dissipating fin and the cooling fan is large, and the size of the micro electromagnetic pump can reach 15 mm*10 mm*3 mm. In this paper, the theoretical and experimental research on the micro-heat dissipation structure is carried out. The structure and parameters of the research object are shown in table 1 and figures 1 and 2. The physical parameters of liquid metal used in the study are shown in table 2.

Table 1. Research object.

	System A	System B
Substrate material	stainless steel	stainless steel
Flow section	0.5 mm*10 mm	1.0mm*15.0 mm
Runner length	388.5 mm	710.2 mm
Fan parameter	Size is 45*45*6 m	Size is 60*60*8 m
	Maximum air volume is 3.8 CFM	Maximum air volume is 4.0 CFM
Heat sink parameter	Fin size is 50 mm*11 mm	Fin size is 60 mm*36 mm
	Wing height is 6 mm	Wing height is 10 mm
	Wing thickness is 0.2 mm	Wing thickness is 0.2 mm
	Gap is 0.8 mm	Gap is 1.3 mm

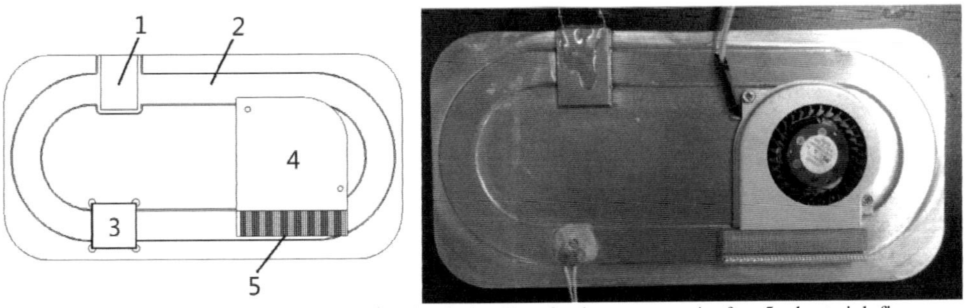

1—miniature electromagnetic pump; 2—circulation line; 3—heat source; 4—fan; 5—heat sink fin

Figure 1. System A.

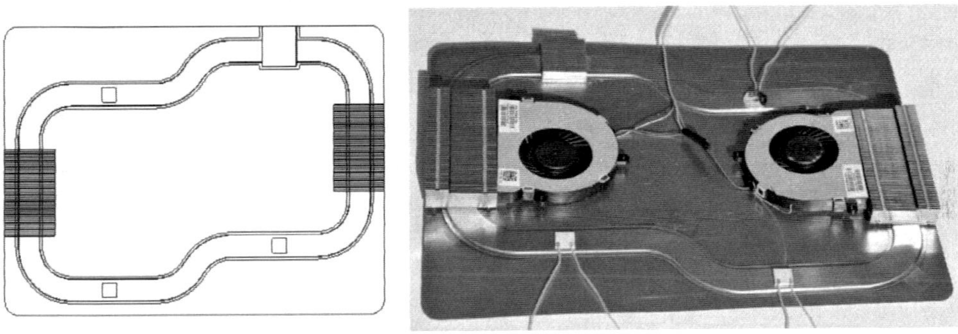

Figure 2. System B.

Table 2. Physical parameters of liquid metal.

Melting Point /°C	Density /kg/m^3	Viscosity /Pa*s	Thermal Conductivity /W/(m* °C)	Specific Heat Capacity /J/(kg* °C)	Pr Number
-5	6500	0.002	30	400	0.0208

2. Experiment

The sample was tested in this paper. The results obtained are shown in tables 3 and 4. A ceramic heater was used to simulate the heat source and a K-type thermocouple was used to measure the temperature. The flow rate of the fluid is estimated by using the value of the temperature rise after the fluid flows through the superheat source. The experimental errors mainly include sample size error, test equipment error, experimental condition error, and micro electromagnetic pump driving force error.

Table 3. Experimental result of system A.

Heatsource power/W	Electromagnetic pump head/kPa	Heat source temperature /°C	Heat source inlet temperature /°C	Ambient temperature /°C	Flow /ml/min	Reynolds number
4.55	0.57	47.7	28.4	22	5.44	57.5
9.5	0.95	54.4	29.2	22	8.70	92.0
11	0.95	60.3	30.0	22	8.37	88.5
13.2	1.14	62.9	31.5	22	9.70	102.6

Table 4. Experimental result of System B.

Electromagnetic pump lift/kPa	Temperature of heat source 1/°C (30 W)	Temperature of heat source 2/°C (15 W)	Temperature of heat source 3/°C (15 W)	Ambient temperature /°C	Flow /ml/min	Reynolds number
0.9	52.5	50.5	43.5	18.0	47.35	328.8
1.8	48.7	46.6	41.5	18.0	86.08	597.8

3. Simulation calculation

The flow and heat transfer simulation analysis of the system was carried out using the finite element software ANSYS. Assuming that the fluid motion state is laminar flow, according to experimental data and simulation analysis results, the structure Reynolds number is less than 2300. Since the temperature of the system is not high and the forced air cooling is used, the influence of radiation heat transfer is finite, so the radiation heat transfer of the structure is neglected in the simulation.

3.1. System A

The results of the system simulation analysis are shown in table 5 and figure 3, The error between the simulation results and the experiment is no more than 5%, the experimental results are reliable and the simulation results are credible.

Table 5. Simulation results at an ambient temperature of 22°C.

Heat source/W	Electromagnetic pump lift/kPa	Heat source temperature /°C	Heat source temperature comparison	Flow /ml/min	Flow comparison	Reynolds number
4.55	0.57	46.5	2.6%	5.19	4.8%	54.9
9.5	0.95	56.2	-3.2%	8.46	2.8%	89.5
11	0.95	61.7	-2.3%	8.46	-1%	89.5
13.2	1.14	64.3	-2.2%	10.1	-4.0%	106.8

As can be seen from the above results, when the thickness of the microcirculation channel is 0.5 mm and the overall thickness is 0.9mm, the flow rate of the low melting point metal of system A can reach 9.7 ml/min. According to formula $Q = c\rho V \cdot \Delta T$, the heat carrying capacity of the liquid metal in the structure is $Q = 0.42\Delta T$. In the above formula, Q is heat W; c is specific heat capacity, ρ is density, V is volume flow and ΔT is— fluid temperature difference.

Heat source temperature comparison and flow comparison means the difference between the experimental value and the theoretical value.

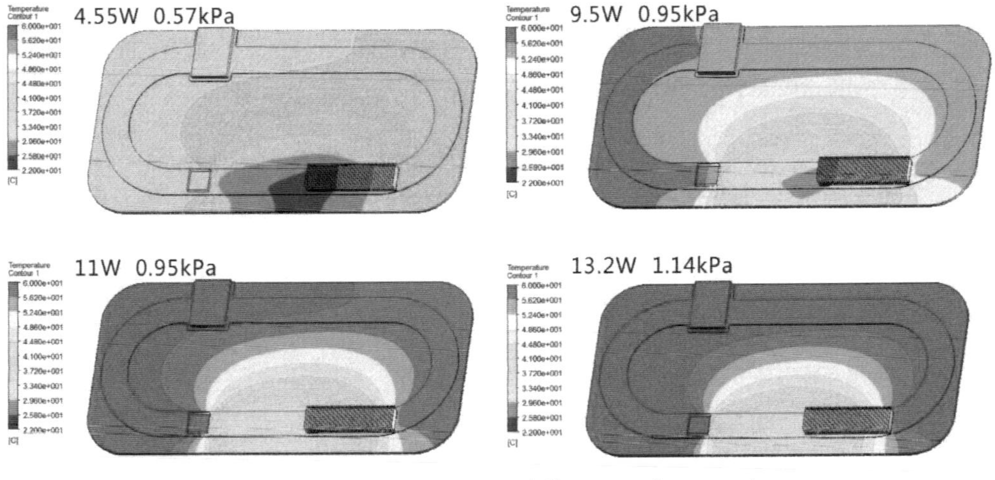

Figure 3. Temperature cloud diagram of system A.

3.2. System B

The results of the system simulation analysis are shown in table 6 and figure 4. The error between the simulation results and the experiment is no more than 5%, the experimental results are reliable and the simulation results are credible.

Table 6. Results at an ambient temperature of 18°C.

	Electromagnetic pump lift/ kPa	Temperature of heat source 1/°C (30 W)	Temperature of heat source 2/°C (15 W)	Temperature of heat source 3/°C (15 W)	Flow /ml/min	Reynolds number
Theoretical value	0.9	50.2	50.6	43.0	45.2	313.9
Experimental error	0.9	4.6%	-0.2%	1.2%	4.8%	
Theoretical value	1.8	47.3	46.9	42.8	84.6	587.5
Error	1.8	3.0%	-0.6%	-3.0%	1.7%	

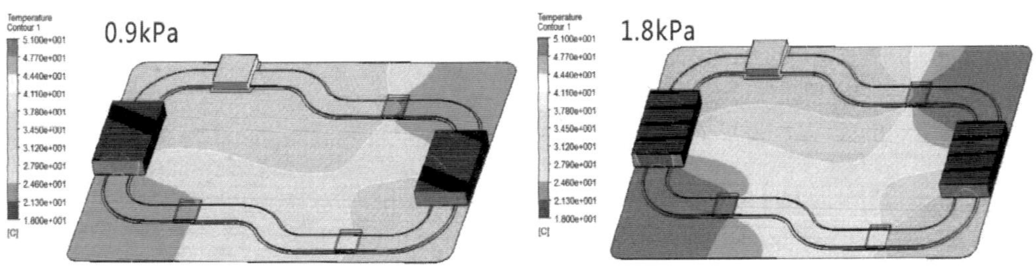

Figure 4. Temperature cloud diagram of system B.

As can be seen from the above results, when the thickness of the microcirculation channel is 1.0 mm and the overall thickness is 1.2 mm, the flow rate of the liquid metal of system A can reach 87.0 ml/min. According to formula $Q = c\rho V \cdot \Delta T$ the heat carrying capacity of the liquid metal in the structure is $Q = 3.77\Delta T$. In the above formula, Q is heat W; c is specific heat capacity, ρ is density, V is volume flow and ΔT is— fluid temperature difference.

4. Theory and result analysis

The system resistance analysis in this paper is shown in table 7 and figure 5. According to the data in the research, the following conclusions can be drawn.

Table 7. System flow resistance analysis.

	Flow section	Hydraulic diameter/ mm	Total length of the runner/mm	System resistance/ kPa	Flow/ ml/min	Reynolds number
SystemA	0.5mm*10mm	0.952	388.5	0.57	5.19	54.9
				0.95	8.46	89.5
				1.14	10.1	106.8
SystemB	1mm*15mm	1.875	710.2	0.9	45.2	313.9
				1.8	84.6	587.5

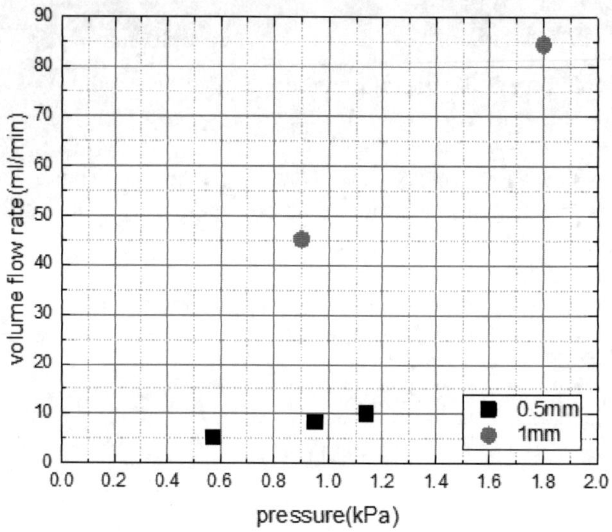

Figure 5. Volume flow and pressure test results.

4.1. Theoretical analysis of system flow

The simulation of the system uses the conventional scale fluid flow theory. The paper compares the simulation and experimental results of the liquid metal flow. The results are in good agreement, indicating that the hydraulic diameter is 0.952 mm and the flow section thickness is 0.5 mm. The flow law conforms to the conventional scale flow heat transfer theory. As can be seen from the data in the table, the smaller the flow path size, the greater the structure resistance, the slower the flow rate, the smaller the size and flow rate reduction, and the smaller the size, the more the flow rate is reduced. On the basis of this paper, the runner size is further reduced, the structure resistance will be greatly increased, the flow rate will be greatly reduced, and the heat dissipation capacity will be greatly reduced. The system will lose its obvious advantages compared with the traditional heat dissipation method, and the processing difficulty will increase. The precision is high and the system performance is poor.

4.2. System size reduction effect

The formula of heat transfer capacity of liquid cooling system has been shown as follows. $Q = c\rho V \cdot \Delta T$ In the above formula, Q is heat W; c is specific heat capacity, ρ is density, V is volume flow and ΔT is— fluid temperature difference.

The heat capacity of the microcirculation system with an overall thickness of 0.9 mm in System A is $Q = 0.42\Delta T$. If a high thermal conductivity material is used, the one-dimensional cylindrical wall thermal conductivity formula is as follows $Q = \frac{2\pi\lambda l(t_1 - t_2)}{\ln(r_2 / r_1)}$. Considering the position of the heat source and the uncertainty of the shape of the heat conducting plate in actual use, taking half of the heat transfer capacity of the one-dimensional cylindrical wall, when the ratio of the heat source of the 0.9 mm copper plate to the heat exchanger plate is 100, the thermal conductivity of the copper plate is $Q = 0.49\Delta T$, which This is equivalent to a liquid metal microcirculation structure. When the overall thickness of the system is further reduced, the heat transfer amount of the high thermal conductive material plate is proportionally reduced, while the flow rate of the liquid metal microcirculation system is accelerated, the heat transfer amount is accelerated, and the liquid metal microcirculation structure loses the performance advantage.

5. Conclusion

In the microchannel flow system, when the flow path size is limited to a certain extent, the fluid cannot be regarded as a continuous medium, and the conventional scale flow and heat transfer theory will not be applicable. The flow path scale is further reduced, and the fluid motion will be dominated by the effects of quantum motion, and the reaction will be free molecular flow motion. This limit is different for gases, liquids, and liquid metals. The critical dimension of liquid is smaller than that of gas, and the critical dimension of liquid metal is smaller than that of water.

Acknowledgments

This work was supported by the New Materials Research Key Program of Tianjin (No. 16ZXCLGX00090), the National Key Research and Development Program of China (No. 2016YFB0303000).

References

[1] Qu W L and Mudawar I 2002 Analysis of three-dimensional heat transfer in micro-channel heat sinks *Int. J. Heat. Mass. Tran.* **19** 3973-85

[2] Lee J and Mudawar I 2008 Fluid flow and heat transfer characteristics of low temperature two-phase micro-channel heat sinks *Int. J. Heat. Mass. Tran.* **17-18** 4327-41

[3] Hsieh S S, Lin C Y, Huang C F and Tsai H H 2004 Liquid flow in a micro-channel *J. Micromech. Microeng.* **4** 436-45

[4] Dong T, Yang Z C, Bi Q C and Zhang Y L 2008 Freon R141b flow boiling in silicon microchannel heat sinks: Experimental investigation *Int. J. Heat. Mass. Tran.* **3** 315-24

[5] Xie Y Q, Yu J Z and Zhao Z H 2005 Experimental investigation of flow and heat transfer for the ethanol-water solution and FC-72 in rectangular microchannels *Int. J. Heat. Mass. Tran.* **8** 695-702

[6] Qu W L and Mudawar I 2004 Measurement and correlation of critical heat flux in two-phase micro-channel heat sinks *Int J Heat Mass Tran.* **10-11** 2045-59

[7] Chen W L, Twu M C and Pan C 2002 Gas-liquid two-phase flow in micro-channels *Int. J. Multiphas. Flow.* **7** 1235-47

[8] Hwang J J, Tseng F G and Pan C 2005 Ethanol-CO_2two-phase flow in diverging and converging microchannels *Int. J. Multiphas. Flow.* **5** 548-70

[9] Ma K Q and Liu J 2007 Liquid metal cooling in thermal management of computer chips *Front of Ener and Power Eng in China* **4** 384-402

[10] Deng Y G and Liu J 2009 Corrosion development between liquid gallium and four typical metal substrates used in chip cooling device *Appl. Phys. A-Mater.* **3** 907-15

[11] Deng Y G and Liu J 2009 Recent advances in direct solar thermal power generation *Jrenew. Sustain. Ener.* **5** 1-23

[12] Deng Y G and Liu J 2010 Hybrid liquid metal-water cooling system for heat dissipation of high powerdensity microdevices *Heat. Mass. Tran.* **11-12** 1327

[13] Tomasz G and Katarzyna B 2018 The applications of Cu substrate in liquid metal cooling systems *Mater Lett* **227** 116-9
[14] Dobosz A, Plevachuk Y, Sklyarchuk V, Sokoliuk B and Gancarz T 2018 The application of liquid metals in cooling systems: A study of the thermophysical properties of eutectic Ga-Sn-Zn with Al additions *Int J Heat and Mass Transfer Part A* **126** 414-20
[15] Carlberg B, Ye L L and Liu J 2012 Polymer-metal nanofibrous composite for thermal management of microsystems *Mater Lett* **75** 229-32

IOP Conf. Series: Materials Science and Engineering **634** (2019) 012014 doi:10.1088/1757-899X/634/1/012014

A study on low-energy light absorption of SWNTs

H Li[1,2], S Y Wu[1], M Y Zhou[1] and X R Liang[1]

[1]School of Basic Sciences for Aviation, Naval Aviation University, Yantai Shandong 264001, China

E-mail: 346521161@qq.com

Abstract. By using semi-empirical method, an approximately one-dimensional model on the basis of tight-binding theory is set up to calculate state density of different sorts of single-walled carbon nanotubes so as to explain the reason for three absorption peaks in the low energy level of light absorption spectrum.

1. Introduction

Since the synthesis of single-walled carbon nanotubes (SWNTs) by Iijima and Ichihashie [1,2], carbon nanotubes have received extensive attention as a photoelectric functional material with wide application prospects [3-13]. In 1998, Pichler [14] measured the photoconductivity of graphite and SWNTs, and three obvious absorption peaks were found out in the graphite, which were located at 4.5±0.05 eV, 13±0.05 eV and 15±0.05 eV, respectively. There were also three similar and slightly lower absorption peaks in SWNTs, which were located at 4.3±0.1 eV, 11.7±0.2 eV and 14.60.1 eV, respectively. At the same time, it was also found that SWNTs had absorption peaks in the lower energy region, which were located at 0.6±0.1 eV, 1.2±0.1 eV and 1.8±0.1 eV. After that, Kataura, Jost, Jeroen and others [15-17] also found the three characteristic absorption peaks in the low energy region of the SWNTs' light absorption spectrum.

SWNTs can be seen as a graphite plane curled in a certain direction. Graphite is a layered six angle structure and each layer expands in a two-dimensional plane, but its edge is unstable, which is just like a thin piece of paper and its edge is tilted up so as to form a more stable reel structure, that is carbon tube.The ratio of the length and the diameter of the carbon nanotube can reach 10^4-10^5 orders of magnitude. Therefore, we can conclude that the carbon nanotubes are one-dimensional or quasi one-dimensional system physically. According to different ways of winding the graphite surface forming carbon nanotubes, the layers of carbon nanotubes can fall into two categories: non helix type and helix type [18]. The helix type is divided into the "armchair" structure and the "zigzag" structure. The spiral vector of carbon nanotubes is usually expressed as $\vec{C}_h = n\vec{a}_1 + m\vec{a}_2$, in which *n*, *m* are integers, $\vec{a}_1, \vec{a}_2$ are the basis of the graphite lattice as shown in the figure 1. The angle θ between $\vec{C}_h$ and $\vec{a}_1$ is called the chiral angle. When $n=m, \theta = 30^0$,this kind of carbon nanotube is called the chair type carbon nanotube, which is the armchair tube; when $m = 0, \theta = 0^0$,this structure is called the sawtooth carbon nanotube, which is the zigzag tube; when $0^0 < \theta < 30^0$, this structure is called the chiral carbon nanotubes. Carbon nanotubes are commonly characterized by (m, n).

IOP Conf. Series: Materials Science and Engineering **634** (2019) 012014 doi:10.1088/1757-899X/634/1/012014

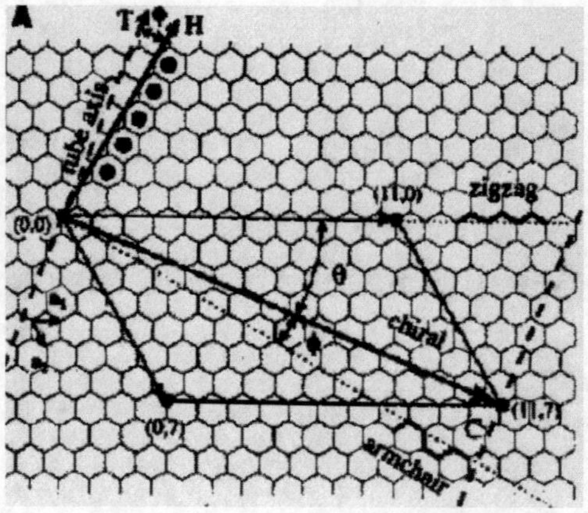

Figure 1. Structural parameters of carbon nanotubes.

2. Model and formula

At present, there are wide and diverse methods to study carbon nanotubes in theory, including the tight binding method [19], the first principle or the AB initio method, the Green function method [20] and so on. We have established a simplified model that is to study non-spiral carbon nanotubes (armchair type SWNTs and zigzag type SWNTs). The armchair type carbon nanotubes are treated as a number of identical trans-polyacetylene chains which are arranged in the axial direction and the zigzag type carbon nanotubes are treated as a number of identical cispolyacetylene chains which are arranged in the axial direction. In the circumferential direction, the same chains are combined by chemical bonds. Given the quasi one-dimensional nature of carbon nanotubes, the two non-helical carbon nanotubes are considered as one-dimensional systems formed by the combination of multiple conjugated polymer chains through interchain coupling in the simplified model. This model is called one-dimensional carbon nanotube model.

The Hamiltonian of the polymer chain described by the extended one-dimensional tight binding model is as follows:

$$H_P = -\sum_{n,\sigma}\left[t_0 - \alpha_P(u_{n+1} - u_n) - t_1\cos(n\pi)\right]\left(a^+_{n,\sigma}a_{n+1,\sigma} + a^+_{n+1,\sigma}a_{n,\sigma}\right) + \frac{1}{2}K_P\sum_n(u_{n+1} - u_n)^2 \tag{1}$$

in which $a^+_{n,\sigma}(a_{n,\sigma})$ is the generation (disappearance) operator of an electron whose spin cis σ at a lattice point n, $t_{n,n+1} = t_0 - \alpha_P(u_{n+1} - u_n) - t_1 cos(n\pi)$ is the transition integral of π electrons in adjacent grid points n and $n+1$, t_0 is the transition integral of electrons in adjacent grid points when lattice points arrange equidistantly, α_P is electroacoustic coupling constant, u_n is the position deviation of the n lattice point (CH group), t_1 is degenerate parameter, and K_P is the elastic coefficient of lattice.

In the axial direction, the Hamiltonian is the sucm of Hamiltonian of all single organically conjugated polymer chains, which is expressed as follows:

$$H_a = -\sum_{j,n,\sigma} t_{j,n,n+1}\left(a^+_{j,n,\sigma}a_{j,n+1,\sigma} + a^+_{j,n+1,\sigma}a_{j,n,\sigma}\right) + \frac{1}{2}K_P\sum_{j,n}(u_{j,n+1} - u_{j,n})^2 \tag{2}$$

in which $t_{j,n,n+1} = t_0 - \alpha_P(u_{j,n+1} - u_{j,n}) - t_1\cos(n\pi)$. j represents the organically conjugated polymer chain of class j, we assume that t_0, α_P, t_1, K_P are the same in each chain.

In the circumferential direction, the Hamiltonian of the coupling term between chains is described as:

$$H_i = - \sum_{j,(n_j,n_{j+1}),\sigma} t_\perp \left(a^+_{j,n_j,\sigma} a_{j+1,n_{j+1},\sigma} + a^+_{j+1,n_{j+1},\sigma} a_{j,n_j,\sigma} \right) \tag{3}$$

in which (n_j, n_{j+1}) represents the nearest neighbor lattice point of two adjacent chains, and $t_\perp$ represents the electron transition integral between the nearest two lattice points in the circumferential direction, which shows the intensity of interchain coupling.

By solving the Schrodinger equation

$$\left(\sum_j H_j + H_i \right) \psi_\mu = \varepsilon_\mu \psi_\mu, \quad \psi_\mu = \sum_{j,n} Z_{\mu,j,n} |j,n\rangle \tag{4}$$

we can get the eigenvalue ε_μ and the eigenfunction ψ_μ.

Total energy

$$E = \sum_{OCC} \varepsilon_\mu + \frac{1}{2} K_P \sum_{j,n} \left(u_{j,n+1} - u_{j,n} \right)^2 \tag{5}$$

The equilibrium equation of the lattice shape is

$$\phi_{j,n} + \phi_{j,n+1} = (-1)^n \pi\lambda \left[\sum_{\mu,\sigma} Z_{j,\mu,n,\sigma} Z_{j,\mu,n+1,\sigma} - \frac{1}{N} \sum_{m=1}^{N} \sum_{\mu,\sigma} Z_{j,\mu,m+1} Z_{j,\mu,m} \right] \tag{6}$$

in which $\phi_{j,n} = (-1)^n (\alpha_P / t_0) u_{j,n}$, preferences [21], t_0=2.5 eV, α_P=4.1 eV/Å, t_1=0.05 eV, K_P=21 eV/Å, $\lambda = 2\alpha_P^2 / \pi K_P t_0$. $\sum_{OCC}$ represents the sum of the occupied energy levels of all electrons. The periodic boundary condition applies in the model.

3. Discussion and examples

The energy level distribution and density of states of two non-spiral carbon nanotubes (armchair type SWNTs and zigzag type SWNTs) are calculated by one-dimensional carbon nanotube model, so as to explain the formation of three absorption peaks in the low energy phase of the carbon nanotube experiment [3].

The intrinsic states of electrons in the atom form a series of discrete energy levels, and we can explain their specific distribution by identifying each energy level. However, the distribution of electrons is very dense in the solid, it is meaningless to indicate each energy level by forming the energy level of quasi continuous distribution. Thus, we introduce the concept of "energy state density" in order to generalize the distribution of energy levels in this case.

The "energy state density" is defined as $N(E) = \lim \frac{\Delta N}{\Delta E}$, in which ΔN is the number of energy state, and ΔE is the energy bite. We calculate the energy state density through Gauss expansion in order to analyse the electronic states near the Fermi level.

$$N(\varepsilon) = \sum_\mu \frac{n_\mu}{(2\pi\sigma)^{\frac{1}{2}}} \exp\left[-\frac{(\varepsilon - \varepsilon_\mu)}{2\sigma^2} \right] \tag{7}$$

in which ε_μ is the eigen energy level, n_μ is the energy level degeneracy, and σ is Broadening parameter.

We can get all the energy eigenvalues by diagonalizing Hamilton of equation (4).

Density of states (DOS) will obtain through Lorentz broadening:

$$\rho(\varepsilon) = \frac{n_\mu}{(2\pi\sigma)^{\frac{1}{2}}} \exp\left[-\frac{(\varepsilon - \varepsilon_\mu)}{2\sigma^2} \right] \quad (8)$$

which shows the number of states in unit energy intervals. $N(\varepsilon)$ shows the sum of the all DOS.

Singular points of DOS mean the existence of quantized energy levels. By studying the DOS near the Fermi level, we can find the electron transition between the energy levels corresponding to the singular points, then figure out the light absorption of the carbon nanotube. First, we calculate the DOS of several zigzag-type SWNTs (*n*, 0) near the Fermi level.

The diameter of zigzag-type SWNTs (14, 0) is 1.10 nm. Figure 2 is the change diagram of the DOS of zigzag-type SWNTs (14, 0). The diameter of zigzag type SWNTs (16, 0) is 1.25 nm. Figure 3 is the change diagram of the DOS of zigzag-type SWNTs (16, 0).

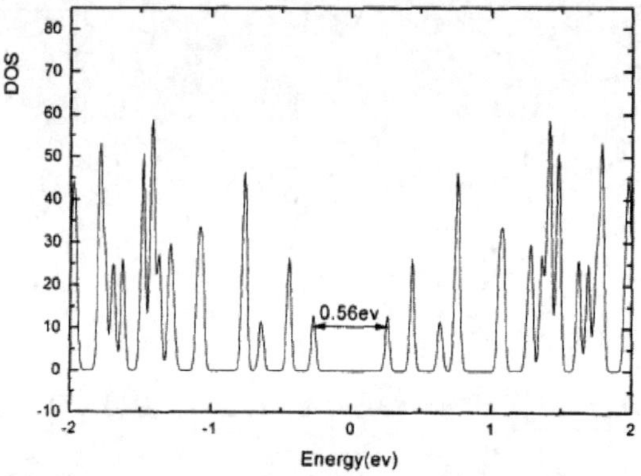

Figure 2. The change diagram of the DOS with the energy of zigzag type SWNTs(14, 0).

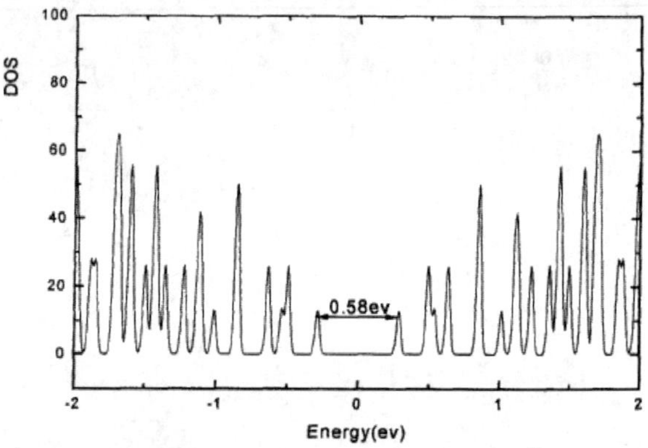

Figure 3. The change diagram of the DOS with the energy of zigzag type SWNTs(16, 0).

In the above two diagrams, the plane that energy equals zero is the Fermi plane, and each peak of energy is called singular point. We can draw the conclusion that zigzag-type SWNTs (14, 0), (16, 0) are semiconducting carbon nanotubes, because around Fermi level, their DOS are almost zero. This means that the Fermi level is formed around the band gap, which is characteristic of the semiconductor. Energy level transition between the first pair of the singular points of two carbon nanotubes is within the range of the first absorption peak in the low energy region of Pichler experimental optical absorption spectrum [17]. Table 1 shows the energy difference between the first pairs of the singular points of more semiconducting carbon nanotubes.

Table 1. Energy difference between the first pair of the singular points of zigzag-type carbon nanotubes with different diameters.

zigzag	(13,0)	(14,0)	(16,0)	(17,0)	(19,0)	(20,0)	(22,0)
Diameter(nm)	1.02	1.10	1.25	1.33	1.49	1.57	1.72
ΔE(eV)	0.66	0.56	0.58	0.49	0.52	0.38	0.40

As you can see from the table above, the diameters of the five zigzag-type carbon SWNTs (13,0), (14,0), (16,0), (17,0) and (19,0) are distributed between 1-2 nm, and the energy level transition between their first pair of the singular points is within the range of the first absorption peak obtained from the experiment. The singular points indicate the existence of the quantum energy level. By calculating the energy levels, We can get the energy corresponding to the first pair of singular points.

In the change diagram of the DOS, the first singular point on the left side of the Fermi plane corresponds to two energy levels, N/2 and N/2-1, the two energy levels are energy merger. Similarly the first singular point on the right side of the Fermi plane corresponds to two energy levels, N/2 and N/2-1, the two energy levels are energy merger. In which N is the total number of lattice. By calculating the electron transition probabilities between these four levels we can arrive at such a conclusion that electronic transition may occur between the first pair of the singular points in the five carbon nanotubes above, but the energy levels of the electron transition are different. The electron transition of SWNTs (13,0), (16,0) mainly occur between N/2 - N/2+2, N/2-1 - N/2+1 energy levels; the electron transition of SWNTs (14,0), (19,0) mainly occur between N/2 - N/2+1, N/2-1 - N/2+2 energy levels; however, in the case of SWNTs (17,0), all four situations are likely to occur, and N is the total number of lattice points. Therefore, the first absorption peak obtained in the experiment is caused by the electron transition between the first pair of the singular points of semiconducting carbon nanotubes.

Next, we will study the DOS of several armchair-type SWNTs (n, n). The diameter of armchair type SWNTs (9, 9) is 1.22 nm. Figure 4 shows its DOS. The diameter of armchair-type SWNTs (10, 10) is 1.36 nm. Its DOS is shown in figure 5.

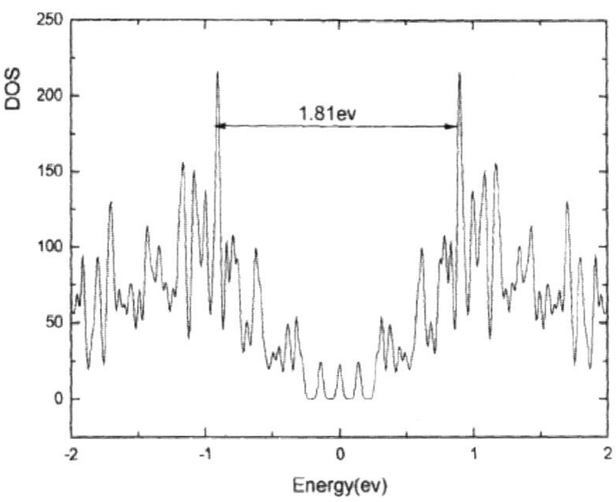

Figure 4. DOS of armchair-type SWNTs (9, 9).

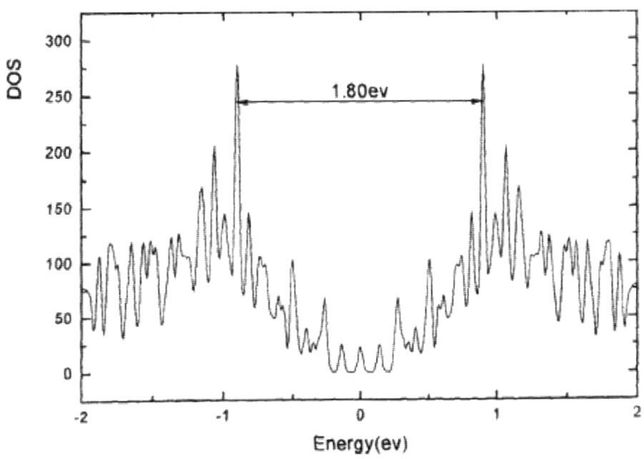

Figure 5. DOS of armchair-type SWNTs (10, 10).

From the above figure 5 we see that there is a minimum of DOS near the Fermi level of the armchair-type SWNTs. It shows that it is a type of metallic carbon nanotube. We take two very obvious singular points as an example, which are the two peaks pointed by the arrow in the figure 5. The two peaks indicate that there is a quantum energy level near the singular points, and the DOS around the quantum energy level is large, so the electron transition is easy to occur. The energy difference between the two singular points is about 1.81 eV, which indicates that if there is an electron transition between the two energy levels, the photon energy needed is about 1.81 eV, which requires the absorption of the photon energy. This value is within the range of the third absorption peak in the low energy region of the experiment.

Table 2. Energy difference between the first pair of the singular points of armchair type carbon nanotubes with different diameters.

Armchair	(7,7)	(8,8)	(9,9)	(10,10)	(11,11)
Diameter(nm)	0.95	1.08	1.22	1.36	1.49
ΔE(eV)	1.80	1.82	1.81	1.80	1.80

IOP Conf. Series: Materials Science and Engineering **634** (2019) 012014 doi:10.1088/1757-899X/634/1/012014

When calculating the DOS of several other armchair-type SWNTs, we found such pair of singular points always exists, and their energy difference is about 1.80 eV, which is shown in table 2.

By calculating the transition probability between the singular points in the armchair-type SWNTs, it is confirmed that there is electron transition between the singular points in the several kinds of armchair-type SWNTs above. The transition probability between the energy levels is different, which is the same case with zigzag-type SWNTs. Therefore, the third absorption peak is formed by the electron transition of metallic carbon nanotube in the experiment.

4. Conclusion

By calculating the DOS of several carbon nanotubes in the diameter range of the experiment we found that energy difference between the first pair of singular points near Fermi lever of the semiconducting zigzag-type SWNTs is basically consistent with the size of the absorption peak of 0.6±0.1 eV obtained in the experiment. We also got the conclusion that the electron transition can occur between the energy levels of the singular points. Therefore, the first absorption peak in the low energy region of Pichler experimental optical absorption spectrum of SWNTs is caused by the electron transition between the first pair of the singular points of the semiconducting carbon nanotubes. From the DOS of two metallic armchair type SWNTs we can get the third absorption peak, 1.8±0.1 ev, which is caused by the electron transition between their two most obvious singular points, with the size basically the same. Several other armchair carbon nanotubes also exhibit this property. The reason for the second absorption peak is not clear. It may be caused by the optical absorption of spiral metallic carbon nanotubes, or caused by the electron transition between the second pair or the third pair of singular points of the semiconducting carbon nanotubes.

References

[1] Iijima S and Lehihashi T 1993 *Nature* **363** 603
[2] Bethune D S, Kiang C H and Devries M S 1993 *Nature* **363** 605
[3] Lawrence T 2017 *Scott Pure and Applied Chemistry* **6** 2016
[4] Yang H P 2018 *Chinese Physics Letters* **8** 375
[5] Cheng H M and Liu C 2018 *Chinese Physics Letters* **8** 78
[6] Zhang X G 2017 *Journal of Polymer Engineering* **1**
[7] Li N N 2017 *Journal of Polymer Engineering* **4** 124
[8] Vasyl M H 2017 *Toxicology Letters* **3** 16
[9] Elisabeth A 2017 *Journal of Nuclear Materials* **8** 32
[10] Abbas P 2018 *Chemical Engineering Science* **7** 21
[11] Iqra S 2017*Computer Methods and Programs in Biomedicine* **10** 16
[12] Marcelina A 2017 *Journal of Fluorine Chemistry* **6** 18
[13] Shang S S 2017 *Applied Mechanics and Materials* **858** 171
[14] Piehler T, KnuPfer M and Golder M S 1998 *Phys. Rev. Lett.* **80** 4729
[15] Kataura H, Kamazawa Y and Maniwa Y 2002 *Appl. Phys. Lett.* **5228** 35
[16] Ömer Güler 2016 *Science and Engineering of Composite Materials* **4** 197
[17] Jost O and Capes L 2004 *Appl. Phys. Lett.* **2004** 3572
[18] Dresselhaus M S, Dresselhaus G and Eklund P C 1996 (San Diego Academic Press) **56** 121
[19] Su W P, Schrieffer J R and Heeger A J 1979 *Phys. Rev. Lett.* **42** 1698
[20] Rohlfing M and Louie S G 1999 *Phys. Rev. Lett.* **82** 1959
[21] Heeger A J, Kivelson S and Schrieffer J R 1988 *Rev. Mod. Phys.* **60** 209

IOP Conf. Series: Materials Science and Engineering **634** (2019) 012015 doi:10.1088/1757-899X/634/1/012015

Development of a geotextile coated with latex and used as a protection element on heated surfaces

C S souza[1,2] and J U L Mendes[1]

[1]Universidade Federal do Rio Grande do Norte, Brazil

E-mail: clarasilvestr@gmail.com

Abstract. A lot has been researched today about new energy resources and sustainable materials aiming to decrease the impacts caused on the environment. Materials waste that were previously destined for the trash, are now seen in a different way. This present paper had as purpose the confection of a geotextile using banana tree leaves, in particular, fibers extracted from the stem of the banana tree leaves, which presently, are recognized as one of the strongest cellulosic fibers. As a natural geotextile matrix, latex was used and, subsequently, possible applications for the geotextile were analysed through previous material characterization based on literature. The samples were weaved and covered by latex, thereafter, analysis and verification were conducted on the main physical-chemical properties and also mechanical and thermal properties. Such results were obtained through tests, as: tensile properties, flammability, thermogravimetric analysis, water and humidity absorption, moisture resistance, friction, density, weight, biodegradation under environmental conditions. and thermal conductivity. Based on the obtained results, it was possbible to conclude that geotextile can be used as surfaces coating, because beside its high tensile strength, even when heated, the goetextile presentes a low thermal conductivity (0.214 W/mK). That fact means that the material has good thermal characteristics, quality of an isulator. Also, it was conclude that the natural composite has combined properties from banana tree fibers (as thermal properties and mechanical strength) and from thermoplastic elastomer latex. Such result did not compromise the hygroscopicity of the banana fiber. Finally, it could be concluded that the material is sustainable, does not damage the environment and fulfil the coating function, as expected.

1. Introduction

A lot has been researched today about new energy resources and sustainable materials aiming to decrease the impacts caused on the environment.

Based on this new sustainable view, we aimed to change how natural geotextile are made, using materials from natural fibers and used in association with soil. These products are being manufactured using textile fibers or using a hybrid combination with synthetic fibers in its manufacture.

According to Raghavendra *et al* [1], sustainable materials made from natural fibers are renewable, have a low cost, they are partially or completely biodegradable materials, and also according to Ramesh *et al* [2], fibers are abrasive, they have a good net calorific value, excellent mechanical properties, low density and they are ecological materials. Fibers containing rich amounts of ligno-celluose such as banana, sisal, bamboo etc, are frequently used as a reinforcement in composites or geotextiles. The geotextile longevity is dependent upon the amount of cellulose and lignin on the fiber, therefore, its chemical composition is extremely important.

The possibility to utilize the waste disposed on plantations, emanating from banana fruit farming (banana trees leaves) to provide the basis to manufacture a natural geotextile has motivated this research.

A natural elastomer matrix, latex, was used. Derived from *Hevea Brasilliensis* tree, which brazilian production is approximately 90.000 ton/year and whose characteristics are low humidity absorption, high elasticity and biodegradability [3].

The geotextile manufacturing process using the reject from the banana tree as raw material, aim to recycle considering the social, economical and environmental aspects and, as a consequence, it contributes to an environmental preservation global policy.

2. Methodology

To create the geotextile, banana fibers from banana tree leaves were used as a reinforcement. And, to complete the biotextile, natural rubber latex pre vulcanized as an agglomerating agent. The fibers were peeled from the central Pseudopeciolo (stem) from the banana tree leaf (figure 1). The fibers presented from 1.0 to 4.0 meters length and an average diameter of 1.32 milimeters.

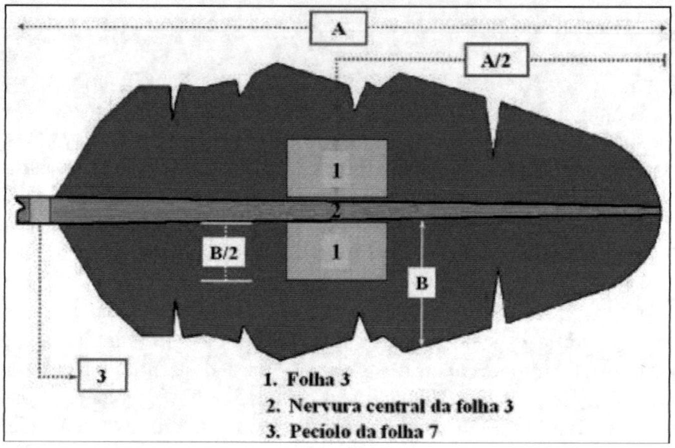

Figure 1. Banana tree leaf divided in 1) Leaf; 2) Leaf central vein; 3) Petioles of the leaf. The natural rubber latex used was pre vulcanized. A is the length of the fiber after being cut from the tree.

To manufacture the geotextile, banana tree leaves were cut from the trees, using a knife and, only cutting only the old leaves. The leaves had been cut on the Petioles to respect the length of the fiber, figure 1.

After the cut, the green surfaces of the leaf, both left and right (figure 1B), were removed, remaining only the central vein of the leaf, where all the fibers were manually extracted. Once the fibers had been removed from the leaf, all the fibers were placed on a marble table aiming to dry the plant's sap. The drying process occurred in a climate-controlled environment that provided a room temperature of (25±3)ºC and humidity 60%.

After a 12-hour period, the leaves had dried and they were cut. All fibers cut in 30 centimeters long for later characterization.

The analysis for fiber and geotextile fabric were conducted in University Federal do Rio Grande do Norte. First, banana fibers without a latex cover were analyzed and, subsequently, after the latex cover. The tests were also conducted with the geotextiles (with and without latex), the tests were: tensile tests of textiles and yarns (strength), Thermogravimetric analysis (TG) and used scanning electron microscopy (SEM).

Aiming to produce a fabric made from vegetable fibers, a manual weaver's loom (25 centimeters x 25 centimeters) was used (figure 2). The yarns disposed as weft and warp, where the weft fibers were

IOP Conf. Series: Materials Science and Engineering **634** (2019) 012015 doi:10.1088/1757-899X/634/1/012015

stronger than the warp fibers that were more malleable, fitting better the weaving process.

Figure 2. Manual weaver's loom.

Manual weaving process lasted on average only five hours. After the process of manufacture of thread and fabric, we removed the tip waste and the base for the geotextile was now concluded (figure 3).

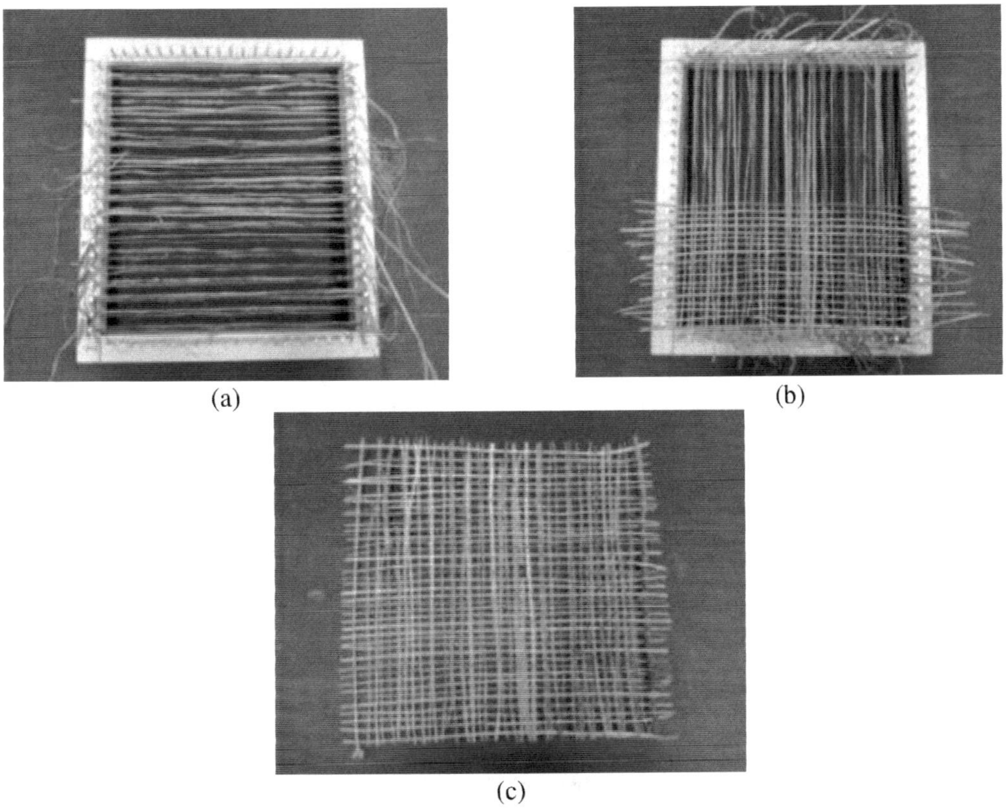

Figure 3. Steps of the process of manufacture of thread and fabric to create the geotextile. (a) Weft, (b) Weft and Warp, (c) Fabric, geotextile base.

The geotextile fabric consisted of a banana fiber fabric linked by a prevulcanized natural latex. Matrix and reinforcement combination took place in a glass container, figure 4, 18 cm wide x 6 cm deep. The recent weaved fabric on the bottom of the container, received a latex bath. The process also used a brush to facilitate its appliance. All curing process lasted average only four hours and, at the end of the process, a biotextile, 25 cm^2 and approx. 0.3 cm thick.

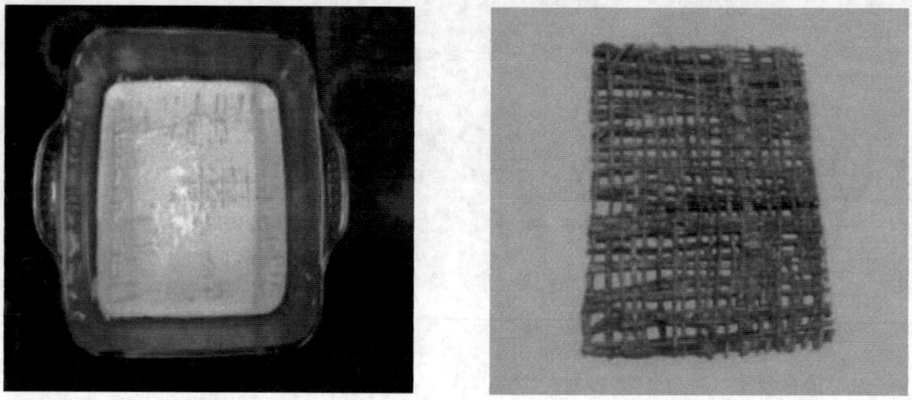

(a) (b)

Figure 4. (a) Fabric in a latex bath and (b) geotextile.

As the main goal of this project was an easy replication and a sustainable geotextile, the substrate received a simple and easy replication coating. For the same reason, in natural fibers have not undergone to any cleaning processes or even have received any chemical component. This also proved that without a special fiber treatment, fibers showed a good bond interaction with latex.

For characterization of the geotextile comparative tests have been made with fabrics, with latex and without latex, to have a better overview of the geotextile qualities and defaults. The tests also took place in University Federal do Rio Grande do Norte. The tests were also: tensile tests of textiles and yarns (strength), Thermogravimetric analysis (TG) and used scanning electron microscopy (SEM), flammability, the absorption of humidity, density (number of hairs per cm^2), abrasion-resistant, environmental degradation, solar exposure, thermal conductivity.

2.1. *Tensile tests of textiles and yarns (strength)*

All tensile properties tests were conducted in University Federal do Rio Grande do Norte at Textile Engineering Laboratory (LabCTex DET/UFRN) using a Tensolab 300 Mesdan. Analysis were carried out according to the technical standard ASTM D5034.

Four samples were cut in 15 centimeters length and 5 centimeters wide. All analysis were carried out under a 15 N/s.

2.2. *Thermogravimetric analysis (TG)*

All analysis were carried out in University Federal do Rio Grande do Norte at NUPEG (Oil and Gas Laboratory), using the equipment Netzsch STA 449F3. All the tests used the manufacturer-specific (proprietary) standard.

Samples (in natural and covered in latex) were cut in 1,5 centimeters length and tested as they were heated until 650°C.

2.3. *Scanning electron microscopy (SEM)*

Aiming a physical characterization of the banana leaf fiber, in natural samples and latex covered samples were cut in 1.2 centimeters length and enlarged from 50 to 100 times on the Hitachi TM3000 Microscope.

All tests were conducted in University Federal do Rio Grande do Norte at DEMAT (Materials Laboratory).

2.4. Flammability
As TG test, flammability analysis took place in NUPEG according to technical standard UL 94 (2013), item 7. Tests were conducted using 3 samples of geotextile and 3 samples of in natural fabric, both 15 centimeters length and 3 centimeters wide.

As the technical standard required, all samples burned for 30 seconds.

2.5. Humidity absorption
According to the manufacturer-specific (proprietary) standard. all samples were cut in a circular shape (in natural samples 1,0 grams and latex samples 3,9 grams) and then, heated until 110,5°C.

All tests were conducted in University Federal do Rio Grande do Norte at Laboratory of fluid mechanics using an ID-200.

2.6. Density
The test conducted in University Federal do Rio Grande do Norte at Textile Engineering Laboratory (LabCTex DET/UFRN) using a very simple thread counter. Analysis were carried out according to the technical standard ASTM D3801/ ISO 2060 (2015).

2.7. Abrasion-resistant
Also at Textile Engineering Laboratory (LabCTex DET/UFRN), analysis were carried out using a Crock Meter able to run 2000 cycles.

The samples used were geotextiles able to stand the endurance test.

2.8. Environmental degradation and solar exposure
According to technical standard EPA 9090A, all samples (in natural fabric and geotextile) were buried in a simulated ground for 2280 hours.

The simulated ground was divided in 23% clayey loam, 23% organic material (manure), 23% sand and 31% distilled water (mass).

2.9. Thermal conductivity
Using a conductive meter KD2 Pro at Laboratory of fluid mechanics, analysis were carried out.

Samples (in natural and covered in latex) were created in a cylindrical shape (50 millimeters thick x 190 millimeters high) and, using the KD2, all sides of the cylinder were tested.

3. Results
Each study followed the technical standard requirement which refers to every analysis, therefore, humidity, temperature and other parameters might vary according to the Standards specifications

3.1. Fiber
The first test conducted on the fibers was scanning electron microscopy (SEM). SEM characterization of the fibers aimed to identify the physical characteristics of the fibers before and after receiving the latex bath and, subsequently, to support the analysis of the geotextile behavior based on its properties, figure 5.

It may be noticed, figure 5, that the banana tree leaf fiber presented a unique microstructure with high porosity due to its natural vegetable fiber characteristic. High amounts of cellulose linked to hemicellulose walls provided to the fiber a large quantity of fibrils. That explains the high level capacity of absorption of humidity and liquids in general, or high hygroscopicity. When covered in latex, it is not evident such characteristic, figure 6.

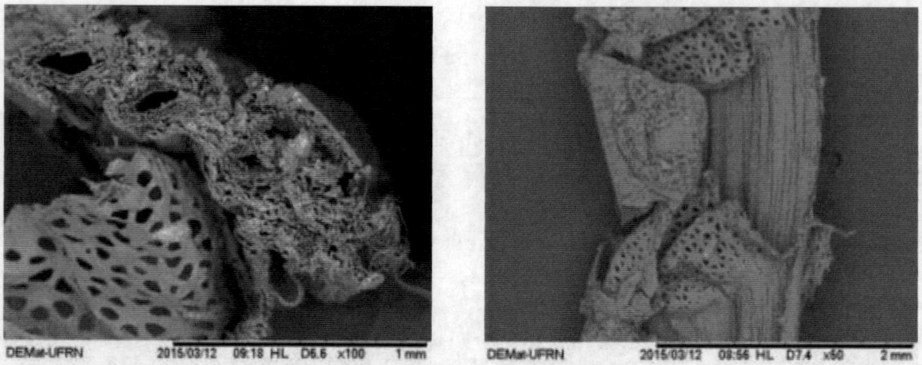

Figure 5. SEM from a banana tree leaf recently removed from the central vein of the leaf. View from the top and lateral view.

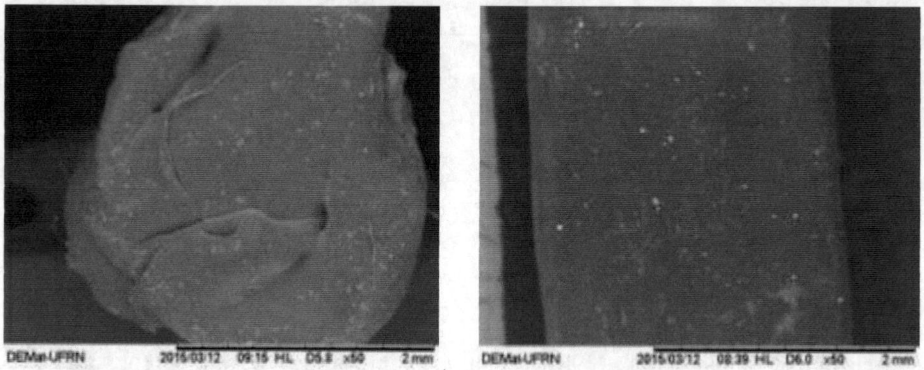

Figure 6. Banana tree leaf fiber covered by latex. View from the top and lateral view.

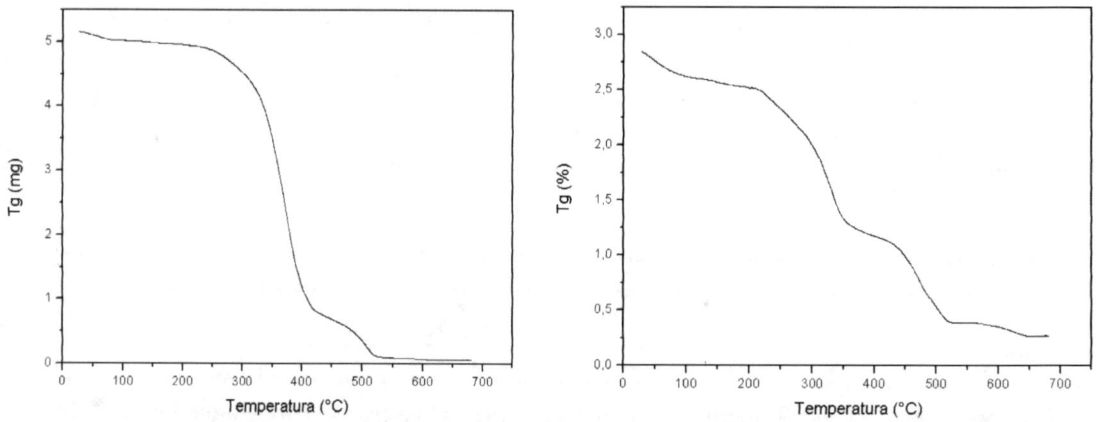

(a) (b)

Figure 7. TGA graphs for fibers a) without latex, b) covered by latex.

Next test made looked for establish the relation between mass loss and heat. That relation could be obtained from continuous measurements of weight in function of temperature along time using thermogravimetric analysis (TGA).

The fibers behavior (with latex and no latex) showed differences between each of them, even though they have the same composition underneath. That is explained by the fact that latex covered up all the small fibrils which covered the fiber. Having no elastic, coating, a considerable mass loss happened due to a higher amount of short fibers and fibers on the fiber surface.

Because of the coating on fibers surface, latex filled up all the empty spaces previously filled by air, therefore, there was less mass loss, figure 7(b).

It could be concluded that latex provided to the natural fiber, a higher stability and resistance to mass variation during the heating.

3.2. Geotextile

The tests were conducted for both fabrics, with and with no latex. As mentioned, all analyzes had the same purpose, which was test the resistance to heat, tearing and traction (tensile strength).

The first test aimed to verify the tensile strength. Geotextiles showed a better and longer resistance, such as more elongation at break and a higher tensile strength than the fabric without the latex coat, since, they received a latex cover as a matrix, they also were weaved as a warp and weft combination as a reinforcement, figure 8.

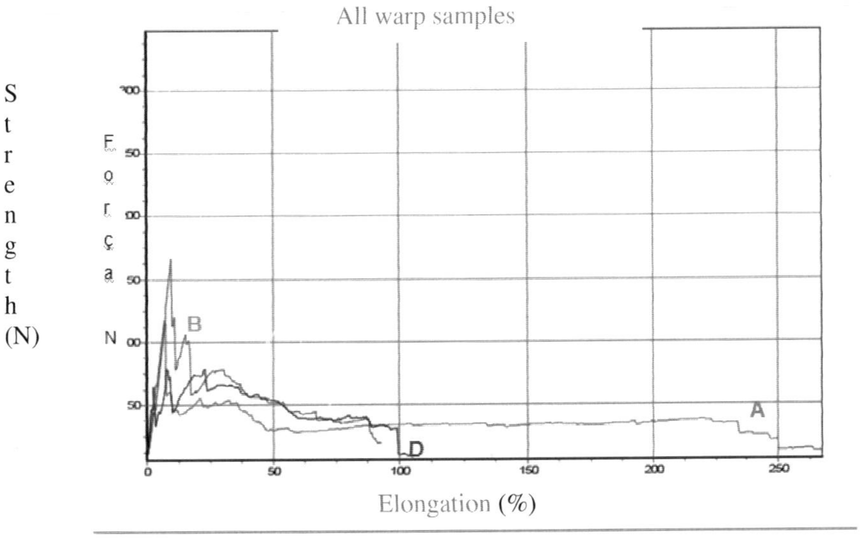

Figure 8. Graph showing 4 samples of Geotextile coated by látex.

In regards to Flammability, it became evident that the behavior observed during previously thermogravimetric analysis (TGA) was attested, which means, latex substrates burned faster than the substrates with in natural fibers. This is due to latex highly flammable characteristics.

With respect to the textiles properties, it became apparent that the geotextile presented a high resistance to abrasion since the material resisted to 380 of the 2000 cycles by the Taber Abraser. The test stopped only because the test tissue tear. In terms of density, the measurement, made by a thread counter, was considered high, being 4 thread/cm². As for a conventional fabric, this density is considered high because of natural porosity of fabrics, but as a geotextile material, it was considered acceptable and an important characteristic which permits heat exchange with the environment.

The procedure related to humidity absorption was conducted using a Moisture Test ID-200, in accordance with the manufacturer's technical specifications. As they were introduced in ID-200, all samples were exposed to high temperatures rising until 110°C and, after the analyzes, they showed different results. It was possible to identify that, after all the heating process, the natural fibers samples presented a capacity of absorbing humidity equal to 12.8%. While latex samples did not show the same capacity, absorbing only a 3.5% of humidity. It occurs due to latex treatment. Receiving latex bath as a plastic matrix, reduces the capacity of absorb air humidity, becoming less hygroscopic. Latex

IOP Conf. Series: Materials Science and Engineering **634** (2019) 012015 doi:10.1088/1757-899X/634/1/012015

filled all of the empty pores, cellulosic fibers characteristics (SEM), upon the fibers, that prevented fibers fibrils to absorb the right amount of moisture without buildup.

In a similar way, water absorption test showed that in natural fabric absorbed more distilled water prior than the geotextile fabric, table 1.

Table 1. Values for water absorption for in natura fabric and geotextile.

In Natura Fabric	Dry Weight	Humid Weight	Weight Gain %
Sample 1	0,21g	1,53g	628,57%
Sample 2	0,23g	1,58g	586.96%
Sample 3	0,24g	1,49g	520,83%
Geotextile	Dry Weight	Humid Weight	Weight Gain %
Sample 1	1,48g	3,26g	120.27%
Sample 2	1,46g	2,64g	80,82%
Sample 3	1,45g	2,61g	80%

There was a large increase in weight when considering both samples, however a larger increase value could be observed on in natural samples, table 1. This happened due to the fact that in natural fabric had no alterations to its physical or chemical composition. The same is not true about the geotextile which absorbed 93% of its original weigth. Even having changes in characteristic of the fiber, the geoxtile as still considered a hygroscopic material.

In accordance to what has been already proved, the geotextile absorbed a large amount of water. This fact propitiated a higher interaction with the soil and the environment where the sample was exposed. It was evidenced by the fact that those samples, in nátural fabric and geotextile, were completely buried in the ground for 120 days

The samples were completely degraded in the natural environment. Similarly, latex composites presented a partial decomposition, figure 9.

Figure 9. Latex specimen partially degraded after 120 days buried.

Something similar happened when the specimens were degraded by the surronding environment. The samples interacted very slowly in the environment and accumulated in their bodies: insects, leaves, sand, etc. Once more, latex has proved to increase samples useful life. Thermoplastic elastomer provided resistance to the environment's degradation, increasing the geotextile quality.

Finally, thermal properties analysis were conducted aiming to check samples measurements for: conductivity, specific heat and thermal diffusivity. Analysis were performed on superior, inferior and lateral sides of both cylindrical specimens, in natural and reinforced by latex.

According to Wang [4] a composite is a thermal insulator when K<0.25 W/mK. As seen in table 2, it can be evidenced that in natural sample is an insulating substrate. After being covered by latex, the substrate loses its insulating capability, however, K remains lower than 0.25 W/mK.

Table 2. Average Measurements of both cylindrical specimens, in natural and reinforced by latex.

Average Measure	K Thermal Condutivity (W/mK)	Cp Specific Hear (J/m³K)	A Thermal Diffusivity (mm²/s)
No Latex	0,146	7816,67	0,188
Latex	0,214	1379,2	0,154

Inserting latex to fill empty pores on the fabric surface, has led to an increase over thermal conductivity measurement values. It can be said that lower than 0.25 W/mK is acceptable for an insulating composite, therefore, geotextile showed lower values, that proves it can be used as an insulating material.

As regards the specific heat, according to Callister, 2005, specific heat of polymers can vary between 1300 and 2000 J/m³K, which relates to the results obtained.

Knowing that, as latex filled the pores on the fibers, what increases density because all the air of empty pores and spaces were fulfilled by a heavy material. Although, this procedure has not changed significantly the thermal difusivity values. This can all be seen in table 2 (α 0,154 mm²/s), in other words, insulating materials characteristics [5].

References

[1] Raghavendra S, Lingaraju, Shetty P B and Mukunda P G 2013 Mechanical properties of short banana fiber reinforced natural rubber *composites International Journal of Innovative Research in Science, Engineering and Technology* **2**

[2] Ramesh M *et al* 2014 Processing and mechanical property evaluation of banana fiber reinforced polymer composites *Procedia Engineering* 563-72

[3] Agostini and Deuber Lincon D S 2009 Caracterização dos constituintes do látex e da borracha natural que estimulam a angiogênese: Presidente Prudente: Unesp p 101

[4] Wang Q Y and Li J P 1998 *Insulating (Thermal) Material Application Engineering Standards Compilation* (Standards Press of China) vol 6 pp 44-8

[5] Callister J R 2005 *W. D. Materials Science and Engineering: An Introduction* 5th ed. (New York: J. Wiley & Sons) p 623

IOP Conf. Series: Materials Science and Engineering **634** (2019) 012016 doi:10.1088/1757-899X/634/1/012016

Fabrication of super-hydrophobic surface on aluminium substrate and a study of surface frosting behaviours

Z J Yu[1,2], Y Y Zhou[1] and S Wang[1]

[1]School of Chemical Engineering, Dalian University of Technology, Dalian 116024, China

E-mail: yuzhijia@dlut.edu.cn

Abstract. Frosting is common found natural phenomenon in late autumn or early spring, which sometimes brings unwanted effects or even disaster to human. With the development of science and technology, frosting may to some extend be mitigated by hydrophobic surfaces. Micro-Nanostructure was gained via chemical etching on aluminium surface. The surface was further modified by low surface energy polymer coating to form a very thin nano-layer on the surface. A kind of fluoroalkylsilane is used as the coating polymer. After fluorination treating, the aluminium surface become super hydrophobic with water contact angles more than150°. The surface was observed to be configured with plateaus and caves of micro-Nanoscale. In subcooled environments, frosting experiment was carried out on the surface of aluminium rods and aluminium sheets using humidifier to supply the water droplets. The comparison of the quantity of frost formed on the super-hydrophobic and untreated aluminium surfaces was conducted in the present work. And the quantity of frost formed on aluminium surfaces of different contact angles was compared. The results show that super-hydrophobic aluminium surfaces have obvious advantages in inhibiting frost. Besides, the contact angle has evident influences on restraining frost. Therefore, super-hydrophobic surfaces may serve as an effective means in decreasing the accumulation of frost on outside solid surfaces.

1. Introduction

Wettability is one of fundamental properties of solid surface. Surfaces with strong water repellency were studied with great effort in the past two decades, such as the self-clean lotus leaves [1], the legs of water striders which walk and jump freely on water [2]. Researchers found hydrophobic surfaces may find great perspective utilizations in many fields [3-6]. Frosting is common found natural phenomena in late autumn or early spring, which sometimes bring unwanted effects or even disaster to human. Frost is commonly formed on home refrigerators resulting in further consideration for the convenience uses. Industry heat exchangers for cooling also face with the problem of frost layer on the cooling surface [7-9], which significantly decreases the heat transfer efficiency [10,11]. With the development of science and technology, frosting may to some extend be mitigated by hydrophobic surfaces owing to the strong water repellency of the surface.

In recent years, many works were done on the fabrication on water repellent surfaces [12-15] among which the method of chemical etching followed by polymer coating may be the feasible means for industry application. Ice inhibition was studied on hydrophobic surfaces [15-18] and ice forming process was analyzed [19,20]. Relatively few work was done on the mitigation of frost on hydrophobic surfaces [21,22]. Since the process of ice formation and frost formation are complex, further studies are necessary for the development for ice or frost resistant surfaces.

In present work, super hydrophobic surfaces(SHS) with micro-nano hierarchical roughness were fabricated via the two steps of chemical etching and organic polymer coating. The hydrophobicity of the surface was examined by means of water contact angle (WCA) measurement. Frosting experiments were carried out to study the anti-frosting effect of SHS.

2. Preparation of SHSs on aluminium substrate

The preparation of SHSs in the present work includes two procedures: chemical etching and surface fluorination.

2.1. Chemical etching

Hydrochloric acid is adopted as the etchant. Specimen cuts were immersed in hydrochloric acid of different concentrations from 1 to 5 M. Etching time was changed from 4 to 70 min. The specimens were immediately rinsed ultrasonically with water after etching, and dried at 50°C in air. A kind of hierarchical roughness of micro-nano scale was formed on the surface of the samples as shown in figure 1. In large micrometer caves stand nano-sized plateaus.

Figure 1. SEM image of SHS at 5000 magnifications.

2.2. Surface coating

Naked aluminum surface is hydrophilic. When etched properly, it becomes super hydrophilic. The surface can be turned super hydrophobic further by low energy material coating. In present work, 1% (wt) tridecafluoroctyltriethoxysilane (FAS) ethanol solution was used as the coating agent. The etched aluminium substrates were put into the FAS solution for 1 hour, and then kept at 130°C heat treatment for about an hour. A thin film with low surface energy of was formed via the following chemical reaction:

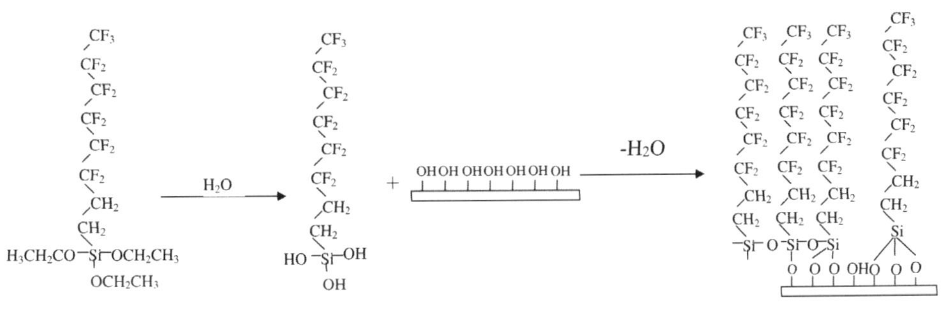

2.3. Characterizations.

The wettability of a solid surface is often characterized by water contact angle measurement. In the present study, WCA were measured with a CA meter. The Equilibrium contact angle values are averages of 5 measurements made on different sites of the surfaces. After fluorination, the surface exhibit super hydrophobicity with WCA up to 158°. Water droplet on the surface looks as a perfect sphere as shown in figure 2.

(a) (b)

Figure 2. The image of a water droplet on SHS. (a) photograph and (b) WCA measurement image with WCA about 158°.

3. Frosting experiment

If the surface temperature of a solid material is below both the dew point of humid air and the freezing point of water, water vapour in humid air will form a frost layer on the surface. Frosting experiments were conducted in a closed cabinet in which temperature can be automatically controlled and kept constant with a refrigeration system. An ultrasonic humidifier was used to generate fog in the air. For comparison, untreated and super hydrophobic samples of aluminium in the shape of rods and sheets were tested for frosting behaviours which are depicted both visually and by means of mass increases of the tested samples. Figure 3 shows a typical visual comparison of the frost formed on an untreated aluminium surface and on an aluminium SHS.

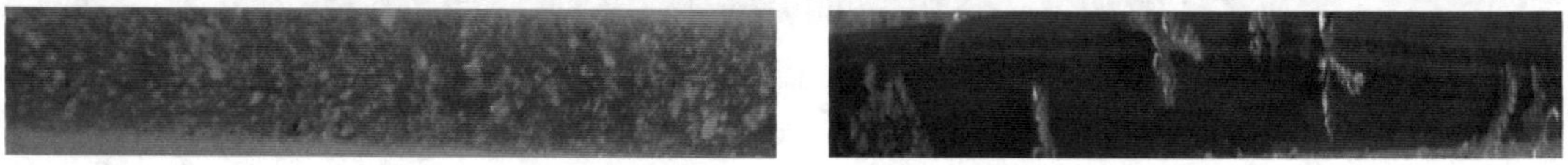

(a) (b)

Figure 3. visual comparison of the frost formed on an untreated aluminium surface and on an aluminium SHS (subcooling: 25 K, time: 30 min). (a) untreated aluminium surface and (b) aluminium SHS.

It can be seen from figure 3 that the frost on SHS is obviously less than on untreated surface. It can retard the starting time of frosting evidently. For longer frosting time, frost can also cover the whole area of SHS, being no sharp visual difference from that of untreated surface, but the amounts of frost (in grams) formed on the two kinds of surface are different as shown in figure 4.

The tested specimens can be etched under different conditions, say, different etchant concentrations and different etching time, resulting in different water repellence with different WCAs. The influence of WCA on frost formation is shown in figure 5, illustrating the frost mitigation effect of the hydrophobic surfaces.

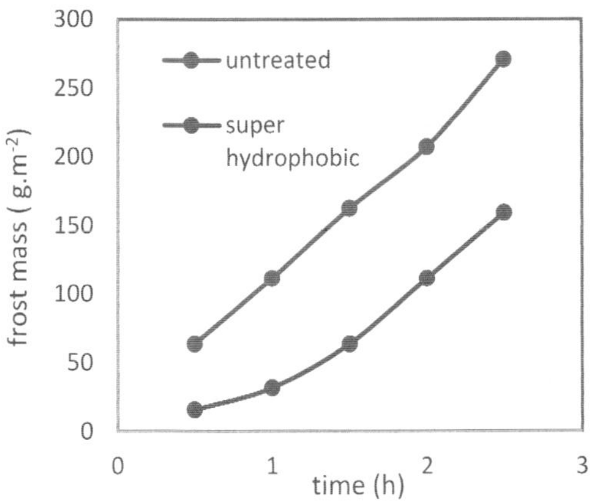

Figure 4. A comparison of frost formation on a SHS and that on an untreated aluminium surface (25 K subcooled).

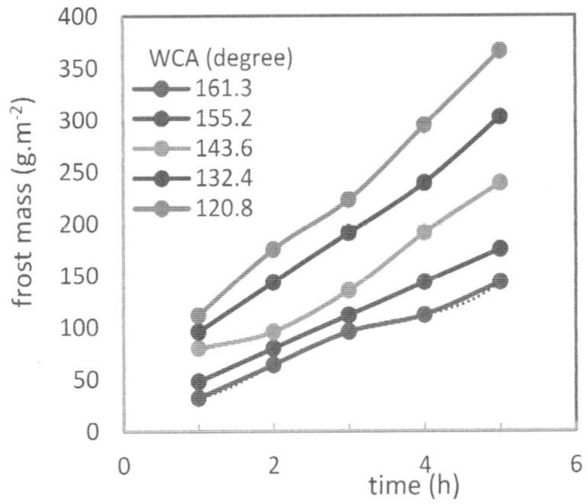

Figure 5. Frost formation on hydrophobic surfaces with different water repellence (20 K subcooled).

4. Conclusion

Micro-Nanostructure was gained via chemical etching on aluminium surface. With fluorine treatment, the surface shows ultra-hydrophobic properties with a WCAs of more than 150°. In the subcooled environments, frosting experiment was carried out on the surface of aluminium rods and aluminium sheets. The results show that aluminium SHSs have obvious advantages in inhibiting frost. Therefore, SHSs may serve as an effective means in decreasing the accumulation of frost on outside solid surfaces.

Acknowledgment
The authors are grateful for the financial support from the National Natural Science Foundation of China (grant No.: 20476014, 51376030).

References

[1] Barthlott N C 1997 Characterization and distribution of water-repellent, self-cleaning plant surfaces *Ann. Bot-London* **79** 667-77
[2] Gao X and Jiang L 2004 Water-repellent legs of water strider *Nature* **432** 36
[3] Kim K, Lichtenhan J D and Otaigbe J U 2019 Facile route to nature inspired hydrophobic surface modification of phosphate glass using polyhedral oligomeric silsesquioxane with improved properties *Appl. Surf. Sci.* **470** 733-43
[4] Liu J, Wang Q, Shan H *et al* 2019 Surface hydrophobicity based heat and mass transfer mechanism in membrane distillation *J.Membrane Sci.* **580** 275-88
[5] Jiang L 2005 Super-hydrophobic nanoscale interface material *Sci. Tech.* **23** 4-8
[6] Guo Y, Zhao E, Guo X *et al* 2019 Fabrication of self-assembled hydrophobic fluorinated silica particulate film *J. Fluorine Chem.* **218** 27-35
[7] Kim M H, Kim H, Lee K S *et al* 2017 Frosting characteristics on hydrophobic and superhydrophobic surfaces: A review *Energ. Convers Manage* **138** 1-11
[8] Kim K, Kim M H, Kim D R *et al* 2014 Thermal performance of microchannel heat exchangers according to the design parameters under the frosting conditions *Int. J. Heat Mass Transfer* **71** 626-32
[9] Kim M H, Kim H, Kim D R *et al* 2016 A novel louvered fin design to enhance thermal and drainage performances during periodic frosting/defrosting conditions *Energ. Convers Manage* **110** 494-500
[10] Song M, Xia L, Mao N *et al* 2016 An experimental study on even frosting performance of an air source heat pump unit with a multi-circuit outdoor coil *Appl. Energy* **164** 36-44

[11] Wang F, Wang Z, Zheng Y *et al* 2015 Performance investigation of a novel frost-free air-source heat pump water heater combined with energy storage and dehumidification *Appl. Energy* **139** 212-9

[12] Qian B and Shen Z 2005 Fabrication of superhydrophobic surfaces by dislocation-selective chemical etching on aluminum, copper, and zinc substrates *Langmuir* **21** 9007-9

[13] Fu X and He X 2008 Fabrication of super-hydrophobic surfaces on aluminum alloy substrates *Appl. Surf. Sci.* **255** 1776-81

[14] Kwon M H, Shin H S and Chu C N 2014 Fabrication of a super-hydrophobic surface on metal using laser ablation and electrodeposition *Appl. Surf. Sci.* **288** 222-8

[15] Liao R, Zuo Z, Guo C *et al* 2014 Fabrication of superhydrophobic surface on aluminum by continuous chemical etching and its anti-icing property *Appl. Surf. Sci.* **317** 701-9

[16] Jung M, Kim T, Kim H *et al* 2015 Design and fabrication of a large-area superhydrophobic metal surface with anti-icing properties engineered using a top-down approach *Appl. Surf. Sci.* **351** 920-6

[17] Rodič P and Milošev I 2019 One-step ultrasound fabrication of corrosion resistant, self-cleaning and anti-icing coatings on aluminium *Surf. Coat. Tech.* **369** 175-85

[18] Liu Y, Li X, Yan Y *et al* 2017 Anti-icing performance of superhydrophobic aluminum alloy surface and its rebounding mechanism of droplet under super-cold conditions *Surf. Coat. Tech.* **331** 7-14

[19] Zuo Z, Song X, Liao R *et al* 2019 Understanding the anti-icing property of nanostructured superhydrophobic aluminum surface during glaze ice accretion *Int. J. Heat Mass Transfer* **133** 119-28

[20] Lei S, Wang F, Fang X *et al* 2019 Icing behavior of water droplets impinging on cold superhydrophobic surface *Surf. Coat. Tech.* **363** 362-8

[21] Yang W, Yuan Y, Liu G *et al* 2018 The anti-icing/frosting aluminum surface with hydrangea-like micro/nano structure prepared by chemical etching *Mater. Lett.* **226** 4-7

[22] Zuo Z, Liao R, Zhao X *et al* 2017 Anti-frosting performance of superhydrophobic surface with ZnO nanorods *Appl. Ther. Eng.* **110** 39-48

Study on the electrical properties of polypyrrole nanowires/silica composites

Z J Weng[1], Z W Zhao[1,3], Y Fang[1], J D Wu[1] and H L Jiang[2]

[1]School of Electronic Science and Engineering, Southeast University, Nanjing 210096, P. R. China
[2]State Key Laboratory of Lake Science and Environment, Nanjing Institute of Geography and Limnology, Chinese Academy of Sciences, 73 East Beijing Road, Nanjing 210008, P. R. China

E-mail: Zhao_zw@seu.edu.cn

Abstract. In recent years, conductive composites have been attracted more attentions, but their applications and developments are restricted by the deficiency of the research on conductive mechanism. The electrical conductivity and percolation threshold of silica (SiO_2) matrix with different particle sizes using polypyrrole nanowires (PPy-NWs) as fillers are experimentally investigated. The results show that the electrical properties of PPy-NWs/SiO_2 composites typically exhibit a non-linear percolation behavior. PPy-NWs/SiO_2 (30 nm) nanocomposites need more filler content to achieve percolation threshold compared to that of PPy-NWs/SiO_2 composites with 2 μm SiO_2 particles as matrix. The conductive behavior of these composites is the result of the interaction and competition of various conductive mechanisms. The difference in electrical properties of PPy-NWs/SiO_2 composites with different particle size SiO_2 as matrix is attributed to the effect of matrix size on tunnelling conduction and the establishment of conductive pathways. Moreover, a two-dimensional conductive network model filled with PPy-NWs and an effective electrical resistivity model of PPy-NWs/SiO_2 composites were established. The simulation models can reflect the electrical properties of conductive composites, and are well agreement with the experimental data.

1. Introduction

Conductive polymers, with superior properties that are suited to a broad range of applications, have gained considerable attentions [1-3]. The advantageous properties of conductive polymers can be combined with those of inorganic materials, and these materials are referred to as "conductive polymer/inorganic composites" [4-6]. Conductive polymer/inorganic composites are highly functional materials that enabled a plethora of new applications ranging from electronics to bioengineering [7, 8]. It has long been pursued to optimize the relationship between the properties of conductive polymer and inorganic materials to satisfy specific applications.

There is an increasing interest in the development of conductive polymer/inorganic composites due to the presence of functionalities, such as their excellent electrical properties [8-11]. However, the conductive mechanism of conductive polymer/inorganic composites is not well understood, and the challenge to optimize the proportion of conductive polymer for achieving better efficiencies still remains. In previous work, typical conductive theories of polymer composites such as percolation theory [12] and quantum tunneling theory [13], were used to study the insulator-to-conductor transition and percolation network formation using randomly-orientated anisotropic fillers in disordered mixtures [14-16]. Moreover, several researchers have attempted to extend the application of percolation theory to predict the critical volume fraction of fillers required to achieve electrical conduction [17, 18].

In this work, the matrix size-dependency of the electrical resistivity of polypyrrole nanowires/SiO_2 (PPy-NWs/SiO_2) composites and the proportion of PPy-NWs that participate in electrical conduction are studied over a wide range of PPy-NWs content, accompanied by analysis of the conductive behavior combined with previous typical conductive theory of polymer composites. Moreover, an extension of this work was also carried out by establishing a two-dimensional (2D) percolation-based model and an effective electrical resistivity model to help study the conductive behavior of PPy-NWs/SiO_2 composites.

2. Experimental

2.1 Materials

All chemicals were of analytical grade and used without further treatment except additional description. The composites matrix used in this study was pure SiO_2 particles specified with micro-size (2 μm) and nano-size (30 nm) and obtained from Nanjing Hongde Company (China) and Shandong Xiya Reagent Company (China), respectively. Pyrrole (Py) with 98% purity and $LiClO_4$ were both purchased from Sinopharm Chemical Reagent Co., Ltd. Na_2HPO_4 was purchased from Nanjing Ningshi Chemical Reagent Co., Ltd. Deionized water was used for all aqueous solutions.

2.2 Preparation of PPy-NWs/SiO_2 composites

PPy-NWs were fabricated by template-free electrochemical polymerization using $LiClO_4$ as dopant as described in the reference [19]. Briefly, 0.2 M Na_2HPO_4, and 0.05 M $LiClO_4$ was added in deionized water. Py (0.15 M) was then added to the above solution and subjected to vigorous magnetic stirring at room temperature for 30 min. The room-temperature electropolymerization was performed in a three-electrode electrochemical cell including a working electrode (Ni sheet), a reference electrode (saturated calomel electrode, SCE) and a counter electrode (Pt sheet). A CHI600D electrochemical working station system (Shanghai Chenhua Instrument Co., Ltd. China) was used for controlling the current applied on the working electrode. Upon a constant potential of 0.8 V applied, the PPy-NWs were formed. The PPy-NWs sample was then carefully rinsed by deionized water to remove the absorbed species and air-dried at room temperature.

PPy-NWs/SiO_2 composites were prepared by simple mechanical mixing with different PPy-NWs loading and different SiO_2 size. At first, pure SiO_2 with average diameter of 2 μm and PPy-NWs were air dried at 100°C for 1 h to remove the moisture. After that, PPy-NWs/ SiO_2 (2 μm) composites with different PPy-NWs loading (≈ 5, 10, 15, 20, 25, 30, 35, 40, 45 wt%) were prepared by simple mechanical mixing using constant speed solid mixer for 15 min and compression molded at 10 MPa pressure. PPy-NWs/SiO_2 (30 nm) nanocomposites with PPy-NWs loading (≈ 5, 10, 15, 20, 25, 30, 35, 40, 45, 50 wt%) series as above but respective different SiO_2 size were also prepared using the same method.

2.3 Characterizations

Surface morphologies of the composites were examined using scanning electron microscope (SEM) (Quanta 200, FEI Ltd, US). The elemental analysis was conducted by energy dispersive X-ray (EDX). Fourier transform-infrared (FT-IR) measurement was employed to characterize the structure of the composites. Electrical resistivity measurements were carried out by 4-probe method using high resistance meter (ST2254) and powder meter (ST2722) (Suzhou Jingge Electronic Co., Ltd, China).

3. Simulation method

To understand the electrical behavior and investigate how different parameters (e.g., intrinsic resistance and concentration) affect electrical properties of the composite system, a two-dimensional

(2D) percolation-based model and an effective electrical resistivity model were created.

Firstly, a 2 D straight PPy-NWs network was generated in the x-y plane by considering each PPy-NW to be a line segment with a length l randomly distributed in the region of 20 μm × 20 μm. The line segments are identified by three parameters: central point coordinates (x_0, y_0), predefined length l, and direction angle θ (defined as the angle from the x-axis to the line segment counterclockwise). Then the occurring of electrical percolation behavior as a result of the formation of a continuous conductive network [17, 18] was evaluated as a function of PPy-NW concentration. Besides, the composite system was set up with random state of the components (e.g., PPy-NWs, SiO_2) by mathematical description. Then an equivalent electrical circuit of the composite system was obtained based on the idea that the resistance through the composite system is the result of a series of a large number of resistors combined in series and parallel [20]. The conductive behavior of the circuit of the composite system could be simulated using HSPICE software and the model-predicted results on the effective electrical resistivity of composites were thus obtained.

4. Results and discussion

4.1 Microstructure of the composites

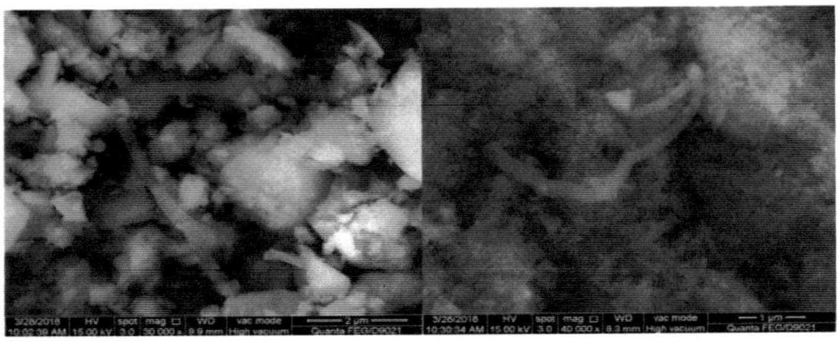

Figure 1. SEM images of PPy-NWs (30 wt%)/SiO_2 composites with micron-sized SiO_2 particles as matrix (left) and nano-sized SiO_2 particles as matrix (right).

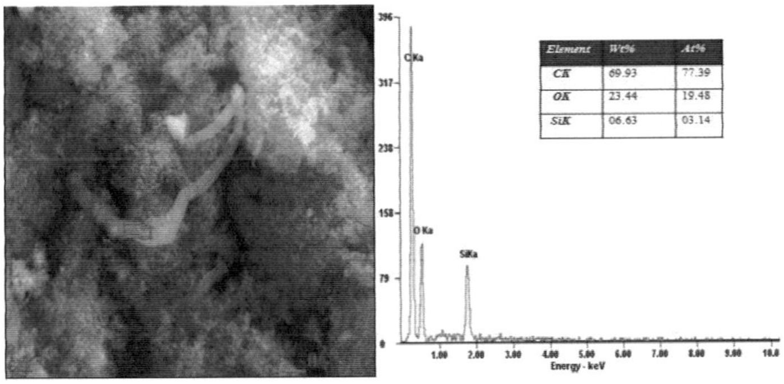

Element	Wt%	At%
CK	69.93	77.39
OK	23.44	19.48
SiK	06.63	03.14

Figure 2. EDX results of the PPy-NWs (30 wt%)/SiO_2 (nano-sized) composites.

Figure 1 shows the SEM images of the PPy-NWs (30 wt%)/SiO_2 composites with micron-sized SiO_2 particles (left) and nano-sized SiO_2 particles (right). The length of the separate nanowires without any branching was on the order of microns and the diameter varied in the range from dozens of nanometers to two hundred nanometers, which is similar to Chen's work [19]. As can be seen in figure 1, the micron-sized SiO_2 particles and nano-sized SiO_2 particles exhibited distinctly different distribution states. Micron-sized SiO_2 exhibited much denser agglomerate with less compressible space and nano-sized SiO_2 was more evenly distributed around the PPy-NWs. The formation of the

PPy-NWs/SiO_2 composites consisting of nano-sized SiO_2 particles and 30 wt% PPy-NWs loading was also confirmed through EDX experiments as shown in figure 2.

Figure 3 shows the typical FT-IR spectra of the composites consisting of nano-sized SiO_2 particles and 30 wt% PPy-NWs loading. The characteristic peak at 3448 cm^{-1} of the spectrum that could be assigned to N-H stretching bending vibration was observed [19, 21]. The peak around 1633 cm^{-1} could be attributed to fundamental vibration bond of Pyrrole rings [22]. The peak at 1107 cm^{-1} was assigned to the C-N stretch, the in plane C-H deformation vibration [2, 22] and the Si-O-Si antisymmetric stretching vibration [23]. Moreover, the peak around the 804 cm^{-1} and 471 cm^{-1} could be attributed to the =C–H out of plane bending of Pyrrole moiety in PPy [22] and symmetrical stretching vibration of the Si-O bond [23], respectively. The presence of these absorption peaks indicated that the composites only contain PPy-NWs and SiO_2 particles.

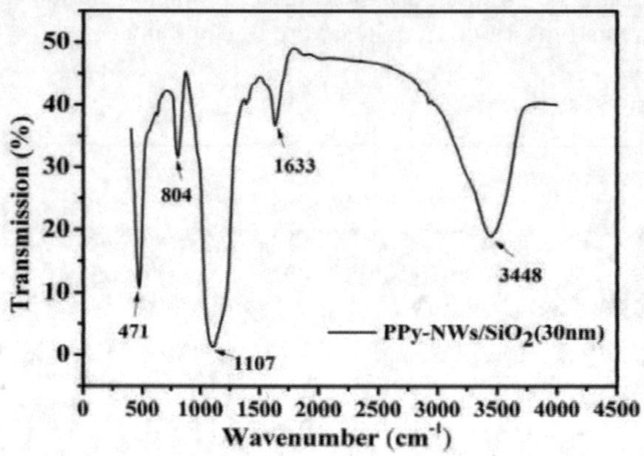

Figure 3. FT-IR spectra of PPy-NWs (30 wt%) /SiO_2 (30 nm) nanocomposites.

4.2 Electrical resistivity

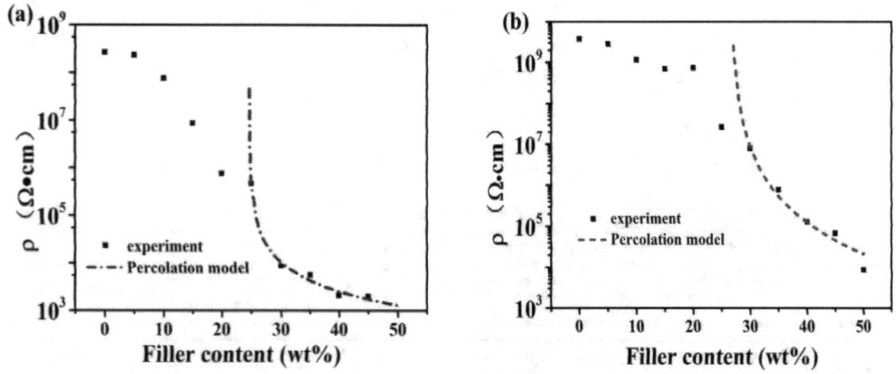

Figure 4. Electrical resistivity of (a) PPy-NWs/SiO_2 (2 μm) composites and (b) PPy-NWs/SiO_2 (30 nm) composites at different PPy-NWs loadings.

Figure 4(a) shows the electrical resistivity of PPy-NWs/SiO_2 (2 μm) composites. There is a general trend of decrease in the electrical resistivity with increasing PPy-NWs. The electrical resistivity decreases much more dramatically from 7.62×10^7 Ω·cm at 10 wt% PPy-NWs loading to 8.65×10^3 Ω·cm at 30 wt% PPy-NWs loading. The electrical resistivity is stable up to 40 wt% PPy-NWs loading and the ρ value of 2.08×10^3 Ω·cm was achieved. This trend is consistent with the non-linear percolation theory. The effect of filler content on PPy-NWs/SiO_2 (30 nm) nanocomposites electrical resistivity is shown in figure 4(b). The resistivity variation of the PPy-NWs/SiO_2 (30 nm) nanocomposites is similar to that of PPy-NWs/SiO_2 (2 μm) composites, but an abrupt decrease occurs when much more PPy-NWs are introduced into the PPy-NWs/SiO_2 (30 nm) nanocomposites. The increase of PPy-NWs content in the PPy-NWs/SiO_2 (30 nm) nanocomposites from 15 wt% to 50 wt% led to a dramatic decrease in the electrical resistivity from 6.98×10^8 Ω·cm to 8.62×10^3 Ω·cm.

The percolation model [12] predicts that, near a metal-insulator (or conductor-perfect conductor) phase transition, the conductivity (or resistivity) will be given by the percolation equation:

$$\sigma_m = \sigma_h(\phi - \phi_c)^t \quad (1)$$

Where σ_m is composite conductivity (S/cm), σ_h is filler conductivity, ϕ is filler volume fraction ($\phi>\phi_c$), ϕ_c is percolation threshold and t is critical exponent.

The values of ϕc and t were determined by fitting the experimental data to equation (1). The estimated values of ϕc and t for PPy-NWs/SiO_2 (2 μm) composites are ≈ 24.69 wt% and 1.33. For PPy-NWs/SiO_2 (30 nm) nanocomposites, the estimated values of ϕ_c and t are ≈ 26.38 wt% and 3.21, respectively. The difference in ϕ_c and t of PPy-NWs/SiO_2 (2 μm) composites and PPy-NWs/SiO_2 (30 nm) nanocomposites can be attributed to the particle size of matrix SiO_2 and its interaction with PPy-NWs.

Basically, the change in electrical resistivity of PPy-NWs/SiO_2 composites with different PPy-NWs loadings could be explained by typical conductive theory. Initially, the electrical resistivity of the pure SiO_2 particles was not affected with small amount of PPy-NWs in the composites. The electrical resistivity in the composites might be decreased with further increasing of PPy-NWs loading because of tunneling mechanism effect [13, 14]. Electrically conductive PPy-NWs are randomly dispersed into the SiO_2 matrix in a margin of nm apart. Here, tunneling of the electrons happened among the adjacent PPy-NWs through SiO_2 matrix with high resistivity or even they are separated by insulating gaps [13]. During this process, tunneling is sensitive to the distance between the PPy-NWs [14], thus highly affected by the SiO_2 matrix size. From SEM images in figure 1, PPy-NWs were partly covered by nano-SiO_2 matrix with high resistivity and more difficult to contact with each other exactly compared with PPy-NWs/SiO_2 (2 μm) composites. The electrical contact between PPy-NWs will be less and contact resistance among them will be high for composites consisting of PPy-NWs and nano-size SiO_2. Thus, It was observed that by using SiO_2 nanoparticles as matrix, the percolation threshold of PPy-NWs/SiO_2 (30 nm) nanocomposites would move towards an increased filler content and exhibit higher resistivity. With PPy-NWs loading at critical concentration, known as percolation threshold [12], the rapid change in electrical resistivity by several orders of magnitude was coincided with the formation of continuous interconnected conductive network chains of PPy-NWs throughout the composites [12, 17, 18]. At this time PPy-NWs are supposed to be connected and PPy-NWs/SiO_2 composites show lower electrical resistivity.

Figure 5 shows the simulation results of PPy-NWs/SiO_2 composites. The percolation probability simulation results are shown in figure 5(a). State 1 consists of very few PPy-NWs randomly dispersed into the SiO_2 matrix. State 2 is the indication for the formation of continuous interconnected conductive network chains of PPy-NWs throughout the composite system. The 2D percolation-based model predicted that the percolation probability increase with increasing number of PPy-NWs. As shown in figure 5(b), the simulated resistivity value of PPy-NWs/SiO_2 (2 μm) composites decreased proportionally with the amount of PPy-NWs. It was noteworthy that the increase of PPy-NWs content from 20 wt% to 40 wt% led to a dramatic decrease in the simulated resistivity from 1.58×10^8 Ω·cm to 9.68×10^3 Ω·cm. The resistivity variation of the composites obtained by simulation is generally consistent with that obtained by experimental measurement, but the percolation threshold value by the simulation is higher than that obtained by experiment due to the super length-diameter ratio of PPy-NWs which is conducive to the formation of conductive network.

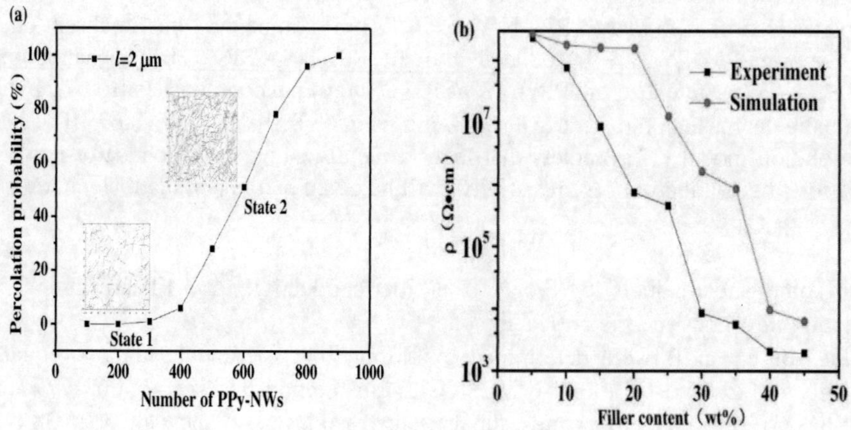

Figure 5. Simulation investigation of electrical properties of PPy-NWs/SiO_2 composites. (a) The percolation probability as a function of PPy-NWs concentration. (b) Electrical resistivity of PPy-NWs/SiO_2 (2 μm) composites from simulation and experiment.

5. Conclusions

The electrical resistivity and percolation threshold of PPy-NWs/SiO_2 (2 μm) composites and PPy-NWs/SiO_2 (30 nm) nanocomposites were studied. The corresponding percolation threshold of PPy-NWs/SiO_2 (2 μm) composites and PPy-NWs/SiO_2 (30 nm) nanocomposites were determined according to experimental results using a percolation model. The electrical resistivity of PPy-NWs/SiO_2 composites strongly depends on the concentration of PPy-NWs. Compared with PPy-NWs/SiO_2 (2 μm) composites, PPy-NWs/SiO_2 (30 nm) nanocomposites has higher resistivity and higher percolation threshold. This difference can be attributed to the effect of SiO_2 matrix size, which is evidenced by SEM measurement. In addition, the 2D conductive network simulation model filled with PPy-NWs can reflect the effects of the number of PPy-NWs on the structure of conductive network and predict the percolation probability of composites well. The electrical resistivity predicted by the effective electrical resistivity model shows the similar trend as experimental results and the effective electrical resistivity model gives an estimation of percolation threshold.

Acknowledgments

This work was financially supported by the National Natural Science Foundation of China (No.: 51879042, 51839011).

References

[1] Kumar G M, Ilanchezhiyan P, Jin K, Park J and Jayavel R 2013 *Sensor Actuat A-Phys.* **199** 283-8
[2] Zhao J X, Wang J, Deng H M, Tang S, Lan Q, Liang J Y, Liu C, Cao Y C and Cheng S J 2019 *J Nanosci. Nanotechno.* **19** 5656-60
[3] Ramirez A M R, Gacitua M A, Ortega E, Diaz F R and del Valle M A 2019 *Electrochem. Commun.* **102** 94-8
[4] Zhitomirsky I 2007 *J. Alloy Compd.* **434** 823-5
[5] Xiaofeng L U, Chao D, Chen J, Zhang W and Wei Y 2006 *MatL.* **60** 2851-4
[6] Anakli D, Cetinkaya S and Karakisia M 2018 *Materials Focus.* **7** 515-8
[7] Chen P P, Hao L, Wu W J, Li Y F and Wang J T 2016 *Electrochim. Acta.* **212** 426-39
[8] Ju H and Kim J 2016 *Chem. Eng. J.* **297** 66-73
[9] Granström M and Inganäs O 1995 *Poly.* **36** 2867-72
[10] Cai Z and Martin C R 1989 *J. Am. Chem. Soc.* **111** 4138-9

[11] Maiti S, Suin S, Shrivastava N K and Khatua B B 2014 *Rsc Advances.* **4** 7979
[12] Kirkpatrick S 1973 *RvMP.* **45** 574-88
[13] Balberg I 1987 *PhRvL.* **59** 1305-8
[14] Maiti S, Shrivastava N K, Suin S and Khatua B B 2013 *Express Polym. Lett.* **7** 505-18
[15] Jagota M and Tansu N 2015 *Sci. Rep.* **5** 10219
[16] Langley D P, Lagrange M, Giusti G, Jimenez C, Brechet Y, Nguyen N D and Bellet D 2014 *Nanoscale.* **6** 13535-43
[17] Han F, Maloth T, Lubineau G, Yaldiz R and Tevtia A 2018 *Sci. Rep.* **8** 17494
[18] Lee B M and Loh K J 2015 *JMatS.* **50** 2973-83
[19] Chen Y Q, Zhao Z W and Wang C 2013 *Nanosci. Nanotech. Let.* **5** 186-90
[20] Ruschau G R, Yoshikawa S and Newnham R E 1992 *JAP.* **72** 953-9
[21] Wan X B, Zhang W, Jin S, Xue G, You Q D and Che B 1999 *J. Electroanal. Chem.* **470** 23-30
[22] Sankir N D and Dogan B 2010 *J. Mat. S.* **45** 6424-32
[23] Chen H S, Sun Z Y and Shao J C 2011 *B. Chin. Ceram. Soc.* **04** 934-7

Distribution of fillers and reinforcements in injection moulded thermoplastic composites: Case study of glass bubbles

A Elsabbagh[1,3], T Attia[1] and A Ramzy[2]

[1]Design and Production Engineering Department, Ain Shams University, Cairo, Egypt
[2]German University in Cairo, Cairo, Egypt

E-mail: Elsabbagh.ahmed@eng.asu.edu.eg

Abstract. Apart from the complex nature of natural fibers, studying glass bubbles, as reinforcement phase, is considered a good start to understand the flow behaviour of fillers or reinforcements and hence their distribution in the final products after injection moulding. Glass bubbles are thermally stable within the processing temperatures of Polypropylene. They are also void from the effect of fibre aspect ratio. This work explores the uneven glass bubbles content along injection moulded spirals out of polypropylene (PP). Glass bubbles are experimented at 10 and 30% volume fraction Vf levels. The processing temperature during the injection moulding varies in the range of 180-200°C. Injection pressure is changed from 250 to 1000 bar. Flow speed is adjusted at two levels of 60and 100 cm^3/s. The response measurement is the distribution of the glass bubbles across the spiral at different distances from the injection point. The experimental results show the significance of the nominal Vf, pressure, temperature and flow speed on the distribution homogeneity where the bubbles move further in case of 30% at high pressure of 1000 bar and higher temperature of 200°C that eases the flow. At 10% Vf, the fluctuations in the measured Vf is minimal. The reason of the fluctuation in case of 30% Vf at high temperature and pressure is the multi-phase flow where the investigation shows that bigger size of bubbles acquires greater momentum to flow. Scaling-up of this study to the synthetic fibres will cover fragmentation. More up-scaling to the case of the natural fibre thermoplastic composites NFTC can deal with other factors like thermal stability and fibre splicing or branching.

1. Introduction

NFTC have an increasing demand in European market. Natural fibre reinforced polypropylene composite (NF/PP) is significantly represented in the injection moulded NFTC of Europe with 92% of the 15000 ton total production volume of NFTC [1]. However, one concern of these composites is the strong fluctuation of the fibre content in comparison to the nominal fibre content [2]. As a direct result, a product with non-uniform mechanical and physical properties is produced. Accordingly, this will induce negative effect on both the production efficiency and the product cost.

The instability of the fibre content values in the injection moulded product is analysed. The reason is due to the separation flow phenomenon between the phases of fibres and the carrying polymer melt during injection into the mould [2]. This separation between phases is attributed to the difference in density between the fibre and the polymer. Also, the shape effect of the fibres whether it is straight or branched. Kubat *et al* [3] reported the same observations during the injection of flowability spiral mould using low density polyethylene (LDPE) reinforced with 25.7% glass fibres. Phase separation, as reported by [3] increases directly with increasing the injected spiral length as well as the polymer

IOP Conf. Series: Materials Science and Engineering **634** (2019) 012018 doi:10.1088/1757-899X/634/1/012018

viscosity. The glass fibre content decreases obviously after the injection point. Then it jumps again along the length of the spiral. Helger *et al* [4] studied the filler homogeneity in injected samples of dumb-bell form under different processing parameters. The significant factors affecting the homogeneity and the stable distribution of the glass bead fillers are namely: filler geometry, mould shape, temperature, injection time, filler content and polymer material. Helger [4] explained the phase separation in terms of the filler inertia along the flow direction. Consequently, fillers accumulate near to the axis with higher inertia leading to higher filler content at the far end of the injection moulded samples.

Peltola *et al* [5] considered the phenomenon of changing fibre content, qualitatively, in case of natural fibres namely hemp and flax. Effects of processing melt viscosity and fibre type on the fibre length were investigated. Thermal processing, during injection, reduced the fibre length. This reduction in fibre length is attributed to the increasing shearing force in compounding. Hemp fibres keep higher aspect ratio due to the more fibrillation compared to flax. Higher aspect ratio results in higher reinforcing, Ramzy *et al* [6] studied the flowability and fibre distribution of NFTC in case of hemp and sisal in spiral injection moulds. Increase of polymer temperature, mould temperature and injection pressure led to increased flowability (longer spiral length). The fibre content, similar to [3], drops after the injection point and then goes up again along the spiral length. The fluctuation in sisal was less than that measured in hemp because of the more agglomeration spots found in hemp case.

Table 1. Degradation modes of synthetic and natural fibres [6].

Mode	Description	Synthetic fibre	Natural fibre
Fragmentation	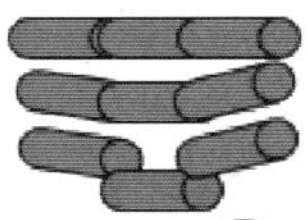	X	X
Thermal degradation	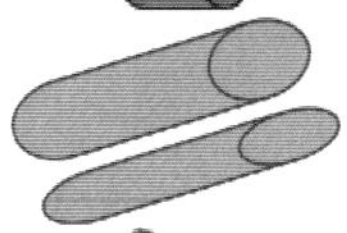		X
Aspect ratio reduction		X	X
Stiffness		X	X
Branching			X
Splicing			X

Studying the fluctuation of natural fibre is more complex than glass fibres and even more complex than glass bubbles as illustrated in table 1. Synthetic fibres undergo fragmentation during processing. Fragmentation, in turn, reduces the aspect ratio and the expected strength especially when the aspect ratio fails to satisfy the critical ratio. Stiffness of the fibre [7] is a function of temperature. On the other hand, natural fibre suffers from all the above-mentioned drawbacks as well as other points. Low thermal stability leads to negative change in mass, size and diameter. Another characteristic of natural fibre is the branching and splicing.

Empirical modelling of NFTC during injection needs more understanding of the fibre degradation in shape and size on the flow behaviour [8]. Analytical modelling in literature addresses mostly the synthetic fibres and their flow behaviour without considering the problem of fibre uneven distribution across and along the injection moulding product [9]. Therefore, this work deals with the problem of filler content fluctuation by starting with glass bubbles and not the complex natural fibres. Glass bubbles are characterised with two advantages namely; thermal stability where no change in mass is accompanied, and unity aspect ratio where no effect of rotational moment is occurring.

The objective is to observe and understand the flow behaviour of glass bubbles in spiral flow-ability moulds during injection moulding. The work is limited in this study to the qualitative description. The distribution of the glass bubbles is measured within the cross section and along the spiral length. Effect of the processing parameters (Pressure, temperature, speed) in a well-designed plan is considered in this study which is not thoroughly done in previous work [3].

2. Experimental work

The carrying polymer matrix is a high flow PP of trade name Moplen EP 500 V developed by Lyondell Basell Industries. The glass bubbles are supplied by 3M GmbH with average size of 15 µm and bulk density of 460 kg/m^3. Spirals are injected per a design of experiments (DoE) Plan using JMP software as listed in table 2.

Table 2. Parameters used in the study.

	Temperature	Pressure	Flow speed	Fibre content
	°C	bar	ccm/s	Vf%
1	180	250	60	10
2	180	250	100	10
3	180	250	100	30
4	180	1000	100	30
5	180	1000	60	30
6	180	1000	100	10
7	200	250	60	10
8	200	250	60	30
9	200	250	100	30
10	200	1000	60	10
11	200	1000	100	10
12	200	1000	100	30
13	180	1000	60	30

Figure 1 presents the injected spiral and the positions of the segments to be cut and investigated. The spiral mould is an Archimedean spiral. The outer diameter is 250 mm and the step is 9.5 mm. The spiral flow cavity is 6 mm wide and the curvature denoted by the radius R=5 mm.

For each condition, almost 15 spirals are injected, and their lengths are measured for the sake of flow-ability. To measure the glass bubble content as well as its size, segments are cut at the injection point and then every 10 cm. Glass bubbles are extracted by dissolving the PP matrix (using dacaline solvent) at 160°C/ 2 hours. The extracted glass bubbles are then dried and preserved for weighing and further investigation of the glass bubble size development (length and diameter). The extraction process is carried out three times for each compound for three different samples of spirals. The average value of the glass bubble content and the standard deviations are then calculated.

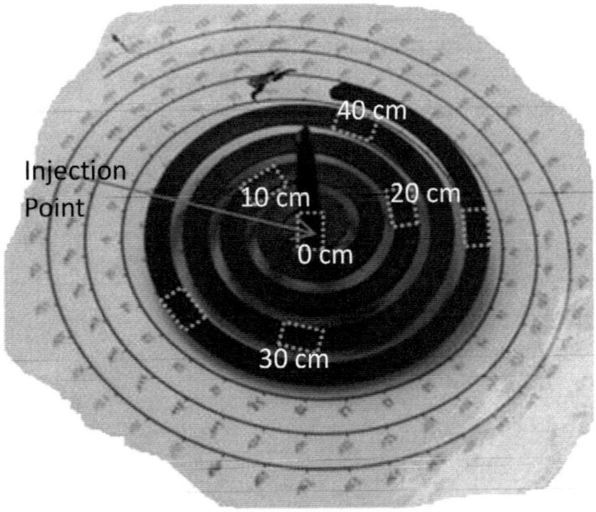

Figure 1. Spiral test for injection moulding flow-ability and the places where the glass bubbles are extracted [Ram18].

3. Results and discussion

The results shown in Figure 2 for Vf=30% match the behaviour reported in [1] for NFTC. The scale of deviation in Vf is small [25.5-29%] relatively in comparison with that in natural fibre case (>10%) as in [1]. Effect of temperature is not observable because Vf fluctuation is reduced under mainly due to the pressure effect. On the other side, it is obvious that low pressure results in less flow-ability. Flow speed is significant where less contents of glass bubbles are measured indicating that portion of the bubbles is hindered before the injection. Also, it is obvious that high pressure increases the Vf fluctuation. This can be attributed to the accompanying increase in spiral length.

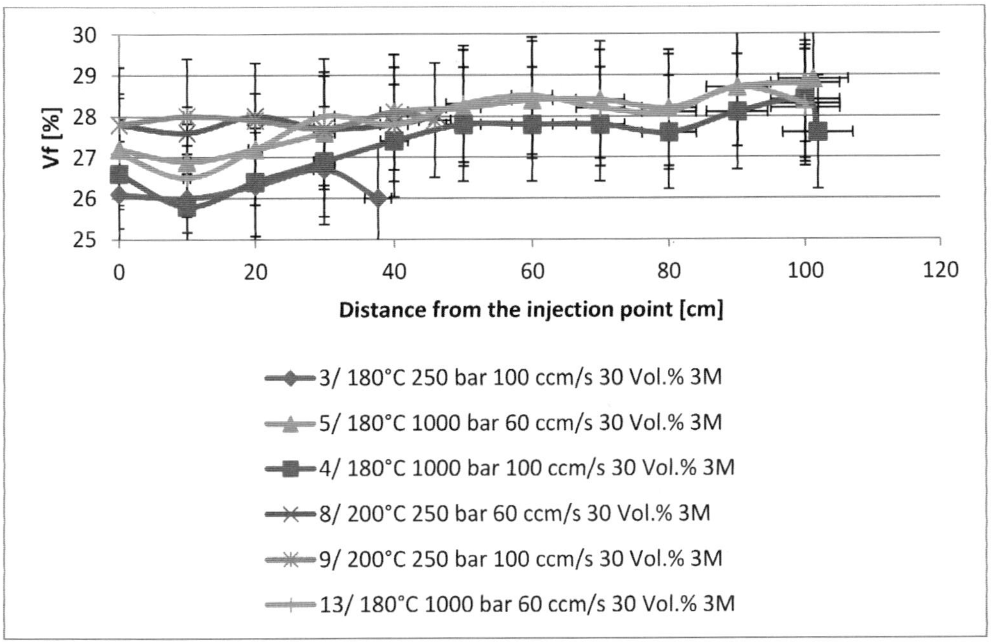

Figure 2. Glass bubbles distribution along spiral length at different processing parameters for 30% Vf of glass bubbles.

Reproducibility of the results is ensured by comparing the very close behaviours of samples 5 and 13 shown in Figure 2. Most important is that the measured Vf along the spiral is close to parabolic nature where the bubbles content is reduced just after the injection start and then increases again. The reason is either that more bubbles are flowing separately faster than the hosting polymer, or the bubbles of greater size have more momentum and got separated from bubbles with less diameter.

This assumption will be investigated microscopically by samples taken from the spiral at different locations. Figure 3 shows the behaviour at 10% Vf, where the measured Vf show almost constant trend regardless the temperature and the flow speed. Error bars are removed for better illustration.

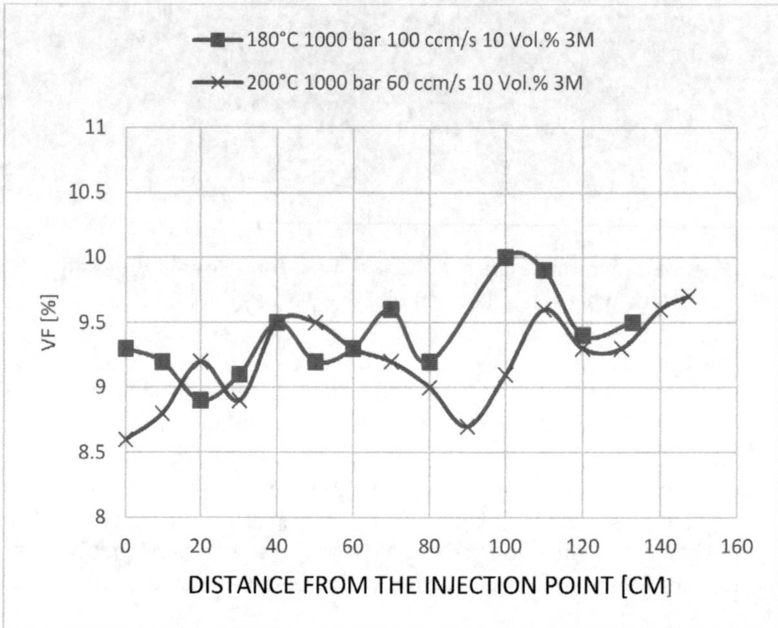

Figure 3. shows the behaviour at 10% Vf, where the measured Vf show almost constant trend regardless the temperature and the flow speed.

Figure 3 Glass bubbles distribution along spiral length at 180-200°C injection temperature, 1000 bar with different flow speeds 60 and 100 ccm/s different processing parameters for 10% Vf of glass bubbles. Figure 4 shows the sites taken for microstructure investigation. Three places were cut from the spiral namely the injection start point, the middle of the spiral length and at the end tip. Then at each one of the three places, the glass bubbles were measured and counted at the upper surface, the cross section middle point and at the bottom curved region.

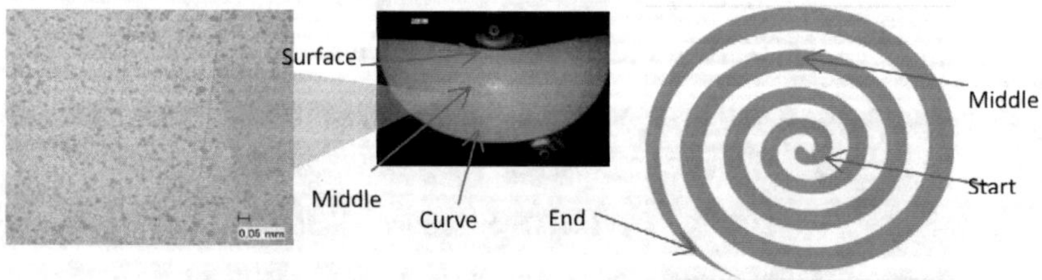

Figure 4. Places of microscopic investigation for glass bubbles diameter along and across the spiral.

Table shows the significance of the processing parameters namely Vf and pressure on the measured diameters of the glass bubbles across the spiral. This result will be helpful in explaining the effect of processing parameters with more complex shapes like synthetic fibres and finally branched flexible natural fibres. As seen in figure 5, the larger glass bubbles in diameter concentrate at the end along the flow. Also, relatively bigger bubbles accumulate at the curve region. This cannot be attributed to gravity because the mould is installed in a vertical position. However, the flow has less resistance at the curve than at the flat surface due to the geometrical effect on the computational fluid dynamics of the polymer/ glass bubbles multi-phase flow.

Table 3. Scales estimate.

Source
Vf [%](10,30)
Pressure [bar]*Vf [%]
Pressure [bar]*Flow speed [ccm/s]
Flow speed [ccm/s]*Vf [%]
Temperature [°C]*Flow speed [ccm/s]
Flow speed [ccm/s](60,100)
Pressure [bar](250,1000)
Temperature [°C]*Vf [%]
Temperature [°C](180,200)
Temperature [°C]*Pressure [bar]
Vf [%]*Vf [%]

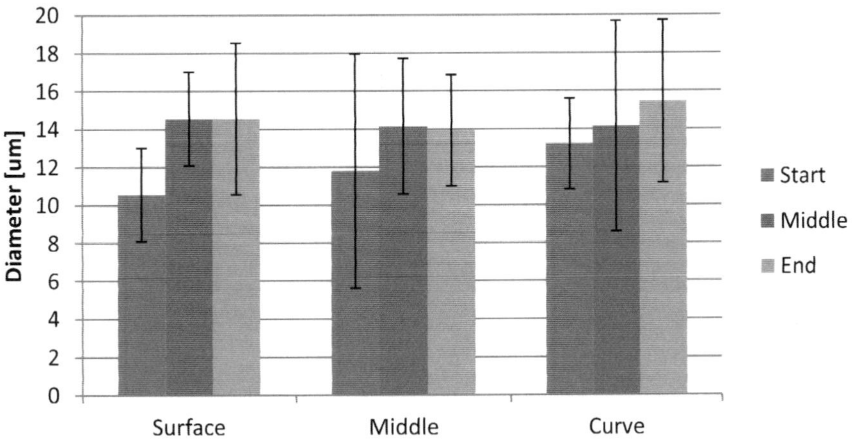

Figure 5. Glass bubble diameter along the spiral length in terms of its position in the cross section.

4. Conclusion and final remarks

The experimental results prove that the distribution of unity aspect ratio particles (glass bubbles) show phase separation. This separation is influenced by the nominal Vf, temperature, pressure and flow speed. The distribution of the glass bubbles behaves parabolically specially at the 30% Vf.

The increase of the measured Vf away from the injection point is interpreted in terms of the size of the bubbles where the glass bubbles of bigger diameter got separated and move further. Diameter of the bubbles across the spiral is influenced by the nominal Vf and pressure. This study can be used further in the analysis of NFTC to exclude the effect of the filler or the fibre size and focus on the other factors like thermal degradation and fibre aspect ratio.

Acknowledgments

Thanks are due to Prof. Gerhard Ziegmann and Dr Leif Steuernagel due to the lightful discussions and facilities given in Clausthal university of Technology/ Institute of polymer materials and plastics

IOP Conf. Series: Materials Science and Engineering **634** (2019) 012018 doi:10.1088/1757-899X/634/1/012018

engineering.

References

[1] Ramzy A, Elsabbagh A, Steuernagel L, Ziegmann L G and Meiners D 2013 Rheology of natural fibers thermoplastic compounds: Flow length and fiber distribution *J. Appl. Poly. Sci.* 1-8
[2] Ramzy A, Elsabbagh A, Steuernagel L, Ziegmann L G and Meiners D 2016 Flow ability and fiber content homogeneity of natural fiber polypropylene composites in injection molding *AIP Conference Proceedings* **1779** 060010
[3] Kubat J and Szalanczio A 1974 Polymer-Glass separation in the spiral mold test *Poly. Eng. and Sci.* **14** 873-7
[4] Helger R P and Menning G 1985 Phase separation effects in processing of glass-bead and glass-fiber-filled thermoplastics by injection molding *Poly. Eng. Sci.* **25** 395-405
[5] Peltola H, Madsen B and Joffe R 2011 Experimental study of fiber length and orientation in injection molded natural fiber /starch acetate composites *Adv. Mater. Sci. Eng.* **2011** 1-7
[6] Ramzy A 2018 Recycling aspects of natural fiber reinforced polypropylene composites
[7] Navin Chand S and Hashmi A R 1993 Mechanical properties of sisal fibre at elevated temperatures *J. Mater. Sci.* **28** 6724-8
[8] Elsabbagh A, Ramzy A, Steuernagel L and Ziegmann G 2019 Models of flow behaviour and fibre distribution of injected moulded polypropylene reinforced with natural fibre composites *Compos. Part B - Eng.* **162** 198-205
[9] Londono-Hurtado A, Osswald T A and Hernandez-Ortiz J P 2011 Modeling the behaviour of fiber suspensions in the molding of polymer composites *Journal of Reinforced Plastics and Composites* **30** 781-90

IOP Conf. Series: Materials Science and Engineering **634** (2019) 012019 doi:10.1088/1757-899X/634/1/012019

Improved wear resistance and mechanical properties of multifunctional polymer nanocomposites for advanced engineering applications

U O Uyor[1,3], A P I Popoola[1] and O M Popoola[2]

[1]Department of Chemical, Metallurgical and Materials Engineering, Tshwane University of Technology, Pretoria West, Private Bag X680. Pretoria, South Africa
[2]Department of Electrical Engineering, Tshwane University of Technology, Pretoria West, Private Bag X680. Pretoria, South Africa

[3]E-mail: UyorUO@tut.ac.za

Abstract. Improvement of wear behaviour and mechanical properties of polymers as bulk and surface coating materials for advanced engineering applications have been research interest. Improved wear resistance and mechanical properties were achieved in this study by incorporation of surface modified graphene and titanium dioxide nano-powder in poly (vinylidene fluoride) matrix via solution blending and melt compounding. The morphology of the nanocomposites was studied using scanning electron microscope (SEM). The nanocomposites showed significant reduction in wear volume and enhanced mechanical properties due to the presence of surface modified nanofillers. About 76.5% reduction in wear volume, 52% and 186.5% increase in tensile strength and Young modulus respectively were achieved in this study. The surface modification of the nanofillers was essential in dispersion of the nanofillers in the polymer matrix and the improved properties recorded. Such nanocomposites can find applications as protective surface coating materials for high frictional condition and bulk materials for automobile and aerospace components.

1. Introduction

Surface modification of materials without changing the bulk properties of the materials to protect them against aggressive environments such as corrosive media and frictional conditions has drawn research and industrial interests. Polymeric materials are often used in such aggressive environments because of their excellent performance in resistance to chemical attacks and self-lubrication in the case of friction [1]. Polymeric materials have good properties required for various engineering applications such as protective surface coating, solid lubricant, electronic, electrical, mechanical and structural applications. Their wide applications are due to their distinct properties such as light weight, flexibility, toughness, easy processing etc. For instance, polytetrafluoroethylene among others have the capability to provide low frictional coefficient under sliding condition. Notwithstanding, they are associated with challenge of high wear rate, which hinders their wide applications in high frictional conditions [2]. Therefore, using such materials as protective surface coating in high frictional environment became industrial challenge despite their high corrosion resistance. Addressing this challenge associated with polymers has drawn attentions of researchers. In the past years, ceramic powders have been incorporated in polymer matrix on the efforts to overcome this shortcoming [3-6].

The recent discovering of carbon-based nanofillers such as graphene and carbon nanotubes has created pathways for improving wear resistance of polymers, while maintaining their self-lubrication and low frictional coefficient properties. For instance, it has been demonstrated that introduction of graphene nanofillers in polymer matrix improves its wear properties [7-10]. About 40% reduction in wear rate by introduction of graphene oxide in polymer matrix has been achieved [11]. In a study carried out by Han *et al* [12], approximately 49.4% enhanced wear performance was recorded with reduced graphene oxide in polymer. Also, Shen *et al* [13] measured high reduction in wear rate when graphene oxide was incorporated in polymer matrix. These results showed the capability of graphene nanofillers in addressing the problem of high wear rate associated with polymers.

Friction and wear mechanisms are well known to be surface phenomenon. Understanding the full detail principles of wear is a challenge, snice it varies from one material to another. Understanding the principles of wear mechanisms will aid in addressing challenges associated with friction and wear not only for bulk materials, also for protective coated surfaces. Basic wear mechanisms are by initial removal of materials from a bulk material surface by the countersurface, followed by subsequent detachment of the material on the countersurface [2]. Comparing materials' removal from polymer composites filled with micro and nano fillers, nanofillers have insufficient size required for low wear rate in polymer matrix [14]. 2D materials such as graphene nanofillers offers large in-plane dimension when incorporated in polymer matrix due to their large surface area. Therefore, obtaining appreciable thickness of such nanofillers will address the challenge of insufficient thickness level and improve wear performance of polymers containing such nanofillers for protective surface coating and bulk materials applications.

This study attached titanium dioxide (TD) nano-powder on surfaces of 2D graphene nanofillers and incorporated into poly (vinylidene fluoride) matrix. During wear operation, there was initial removal of TD attached on graphene surfaces, then removal of graphene sheets from the polymer matrix. This promoted reduction in wear volume of the nanocomposites. Also, the rough surfaces of graphene sheets due to attachment of TD promoted interlocking of the polymer molecular chains, which resulted to improved mechanical properties. Good interfacial interaction between the two nanofillers and the polymer matrix was necessary in achieving the enhanced wear and mechanical performance. This was achieved by surface modification of graphene and TD via functionalization and hydroxylation respectively. There was significant increase in wear resistance and mechanical properties of the developed nanocomposites. The materials combination and methodology demonstrated in this study will be able to address the industrial challenges of high wear rate and poor mechanical properties associated with polymeric materials. For instance, such developed nanocomposites can find application as surface coating materials in high frictional environment. It can be used as solid lubricant where liquid lubrication is not suitable such as high temperature condition. Aerospace and automobile industries where materials with good flexibility, light weight and high mechanical strength are required can be beneficiaries of such fabricated nanocomposites.

2. Materials and methodology

The following materials were sourced from Sigma-Aldrich and were used in the course of this study; poly (vinylidene fluoride) (PVDF) powder, slightly oxidized graphene nanoplatelets (G), anatase titanium dioxide nano-powder (TD), 3-hydroxytyramine hydrobromide (HTHB), N, N-dimethyl formamide (DMF), ammonia solution, hydrogen peroxide (H_2O_2) and nitric acid (HNO_3).

Graphene (G) was functionalized with HTHB to promote its dispersion in PVDF matrix. First, 50mg of G was added in 50mL of distilled water and ultrasonicated at 80°C for 1 hour. Then, 40mg of HTHB was introduced into the suspension at the same condition. Ammonia solution was added in drops to activate the reaction process. The suspension was kept under ultrasonication and mechanical stirring for 6 hours at 80°C. Then, the particles were severally washed with distil water and DMF to obtain functionalized graphene (fG). H_2O_2 and HNO_3 were used in hydroxylation of TD to improve its wettability on fG in the polymer matrix.

In fabrication of the nanocomposites, certain amount of fG and TD were introduced in PVDF matrix by solution blending. fG and TD were first mixed in DMF. The blended nanofillers were added into pre-dissolved PVDF. The suspension was homogenously mixed with the help of ultrasonicator and mechanical stirrer for 1 hour at 80°C. Then, the mixture was dried overnight in an oven at 80°C. The nanocomposites were then grinded and melt-mixed using HAAK rheomix 600 OS for 15 min at 200°C and 150 rpm. The nanocomposites were finally hot pressed to desired shapes for analysis using carver press moulder at 200°C and 10 MPa for 5 min.

Attachment of TD on fG surfaces was investigated using Transmission Electron microscope (TEM) as reported in our previous work [15]. Morphology of the nanocomposites were investigated using Scanning Electron Microscope (SEM). Wear performance of the nanocomposites was obtained using tribometer (Anton Paar). This was carried out in accordance with ASTM G99-95 standard at room temperature and speed of 200rpm with steel ball of 0.3cm radius and 0.03μm roughness (Ra). Each sample was subjected to 1 hour wear test under applied normal load of 10N. Wear volume of the nanocomposites was obtained from their respective weight losses. Average of three tests were taken for each sample. Mechanical properties of the nanocomposites were determined using universal testing machine at ambient temperature. The tests were conducted according to ASTM D638 (Type V) standard. Crosshead rate of 5 mm/min was set. Each nanocomposite was tested five times and the average is presented as previously reported by Le, Huang [16].

3. Results and discussion

3.1. Morphological study of the nanofillers and nanocomposites

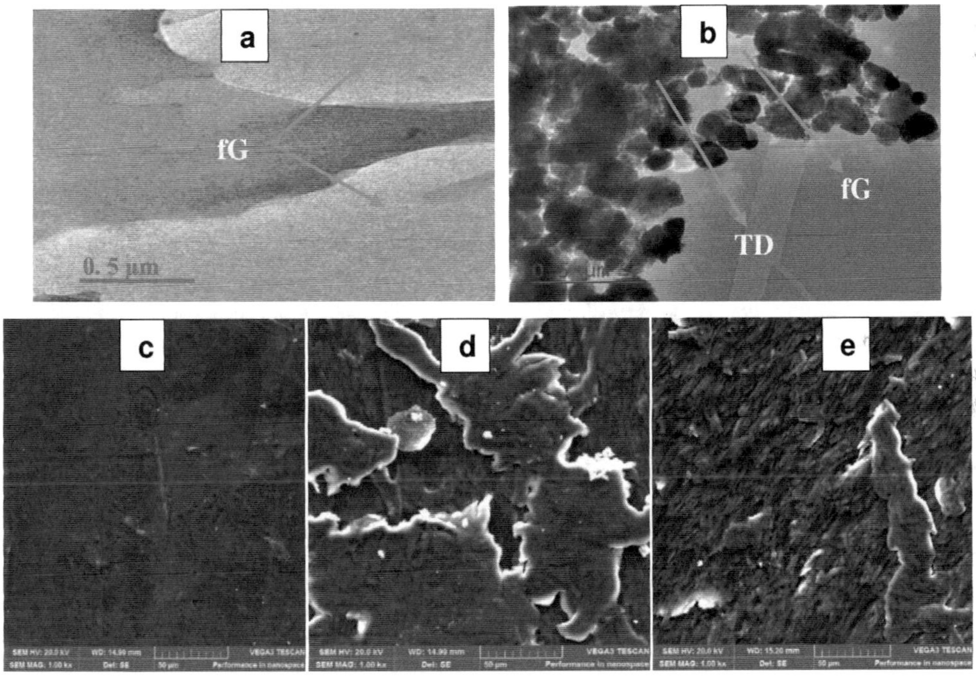

Figure 1. TEM micrographs of (a) G (b) G attached TD [15] and SEM images of (c) Pure polymer (d) 3.34wt%fG/10wt%TD and (e) 6.67wt%fG/10wt%TD nanocomposites.

Figure 1 shows successful attachment of TD on fG surfaces. Large surface area of G can be seen from figure 1(a). After blending with TD, it was noted that TD was attached on the surfaces of fG as shown in figure 1(b). The attachment of TD on fG gave the nanofillers appreciable thickness, which helped in improving wear resistance of the polymer. Again, there was good wettability between fG and TD since there was no noticeable segregation of the nanofillers. This resulted from the surface modification of the nanofillers, which was essential in good compatibility of nanofillers. On the other hand, figure 1(c) shows smooth microstructure of PVDF because it was still a homogenous material. On addition of 3.34 wt%fG/ 10 wt%TD in the matrix, the morphology changed as shown in figure 1(d). fG layers can be seen with dot-like surfaces, indicating the presence of TD on fG surfaces. As fG increased to 6.67wt%, the presence of TD supressed in the matrix due to high content of fG as shown in figure 1(e). Notably, there was no significant agglomeration of the nanofillers in the matrix. This indicates good bonding between the constituents of the nanocomposites due to surface functionalization of G and hydroxylation of TD.

3.2. Wear performance of the nanocomposites

The wear test conducted showed that there was significant reduction in wear volume of the nanocomposites when compared to pure polymer as shown in figure 2. The reduction in wear volume was due to the presence of nanofillers in the polymer matrix. The attachment of TD on fG played a vital role in reduction of wear volume of the nanocomposites by increasing the thickness of fG. It is believed that during wear process, TD on fG are first removed by the sliding countersurface, before detaching the large in-plane fG from the bulk material. Again, the initial removal of TD formed debris, which was transferred on the sliding countersurface. When this happened, it took longer duration before detaching fG, thereby increasing wear resistance of the nanocomposites. Due to the large dimension of fG in the matrix [2], it was possible for the countersurface to slide across fG without easy detachment. This is because after initial transfer of TD to the sliding countersurface, the wear operation continued by material to material contact instead of bared sliding countersurface to material contact. Therefore, reduction in wear volume was noted compared to pure polymer. For instance, wear volume reduction of about 69% was measured for 1.67wt%fG/10wt%TD nanocomposite when compared to pure polymer. Further increase in the nanofillers to 3.34wt%fG/10wt%TD, further reduced the wear volume to about 76.5%. At this concentration, there was thinner material transfer to the countersurface, uniform covering of countersurface and good bonding of transferred material on the countersurface. These conditions favour self-lubrication and low wear rate [1]. Beyond 3.34wt%fG/10wt%TD content, the wear resistance started decreasing as shown in figure 2(b) but was still better than pure polymer and 1.67wt%fG/10wt%TD nanocomposite. This can be attributed to high wt% of the nanofillers, especially fG as shown by the SEM image in figure 1(e). This might have reduced polymeric resin binding the nanofillers, resulting to easy detachment of the nanofillers with increasing wear volume when compared to 3.34wt%fG/10wt%TD nanocomposite. The significant reduction in wear volume shows that such nanocomposites are potential materials for high frictional environment.

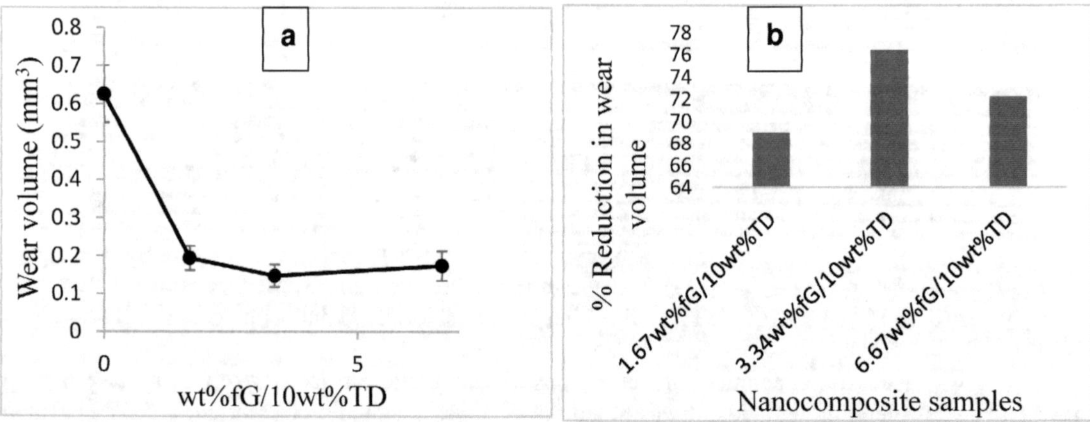

Figure 2. (a) Wear volume and (b) Percentage reduction in wear volume of the nanocomposites.

3.3. Mechanical properties of the nanocomposites

Figure 3 shows tensile strength and Young modulus of the nanocomposites. The tensile strength of the nanocomposites revealed improvement when compared with pure polymer. The improved strength was due to well distribution of the nanofillers and strong interfacial interaction with the polymer matrix. This often result to improved tensile strength [16]. With the presence of fG in the polymer matrix, there was formation of network structures by interconnection of the nanofillers. This resulted to restriction in mobility of the polymer chains when load was applied and increase in strength of the nanocomposites. It was noted that tensile strength was increasing continuously with increasing nanofillers content as shown in figure 3(a). As revealed in figure 3(b), tensile strength percentage increment of about 15.5%, 25.8% and 52% were respectively recorded for 1.67wt%fG/10wt%TD, 3.34wt%fG/10wt%TD and 6.67wt%fG/10wt%TD nanocomposites compared to the pure polymer. These indicate that there was good load transfer from the matrix to the nanofillers when external load was applied. This was due to good entanglement between the nanofillers and the polymer's molecular chains. This resulted from surface modification of the nanofillers, which enhanced interaction with the polymer matrix.

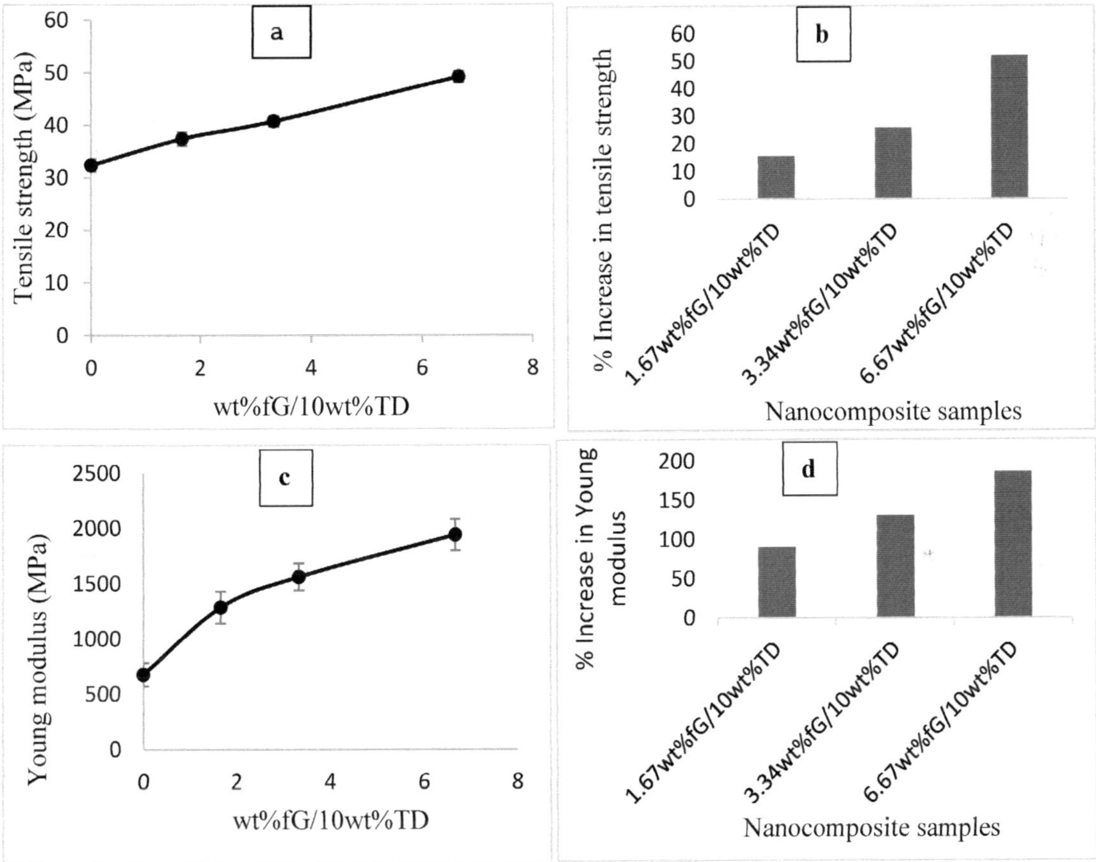

Figure 3. (a) Tensile strength (b) Percentage increase in tensile strength (c) Young modulus and (d) Percentage increase in Young modulus of the nanocomposites.

On the other hand, the Young modulus of the nanocomposites showed significant increase compared to the pure polymer as shown in figure 3(c). There was steady increase in Young modulus with increase in nanofillers content. The increase in Young modulus was due to increase in stiffness of the materials, which resulted from good interfacial bonding between the nanocomposites' constituents [17,18]. Also, the dot-like surfaces of fG in the polymer matrix due to the attachment of TD as revealed by SEM image in figure 1(d) promoted mechanical interlocking of the polymer molecular chains. This resulted to reduction in interfacial sliding between the nanofillers and polymer's chains. Hence, there was increase in stiffness and Young modulus of the nanocomposites compared to the pure polymer. High percentage increase in Young modulus of about 90%, 130.5% and 186.5% were obtained for various nanocomposites concentrations as shown in figure 3(d). Therefore, such nanocomposites are potential materials for various advanced engineering applications.

4. Conclusion

This study has been able to improve wear and mechanical properties of polymer using hybrid mixture of nanofillers. The nanofillers were noted to be well distributed in the polymer matrix due to surface modification. The nanocomposites showed significant increase in wear resistance. About 76.5% reduction in wear volume was recorded. Also, mechanical properties of the nanocomposites were improved. About 52% and 186.5% increase in tensile strength and Young modulus were measured respectively. With the improved wear resistance and mechanical properties, such nanocomposites can find application as protective surface coating materials for high frictional environment and mechanical components for automobile and aerospace industries.

Acknowledgment

This work is based on the research supported in part by the National Research Foundation of South Africa (Grant Numbers: 112238).

References

[1] Bahadur S and Schwartz C 2013 The effect of nanoparticle fillers on transfer film formation and the tribological behavior of polymers tribology of polymeric nanocomposites *Friction and Wear of Bulk Materials and Coatings* 23-48

[2] Kandanur S S, Rafiee M A, Yavari F, Schrameyer M, Yu Z-Z, Blanchet T A and Koratkar N 2012 Suppression of wear in graphene polymer composites *Carbon* **50** 3178-83

[3] Wang Y, Lim S, Luo J and Xu Z 2006 Tribological and corrosion behaviors of Al_2O_3/polymer nanocomposite coatings *Wear* **260** 976-83

[4] Wang Q, Xue Q, Shen W and Zhang J 1998 The friction and wear properties of nanometer ZrO2-filled polyetheretherketone *J. Appl. Polym. Sci.* **69** 135-41

[5] Schwartz C J and Bahadur S 2000 Studies on the tribological behavior and transfer film–counterface bond strength for polyphenylene sulfide filled with nanoscale alumina particles *Wear* **237** 261-73

[6] Bahadur S, Zhang L and Anderegg J 1997 The effect of zinc and copper oxides and other zinc compounds as fillers on the tribological behavior of thermosetting polyester *Wear* **203** 464-73

[7] Li Y, Wang Q, Wang T and Pan G 2012 Preparation and tribological properties of graphene oxide/nitrile rubber nanocomposites *J. Mater. Sci.* **47** 730-8

[8] Li C, Xiang M and Ye L 2017 Intercalation structure and highly enhancing tribological performance of monomer casting nylon-6/graphene nano-composites *Compos. Appl. Sci. Manuf.* **95** 274-85

[9] Thangavel E, Ramasundaram S, Pitchaimuthu S, Hong S W, Lee S Y, Yoo S-S, Kim D-E, Ito E and Kang Y S 2014 Structural and tribological characteristics of poly(vinylidene fluoride)/functionalized graphene oxide nanocomposite thin films *Compos. Sci. Tech.* **90** 187-92

[10] Bhargava S, Koratkar N and Blanchet T A 2015 Effect of platelet thickness on wear of graphene–polytetrafluoroethylene (PTFE) composites *Tribol. Lett.* **59** 17

[11] Tai Z, Chen Y, An Y, Yan X and Xue Q 2012 Tribological behavior of UHMWPE reinforced with graphene oxide nanosheets *Tribol. Lett.* **46** 55-63

[12] Han P, Fan J, Jing M, Zhu L, Shen X and Pan T 2014 Effects of reduced graphene on crystallization behavior, thermal conductivity and tribological properties of poly (vinylidene

fluoride) *J. Compos. Mater.* **48** 659-66
[13] Shen X-J, Pei X-Q, Fu S-Y and Friedrich K 2013 Significantly modified tribological performance of epoxy nanocomposites at very low graphene oxide content *Polymer* **54** 1234-42
[14] Uyor U O, Popoola A P I, Popoola O M and Aigbodion V S 2019 Effects of titania on tribological and thermal properties of polymer/graphene nanocomposites *J. Thermoplast. Compos. Mater.* 0892705718817337
[15] Uyor U O, Popoola A P I, Popoola O M and Aigbodion V S 2018 Mechanically insulated graphene/polymer nanocomposites with improved dielectric performance and energy storage capacity *Polym. Sci. Ser. A Polym. Phys.* **60** 875-85
[16] Le M T and Huang S C 2015 Thermal and mechanical behavior of hybrid polymer nanocomposite reinforced with graphene nanoplatelets *Materials (Basel)* **8** 5526-36
[17] Maity N, Mandal A and Nandi A K 2015 Interface engineering of ionic liquid integrated graphene in poly(vinylidene fluoride) matrix yielding magnificent improvement in mechanical, electrical and dielectric properties *Polymer* **65** 154-67
[18] Ahmadi-Moghadam B, Sharafimasooleh M, Shadlou S and Taheri F 2015 Effect of functionalization of graphene nanoplatelets on the mechanical response of graphene/epoxy composites *Mater. Des.* **66** 142-9

IOP Conf. Series: Materials Science and Engineering **634** (2019) 012020 doi:10.1088/1757-899X/634/1/012020

Effect of the combination of hydrophilic nano-SiO_2 and gaffix VC-713 polymer on CH_4 hydrate formation: Insight from the experimental study

R Wang[1,2,3], X G Xu[1,2], X M Shi[1,2], Y Y Li[1,2], J Zhang[1,2] and Z Yang[1,2]

[1]CNPC Engineering Technology R & D Company Limited, Beijing 102206, China
[2]National Engineering Laboratory for Oil & Gas Drilling Technology, Beijing 102206, China

E-mail: wangrdr@cnpc.com.cn

Abstract. In this paper, the effect of the combination of hydrophilic nano-SiO_2 with a particle size of 80 nm and gaffix VC-713 polymer (VC-713) on their hydrate inhibition was investigated at 4.0°C and 6.0 MPa, and the underlying influence mechanism was analysed by a multiple scales of the meso-structural observation experiments. Experimental results indicated that the combination doesn't exhibit any synergistic inhibition effect on the formation of CH_4 hydrate, but it weakens the hydrate inhibition abilities of both components. Moreover, with the increase of the VC-713 content to 0.75 wt% and 1.00 wt%, the inhibition of hydrophilic nano-SiO_2 completely disappears.

1. Introduction

As a potential alternative energy source, drilling operations are an important part of natural gas hydrate exploration and exploitation [1-3]. In order to maintain the borehole stability and safety, The gas hydrate-bearing sediment (GHBS) drilling operations are performed under overpressure conditions, and drilling fluid would permeate into the GHBS due to the pressure difference [4]. Research work has demonstrated that to eliminate the negative effects and keep the borehole stable, drilling fluid loss should be minimized, and the application of nanoparticles is an effective way to reach such goal [5]. Wang *et al* has reported that 4.0 wt% of hydrophilic nano-SiO_2 with a particle size of 80 nm can effectively inhibit the formation of hydrates in a simulated drilling fluid under different driving force conditions [6]. Liu *et al* has reported that the addition of 2% hydrophilic nano-SiO_2 into the hydrate drilling fluid system can not only optimize the properties such as rheology, depressing filter loss, and shale inhibition, but also ensure the hydrate inhibition of the low-cost drilling fluid system [7].

Although it has been reported that the application of hydrophilic nano-SiO_2 in hydrate drilling fluids has achieved some progresses, these studies just involved simple nano-fluids and mature drilling fluid systems. In addition, the effect of the composite of hydrophilic nano-SiO_2 and hydrate kinetic inhibitors on their hydrate inhibition was not revealed, and the underlying influence mechanism was also not explained. In this study, the combination was prepared using 4.0 wt% of hydrophilic nano-SiO_2 with a particle size of 80 nm (studies have reported that hydrophilic nano-SiO_2 under this condition exhibit the optimum inhibitory effect on methane hydrate formation [8]) and the hydrate kinetic inhibitor VC-713 with good inhibition and widely used (the concentrations of VC-713 were

0.25 wt%, 0.50 wt%, 0.75 wt%, and 1.00 wt%). In addition, the above issues were investigated in detail by multiple perspectives using various methods, including simulations of methane hydrate formation and meso-structural observations of simulated drilling fluid. The results obtained from this study can provide important reference values for the industrial applications of nano materials for hydrate field and the development on new types of conventional oil as well as gas drilling fluid and hydrate drilling fluid systems.

2. Experimental section

2.1. Materials and instruments

The purity of CH_4 gas used in this work is 99.9%, provided by Wuhan Niuruide Trading Co., Ltd., China. Hydrophilic silica, grain size 80 nm, provided by Qinhuangdao Nano products Co., Ltd., China. VC-713 is produced by International Special Products (ISP) Co., Ltd., America. Ultrapure water (abbreviated as UPW) resistivity more than 18.25 MΩ·cm is made up of a type UPW-S Millipore unit. In order to uniformly disperse the hydrophilic nano-SiO_2 in simulate drilling fluids, high speed stirring at 20000 r/min was used in our experiments.

The schematic of the experimental apparatus for hydrate formation is shown in figure 1. In order to conduct meso-structure observations, initially the lyophilized samples from simulate drilling fluid were prepared using the Four-Ring LGJ-10D freeze dryer. Then, the samples were observed by the Quanta 200F field emission scanning electron microscope (SEM).

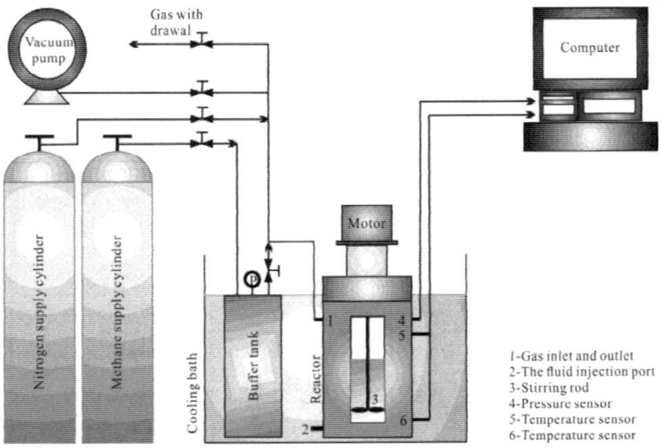

Figure 1. Schematic of the experimental setup.

2.2. Methodology

2.2.1. Experiments on CH_4 hydrate formation. Hydrate formation process was simulated by constant volume method [9]. To ensure the repeatability of the experiments, each experiment was repeated three times. The specific experimental steps are as follows.

- Use a clean 650 ml reactor and check its tightness.
- The liquid sample of 250 mL was injected into the reactor and vacuum continuously for 10 min. CH_4 gas was injected into the buffer tank until the pressure reached 10.0 MPa. Then turn on the temperature control system to lower the temperature. To avoid the aggregation and precipitation of the hydrophilic nano-SiO_2, the liquid sample was stirred at a speed of 300 r/min.
- After the temperature of the system was stabilized at 4.0°C, the CH_4 gas was injected into the reactor through the upper air inlet until the pressure reached 6.0 MPa. Then, the stirring speed was increased to 600 r/min and the monitoring software was turned on to collect the temperature and pressure data (the frequency of data acquisition is one time per 2 seconds).

2.2.2. Observing the mesoscopic structure. It has been reported that the meso-structural features of simulate drilling fluid can analyse its underlying influence mechanism of hydrate formation [10,11]. The experimental procedure is shown as following:

- Use the dropper to take a small amount of simulated drilling fluid, drop onto a mica film, and place it in liquid nitrogen for 2.0 min.
- The frozen sample was lyophilized in a freeze-drier for 20.0 h.
- The mesoscopic structure of the sample was observed by SEM after preparation.

3. Results

3.1. Evaluation of the hydrate inhibition

3.1.1. Data processing. During the evaluation of hydrate inhibition, the induction time, amount and rate of hydrate formation will be comprehensively analyzed. Taking the curves of temperature and pressure during the formation of hydrates in UPW and the nano-fluid with concentration of 4.0 wt% as an example (figure 2), the induction time t_1, the stopping time of hydrate formation t_2, the starting pressure P_1 and the stopping pressure P_2 of hydrate formation can be obtained.

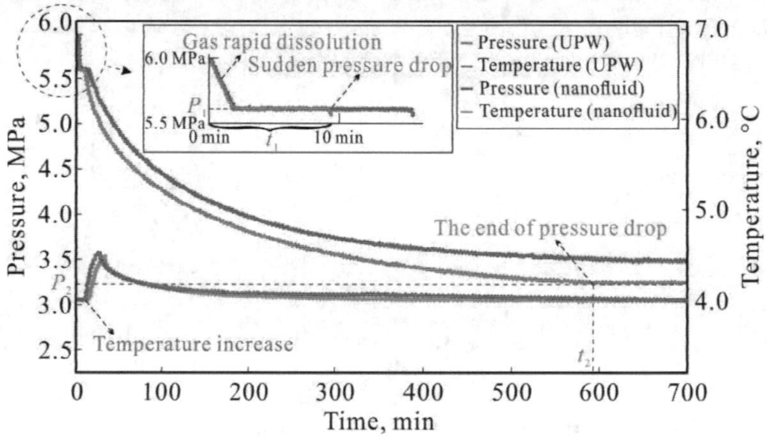

Figure 2. Variation in temperature and pressure with time for each experiment.

Since the guest gas is high-purity CH_4, its consumption represents the formation amount of hydrates. Therefore, the amount of hydrate formation can be calculated using the following equation [12-14]:

$$\Delta n = (\frac{P_1}{Z_1} - \frac{P_2}{Z_2}) \cdot \frac{V}{RT} \tag{1}$$

where Δn is the amount of methane consumption, mol; R is gas constant, equal to 8.31441 J/(mol·K); T is gas temperature, K; V is gas volume, m^3; and Z_1, Z_2 are the initial compressive factor and final compressive factor, respectively. The compressive factor can be calculated as follows [15]:

$$Z = 1 + (A_1 + \frac{A_2}{T_{pr}} + \frac{A_3}{T_{pr}^3}) + (A_4 + \frac{A_5}{T_{pr}})\rho_r^2 + (\frac{A_5 A_6}{T_{pr}})\rho_r^5 - \frac{A_7}{T_{pr}^3}\rho_r^2(1 + A_8\rho_r^2)\exp(-A_8\rho_r^2) \tag{2}$$

where A_i is given parameters (A_1=0.31506237, A_2=−1.0467099, A_3=−0.57832729, A_4=0.53530771, A_5=−0.61232032, A_6=−0.10488813, A_7=0.68157001, A_8=0.68446549) [14]; P_{pr} is comparative pressure; T_{pr} is comparative temperature; ρ_r is methane comparative density. P_{pr}, T_{pr} and ρ_r are calculated as follows:

$$P_{pr} = \frac{P}{P_c},\ T_{pr} = \frac{T}{T_c},\ \rho_r = 0.27(\frac{P_{pr}}{ZT_{pr}}) \tag{3}$$

where P is methane gas pressure, MPa; P_c is critical pressure of methane gas, about 4.60 MPa [12]; T is methane temperature, K; T_c is critical temperature of methane gas, about 191.11 K [12]. Substituting the experimental data into equations (1), (2), and (3), the consumption of methane gas can be calculated.

The rate of hydrate formation can be characterized by the methane consumption per unit time, which is the average consumption rate of methane between the period of starting and stopping hydrate formation. The calculation formula can be deduced based on equation (1) and is expressed as follows.

$$v = \frac{\Delta n}{(t_2 - t_1)} \tag{4}$$

where v is the average consumption rate of gas, mol/min.

3.1.2. Data obtained. According to the various parameters shown in figure 2, when 4.0 wt% of hydrophilic nano-SiO_2 with a particle size of 80 nm was uniformly dispersed in the system, the induction time of CH_4 hydrate formation can be calculated to be prolonged by 70.4%, the formation amount decreased by 9.7%, and the formation rate decreased by 14.3% compared to the UPW experiment (induction time of CH_4 hydrate formation was 9.06 min; gas consumption was ~0.51 mol; and average gas consumption rate was ~0.88×10^{-3} mol/min). Results indicated that hydrophilic nano-SiO_2 effectively inhibits the nucleation and growth of hydrates under this experimental condition.

According to the method mentioned in Section 3.1.1, the induction times, amounts and rates of CH_4 hydrate formation in the VC-713 aqueous solutions with concentrations of 0.25−1.00 wt% and the groups of fluids with the addition of hydrophilic nano-SiO_2 can be calculated (figure 3). It can be known from figure 3 that the addition of hydrophilic nano-SiO_2 can decrease the induction times for the formation of CH_4 hydrate and increase the formation rates compared to the CH_4 hydrate formation in VC-713 aqueous solutions with different concentration. When the additions of VC-713 in the nano-fluids were 0.25 wt% and 0.50 wt%, the amounts of the formed CH_4 hydrates decreased, but the amounts were slightly greater than those in the nano-fluids prepared with hydrophilic nano-SiO_2 and UPW under the same experimental conditions. The addition of a kinetic inhibitor will not change the amount of hydrate formation [9]. Therefore, with the increase of the content of VC-713 in the tested fluid samples, the inhibition ability of hydrophilic nano-SiO_2 on the hydrate formation is gradually weakened. With the VC-713 content being greater than or equal to 0.75 wt%, the inhibition effect of hydrophilic nano-SiO_2 was completely lost.

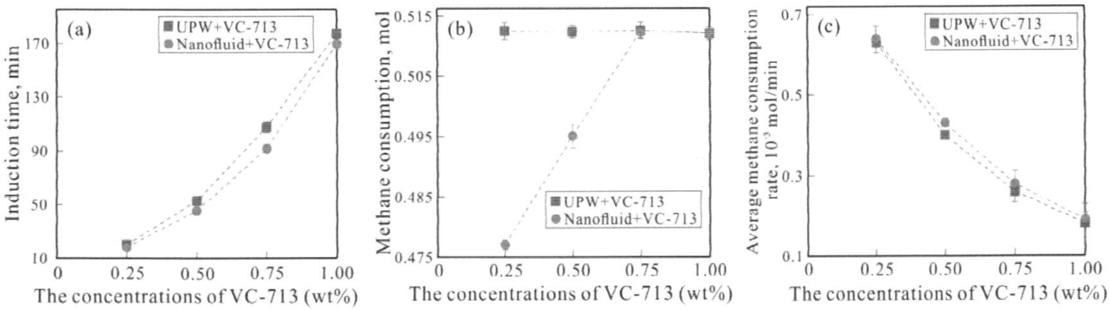

Figure 3. Hydrate formation (a) induction time, (b) amount, and (c) rate in each experiment.

3.2. Mesoscopic structural characteristics of the system

The meso-structural characteristics of the freeze-dried samples from two groups of fluids (i.e., UPW + 1.00 wt% VC-713 and UPW + 1.00 wt% VC-713 + 4.0 wt% nano-SiO_2, respectively) observed under different magnifications are shown in figure 4. Using 1.00 wt% of the VC-713 aqueous solution, hydrated VC-713 molecular aggregates were observed as thin rods (diameter of 0.5–1.0 μm) and smooth thin sheets (thickness of ~0.5 μm and diameter of 40–100 μm), comprising a skeleton structure with a dense, complex spatial distribution (figures 4(a), 4(b), and 4(c)). With the addition of 4.0 wt% of hydrophilic nano-SiO_2, the hydrophilic characteristic of hydrated VC-713 molecules, small-sized nanoparticles and nanoparticle aggregates adhered to the surface of the skeleton structure, clearly changed the original morphology and spatial structure of the hydrated VC-713 molecular aggregates. The diameter of the thin-rod-like hydrated molecular aggregates slightly increased (around 1.0–3.0 μm), while the amount of thin-sheet-like hydrated molecular aggregates dramatically decreased with the increase of the thickness to 1.5–3.0 μm and decrease of the diameter to 30–70 μm.

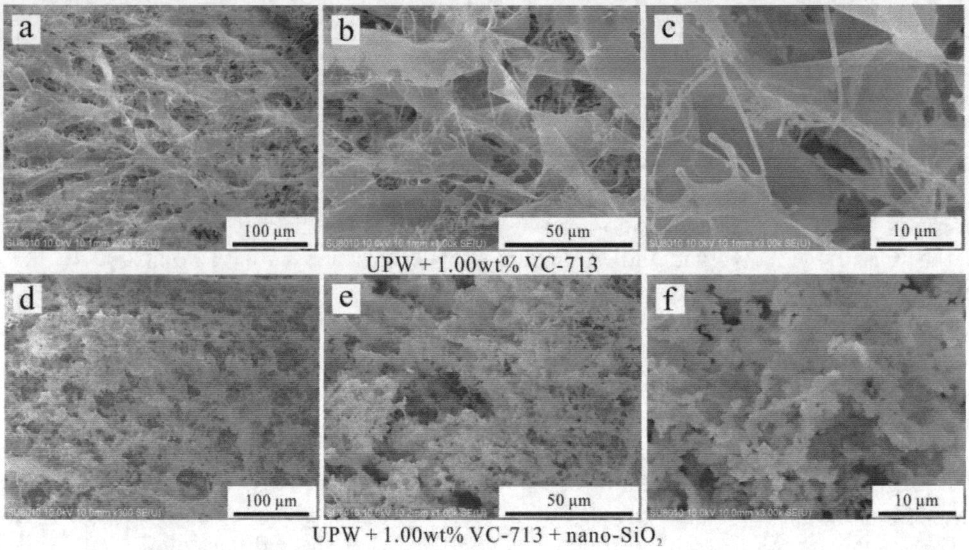

Figure 4. Meso-structures of freeze-dried samples from two groups of fluids under different magnifications.

4. Analysis and discussion

After uniformly dispersing 4.0 wt% of hydrophilic nano-SiO_2 with a particle size of 80 nm in UPW, because of its hydrophilic functional groups on the surface, water molecules near the nanoparticles were bound, and the formation of a cage-like structure from water molecules was inhibited, thereby prolonging the induction time of hydrate formation and decreasing the formation rate of hydrate. Moreover, the water molecules wetted on the hydrophilic nano-SiO_2 surface further formed a water film layer, which did not involve in the hydrate formation, leading to the decrease in the amount of methane hydrate formation (figure 5(a)) [6,8]. Compared to the effect of the hydrate inhibition between hydrophilic nano-SiO_2 and VC-713, the hydrate inhibition of VC-713 was considerably better than that of hydrophilic nano-SiO_2. Hence, the inhibition effect on methane hydrate formation is dominated by VC-713 under the condition of the tested fluid sample containing hydrophilic nano-SiO_2 and VC-713. On the other hand, hydrophilic nano-SiO_2 played a marginal role in decreasing the amount of the formed methane hydrate with a relatively low amount of VC-713 (e.g., 0.25 wt% and 0.50 wt%, figure 5(b)).

Due to the addition of hydrophilic nano-SiO_2 into the VC-713 aqueous solution, hydrophilic nano-SiO_2 adhered to the surface of hydrated VC-713 molecular aggregates and inhibited the migration of hydrated VC-713 molecules (figure 4). Since VC-713 inhibited the nucleation and growth of hydrates via the adsorption of the unclosed cage-like structure on water molecules and the formed hydrate crystalline particles [16,17]. Thus, the inhibition of the migration of hydrated VC-713 molecules by hydrophilic nano-SiO_2 possibly weakened the inhibition ability of VC-713 to the nucleation and growth of hydrates, leading to the decrease of the induction time for methane hydrate formation during the experiment and the increase of the formation rate (figure 3). Meanwhile, lactam-containing polymers (e.g., polyvinyl pyrrolidone and VC-713) exhibit a strong film-forming property [17]. Because of the addition of high contents of VC-713 in the fluid samples (i.e., 0.75 wt% and 1.00 wt%), hydrophilic nano-SiO_2 uniformly dispersed in the system was possibly wrapped by VC-713, which was similar to the wrapping mechanisms of polyvinyl pyrrolidone and VC-713 on hydrate crystalline particles. Therefore, the surface of hydrophilic nanoparticles cannot form a water film that does not participate in the formation of hydrate (figure 5(c)). This result in turn leads to hydrophilic nano-SiO_2 not being able to decrease the hydrate formation amount.

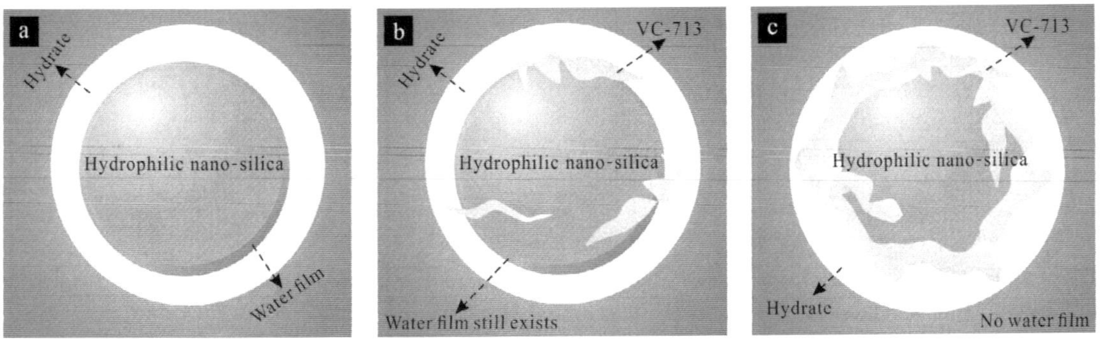

UPW + Nano-SiO_2 UPW + Nano-SiO_2 + VC-713 (0.25wt%, 0.50wt%) UPW + Nano-SiO_2 + VC-713 (0.75wt%, 1.00wt%)

Figure 5. The possible mechanism.

5. Conclusions and prospects

In this study, the effect and its mechanisms of the combination of hydrophilic nano-SiO_2 and VC-713 on their hydrate inhibition was discovered based on the observation of the meso-structural characteristics of the freeze-dried simulated drilling fluids. The results prove that the adsorption and wrapping effect occurred between hydrophilic nano-SiO_2 and VC-713 owing to their hydrophilic characteristic and the film-forming property of VC-713, thereby leading to weakened hydrate inhibition of VC-713. Moreover, with the increasing VC-713 content, the hydrate inhibition of hydrophilic nano-SiO_2 decreased. When the concentration of VC-713 is greater than or equal to 0.75 wt%, the hydrate inhibition of hydrophilic nano-SiO_2 completely disappeared. Hence, the combination of hydrophilic nano-SiO_2 and VC-713 does not exhibit a synergistic inhibition effect on the formation of CH_4 hydrate, but instead, it weakens the hydrate inhibition ability of each component. This study provides an important research idea for the development of conventional oil as well as gas drilling fluid and natural gas hydrate drilling fluid systems.

Acknowledgments

This work was supported by the China Postdoctoral Science Foundation (206402) and the Research Project of CNPC Engineering Technology Research & Development Company limited (CPET201807).

References

[1] Mcconnell D R, Zhang Z and Boswell R 2012 *Mar. Petrol. Geol.* **34** 209-23
[2] Wang R, Ning F L, Liu T L, Zhang L, Sun H C, Peng L, Guo D D and Jiang G S 2017 *Acta. Petrol. Sin.* **38** 963-72
[3] Wang R, Sun H C, Sun J S, Qu Y Z, Zhang J, Shi X M, Zhang L and Guo D D 2019 *J. Chem-NY* **Online.** https://doi.org/10.1155/2019/3547518
[4] Ning F L, Zhang K N, Wu N Y, Zhang L, Li G, Jiang G S, Yu Y B, Liu L and Qin Y H 2013 *Geophys. J. Int.* **193** 1370-84
[5] Zakaria M, Husein M M, and Harland G 2012 Novel nanoparticle-based drilling fluid with

improved characteristics *In SPE International Oilfield Nanotechnology Conference and Exhibition* (Noordwijk, Netherlands) p6

[6] Wang R, Liu T L, Ning F L, Ou W J, Zhang L, Wang Z, Peng L, Sun J X, Liu Z C, Li T S, Sun H C and Jiang G S 2019 *J. Energy. Chem.* **30** 90-100

[7] Liu T L, Jiang G S, Zhang P, Sun J X, Sun H C, Wang R and Zheng M M 2016 *J. Nat. Gas. Sci. Eng.* **33** 934-41

[8] Wang R 2017 Study on the effects of nano-SiO_2 and hydrate inhibitors on hydrate formation in drilling fluid (Wuhan: China University of Geosciences)

[9] Sloan E D and Koh C A 2007 *Clathrate Hydrates of Natural Gases* 3rd ed. (Boca Raton: Crc Press Inc.)

[10] Zhang L, Sun H C, Han B, Peng L, Ning F L, Jiang G S and Chehotkin V F 2016 *J. Unconv. Oil Gas Resour.* **14** 86-98

[11] Wang R, Sun H C, Shi X M, Xu X G, Zhang L and Zhang Z L 2019 *Energies* **12** 2026

[12] Li Z H and Miao L M 2014 *Shanghai Chem. Ind.* **39** 7-10

[13] Sun H C, Wang R, Xu X G, Wang J Ha, Jiang G S, Zhang L, Liu T L and Ning F L 2018 *J China Uni. Pet.* **42** 81-7

[14] Wang R, Sun H C, Xu X G, Zhang J, Wang J H, Yang Z, Zhang Z L, Lai X Q, Liu T L and Jiang G S 2018 *Energ. Source Part A* **40** 2257-64

[15] Ahmed T H **1989** *Hydrocarbon Phase Behavior* (Texas: Guif Publishing Company)

[16] Koh C A, Westacott R E, Zhang W, Hirachand K, Creek J L and Soper A K 2002 *Fluid Phase Equilibr.* **194** 143-51

[17] Peng B Z, Sun C Y, Liu P, Liu Y T, Chen J and Chen G J 2009 *J. Colloid Interf. Sci.* **336** 738-42

Dielectric and piezoelectric properties of (0.10-*x*) Pb $(Mn_{1/3}Nb_{2/3})O_3$ -*x* $Pb(Ni_{1/3}Nb_{2/3})O_3$-$0.91Pb(Zr_{1/2}Ti_{1/2})O_3$ ceramics with high Q_m

J Yoo[1,2] and K Kim[1]

[1]Department of Electrical Engineering, Semyung University, Jecheon 390-711, Korea

E-mail: juhyun57@semyung.ac.kr

Abstract. In this paper, for developing the piezoelectric ceramics suitable for ultrasonic surgical instruments application, (0.10-x) $Pb(Mn_{1/3}Nb_{2/3})O_3$ -x $Pb(Ni_{1/3}Nb_{2/3})O_3$- 0.91 $Pb(Zr_{1/2}Ti_{1}/_{2})O_3$ (x=0. 0.015, 0.03, 0.045, 0.06, 0.075) ceramics were manufactured and then their physical characteristics were systematically analyzed. When $Pb(Ni_{1/3}Nb_{2/3})O_3$ and sintered temperature were is 0 mol%, 940°C, respectively, the excellent electrical properties (d33= 236 pC/N, εr= 731.9, kp= 0.5839, Qm= 1982) were obtained for ultrasonic surgical instruments and actuators.

1. Introduction

Recently, $Pb(Zr,Ti)O_3$(PZT) system ceramics has been actively studied for the application of piezoelectric ceramics such as ultrasonic sensor, ultrasonic cleaners, ultrasonic surgical instruments and piezoelectric actuators [1]. Although $Pb(Zr,Ti)O_3$ ceramics have been widely utilized in the various kind of fields, deterioration of piezoelectric characteristics and environmental pollution due to PbO, which is rapidly volatilized at around 1000°C, are problematic.

Therefore, for solving the PbO problems, the studies on the two, three component system PZT, and PMN-PZT ceramics have been actively performed in order to further more increase physical properties together with lowering the sintering temperature of the ceramics. Therefore, to decrease the sintering temperature and to increase the piezoelectric characteristics is to use various kinds of additives as sintering aids [2,3]. Recently, ultrasonic surgical instruments for laparoscopic surgery have been widely used. The ultrasonic surgical instrument uses a piezoelectric vibrator, which generates heat due to high-power driving for a long time. The generated heat promotes thermal expansion of the piezoelectric vibrator and can cause destruction.

Therefore, it is necessary to effectively design the ultrasonic surgical instrument for radiating the heat generated well even if it is used for a long time. And also, it is required to increase Q_m for reducing the heat rise of the piezoelectric vibrator .

The piezoelectric ceramic materials with high Q_m for ultrasonic surgical instrument application may be achieved by developing $Pb(Ni_{1/3}Nb_{2/3})O_3$- $Pb(Zr_{1/2}Ti_{1/2})O_3$- $Pb(Mn_{1/3}Nb_{2/3})O_3$ system. Moreover, in the system ceramics, dielectric and piezoelectric properties can be greatly enhanced by the variations of composition. The additives of Li_2CO_3, CuO, Bi_2O_3 can induce liquid phase state of the system, causing the enhancement of the physical properties [4-7].

Here, in order to develop the system ceramics with high kp and high Qm for ultrasonic surgical instruments and actuators applications,

CuO, Bi_2O_3 and Li_2CO_3 (CLBO) as additives were used to (0.10-*x*) $Pb(Mn_{1/3}Nb_{2/3})O_3$-*x* $Pb(Ni_{1/3}Nb_{2/3})O_3$- 0.91 $Pb(Zr_{1/2}Ti_{1/2})O_3$(*x*=0. 0.015, 0.03, 0.045, 0.06, 0.075) ceramics. Here, *x* was varied for investing the physical characteristics of the manufactured ceramics.

2. Experiments

In this study, the used compositions are as follows;

$$(0.10\text{-}x)\ Pb(Mn_{1/3}Nb_{2/3})O_3\text{-}x\ Pb(Ni_{1/3}Nb_{2/3})O_3\text{-}0.91\ Pb(Zr_{1/2}Ti_{1/2})O_3+0.5wt\%\ PbO$$

$$(x=0.\ 0.015, 0.03, 0.045, 0.06, 0.075) + \text{additives } (Li_2CO_3\text{- } CuO\ \text{-}Bi_2O_3)$$

The PbO, MnO_2, ZrO_2, TiO_2, NiO and Nb_2O_5 materials with purity more than 99% were weighed and then were ball-milled for 24 hours. And then they were also calcined at 850°C for 2hour. And then, Li_2CO_3- CuO-Bi_2O_3 as additives was added and ball-milled again for 24 hour. 5wt% PVA was added and mixed to the calcined ceramic powders. The ceramic powders were molded under a pressure of 15 MPa. And then molded powders were sintered at 940°C for 2 hour. For investigating the physical properties, the ceramic specimens were polished to thickness 1 mm. And then were electrodeposited with silver paste. The poling of specimens was made by 30 kV/cm for 30 minutes in 120°C silicon oil bath. The capacitance was measured by an LCR meter (Instek LCR-819) and the dielectric constant (ε_r) was measured. At the room temperature, the resonant frequency *fr* and anti-resonant frequency *fa* were investigated by Agilent 4294A Impedance Analyzer. And then the Q_m and k_p were measured. Piezoelectric d_{33} constant was measured by d_{33} meter (APC-90-2030).

3. Results and discussion

Figure 1 shows the density of the specimen with *x* (=$Pb(Ni_{1/3}Nb_{2/3})O_3$). The density showed the highest value of 8.24 g/cm^3 at the 3 mol% *x*. It is analyzed that Bi_2O_3, CuO, and Li_2CO_3 with a low melting temperature can react with PbO for forming liquid phase, resulting in the increase of densification.

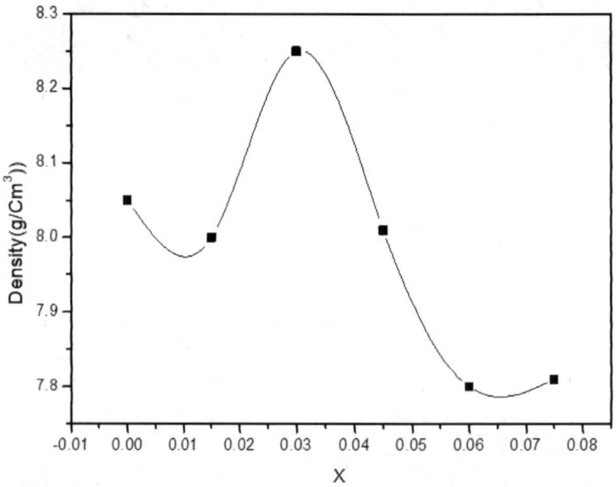

Figure 1. Density of the specimen with *x*.

Figure 2 shows the k_p of the specimen with *x*. The *kp* showed the highest value of 0.62 at 6 mol% *x*, and then gradually decreased. In general, the *kp* is enhanced when the sintered density of the ceramics is increased. However, density was decreased due to excess PNN substitution but k_p showed the trend of increment according to the movement toward MPB composition [8]. Figure 3 shows the piezoelectric d_{33} constant of the specimen with *x*. When *x* was 7.5 mol%, the d_{33} maximum value was 350 pC/N, and then the piezoelectric d_{33} constant decreased.

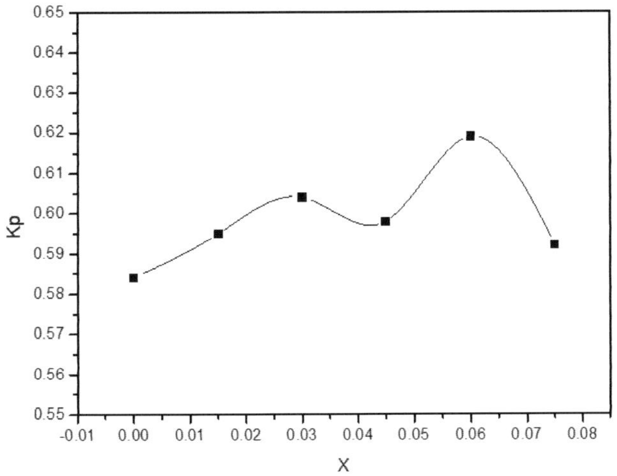

Figure 2. K_p of the specimen with x.

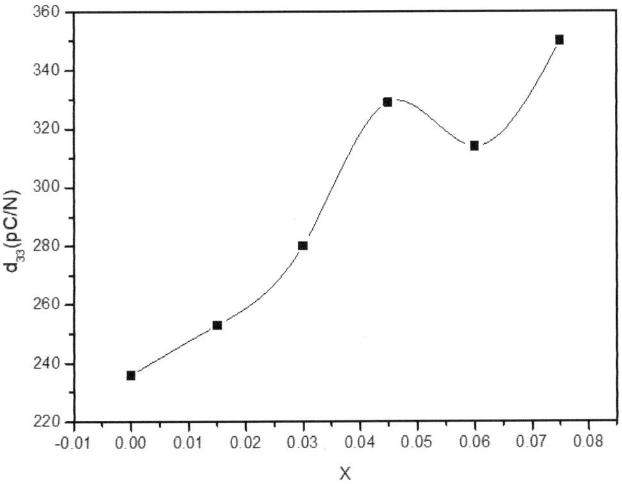

Figure 3. d_{33} of the specimen with x.

In this study, it can be explained by the fact that the d_{33} values were slightly increased due to the increase of x with -120°C Curie temperature.

Figure 4 shows the dielectric constant of the specimen with x carried out at room temperature. The dielectric constant showed the highest value of 1,345 at 7.5 mol% x. This is a similar result with d_{33}.

This result can be also illustrated by the fact that Curie temperature of the specimen is decreased according to the increase of PNN with -120°C Curie temperature. Accordingly, the dielectric constant of the specimen at room temperature is increased.

Figure 5 shows the Qm with x.

According to x substitution, the content of MnO_2 which coexist in the Mn^{3+} state in PZT systems as the substitutes for Zr^{4+}, and Ti^{4+} can be decreased. Thus, the generated oxygen vacancies is decreased, causing a decrease in the Qm . At the 0 mol% x, the highest value of Qm was 1982.

These phenomenon can be illustrated by the fact that the decrease of Mn ratio increases the ratio of x. As the results, As the Mn ratio was small, the Qm was decreased.

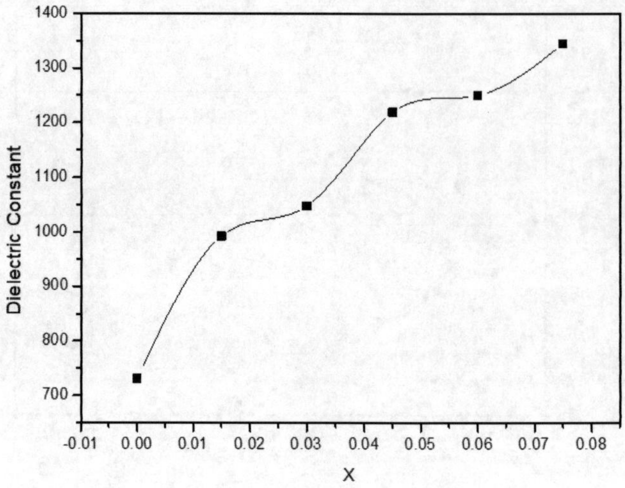

Figure 4. Dielectric constant of the specimen with *x*.

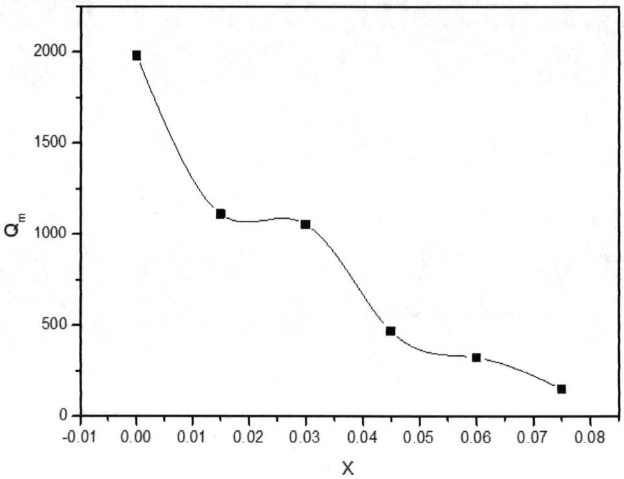

Figure 5. *Qm* with *x*.

Table 1 shows the physical properties of specimens with *x*.

Table 1. Physical properties of the specimen with *x*.

x	ρ (g/cm^3)	k_p	ε_r	Q_m	d_{33} (pC/N)
0	8.05	0.584	732	1982	236
0.015	8.00	0.595	993	1112	253
0.03	8.25	0.604	1048	1058	280
0.045	8.01	0.598	1220	470	329
0.06	7.80	0.619	1251	325	314
0.075	7.81	0.592	1345	150	350

4. Conclusions

In order to develop low temperature sintering piezoelectric ceramics for ultrasonic surgical instruments and actuators, (0.10-*x*) $Pb(Mn_{1/3}Nb_{2/3})O_3$-*x* $Pb(Ni_{1/3}Nb_{2/3})O_3$-0.91 $Pb(Zr_{1/2}Ti_{1/2})O_3$ +0.5wt% PbO were fabricated with the variation of $Pb(Ni_{1/3}Nb_{2/3})O_3$. The piezoelectric and dielectric properties of the manufactured samples were investigated. The following results were obtained:

- At the 3 mol% $Pb(Ni_{1/3}Nb_{2/3})O_3$, the maxinum density of 8.25 g/cm^3 was obtained.
- At 0 mol% $Pb(Ni_{1/3}Nb_{2/3})O_3$ composition, physical characteristics of d_{33}= 236 pC/N, ε_r= 732, k_p= 0.5839, Q_m= 1982 were obtained ,suitable for ultrasonic surgical instruments.

Acknowledgments
This paper was studied by the 2016 National Research Foundation of Korea (N RF).

References

[1] Lee Y H and Yoo J H 2010 *Ferroelectrics* **10** 396
[2] Yoo J H 2011 *Ferroelectrics* **135** 423
[3] Yoo J H, Lee Y H, Kim D H, Lee I H, Kwon J S and Paik D S 2008 *J. Korean Inst. Electr. Electron. Mater. Eng.* **21** 126
[4] Lee D H, Lee S H, Nam S P and Lee Y H 2010 *J. Korean Inst. Electr. Electron. Mater. Eng.* **59** 11
[5] Chao X, Yang Z, Huang X, Ma D and Zeng J 2009 *Curr. Appl. Phys.* **9** 1283
[6] Jeong Y, Yoo J, Lee S and Hong J 2007 *Sens. Actuators A* **135** 215
[7] Zhang Q, Yue Y and Nie R 2017 *Mater. Res. Bull.* **85** 96
[8] Tiwari V and Srivastava G 2015 *Ceram. Inter.* **41** 2774

IOP Conf. Series: Materials Science and Engineering **634** (2019) 012022 doi:10.1088/1757-899X/634/1/012022

Improvement of tribological properties of Al alloy with CNTs and Nb nanopowder for industrial application

C O Ujah[1,3], A P Popoola[1] and O M Popoola[2]

[1]Faculty of Engineering and the Built Environment, department of Chemical, Metallurgical and Materials Engineering, Tshwane University of Technology, Pretoria, South Africa
[2]Faculty of Engineering and the Built Environment, department of Electrical Engineering, Tshwane University of Technology, Pretoria, South Africa

[3]E-mail: omega.ujah@gmail.com

Abstract. Unreinforced Al alloy lacks adequate tribological property required in service. To improve on this property, it must be reinforced with materials having good wear properties. Hence, Al was reinforced with CNTs and Nb and consolidated with spark plasma sintering (SPS). The wear experiment was conducted on a ball-on-disc Tribometer where loads of 10 N, 20 N and 30 N were systematically applied. The coefficients of friction (COF) of the sintered samples were subsequently obtained. The samples were weighed before and after each run to determine their weight losses which were used to compute their wear volumes. The experiment was designed and analyzed with Taguchi and ANOVA. Results showed that Al-8CNTs-8Nb composite had the best wear characteristics. Its COF got improved by 79% which shows that it is suitable for industrial applications.

1. Introduction

This study deals with improvement of Al alloy for industrial applications, especially in high transmission conductor. Ever since the conventional conductor, Aluminium conductor steel reinforced (ACSR) was found deficient in some operational areas, such as in operating temperature, coefficient of thermal expansion (CTE), corrosion, wear, and density [1], research has been on going to find a possible replacement. Al-CNTs-Nb composite is investigated in this study as a possible replacement for the conventional conductor. Traditionally, transmission conductors should be light in weight with high wear resistance, however, the base metal of this work, Al 1000 series alloy powder, is light in weight, corrosion resistant, electrically conductive, cost effective but poor in strength and tribology [2]. Therefore, Carbon nanotubes (CNTs) which have high strength, high thermal conductivity, high wear resistance and light weight [3]; Nb is a superconductor, a grain refiner, a fracture toughening additive and a wear resistant additive [4]. These properties should be able to improve the properties of Al alloy for industrial application. The fabrication process is spark plasma sintering (SPS) which is a non-conventional sintering method that utilizes high heating and cooling rates to minimize grain growth. The presence of plasma heating in SPS helps to vaporize impurities. More so, the low sintering temperature is necessary for energy conservation. And also, the short sintering time makes the process cost effective; while the simultaneous heating with application of pressure helps to achieve a solid compact at a temperature lower than the melting point of the material [5]. Even though SPS is a bulk synthesising technique, it is believed to be a surface improving process as it generates products with wear and corrosion resistance. Tokita [6] produced high wear resistance nozzle with SPS. Many researchers have worked on improving the properties of Al alloy with ceramics, particulates, metal powders and even thermo-mechanical treatments. For instance, Ehsan *et al* [7] reinforced Al alloy with TiC, Loto and Babalola [8] improved Al with SiC using SPS, Zawadzki *et al* [9] improved AA7075 alloy with heat treatment and shot penning, and Kyziol *et al* [10] surface-modified AA2024 by chemical vapour deposition. However, SPS of Al-CNTs-Nb composite aimed at improving the wear characteristics of Al alloy has not been reported anywhere in the literature despite the excellent properties of these reinforcements. This study therefore, is aimed at utilizing the characteristics of the CNTs and Nb additives in one hand and SPS on the other hand to improve the tribological properties of Al alloy for use in transmission conductor.

2. Material and methods

2.1. Material

Aluminium alloy powder (1000 series) of purity (99.9 wt.%) with particle size of 40 nm; multi-walled carbon nanotubes of purity (99.5 wt.%), diameter (10-30 nm), (length 5-20 μm); and Niobium nano powder of purity (99.9 wt.%) and particle size of 20-40 nm, which were supplied by a company in China named Hongwu International Group, constituted the materials used in this study. Masses of the powders required to reflect the Taguchi design of experiment (DOE) in table 1(a) were weighed out and mixed. The CNTs with Young's modulus of 950 -1500 GPa, Poisson ratio of 0.32, thermal conductivity of 3000 W/mK and produced by chemical catalytic vapour deposition (CCVD) were initially agitated to de-agglomerate particles with a Tubular Shaker Mixer at an angular velocity of 69 r/min for 8 h with steel balls incorporated therein. Thereafter, the whole powders were blended together at 110 rpm for 10 hours with balls made of high carbon steel at a ball to powder (B:P) ratio of 2:5 following Chieh *et al*, Ujah *et al* [11, 12].

Table 1. (a) Taguchi DOE (L9 3^2 Orthogonal array) for sintering. (b) Taguchi DOE (L9 3^3 Orthogonal array) for tribology experiment.

(a)					(b)			
Exp no	Al (wt.%)	CNTs (wt%)	Nb (wt%)	Description	Exp no	CNTs	Nb	Load (N)
1	98	1	1	Al-1CNTs-1Nb	1	1	1	10
2	95	1	4	Al-1CNTs-4Nb	2	1	4	20
3	91	1	8	Al-1CNTs-8Nb	3	1	8	30
4	95	4	1	Al-4CNTs-1Nb	4	4	1	20
5	92	4	4	Al-4CNTs-4Nb	5	4	4	30
6	88	4	8	Al-4CNTs-8Nb	6	4	8	10
7	91	8	1	Al-8CNTs-1Nb	7	8	1	30
8	88	8	4	Al-8CNTs-4Nb	8	8	4	10
9	84	8	8	Al-8CNTs-8Nb	9	8	8	20

2.2. SPS technology

Optimized SPS parameters reported elsewhere [12] was used to sinter the samples. They include 630°C 30 MPa, 10 mins and 200°C/mins. The blended powders were singly sintered one after the other with these parameters at a vacuum of 0.605e-3 bar, relative pressure of -500e-3 bar, and absolute pressure of 1.2e-3 bar. Graphite paper was used to fix at the wall of the die to aid in homogenous heating and for easy removal of the sample after sintering. The number of samples produced was 10 and they were cut for analysis. Figure 1 shows an illustration of SPS mechanism and sintered sample. It can be seen that at the start of sintering (figure 1(a)), heat is circulated within the powder mass through plasma heating. But once neck formation takes place, heat is transferred to the adjacent particles via joule heating.

IOP Conf. Series: Materials Science and Engineering **634** (2019) 012022 doi:10.1088/1757-899X/634/1/012022

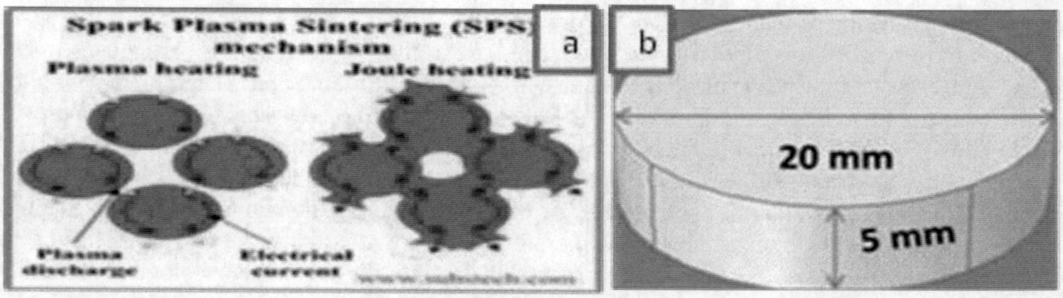

Figure 1. (a) Mechanism of SPS (b) Sketch of sintered sample.

2.3. Tribology test

The 10 samples were prepared for wear test by cutting them into (20 x 20 x 2 mm) sizes followed by metallographical polishing. The test was conducted at room temperature. The Tribometer consists of ball-on-disc reciprocating machine capable of generating the coefficient of friction (COF) of test samples. The ball is made of Grade 25 alloy steel E52100, 6.35 mm in diameter. The stroke length was 5 mm, velocity was 3 mm/s, time was 360 seconds and the loads were 10 N, 20 N, and 30 N. The masses of the sample before and after each test were taken and used to compute the wear volume as shown below:

$$\text{Wear volume (cm}^3\text{)} = (M_1 - M_2) / \rho \quad (1)$$

Where M_1 and M_2 [g] are masses obtained before and after each wear tests respectively, ρ [g/cm^3] is the actual density of the composite material. The tribology test was analyzed with statistics by firstly designing the experiment with Taguchi DOE (L9 3^3 Orthogonal array) and analyzing it with ANOVA. The Taguchi DOE is shown in table 1(b).

3. Results and discussion

The TEM micrograph of the start-up powders are shown in figure 2. It can be seen that the Al particles (20 -100 nm) are spherical in shape (figure 2(a)), CNTs are made of cylindrical tubes entangled to one another (figure 2(b)) and Nb particles (10 – 80 nm) are spherical too with some tiny particles clinging to bigger ones to form some sorts of agglomeration, though may be due to packaging from the industry. The agglomeration of CNTs was due to its high aspect ratio [13].

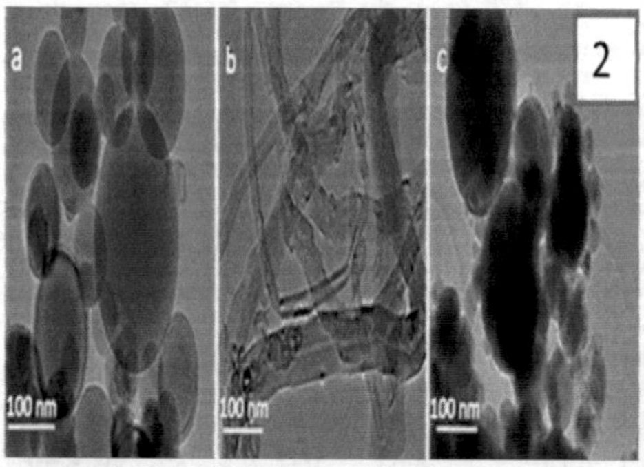

Figure 2. TEM Images of as-received powders (a) Al (b) CNTs (c) Nb.

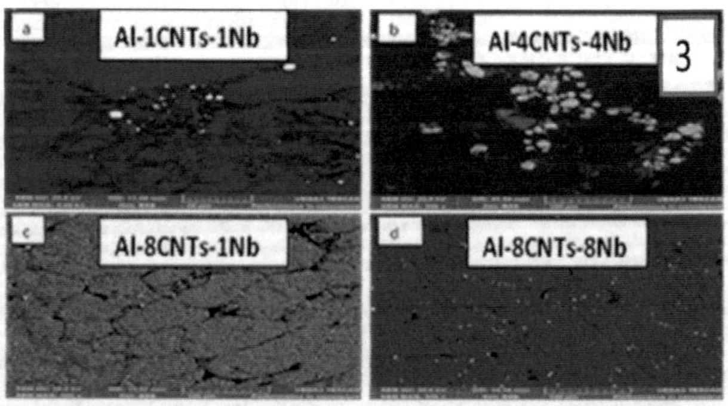

Figure 3. SEM images (a) Al-1CNTs-1Nb (b) Al-4CNTs-4Nb (c) Al-8CNTs-1Nb (d) Al-8CNTs-8Nb.

Scanning electron microscope (SEM) images of sintered samples are shown in figure 3. It can be seen that there was grain refinement, homogenous dispersion of reinforcements and cohesive matrix/reinforcement interface except for figure 3(c) which has pores. The pores would definitely

IOP Conf. Series: Materials Science and Engineering **634** (2019) 012022 doi:10.1088/1757-899X/634/1/012022

affect the sample as could be seen later on.

Figure 4 shows coefficient of friction (COF) and wear volume profiles. It can be seen that the COF decreased as the load increased (figure 4(a)). This could be because higher loads increased the temperature of the surface and subsequently softened it thereby reducing the COF [14]. However, Wang *et al* [15] had a contrary observation where it was seen that COF increased with increase in applied load. The COF of pure Al was higher than that of the reinforced composites and got improved by 79 % (from 0.48 for pure Al to 0.10 for Al-8CNTs-8Nb). The improvement can be attributed to the incorporated reinforcements which induced refinement of the grain sizes as can be seen in figure 3(d), well dispersed reinforcements and compact grain boundaries [4-7]. Figure 4(b) shows wear volume of the sintered samples subjected to wear test. Here it could be seen that unreinforced Al alloy had the highest wear volume while Al-8CNTs-8Nb had the least wear volume. The reduced wear volume in the composites was because the reinforcements had high wear resistance and the SPS processing technique impacted positively on the products [5,6,14]. Also observed was that the higher the load, the higher the wear volume which could be attributed to the fact that the increased load increased the shearing force on the surface thereby removing more materials from the surface [14,15]. The wear volume got improved by 23% (from 12.8 ± 1.2 μm^3 for Pure Al alloy to 9.8 ± 2.0 μm^3 for Al-8CNTs-Nb composite).

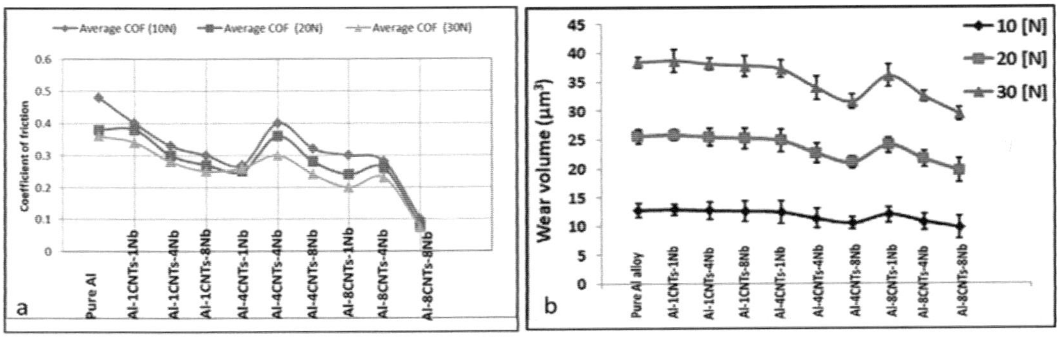

Figure 4. Tribology profiles (a) COF (b) wear volumes.

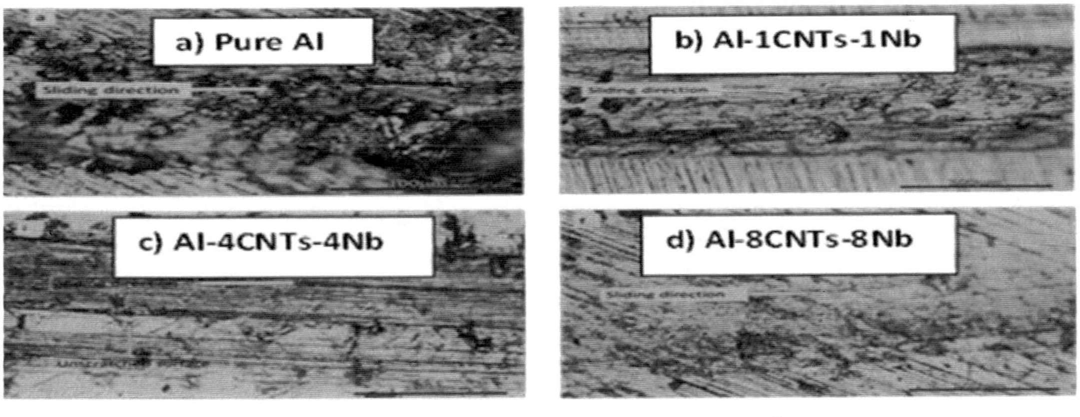

Figure 5. Optical microscopy of some worn surfaces.

Figure 5 shows the optical micrographs of worn surfaces. It can be seen that unreinforced pure Al alloy had more severe wear track than other samples (Figure 5(a)). In Figure 5(d), the wear track was very shallow which concur to the low COF and low wear volume of Al-8CNTs-8Nb composite. Figure 5(b) shows wear track that is severe though less than that of pure Al. In Figure 5(c), the wear track still reduced. The trend shows that the higher the quantity of reinforcement, the shallower the wear tracks which could be interpreted from the angle that the protective films generated by the reinforcements increased with their weight fractions. The wear mode observed in the whole samples was that of abrasion. It was noticed also that the higher the load, the more severe the wear track. Figure 6 shows coefficient of friction (COF) profiles of Al-CNTs-Nb at 20 N loading over time. The outstanding feature of the profile is that there are two main stages, the running-in stage and the steady state stage. The friction rises rapidly from zero mark to the maximum value, and then falls down before it stabilizes.

Figure 7 shows Taguchi-ANOVA analysis of tribology result. In figure 7(a), it could be seen that load 1 (10 N) had the highest influence on the outcome of the coefficient of friction. This was confirmed in figure 7(b) where load had the highest percentage contribution of 67.12% to the outcome of the tribology experiment. Next in significance was Nb reinforcement (level 3, that is 8 wt.%) and then CNTs (level 3, that is 8 wt.%). The interpretation of this is that 8 wt% of both Nb and CNTs reinforcements produced highest effect on the reduction of COF of the composite. This was confirmed in figures 7(c) & 7(d) where 10 N load, 8 wt.% Nb and 8 wt.%CNTs gave the highest mean values and the least signal-to-noise ratios respectively. Hence, Al-8CNTs-8Nb has shown the least COF in this analysis.

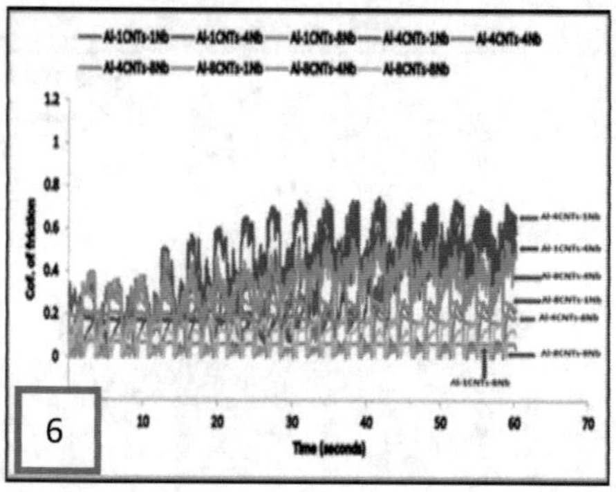

Figure 6. Coefficient of friction profiles of Al-CNTs-Nb samples.

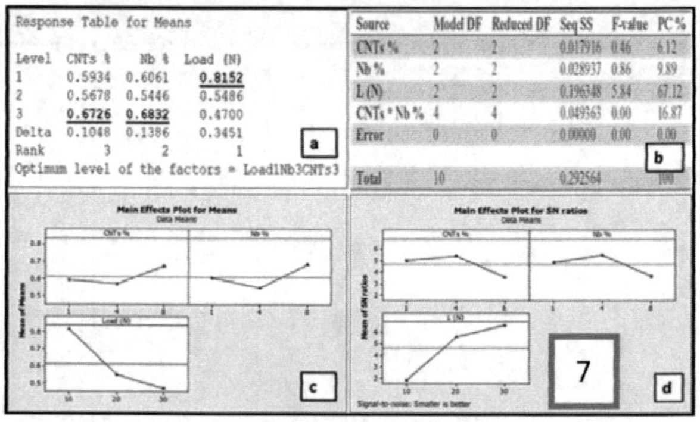

Figure 7. Taguchi-ANOVA analysis of tribology of (a) Responses of the input variables (b) Significance of the variables (c) Plot of the means of the variables (d) Plot of the signal-to-noise ratio of the variables.

4. Conclusion

The experiment has been conducted and findings show that wear resistance of Al composite improved by 79% in terms of COF and 23% in terms of wear volume. The sample with the best wear characteristics was Al-8CNTs-8Nb with a density of 2.72 g/cm^3. Taguchi and ANOVA analyses concurred with this result. These improvements could be attributed to the generation carbon film by CNTs which acted as a solid lubricant and the Nb_2O_5 which was formed by Nb that acted as a protective layer against wear, together with the SPS route which aided in improving the properties of the composite so far developed. So, the properties achieved concur that Al-CNTs-Nb composite is a robust material for industrial application such as in high power transmission conductor and its core.

Acknowledgment

The authors want to appreciate centre for energy and electric power (CEEP) and Tshwane University of Technology (TUT) Pretoria for their financial assistance in the course of this work. There was no conflict of interest in this work.

References

[1] Alawar A, Bosze E J and Nutt S R 2005 A composite core conductor for low sag at high temperatures *IEEE Trans. Power Delivery* **20** 2193-9

[2] Starke Jr E A and Rashed H M M A 2017 Alloys: Aluminium *Reference Module in Mater Sci and Mater Eng*. doi:10.1016/B978-0-12-803581-8.09210-9

[3] George R, Kashyap K T, Rahul R and Yamdagni S 2005 Strengthening in carbon nanotube - aluminium (CNT/Al) composites *J Scripta Mater* **53** 1159-63

[4] Thomson K E, Jiang D, Yao W, Ritchie R O and Mukherjee A K 2012 Characterization and mechanical testing of alumina-based Nano composites reinforced with niobium and/or carbon nanotubes fabricated by spark plasma sintering *J. Acta Mater* **60** 622-32

[5] Delaizir G, Bernard-Granger G, Monnier J, Grodzki R, Kim-Hak O, Szkutnik P D, Soulier M, Saunier S, Goeuriot D and Rouleau O 2012 A comparative study of spark plasma sintering (SPS), hot isostatic pressing (HIP) and microwaves sintering techniques on p-type Bi 2 Te 3 thermoelectric properties *Mater. Res. Bull.* **47** 1954-60

[6] Tokita M 2015 Recent and future progress on advanced ceramics sintering by spark plasma sintering *Nanotechnologies in Russia* **10** 261-7

[7] Ehsan G, Ali F, Masoud A *et al* 2017 Evaluation of microstructure and mechanical properties of Al-TiC metal matrix composite prepared by conventional, microwave and spark plasma sintering methods *Mater* **10** 1255

[8] Loto R T and Babalola P 2017 Corrosion polarization behaviour and microstructural analysis of AA1070 aluminium silicon carbide matrix composites in acid chloride concentrations *Cogent Eng* **4** 1422229

[9] Zawadzki P, Stegliński M, Wójcik R, Klich M, Kyzioł K, Kottfer D and Pawłowski W 2015 The effect of two-stage age hardening treatment combined with shot peening on stress distribution in the surface layer of 7075 aluminium alloy *Arch of Met and Mater* **60** 1993-8

[10] Kyzioł K, Mędala M and Kaczmarek 2014 Influence of plasmochemical modification of Al–Cu–Mg alloys on surface structure and functional properties *Vacuum* **105** 52-8

[11] Chieh K, Te-Tan L, Kuo-Hwa T *et al* 2009 The influences of powder mixing process on the quality of W-Cu composites *J Trans Canada Soc Mech Eng* **33** 3

[12] Ujah C O, Popoola A P I, Popoola O M and Aigbodion V S 2019 Optimisation of spark plasma sintering parameters of Al-CNTs-Nb nano-composite using taguchi design of experiment *Int J Adv Manuf Technol* **100** 1563-73

[13] Ameri S, Sadeghian Z and Kazeminezhad I 2016 Effect of CNT addition approach on the microstructure and properties of NiAl-CNT nanocomposites produced by mechanical alloying and spark plasma sintering *Intermetallics* **76** 41-8

[14] Abou Bakr El A, Kamel B M, Khattab A, Osman T A, Azzam B and Zaki M 2018 Improved mechanical and tribological properties of A356 reinforced by MWCNTs *Fullerene, Nanostructures and Carbon Nanostructures* **26** 185-94

[15] Wang Y Q, Afsar A M, Jang J H, Han J I and Song J I 2010 Room temperature dry and lubricant wear behaviors of Al2O3f/SiCp/Al hybrid metal matrix composites *Wear* **268** 863-70

Assessment of the physical properties of banana pseudo stem/ ABS composites

M S Ahmed[1,2], T Attia[1], A A Abd El-Wahab[1], R Elgamsy[1] and M H Abd El-latif[1]

[1]Designs and Production Department, Faculty of Engineering, Ain Shams University, Cairo, Egypt

E-mail: mohsamehahmed23@gmail.com

Abstract. The physical properties of Banana Pseudo Stem (BPS) reinforced Acrylonitrile Butadiene Styrene (ABS) composites with different fiber weight percentages were studied. BPS fibers were pretreated by using Sodium hydroxide solution with 5% concentration for 24 hours. ABS/BPS composites were prepared with 0, 10, 20 and 30 wt. % fibers. The Moisture Content (MC %) and Water Absorption (WA %) for BPS fiber were studied. The water absorption (WA %) and Thickness Swelling (TS %) of the composites were investigated. The scanning electron microscope (SEM) observation for fracture surfaces was also studied. The Thermogravimetric_analysis (TGA) and Differential Scanning Calorimetry (DSC) analyses were used to measure the thermal stability and glass transition temperature (Tg) of BPS/ABS composites. Moreover, the melt flow index (MFI) and dimensional stability tests for the composites were studied.

1. Introduction

Polymeric matrix composites play an important role in our life. Composites share in manufacturing of many applications such as aerospace, transportation, wind energy, construction, automotive and marine applications [1]. The global composite materials market is expected to reach an estimated $40.2 billion by 2024 and it is forecast to grow at a CAGR of 3.3% from 2019 to 2024 [2]. The major motives for growth in composites market are increasing demand for lightweight materials in the aerospace, defense and automotive industry corrosion and chemical resistance materials demand in construction and pipe & tank industry [2].

ABS is one of the important copolymers which consists of Acrylonitrile, butadiene and styrene. It is used in many applications such as polymeric wood composites. It has a good physical property when compared with other plastics, ABS is the most widely used material in fused deposition modelling (FDM) applications such as 3D- Printing technology [3]. The global market size of ABS was USD 20.22 billion in 2014. Its market share is expected to be USD 34.88 billion by 2022. This is due to rising the use of the product in numerous applications including automotive ancillaries, consumer goods, and construction materials [1]. Importance of light weight automobiles on account for a better fuel efficiency is expected to increase products application scope. Recent developments in the Corporate Average Fuel Economy (CAFE) have forced automobile manufacturers to seek new ways to reduce fuel consumption which will have a strong impact on the industry over the forecast period [4].

Due to environmental pollution and increase of harmful emissions, that comes from using non-natural compounds in the industry field and causes many different health and environmental problems.

Natural fibers became good candidates to invade the field of industry as an alternative to synthetic fibers. Due to low cost, low density and good properties compared with synthetic fibers [5]. Nowadays, the utilization of agriculture wastes (AGW) is combined between past tradition and present technology. By using them as composting, biogas, electricity and animal foods. They are also a big source of natural fibers to be used as filler or reinforce in polymeric matrix composites (PMCs) [5]. The amount of (AGW) in Egypt were reached 35 million tons / year (23 million tons of vegetarian wastes and 12 million tons of animal wastes per year). Almost 21 million tons/year of AGW haven't been utilized till now [6]. During last years, different industries tried to decrease usage of composites from petroleum due to harmful impact on environment. So natural fiber composites (NFCs) can be considered as replacement for petroleum ones. The price and manufacturing consumed energy of natural fibers are low when compared with synthetic fibers [6].

The estimated amount of Banana wastes reached 1.7 million tons /year. Banana residues can be used as a source of banana fiber (BF). The production of banana fruit over 130 countries is nearly 120-150 million tons (as the second major production of fruit) with nearly 16% of total global production of fruit, only 12% of the weight of banana tree is represented as banana fruit and the others (leaves and pseudo stems, etc.) aren't used as food but used in fiber production [7]. BF can be obtained from many parts of banana plant such as leaves, rachis and stems. Many industries such as paper, mats, textiles and composites are improved by BF. Pseudo stems are considered as a cluster of leaf stalk which aggregated together in cylindrical shape. They have many applications in industry like cardboards with high quality, ropes for marine field, currency paper and fabric materials. Banana pseudo stem (BPS) fibers are better than rachis and leaf fibers due to its high mechanical strength [8]. BPS fibers are considered as Bast plant fibers with complex structure of cellulose, hemicellulose and lignin. BPS have high mechanical properties due to high cellulose percentage compared with others [8].

The aim of this paper is to study the physical and thermal properties of BPS/ABS composites through using BPS (fibrils extracted from banana tree wastes) with the ABS material as a thermoplastic composite material. ABS is used as a basic matrix and BPS fibers are used as a reinforcing material with different wt. % of (0-10-20-30%).

2. Methodology

2.1. Materials

Banana Pseudo Stem (BPS) fibers were supplied by Banana Kingdom Co. The chemical analysis and density of banana fibres are shown in table 1 [9]. The banana fiber length and diameter were measured by using FEROX PL Microscope, the average fiber length, diameter, and aspect ratio are listed in table 2.

Table 1. Chemical composition of BPS fibers [9].

Cellulosc [wt.%]	Hemicellulose [wt.%]	Lignin [wt.%]	Others [wt.%]
63	12	5	20

Table 2. Physical Properties of BPS fibers [9].

Density (kg/m^3)	Average fiber length	Average fiber diameter	Average fiber aspect ratio
721	4 ± 0.5 (mm)	163.64 ± 20 (μm)	25 ± 3

Samsung ABS code SD-0150, UN-reinforced General-Purpose ABS granules were used in this research.

2.2. Pre-treatment of BPS fibers

BPS fibers were chemically treated by using (5%) NaOH Solution at room temperature (25°C) for 24 hrs [8]. The treated fibers were washed by running water and immersed into hydrochloric acid HCL solution of (1%) concentration at room temperature (25°C) for 10 minutes [9]. This was followed by washing in running water. Consequently, the fibers were dried in open air under the sun light for 24-48 hrs., then placed in the oven for drying at 80°C for 48 hrs. [9].

2.3. BPS/ABS Composite preparation
BPS/ABS Composites were prepared by placing the ABS granules in thermal mixer machine at 180°C for 15 minutes [10]. Then BPS fibers were added with different weight percentage (0-10-20-30) wt. %. The recommended mixing time is 10 minutes to achieve homogeneous mixing results [10]. These composites were consequently crushed into granules and then injected using injection machine type METAL MECCANICA (65/90 SR).

2.4. Physical and thermal tests

2.4.1. Moisture content test. The moisture content (MC) test was performed for untreated and treated BPS fiber according to DIN EN 322 by using ADAM balance NBL 623e.

2.4.2. Water absorption (WA%) and thickness swelling (TS%) test. The WA% and TS% tests were carried out for pure ABS and BPS/ABS composites at different fiber wt. % according to ASTM D570 b using ADAM balance NBL 623e.

2.4.3. Melt flow index (MFI) test. The melt flow index (MFI) tests were carried out through *a Melt* Flow Index Tester DW5450D for both pure ABS and BPS/ABS composites at different fiber wt.% according to ASTM D1238-04.

2.4.4. SEM observation. The microstructure analysis was made for BPS/ ABS at different wt.% using scanning electron microscope Quanta FEG-250 microscopy.

2.4.5. TGA test. TGA test was done using Thermo gravimetric device with model TA instrument Q500 according to ASTM D 6370.

2.4.6. DSC analysis. DSC analysis was performed by using TA instrument Q2000, according to ASTM D 3417.

2.4.7. Dimensional stability test. The dimensional stability test was performed using Q400 TA Universal according to ASTM D1204.

3. Results and discussion

3.1. Moisture content of untreated & treated BPS fibers
The moisture content of untreated and treated BPS fibres in an ambient with room temperature of 25 oC and humidity of 50%. The moisture content for treated fibres was 14.52 % while, for untreated fibres was 8%. Due to alkaline treatment for BPS fibres, the moisture content increases from 8% to 14% (with an increase of 75%). Therefore, the BPS fibres should be dried in oven for 24-48 hrs. before processing.

3.2. Effect of fibers weight percentage in water absorption (WA%) of BPS/ABS composite
The water absorption percentage of BPS/ABS composite as a function of time at different fiber wt.% and time is presented in figure 1. The WA% of BPS/ABS composites after immersion time of 288 hrs. were 1.21, 1.85, 2.42, and 2.27 % at 0, 10, 20, and 30 % BPS fiber weight percentage.

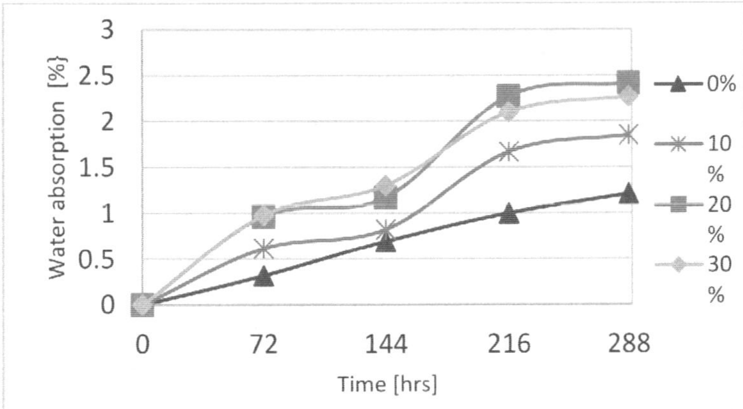

Figure 1. Water absorption as a function of time for BPS/ABS composites.

The results show that the WA% increases with increments of fiber weight percentage wt. %. BPS/ABS composite samples absorb moisture more than ABS matrix sample. This is due to hydrophilic nature of BPS fiber which is responsible for water absorption of composite. However, the increase of voids at fiber - matrix interface influences increasing the water absorption of composites.

3.3. Effect of fiber weight percentage in thickness swelling (TS%) of BPS/ABS composite

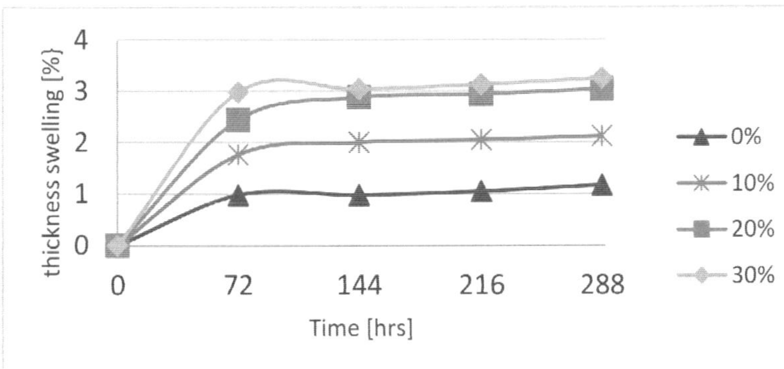

Figure 2. Thickness swelling as function of time for BPS/ABS composites.

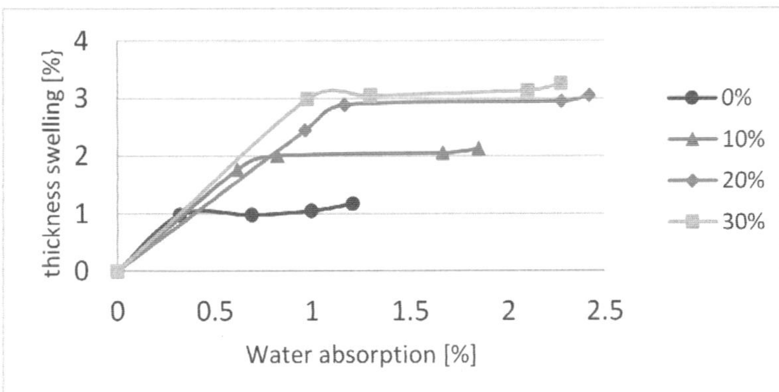

Figure 3. Relationships between water absorption and thickness swelling.

Figure 2 shows the percentage thickness swelling of the TS% of BPS/ABS composite as a function of time at different fibres wt. %. The TS% of BPS/ABS composites after immersion time of 288 hrs., were 1.17, 2.12, 3.04, and 3.25 at 0, 10, 20, and 30% BPS fiber weight percentage. The TS% of BPS/ABS composites increase as the water absorption increases as shown in figure 3. ABS with wt. % of 30% BPS fibers exhibit maximum water absorption and thickness swelling during the whole duration of immersion. Water absorption leads to a build-up of moisture in the fiber cell walls, which then results in fiber swelling. Increasing fiber weight percentage wt. % increases hydrophilic sites which are available for water absorption [11].

3.4. Effect of fiber weight percentage in MFI of BPS/ABS composite

The melt flow index of pure ABS under investigation recorded 20.5 gm/ 10 min, which is concordant with standard MFI of ABS. The effect of fiber weight percentage wt.% on MFI of BPS/ABS composite is presented in figure 4. The MFI of BPS/ABS was 18.5, 17, and 15 gm/ 10 min at 10, 20, and 30% fiber content. The results show decreasing MFI with increasing BPS fiber wt.%. This may be due to BPS fiber misalignment which affects the viscoelasticity dynamics of BPS/ABS melts. Also, the mobility of molecular chains of polymers have been hindered by BPS fibers.

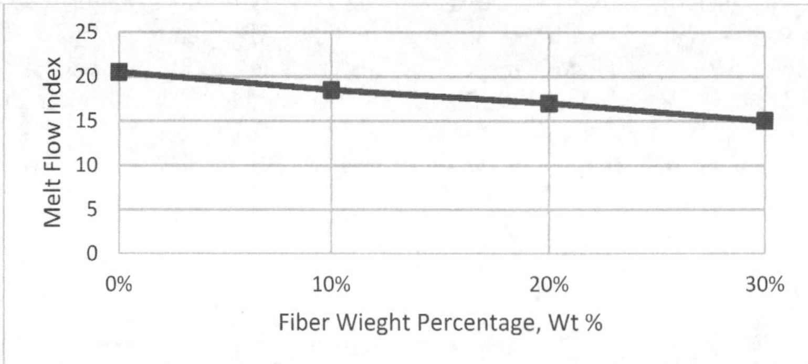

Figure 4. Effect of fiber wt. % on melt flow index MFI of BPS/ABS composite.

3.5. SEM study

SEM micrographs of tensile test fracture surfaces for (BPS/ABS) composites containing 10% fibers weight percentage show uniform fiber distribution and decrease of cavities and voids comparing to 20% and 30% fibers content in figure 5(a). However, as the fiber weight percentage increased to 20% and 30%, voids appeared at the fiber/matrix interface as shown in figures 5(b) and 5(c).

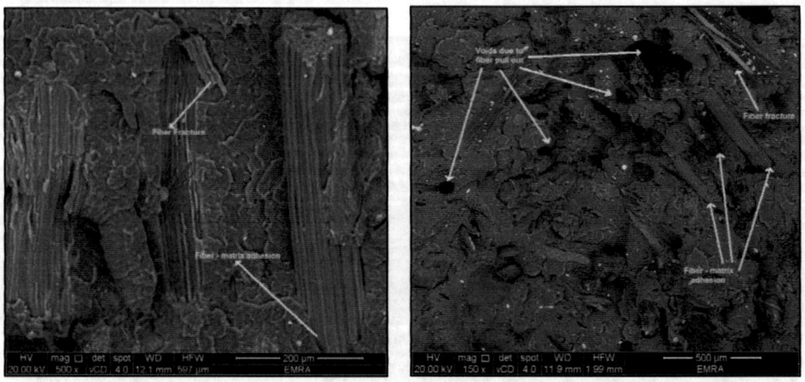

(a) (b)

(c)

Figure 5. The SEM micrograph for composite at fracture surface. (a) 10% BPS/ABS, (b) 20% BPS/ABS, (c) 30% BPS/ABS.

3.6. Effect of fiber weight percentage on thermal stability of BPS/ABS composite

The previous researches indicated that as the total weight residue increases, the thermal stability also increases [9]. Thus, increasing the fiber weight percentage, the total weight residue at 500°C increases dramatically from 2.72% at 0% BPS to 4.1%, 7.2% and 7.29% at 10%, 20% and 30% BPS fiber weight percentage respectively (see figure 6). Corresponding, the thermal stability of BPS/ABS composites increased.

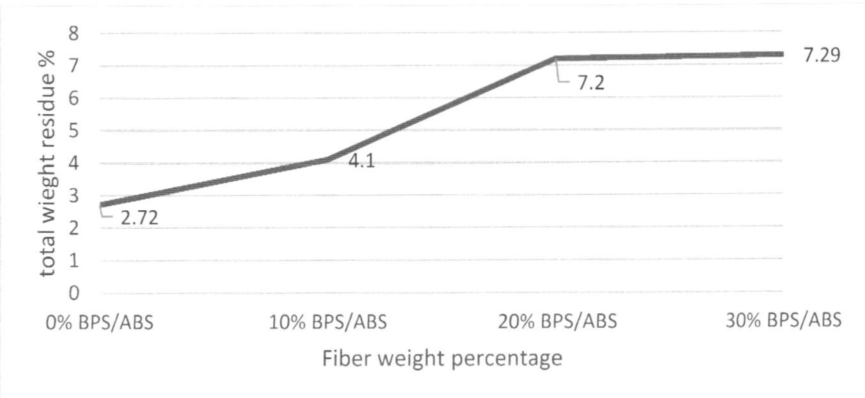

Figure 6. Total weight residue % on 500°C at different fiber weight percentage.

3.7. Effect of fiber weight percentage on Tg of BPS/ABS composite

Figure 7 Shows the effect of fiber weight percentage on Glass transition temperature of BPS/ABS composite. The Tg of pure ABS matrix was 94.8°C. It is observed that glass transition temperature increases with addition of BPS fiber in ABS matrix, also related with decrease in mobility of polymer with the addition of fiber, so Tg is considered as an indicator for interfacial interaction between fiber and matrix. This is an agreement with the loadings of other molecules. Thus, the addition of BPS fiber to ABS matrix increases the Tg of composite to 102.85°C and 103.43°C at 10 and 20% BPS fiber wt.% respectively. The DSC thermogram showed that the glass transition temperature of composites above 20% fiber weight percentage decreased 102°C at 30% fiber weight percentage.

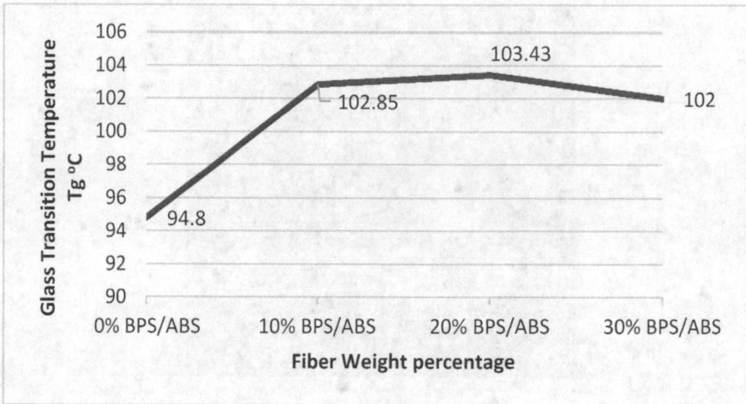

Figure 7. Glass transition temperature Tg at different wt. % fiber weight.

3.8. Effect of fiber weight percentage on dimensional stability of BPS/ABS composite

The coefficient of shrinkage -due to heating at 100°C (24 hrs) in oven- for parallel and transverse directions of injection molding are shown in figures 8(a) and 8(b) respectively. It is noticed that the coefficient of shrinkage (parallel direction of process) were 12.72, 10.399, 9.286, and 7.306% for 0, 10, 20, and 30% fiber weight percentage wt. % respectively. While the coefficient of shrinkage (transverse direction of process) were 23.61%, 20.46%, 19.27%, and 15.54% respectively. Specimens with higher fiber weight percentage wt.% exhibited lower shrinkage. During the injection molding of BPS/ABS composites, most BPS fibers were suspended in molten ABS matrix with its long axis parallel to liquid surface which led to BPS fiber orientation parallel to the finished composite surface. Thus the shrinkage coefficients in parallel direction were less than the others in the transverse direction (fiber orientation). This is an agreement with the work of Shinoj *et al* [11].

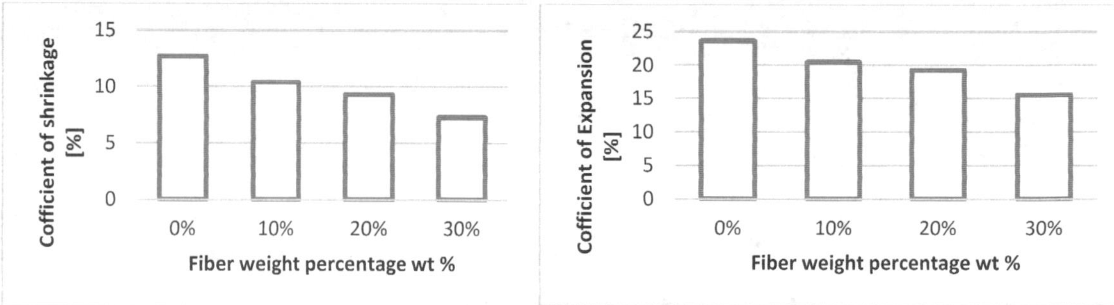

(a) (b)

Figure 8. Coefficient of shrinkage for (a) parallel direction (b) transverse direction of injection molding.

The equivalent shrinkage coefficient of BPS/ABS composite was calculated as an average of linear shrinkage coefficient in both directions and shown in figure 9. As the expansion or shrinkage coefficient decreases, the dimensional stability increases. The addition of BPS fibers in ABS matrix enhances the dimensional stability of ABS composite.

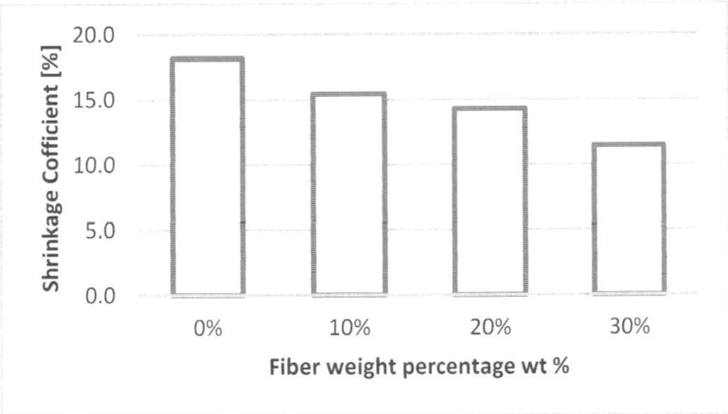

Figure 9. Equivalent shrinkage coefficient of BPS/ABS composite.

4. Conclusion

The following conclusions can be drawn from this work.

- The moisture content for treated BPS fibers of 14.52% while the moisture content of untreated was 8%.
- The MFI of BPS/ABS composite decreased as the fiber wt.% increase, thus the viscosity increased.
- The water absorption WA% and thickness swelling TS% of BPS/ABS composite increased with increasing fiber content from 10% to 30%.
- The addition of BPS fiber in ABS matrix enhances the dimensional stability of ABS composite.
- The thermal stability of BPS/ABS composite increase as the fiber weight percentage increases.
- The dimensional stability increases with the addition of BPS fibers in ABS matrix.

Acknowledgment

Thanks to the faculty of Eengineering (Ain Shams University) and Al-Salam Plastic Company for offering their facilities to complete this experimental work. Also Science and Technology Center of Excellence "STCE" for their support.

References

[1] Sai M K 2016 *Journal of Latest Trends in Engineering and Technology* **6** 129
[2] Dudek P F 2013 *Arch. Metall. Mater.* **58** 1415
[3] Bateni F and Ahmed F 2012 *Thermoplastic Elastomers* 329
[4] Saba N and Tahir M *2014 Polymers* **6** 2247
[5] Abou Hussein S and Sawan M *2010 Journal of Applied Sciences Research* **6** 1116
[6] Hassan H ans ElGebaly 2014 *Journal of Agriculture Research* **3** 592
[7] Reddy N and Yang Y 2015 *Innovative Biofibres from Renewable* 89
[8] Mukhopadhyay S and Fangueiro R 2008 *Journal of Engineered Fibers and Fabrics* 3 592
[9] Bhatnagar R and Gupta G 2015 *International Journal of Scientific & Engineering Research* **6** 49
[10] Elgamsy R *et al* 2018 *Ain Shams Engineering Journal* **9** 3271
[11] Shinoj S *et al* 2010 *Journal of Applied Polymer Science* **117** 1064

Vibration analysis of fiber reinforced composite hydrofoils using finite element method

B-J Joe[1], S-Y Hong[1,4], J-H Song[2] and H-W Kwon[3]

[1]Department of Naval Architecture and Ocean Engineering, Seoul National University, South Korea
[2]Department of Naval Architecture and Ocean Engineering, Chonnam National University, Yeosu, South Korea
[3]Department of Naval Architecture and Ocean Engineering, Koje College, Koje, South Korea

E-mail: syh@snu.ac.kr

Abstract. In this study, the natural responses of fiber reinforced composite hydrofoils are obtained using FEM. It is of great importance to estimate the vibrational responses of fiber reinforced composite hydrofoils correctly due to the flexible characteristics of the material. Modal analysis is performed on a cantilevered hydrofoil which has a cross-section of a modified NACA0009 profile using FEM software Abaqus 2019. Newly developed structured mesh modeling method is used, and the results are compared with the conventional modeling methods and experiment. In the developed mesh model, boundaries of the layers are modified in order to apply a structured mesh. By investigating the stress distribution of the hydrofoil for the bending mode case, it is evident that the developed mesh model results in the most accurate and physical result compared to existing modeling method.

1. Introduction

Fiber reinforced composites, also referred to as Fiber reinforced polymer or plastics (FRP) are currently used in various fields. Due to their high specific strength and resistance to corrosion [1], FRP is thought of as the next generation material which could replace metallic alloys in the field of naval engineering [2,3]. Small vessels such as leisure boats and yachts already put FRP to use as the primary source of structural strength [4]. Innovative designs are starting to be used in the rage of large ships such as commercial bulk carriers and naval vessels [2,3].

Marine propellers provide thrust and enable vessels to navigate through its course. Since propellers are crucial for ships to provide their original purpose, an adaptation of FRP has been delayed due to uncertainties. Until recent registrational changes by the International Maritime Organization (IMO), the main concern for propeller was to provide the appropriate thrust. However, new regulations require propellers to meet noise standards because sound radiated underwater may cause severe damage to the ocean environment.

Flexible properties of FRP provides the propeller the ability to adaptively change shape according to the nearby flow, which in result, reducing noise radiated by the propeller [5,6]. It is of common understanding that fluid-structure interaction (FSI) analysis is required to fully estimate the performances of fiber reinforced composite marine propellers. With a focus on acoustical performances of the FRP propellers, it is crucial to fully understand the structural characteristics and vibrational characteristics of the propeller due to external forces [7]. Thus, the main concern is to correctly estimate the natural responses of FRP propellers while correctly reflecting the structural and material properties of FRP in the simulation.

Three-dimensional solid elements were used to include the effect of shear deformation in the analysis instead of plate or shell elements. Laminated composite theory based 2D elements commonly assume in-plane displacements and corresponding shear deformations [8]. Due to this assumption, 2D elements are unable to incorporate the appropriate amount of shear deformation effects for thick structures such as hydrofoils and marine propellers. Solid elements can be divided into structured and unstructured mesh. For laminated composites, structured mesh stacked in the thickness direction results in the most accurate displacements. [9]

Fiber reinforced composite hydrofoils were selected for analysis to incorporate simplicity of the analysis and modeling of the structure. The goal is to simulate the natural responses of these hydrofoils using 3D solid elements, which can be used similarly on marine propellers in future work. The composite hydrofoil models had complex lamination boundaries where structured mesh cannot be applied. To accurately simulate the responses of the hydrofoil, a new laminated composite structured mesh modeling method was developed in order to apply the structured mesh on composite hydrofoils. The FEM software Abaqus 2019 was used to perform the model analysis on the composite hydrofoils.

2. Methods

2.1. Geometry and lamination properties of hydrofoil

All information regarding the geometry and the lamination properties of the hydrofoil used are from experimental researches present in current literature [10-13]. Figure 1 shows the sectional geometry of the hydrofoil at plane z = 0 mm.

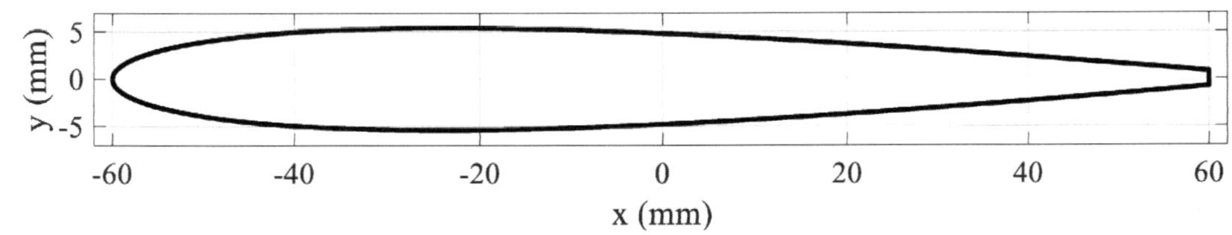

Figure 1. Sectional geometry of the hydrofoil at z = 0 mm. The section has the profile of modified NACA0009 and the chord length is 120 mm.

The profile was modified NACA0009 which can be expressed by equation (1). The modified profile had the same constants for variable x except for the constant for x to the power of four, which was changed from -0.1015 to -0.08890.

$$y = 5t(0.2969x^{0.5} - 0.1260x - 0.3516x^2 + 0.2843x^3 - 0.08890x^4) \quad (1)$$

Variable t represents the thickness to chord length ratio which was 0.09 for NACA0009 profile. Variables x and y are nondimensionalized using the chord length. The profile in figure 1 is the plot of equation (1) shifted in the negative x-direction by half of the chord length. The positive x-direction indicates the trailing-edge direction while the negative x-direction is the leading-edge direction. The detailed profile of the hydrofoil is shown in figure 2.

The hydrofoil had 110 mm of root section where the profile was the same. This region was constructed in order to apply the boundary conditions. The regions where the boundary conditions were applied can be seen in figure 3. The rest of the profile had 300 mm in length in the positive z-direction which is the spanwise direction. The rest of the hydrofoil had a geometry where the chord length linearly decreased from 120 mm to 60 mm with the mid-chord line remaining straight.

Figure 2. The geometry of hydrofoil with all lamination information included.

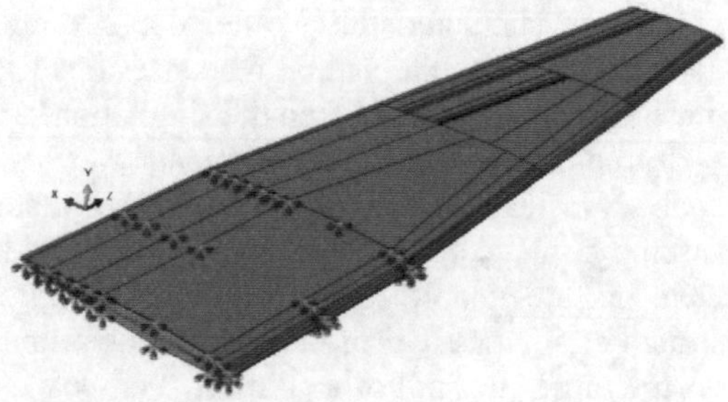

Figure 3. Locations of fixed boundary condition application regions.

Three different lamination configurations were considered. The stacking sequences of the three hydrofoils in question were $[(0^o/90^o)_g, (X)_{5c}, (0^o/90^o)_{2g}, (X)_{4c}]_s$ with values of X as 0°, 30° and -30° for each structure. The reference direction was the span direction which is the same as the z-axis. Subscript 'g' represents glass fiber and epoxy composite layer, whereas subscript 'c' represents carbon fiber and epoxy composite layer. The number with the subscript represents the number of times each sequence is repeated. The last subscript 's' indicates that the plies were set to be laminated symmetrically with respect to the center plane which is the same as the zx-plane. The center of the hydrofoil was filled with glass-mat fabric to construct a core for the hydrofoil. The outermost surface of the hydrofoil was finished with basket weaved glass-epoxy layer. The details of the lamination boundaries are shown in figure 4. Color represents the same composite material and the structured mesh was constructed using the geometry in figure 4(b).

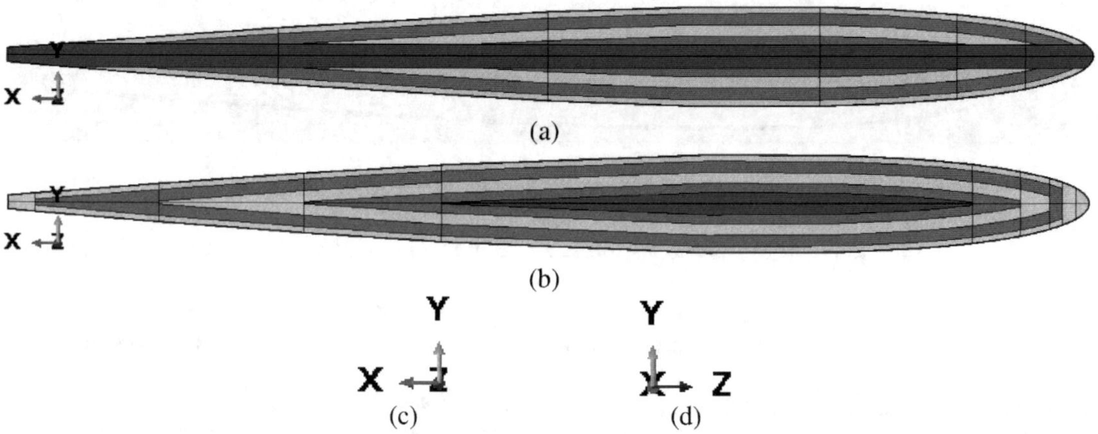

Figure 4. Lamination boundaries and the structural, material coordinated of the hydrofoil. (a) The original geometry of the hydrofoil. (b) Developed structured mesh modeling method applied geometry. (c) Structural coordinated. (d) Material coordinates.

A new laminated composite hydrofoil modeling method was developed in order to apply structured mesh because it is impossible to apply a structured mesh on complicated inner boundaries of the original geometry. The proposed modeling method modifies the lamination boundaries according to two principles. One, only modify regions where structured mesh is inapplicable after partitioning, and two, keep main boundaries in thickness direction and modify only if necessary. Using this developed modeling method, the structured mesh was applicable on the composite hydrofoil.

The material properties of the fiber reinforced composite materials used are given in table 1.

IOP Conf. Series: Materials Science and Engineering **634** (2019) 012024 doi:10.1088/1757-899X/634/1/012024

Table 1. Fiber reinforced composite material properties.

	ρ (kg/m^3)	E_{11}, E_{22}, E_{33} (GPa)	ν G_{12}, G_{23}, G_{31} (GPa)
Carbon-Epoxy [0°]	1590	117.8, 13.4, 13.4	0.253.9, 3.9, 3.9
Glass-Epoxy [0°/90°]	1750	15.1, 19.3, 15.1	0.034.2, 4.2, 4.2
Glass-Basket	1660	15.3, 12.8, 15.3	0.133.0, 3.0, 3.0
Glass-Mat	1370	6.8, 5.0, 5.0	0.302.5, 2.5, 2.5

Anisotropic material properties are defined with respect to the material axes is figure 4(d). Only one shear modulus in the xy-direction was provided in current literature [10-13], thus the rest of the shear modulus was set equal to that of the value in the xy-direction.

2.2. Conditions of FEM modal analysis

Fixed boundary conditions were applied on the root section of length 110 mm as shown in figure 3. This was done in order to mimic the experimental boundary conditions where the root section was camped mechanically using a contraption which had the same profile as the hydrofoil. The rest of the nodes were set to be free. Three eigenvalues and the corresponding eigenfunctions in the range between 0 Hz and 1 kHz were evaluated using the FEM software Abaqus 2019. This range was based on the experimental results where the natural frequencies were within that limit. Lanczos eigensolver was used in the process.

$$[M]\{\ddot{x}\} + [K]\{x\} = \{f\} \tag{2}$$

The modal analysis obtains the eigen values and the eigen vectors of the governing equation of the structure. The governing equation of structured can be expressed as equation (2) where M and K represents the global mass and stiffness matrix respectively and x represents the node displacement vector.

Two kinds of FEM mesh were used. First, unstructured mesh using continuum three-dimensional ten nodes, quadratic tetrahedral elements (C3D10) was constructed. The hydrofoil geometry required 320,169 elements to mesh the whole structure. The geometry used is described in figure 4(a). The second mesh was the structured mesh where the new modeling method was applied in order to construct a structured mesh on pointy shaped regions. Using the continuum three-dimensional twenty nodes, quadratic hexahedral element (C3D20), the total number of elements required for the structured mesh was 436,840. The geometry used to create a structured mesh is shown in figure 4(b). Three eigenfrequencies for each type of mesh and each type of lamination direction was obtained.

3. Results of Hydrofoil modal analysis

Three eigenfrequencies in the range from 0 Hz to 1 kHz were obtained. Modal analysis was performed for each unstructured and structured mesh, and for each lamination sequence, X = 0°, 30°, -30°. The simulation results were compared with the experimental results of the same hydrofoil and same lamination configuration [13]. The results were also compared to that of the same reference were the authors performed a FEM frequency analysis. 3D solid elements and multiple plies per element method were used in the reference [11]. In contrast, in this study, one or more elements per ply method were used. The results of the modal analysis compared to that of the reference results are in table 2.

It is shown in table 2 that using FEM modal analysis method, the eigenfrequency of the FRP hydrofoil can be estimated within a reasonable error boundary. Even though the reference used multiple plies per element method [13], the eigenfrequency which can be thought of as a global response can be obtained with small error. However, observing the result in a local point of view, multiple plies per element method revealed its limits.

Table 2. Modal analysis results and comparison with that of the reference.

Case		**1**			**2**			**3**		
		X = 0⁰			**X = 30⁰**			**X = -30⁰**		
	Mode No.	FEM (Hz)	Exp. (Hz)	Error (%)	FEM (Hz)	Exp. (Hz)	Error (%)	FEM (Hz)	Exp. (Hz)	Error (%)
Ref. [a]	1	118	120	-1.67	72	74	-2.70	78	79	-1.27
Unstructured		125.52		4.60	76.754		3.72	83.388		5.55
Developed Model		121.85		1.54	74.052		0.07	80.425		1.80
Ref. [a]	2	394	428	-7.94	286	305	-6.23	311	338	-7.99
Unstructured		423.85		-0.97	315.87		3.56	338.87		0.26
Developed Model		407.45		-4.80	308.79		1.24	330.79		-2.13
Ref. [a]	3	464	482	-3.73	594	611	-2.78	598	591	1.18
Unstructured		500.82		3.90	633.48		3.68	643.67		8.91
Developed Model		494.44		2.58	613.84		0.46	626.37		5.98

[a] Ref. here indicates the FEM frequency analysis results of that of the reference [13].

Observing the mode shape of the hydrofoil, the deviation from the experimental results when one element represents multiple plies can be seen. Figure 5 shows the eigenmode of the hydrofoil with X = 30° lamination sequence. The results were compared with reference FEM results and reference experiment results [13]. The color legends are neglected because the exact values are of insignificance.

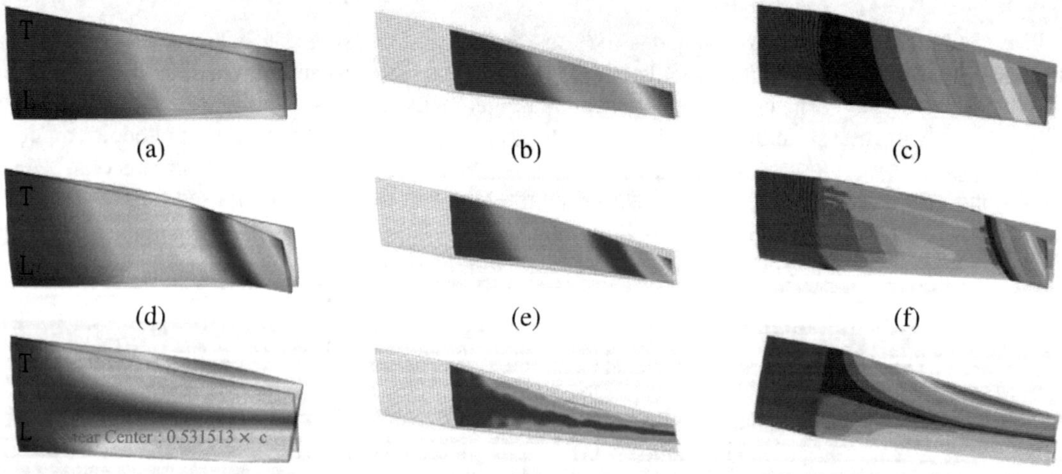

Figure 5. Mode shapes of hydrofoil with X = 30° lamination sequence. 'T' represents the trailing edge and 'L' represents the leading edge. (a), (b), (c) Modeshape of the first mode. (d), (e), (f) Mode shape of the second mode. (g), (h), (i) Third mode shape. The first column parts (a), (d) (g) Results of this paper using the developed model. (b), (e), (h) Reference experiment results [13]. (c), (f), (i) FEM results of the reference [13].

By analyzing the results shown in figure 5, the fact that the first mode is the first bending mode, the second mode is the second bending mode and the third mode corresponds to the first twisting mode is the same. Since exact values, which are the displacements for each node was not given in the reference [13]. Thus, quantitative comparisons between the FEM results and the results of that of the experiments were not possible. However, performing a qualitative comparison between the FEM results and the experimental results, it was found that by using one or more elements per ply method, the location of the shear center of the first twisting mode was correctly estimated. While using multiple plies per element method, which can be found in the reference FEM results, the shear center was not correctly located. Comparing figure 5(h) with figure 5(i), it can be easily picked out that the experimental results in figure 5(h) shows that the shear center is at 53.15% chord length location from the leading edge, which is leaned towards the trailing edge while the reference FEM results in figure 5(i) the shear center is located towards the leading edge.

In order to find the differences between using an unstructured mesh and the structured mesh from the developed modeling method, the stress distributions were analyzed for a bending mode case. Figure 6 shows the interlaminar shear stress (ILSS) for an unstructured mesh and the developed modeling method. The shear stress, S23 is in structural coordinates. The lamination sequence is when X = 0°. The ILSS was plotted instead of all the other stresses because for laminated composites, the estimation of shear stress between individual layers is one of the important objectives of FEM analysis. From ILSS distribution information, further analysis is possible such as fatigue analysis. It is clearly seen in this figure, that the structured mesh using the new modeling method results in more physically reliable stress distribution. While the unstructured mesh shows local peaks in the stress where it is not plausible.

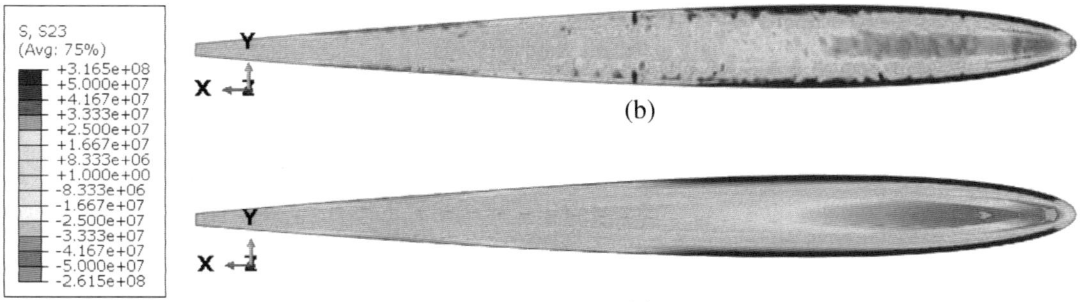

(a) (c)

Figure 6. Interlaminar shear stress (ILSS) plot of the hydrofoil for first bending mode. The lamination sequence is when X = 0°. The location of the section is at the root where z = 110 mm. (a) Color legend which is identical for parts (b) and (c). (b) ILSS plot for an unstructured mesh case. (c) ILSS plot for a structured mesh using developed modeling method.

4. Conclusion

Modal analysis was performed on hydrofoils which had a modified NACA0009 sectional profile. The natural frequencies were compared between the FEM results and experimental results. FEM frequency analysis shows that it can estimate the natural responses within a small error range. Whether multiple plies per element method was used, or one or more elements per ply method was used, did not show any significant difference. In addition, whether an unstructured mesh or a structured mesh was used did not show any meaningful variations in natural frequency.

On the other hand, observing the eigenmode FEM analysis it can be concluded that the one or more elements per ply method must be used to correctly capture the local responses of the FRP hydrofoils. Analyzing the results of the interlaminar shear stress for bending condition, it is evident that a structured mesh combined with laminated composite modeling method is required for an accurate assessment of stress in FRP hydrofoil. The main downside of using FRP materials when constructing a marine propeller is the uncertainty of material failure. Thus, an accurate estimate of the internal stress distribution from the structural simulation is mandatory, and the structured mesh is the only type which can deliver correct results of the stress.

In conclusion, a structured mesh, together with laminated composite modeling method must be applied in order to correctly simulate FRP hydrofoils. Using the developed modeling method, a modified lamination geometry was constructed in order to apply a structured mesh. For future work, it may be required to develop a mesh model which can consider all the lamination information without any simplification.

Acknowledgments
This research was funded by the Education and Research Center for Creative Offshore Plant Engineers. Also, supported by the Future Submarine Low Noise Laboratory.

References

[1] Carlsson L A, Donald F A and Pipes R B 2014 *Experimental Characterization of Advanced Composite Materials 4th edi.* (Boca Raton: CRC Press)
[2] Mouritz A P, Gellert E, Burchill P and Challis K 2001 Review of advanced composite structures for naval ships and submarines *Composite Struc.* **53** 21-42
[3] Koko T S, Shahin K O, Akpan U O and Norwood M E 2012 *Review of composite p propeller developments and strategy for modeling composite propellers using PVAST* (Canada)
[4] Baley C, Lan M, Davies P and Cartié D 2015 Porosity in ocean racing yacht composites: A review *Appl. Composite Materials* **22** 13-28
[5] Raj S S and Reddy P 2014 Bend-twist coupling and its effect on cavitation inception of composite marine propeller *Int. J. Mech. Eng. and Tech.* **5** 306-14
[6] He X D, Hong Y and Wang R G 2012 Hydroelastic optimization of a composite marine propeller in a non-uniform wake *Ocean Eng.* **39** 14-23
[7] Paik B G, Kim G D, Kim K Y, Seol H S, Hyun B S, Lee S G and Jung Y R 2013 Investigation on the performance characteristics of the flexible propellers *Ocean Eng.* **73** 139-48
[8] Kreja I 2011 A literature review on computational models for laminated composite and sandwich panels *Open Engineering* **1** 59-80
[9] Hu T 2017 Optimizing the sandwich composite structure in the cantilever beam *Mechanical Engineering* **7** 127-43
[10] Zarruk G A, Brandner P A, Pearce B W and Phillips A W 2014 Experimental study of the steady fluid–structure interaction of flexible hydrofoils *J. Fluids and Structures* **51** 326-43
[11] Ibrahim M E, Phillips A W, Ditchburn R J and Wang C H 2014 Nondestructive evaluation of mechanically loaded advanced marine composite structures *Advanced Materials Research* **891** 594-9
[12] Herath M T, Prusty B G, Phillips A W and John N S 2017 Structural strength and laminate optimization of self-twisting composite hydrofoils using a genetic algorithm *Composite Struc.* **176** 359-78
[13] Young Y L, Garg N, Brandner P A, Pearce B W, Butler D, Clarke D and Phillips A W 2018 Load-dependent bend-twist coupling effects on the steady-state hydroelastic response of composite hydrofoils *Composite Struc.* **189** 398-418

IOP Conf. Series: Materials Science and Engineering **634** (2019) 012025 doi:10.1088/1757-899X/634/1/012025

Dynamic response analysis of polyoxymethylene hydrofoils using the hybrid pitch mode FSI method

W-S Jang[1], W-S Choi[1], S-Y Hong[1,4], J-H Song[2] and H-W Kwon[3]

[1]Department of Naval Architecture and Ocean Engineering, Seoul National University, Seoul, South Korea
[2]Department of Naval Architecture and Ocean Engineering, Chonnam National University, Yeosu, South Korea
[3]Department of Naval Architecture and Ocean Engineering, Koje College, Koje, South Korea

E-mail: syh@snu.ac.kr

Abstract. In this paper, the objective is to analyze the dynamic responses of 2D pitching polyoxymethylene (POM) hydrofoils in the uniform flow using the hybrid coupling Fluid-Structure Interaction (FSI) method, which takes account of added mass effects and show its robustness by comparing with the experiments. First, FSI analysis using the empirical formula of hydrodynamic forces around hydrofoils was performed and validated by comparing the analysis results of the weak coupling FSI method and the hybrid coupling FSI method with the analytic solution. Next, The FSI method with Computational Fluid Dynamics (CFD) was implemented in the framework of OpenFOAM, the open-source CFD code. Flow analysis and structural analysis were performed for every time step in the calculations. Here, the flow analysis was performed using Large Eddy Simulation (LES) scheme and the structural analysis was performed by solving the equation which considered the pitch mode of deformable hydrofoils. Dynamic responses of 2D NACA0015 POM hydrofoils using the hybrid coupling FSI method well matched with the experimental results qualitatively, where dynamic responses without the hybrid coupling FSI method showed numerical instability. It is shown that the hybrid coupling FSI method considering added mass effects should be considered for analysis of deformable hydrofoils.

1. Introduction

Most ships and submarines generally use rigid propellers made of Nickel-Aluminum-Bronze (NAB) alloy or stainless steel. However, composite propellers and hydrofoils have been a very active area of research recently. There are many advantages of composite materials for propellers and hydrofoils, such as high strength to weight, high stiffness to weight ratios, high Cavitation-Inception-Speed (CIS) and low subcavitating noises [1,2].

Computational Fluid Dynamics (CFD) methods are generally used for predicting performances of rigid hydrofoils. However, it is unable to predict composite hydrofoils with CFD analysis because they suffer from large displacements which leads to instantaneous variations of the flow field. Hence, Fluid-Structure Interaction (FSI) analysis is needed to predict dynamic responses of composite hydrofoils taking account of their large displacements. Fluid-structure interactions represent interactions among hydrodynamic forces, elastic forces, and inertia forces.

Numeric procedures to solve these FSI problems are classified into two approaches, the monolithic and the partitioned approach [3]. The monolithic approach regards coupled fluid-structure system as a whole, solving them simultaneously. Although it has the advantage of ensuring stability and accelerating convergence, order differences in fluid and structural properties may result in the ill-conditioned matrix, and most of all, it is a great challenge to implement completely new analyzing code [4]. On the other hand, the partitioned approach, also known as the staggered approach, solves fluid domain and structure domain sequentially and each result transfers to the other domain as boundary conditions [5]. This approach is not mathematically rigorous or stable compared to the monolithic approach. However, the partitioned approach is still actively used as FSI analysis approach because it has a great advantage of being able to utilize 'legacy' codes used in each area [3].

The partitioned approach is also classified into two coupling methods, the weak (loose) coupling method and the strong (tight) coupling method. The difference between the weak coupling method and the strong coupling method is whether or not there are iterations in the coupling process. The weak coupling method, a coupling method without iterations, takes less computing time while it does not accurately solve situations with large displacements and large added mass. The strong coupling method, on the contrary, a coupling method with iterations, predicts responses more accurately than the weak coupling method while it takes a lot of computation time.

Various attempts are being researched to take large displacements and large added mass effects into account with relatively less computation time. Astorino *et al*, Fernandez *et al* developed a semi-implicit coupling method implementing robin boundary conditions [6,7]. Young and Chae *et al* predicted responses of flexible propellers and hydrofoils using the hybrid coupling method [1,8.9].

In this paper, dynamic responses of 2D pitching polyoxymethylene (POM) hydrofoils in water, which is the case where substantial added mass effects are expected, were predicted using the hybrid coupling FSI method. The simple hybrid coupling FSI analysis was performed using the empirical formula of hydrodynamic forces to validate the hybrid coupling FSI analysis. Next, the hybrid coupling FSI analysis using CFD was performed to predict dynamic responses of pitching POM hydrofoils.

2. Methods

Two-dimensional symmetric hydrofoils (see figure 1) were used for analyses where c denotes the chord length of the hydrofoil and b $(= c/2)$ denotes the mid-chord length. The elastic axis, defined as the locus of shear centers, is positioned at distance ab $(= ac/2)$ from the mid-chord. Moment and pitch angle θ have clockwise positive direction.

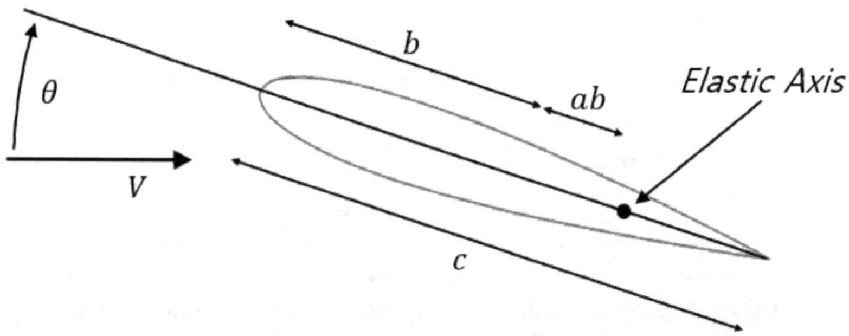

Figure 1. Two-dimensional symmetric hydrofoil undergoing pitch motion.

2.1. Flow analysis

Consider a 2D hydrofoil undergoing pitch motion with frequency ω in the uniform flow. The solution for the moment about the elastic axis is written as in equation (1) [10].

$$M_T = \pi\rho b^2\left[-Vb(0.5-a)\dot{\theta} - b^2(0.125+a^2)\ddot{\theta}\right] + 2\pi\rho V b^2(a+0.5)C(k)\left[V\theta + b(0.5-a)\dot{\theta}\right] \tag{1}$$

where ρ and V denote fluid density and uniform flow velocity each and k $(=\omega b/V)$ denotes reduced frequency. $C(k)$ is Theodorsen function and it is written as equation (2). Theodorsen function represents a relation between quasi-steady and unsteady hydrodynamic forces of hydrofoils undergoing pitch motion at reduced frequency k. Length ratio a, related with elastic axis, contributes to the elasticity because hydrodynamic elastic force $2\pi\rho Vb^2(a+0.5)C(k)V\theta$ is function of a.

$$C(k)=\frac{H_1^2(k)}{H_1^2(k)+iH_0^2(k)} \tag{2}$$

$H_n^{(2)}(k)$ denote Hankel functions of the second kind.

Equation (1) is obtained by the potential flow assumption. According to potential flow theory, flow is incompressible, inviscid and irrotational. However, there are viscous effects like boundary layers in practice and they should be considered to predict dynamic responses accurately. Thus, hydrodynamic forces should be obtained by using CFD because the forces using CFD include forces obtained by equation (1) and also viscous forces.

Without CFD, equation (3) represents an empirical formula for the hydrodynamic force of the forced oscillating NACA 0009 hydrofoils in the uniform flow [11].

$$M_m=-(J_m\ddot{\theta}+\mu_m\dot{\theta}+k_m\theta) \tag{3}$$

Where $J_m=0.125\rho\pi b^4s$, $\mu_m=0.84\rho Vb^3sk^{-0.4}$ $for\ k\leq 4$, $\mu_m=0.08\rho Vb^3sk^{0.6}$ $for\ k\geq 12$, $k_m=2\rho V^2b^2s(0.18k-1.6)$ $for\ k\leq 4$, $k_m=2\rho V^2b^2s(0.13k-1.8)$ $for\ k\geq 12$.
here, s denotes span.

2.2. Structure analysis

Structure analysis considered only pitch motion of hydrofoils. Thus, structure analysis can be performed by equation (3) which is a simple 1-degree-of-freedom equation of motion about pitch angle θ of hydrofoils. I, C and K denote moment of inertia, structural damping parameter and structural stiffness respectively.

$$I\ddot{\theta}+C\dot{\theta}+K\theta=\mathrm{F} \tag{4}$$

2.3. Partitioned hybrid coupling FSI method

In the partitioned approach, there are two boundary conditions for FSI analysis. First, fluid mesh velocity should be equal to the time derivative of structural displacement on the fluid-structure interface. The second boundary condition is that stress at each domain should be equal on the fluid-structure interface. Using the second boundary condition, equation (5) is obtained by coupling equation (4) and the hydrodynamic force of hydrofoils.

$$I\ddot{\theta}+C\dot{\theta}+K\theta=\mathrm{M_{CFD}} \tag{5}$$

M_{CFD} can be replaced by M_M which is the empirical formula of hydrodynamic forces without CFD analysis [11]. If FSI analysis is performed at each time step in this condition, this is the weak coupling approach.

M_{CFD} includes the potential hydrodynamic force in the form of equation (1), which is functions of $\ddot{\theta}$, $\dot{\theta}$, θ. If their effects are reflected in the structure analysis matrix, the dynamic responses of hydrofoils between old time step and new time step can be more accurately predicted during the FSI analysis process. The following equation (6) considers the effects of equation (1) as added mass, added damping and added stiffness and this FSI analysis is called the hybrid coupling FSI method.

$$(I+I_a)\ddot{\theta}+(C+C_a)\dot{\theta}+(K+K_a)\theta=\mathrm{M_{CFD}}-M_T \tag{6}$$

Where $I_a=\pi\rho b^4(0.125+a^2)$, $C_a=\pi\rho b^3V(0.5-a)$ (1- $(2a+1)C(k)$), $K_a=-2\pi\rho V^2b^2(a+0.5)\mathrm{C(k)}$.

If M_{CFD} is replaced by M_M, all terms of equation (6) can be described as a function of $\ddot{\theta}, \dot{\theta}, \theta$. Therefore, it is possible to obtain an exact solution without numerical analysis.

3. Result and discussion

3.1. Non-CFD FSI analysis

FSI analysis predicting dynamic responses of NACA 0009 hydrofoils of the pitching frequency ω were performed and validated. The parameters of the analysis are given in table 1. Runge-Kutta 4th order scheme was used for the time derivative scheme and flow analysis was performed by the empirical formula [11].

Table 1. Parameters for Non-CFD FSI analysis.

Parameter	Rigid hydrofoil	Deformable hydrofoil
I (kg·m)	1.0×10^{-2}	1.428×10^{-3}
C (kg·m /s)	0.253	0.096
K (kg·m/s^2)	1000	1000
c (m)	0.1	0.1
b (m)	0.05	0.05
a (-)	0	0
V (m/s)	5	5
k (-)	2.63×10^{-2}	4.84×10^{-2}
$\omega\ (rad/s)$	2.63	4.84
$\rho\ (kg/m^3)$	1000	1000
M_{forced} *(Nm)*	34.9	34.9
Time step (s)	5.0×10^{-3}	2.70×10^{-3}

This analysis was aimed at forced pitching hydrofoils. Thus, there was an additional moment term $M_{forced}\times sin(\omega t)$ at the right hand side of equations (5) and (6). The weak coupling FSI method and the hybrid coupling FSI method were used for predicting dynamic responses of two different materials and compared with the analytic solution.

The results of non-CFD FSI analysis were given as figure 2. For rigid hydrofoils which have high structural density, both coupling methods showed the same results with the analytic solution. On the other hand, dynamic responses of deformable hydrofoils varied depending on the coupling method. While the response of the weak coupling FSI method diverged, the response of the hybrid coupling FSI method converged and was shown to be the same as the analytic solution.

In the case of deformable hydrofoils, the structural density is relatively low and fluid around them has relatively high density making mass ratio $\rho_{\text{fluid}}/\rho_{\text{structure}}$ close to 1. Thus, the effects of added mass, added damping, added stiffness by the flow field around the hydrofoil become relatively significant. Hence, the hybrid coupling FSI method that considered those added effects can predict dynamic responses of deformable hydrofoils. It is difficult to predict dynamic responses of deformable hydrofoils with the weak coupling FSI method that ignores the added effects. If the weak coupling FSI method is used to predict dynamic responses of deformable hydrofoils, relatively small time step is required, which is inefficient in the respect of computation time.

The results showed the same results with Young's results [12].

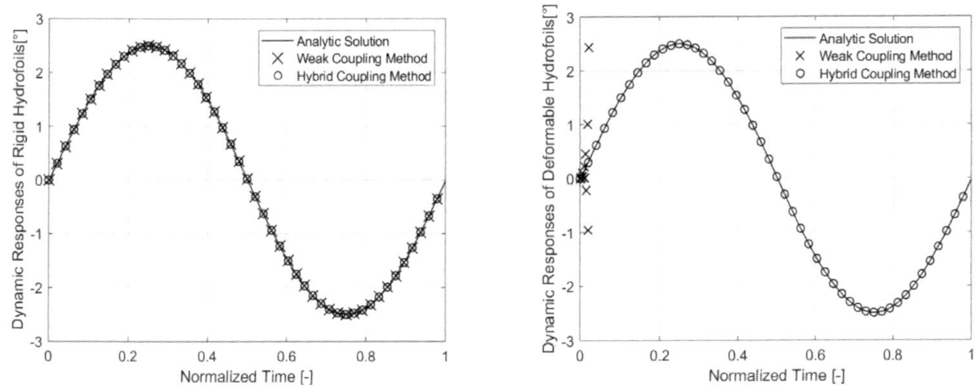

(a) (b)

Figure 2. Dynamic response of hydrofoils oscillating with reduced frequency k with the weak coupling FSI method, the hybrid coupling FSI method and the analytic solution (a) rigid hydrofoils (b) deformable hydrofoils

3.2. FSI analysis

Dynamic responses of POM NACA0015 hydrofoils undergoing only pitch motion in the uniform flow were obtained through FSI analysis using CFD. The FSI analysis was implemented in the framework of OpenFOAM, an open source CFD software. Flow analysis was performed by OpenFOAM's own solver and structural analysis was performed by the same method as in *subsection 3.1*. The hybrid coupling FSI method was used to take the added effects into account and POM which has low density was used for hydrofoils' material to observe dynamic responses of POM hydrofoils and to compare the responses with the results of Chae [8]. Fluid mesh for the FSI analysis was given in figure 3. To precisely obtain the transient characteristics of the hydrodynamic forces, the Large Eddy Simulation(LES) method was selected for the turbulence modeling method and the y+ value of the fluid mesh around the hydrofoil was set to 1 or less. Detailed parameters were given in table 2.

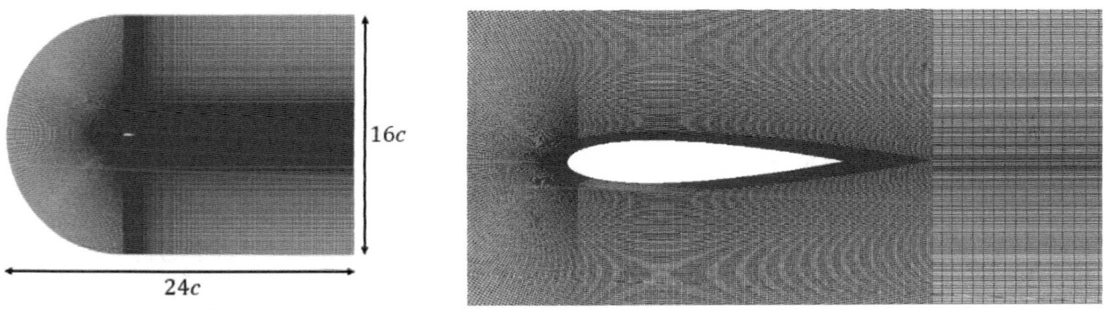

(a) (b)

Figure 3. Fluid mesh for the FSI analysis (a) Full mesh (b) Mesh around NACA 0015.

The results of the FSI analysis compared to the experimental result by Chae was given by figure 4 [8]. Results of the FSI analysis using the weak coupling FSI method diverged. Thus, it is difficult to obtain dynamic responses of POM hydrofoils using the weak coupling FSI method and figure 4 contained only the analysis result by the hybrid coupling FSI method. The result by the hybrid coupling FSI method takes approximately 3 days for parallel calculation using 5 CPUs. The result showed that the order of responses' amplitude and peak frequency of the FSI analysis is equal to the experimental result. Contrary to FSI results, the experimental results had two frequency peaks because the experiments considered motions of all modes. Therefore, the FSI analysis should consider the other modes to obtain more accurate dynamic response predictions of POM hydrofoils.

Table 2. Parameters for the FSI analysis.

Parameter	POM hydrofoil
$I\ (kg{\cdot}m)$	1.28×10^{-4}
$C\ (kg{\cdot}m\ /s)$	1.26×10^{-2}
$K\ (kg{\cdot}m/s^2)$	769.88
$c\ (m)$	0.1
$a\ (\text{-})$	-0.28
$AOA\ (°)$	8
$\rho\ (kg/m^3)$	1000
$Time\ step\ (s)$	1.0×10^{-5}

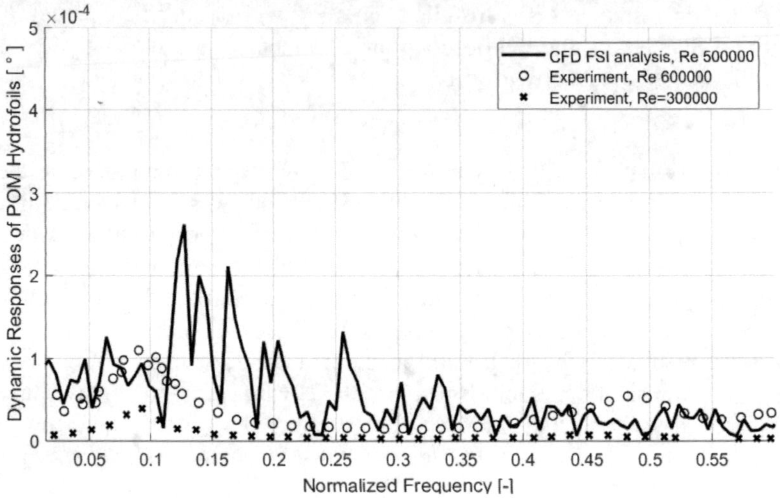

Figure 4. Dynamic responses of POM hydrofoils obtained by the FSI analysis and experiments.

4. Conclusion

To predict dynamic responses of composite hydrofoils, Fluid-structure interaction(FSI) analysis should be performed and in particular, various FSI methods taking account of added mass effects are being researched. In this paper, dynamic responses of 2D pitching POM hydrofoils in water, which is the case where substantial added mass effects are expected, were predicted using the hybrid coupling FSI method.

Non-CFD FSI analysis based on an empirical formula to predict dynamic responses of NACA 0009 hydrofoils of the pitching frequency ω in the uniform flow were performed and validated. Dynamic responses of rigid hydrofoils can be obtained by both the weak coupling FSI method and the hybrid coupling FSI method. However, dynamic responses of deformable hydrofoils can be obtained only by the hybrid coupling FSI method because added mass, added damping, added stiffness have significant effects on the dynamic responses.

Dynamic pitching responses of POM NACA0015 hydrofoils in the uniform flow were obtained through FSI analysis using the hybrid coupling FSI method and compared to experimental results. The order of response amplitude and the peak frequency of FSI analysis are equal to the response of experiments. On the contrary to the weak coupling FSI method, FSI analysis using the hybrid coupling FSI method can predict the tendency of dynamic responses of POM hydrofoils. However, there is a limit that FSI analysis considering only the pitch motion has difficulty in predicting accurate dynamic responses of POM hydrofoils in practice.

For future work, the other modes should be considered in order to obtain accurate dynamic responses of POM hydrofoils. Furthermore, other natural modes can be considered to obtain more accurate dynamic responses.

Acknowledgments
This research was funded by the Education and Research Center for Creative Offshore Plant Engineers. Also, supported by Future Submarine Low Noise Laboratory.

References

[1] Young Y L 2008 Fluid-structure interaction analysis of flexible composite marine propellers *J. Fluid and Struc.* **24** 799-818
[2] Lee H S 2017 A study on the BEM-FEM based hydroelastic analysis of composite marine propellers (Seoul National University)
[3] Hou G, Wang J and Layton A 2012 Numerical methods for fluid-structure interaction: A review *Community of Comp. Phys* **12** 337-77
[4] Hubner B, Walhorn E and Dinkler D 2004 A monolithic approach to fluid-structure interaction using space-time finite elements *Comp. Methods Appl. Mech. Eng.***193** 2087-104
[5] Campbell R L 2010 Fluid-structure interaction and inverse design simulations for flexible turbomachinery (Pennsylvania State University)
[6] AStrorino M, Chouly F and Fernandez M A 2009 Robin based semi-implicit coupling in fluid-structure interaction: stability analysis and numerics *Society for Industrial and Appl. Math.* **31** 4041-65
[7] Fernandez M A, Mullaert J and Vidrascu M 2013 Explicit robin-neumann schemes for the coupling of incompressible fluids with thin walled structures *Comp. Methods of Appl. Mech. Eng.* **267** 566-593
[8] Chae E J 2015 Dynamic response and stability of flexible hydrofoils in incompressible and viscous flow (University Michigan)
[9] Chae E J, Akcabay D T, Lelong A, Astolfi J A and Young Y L 2016 Numerical and experimental investigation of natural flow-induced vibrations of flexible hydrofoils *J. Phys. Fluids.* **28** 075102
[10] Theodorsen T 1949 General theory of aerodynamic instability and the mechanism of flutter: laboratory report 496 (Washington D C: National Advisory Committee for Aeronautics)
[11] Munch C, Ausoni P, Braun O, Farhat M and Avellan F 2010 Fluid-structure coupling for an oscillating hydrofoil *J. Fluids and Struc.***26** 1018-33
[12] Young Y L, Chae E J and Akcabay D T 2012 Hybrid algorithm for modeling of fluid-structure interaction in incompressible, viscous flows *Acta Mechanica Sinica* **28** 1030-41

IOP Conf. Series: Materials Science and Engineering **634** (2019) 012026 doi:10.1088/1757-899X/634/1/012026

Effect of storage environment on the crystallinity and compressive load of starch based biodegradable cup

R A Garalde[1,2], E Orendain[1] and J Urbona[1]

[1]Packaging Technology Division, Industrial Technology Development Institute, Department of Science and Technology, Taguig 1631, Philippines

E-mail: raygaralde@gmail.com

Abstract. The effect of storage environment on the overall crystallinity and compressive load of commercially sold starch based biodegradable cup was determined. Using FTIR and DSC, the components of the cup was identified as starch, polypropylene (PP) and organic acid as compatibilizer. Using TGA, the weight percentage of starch and PP components were 13 and 58%, respectively. The samples were stored 20°C, 80% RH and 40°C, 40% RH for 4 months. Using DSC, the overall crystallinity was evaluated based on the heat of fusion (H_f). Using XRD, the crystalline structure was evaluated based on peaks and full width at maximum height (FWMH). After 4 months, DSC results showed only one endothermic peak with melting point of ≈165°C at both conditions. The H_f was remained ≈50 J/g. Based on XRD results, there were no development of new crystalline peaks and the peak FHWM remained unchanged. Initially, the percentage samples below the set limit of compressive load was ≈20%. The percentage below the set limit was ≈40% for samples stored at 20°C, 80% RH, whereas ≈20% for samples stored at 40°C, 40% RH. The 80% RH at 20°C increased the moisture content from 2% to 3.1%. The overall crystallinity remained unchanged after storage. Water became a plasticizer for the starch component. The increase in moisture content could have decreased the compressive load during storage.

1. Introduction

Thermoplastic starch (TPS) is composed mainly of starch and plasticizer [1,2]. The components are processed to have thermoplastic properties. It can then be converted into films, tray, cup, cutlery, foams and sheets using standard equipment used in conventional polymers. TPS is a very attractive material because its raw material comes from a renewable resource. It is biodegradable and non-toxic. TPS can also be blended with other polymers to improve its mechanical and water barrier properties [1].

Though TPS is very promising material to combat solid waste generation, TPS can become brittle during long term storage [1,2]. The environmental storage conditions can induced recrystallization of starch (i.e. retrogradation) [3] and migration of plasticizer out of the TPS matrix [1,2]. When polymers are stored above its glass transition (T_g), the polymer chains has some mobility and it can form more orderly structures during storage. This phenomenon is called recrystallization. Exposure to high humidity environment will increase the moisture content of TPS. Water is effective in plasticizing TPS because it is capable of hydrogen bonding with the hydroxyl groups of the starch chains [3]. The molecular size of water is also smaller than conventional plasticizers like glycerol, thus, it is more mobile to penetrate between the starch chains. The effectiveness of water as a plasticizer which is coupled with storage temperature above T_g will increase the rate of retrogradation.

The recrystallization of polymers can be evaluated using differential scanning calorimeter (DSC) and x-ray diffraction (XRD) techniques [3,4]. Using DSC technique, the presence of crystalline structures can be detected by the melting peak. The heat of fusion of the melting peak is an indicator of the relative amount crystalline structures that are present [5,6]. Using XRD technique, the peaks of crystalline structures can be detected. Recrystallization can also be detected by formation of a new peak [6,7]. The increase in crystallite size can also be evaluated using the full width at maximum height (FWMH). A decrease in FWHM implies increase of crystalline size [7] as effect of storage.

The changes in crystallinity and moisture content during storage can affect the mechanical properties of TPS [8-10]. For films, the increase in crystallinity will increase the tensile strength and modulus of elasticity and will decrease elongation at break. When the film is stored in ≥ 75% relative humidity (RH), the moisture content will increase and films will become very elastic [10]. There are many ageing studies of TPS materials [3,8-10]. In TPS composites, individual components can be influence by ageing and can affect the mechanical properties of the composite [11]. Ageing studies can be conducted to monitor the properties of the packaging and to assure its service life.

In this study, the commercially sold starch-based biodegradable cup is in its final and functional form. The cup is a takeaway packaging. The cup is subjected to compression when the lid is applied as closure or when covered cups were stacked in food delivery. The body section is more likely to be deformed during compression. The objective was to evaluate the effect of storage environment on the overall crystallinity and compressive load of starch based biodegradable cup.

2. Materials and methods

2.1. *Materials*
Starch-based biodegradable cup (12 oz.) was purchased from a local supplier in the Philippines. The cups with defects such as dents, holes and tear were eliminated.

2.2. *Identification of plastic components*
Fourier Transform Infra-red (FTIR) spectrum of the body section was recorded using Shimadzu IR-Prestige coupled with attenuated total reflectance (ATR). Using 32 accumulated scans, the absorbance was examined between 400 and 4,000 cm^{-1} wavenumber using a spectral resolution of 4 cm^{-1} [12].

Using differential scanning calorimeter (Shimadzu DSC-60), the melting point temperature (T_m) of the body section was determined. The heating and cooling rates were set at 10°C/min based on ASTM D3418 [13]. The sample was heated from 30 to 250°C and then was cooled under nitrogen atmosphere. The measurement was performed in duplicate.

Using thermogravimetric analysis (TGA), the weight percent (%) of the plastic components were estimated. The TGA thermogram was obtained using a STA 6000 (PerkinElmer, USA). About 20 mg sample was used. It was heated from 30 to 900°C at a heating rate of 10°C/min under a nitrogen atmosphere with N_2 flow rate of 50 ml/min.

2.3. *Crystallinity analysis*

2.3.1. Overall crystallinity analysis. The overall crystallinity was estimated based on the heat of fusion *(Hf)* [6,14]. From the DSC result of Section 2.2, the H_f (J/g) was calculated using TA-60WS software.

2.3.2. Crystalline structure analysis. X-ray diffraction (XRD) pattern of the body section was analyzed with Shimadzu Maxima XRD-7000. Sample was scanned at diffraction angles (2θ) from 5 to 35° using a scan rate of 0.02/s. The crystalline peaks and full width at half maximum (FWHM) of each peak were determined using Jade Version 6.0 software (MDI, USA).

2.4. *Compressive load*

The samples were conditioned at 23°C, 50% RH for 2 days prior to compressive load test. The compressive load was measured using a compression tester (Shimadzu EZ-SX) equipped with upper and lower compression plates. The jig plate diameter was 118 mm. The speed was set at 1.3 mm/min based on ASTM D695 [15]. The 15 replicates tested. The sample was positioned facing up wherein the upper jig would be in contact with the cup opening.

2.5. Moisture content
The sample was taken from the body section and cut into 5 cm x 5 cm sheet. The sheet was dried using vacuum oven (France Etuves XFL020, France) at 70°C under – 1 bar for 18 hours. The test was performed in triplicate.

2.6. Effect of storage environment
The samples were stored in two conditions: 1) 20°C, 80% RH and 2) 40°C, 40% RH for 4 months. The incubator used can only control the temperature. The relative humidity was monitored.

The overall crystallinity analysis by DSC, compressive load test and moisture content were monitored every month. The crystalline structure analysis by XRD was conducted at 0 month and after 4 months of storage.

To further evaluate the effect of storage on compressive load, the lower limit of compressive load was set to be the average compressive load subtracted by the standard deviation. The percentage of samples below the set lower limit was also calculated.

3. Results and discussion

3.1. Identification of plastic components
The FTIR spectrum of the body section (no figure) exhibited a broad band between 3600–3000 cm^{-1}. This was the absorbance band of O–H stretching of hydroxyl group [16]. Other absorbance bands were at 2949, 2914 and 2849 cm^{-1} (C–H stretching), 1730 cm^{-1} (C=O stretching of carbonyl group), 1454 and 1375 cm^{-1} (C–H blending), 1016 cm^{-1} (C–O stretching), and 872 cm^{-1} (C–C vibrations) [16,17].

The absorbance bands of O–H stretching of hydroxyl group and C–O stretching were attributed to the starch component. The sharp peaks of C–H stretching with very strong absorbance were attributed to polypropylene (PP) component. The C=O stretching indicated presence of carboxylic acid group which was not a functional group of the starch and PP components. This implied possible usage of organic acids as compatibilizer in TPS - PP blend [18].

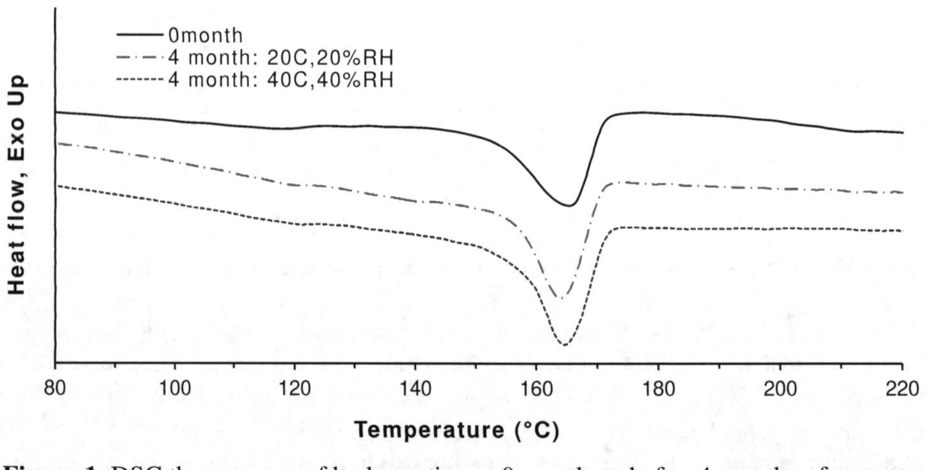

Figure 1. DSC thermogram of body section at 0 month and after 4 months of storage.

Using DSC technique, the heating scan of the body section is shown in figure 1. There was only one endothermic peak with melting point temperature (T_m) of 166°C. The peak T_m indicated a PP component in the composite. One endothermic peak of the TPS – PP composite also indicated a degree of miscibility as an effect of the compatibilizer [18].

DSC and FTIR techniques identified the starch, PP and organic acid components of the starch-based biodegradable cup.

Using TGA technique, the mass loss curve (figure 2) was used to estimate the weight percent (%) of components. The maximum rate of weight loss (T_{max}) was observed at 119, 315 and 464°C [14,19]. These were attributed to water, starch and PP components with weight percent of 1.7, 13 and 58%, respectively. The T_{max} of the compatibilizer was not detected probably because it has degraded in the same temperature range of starch and PP components.

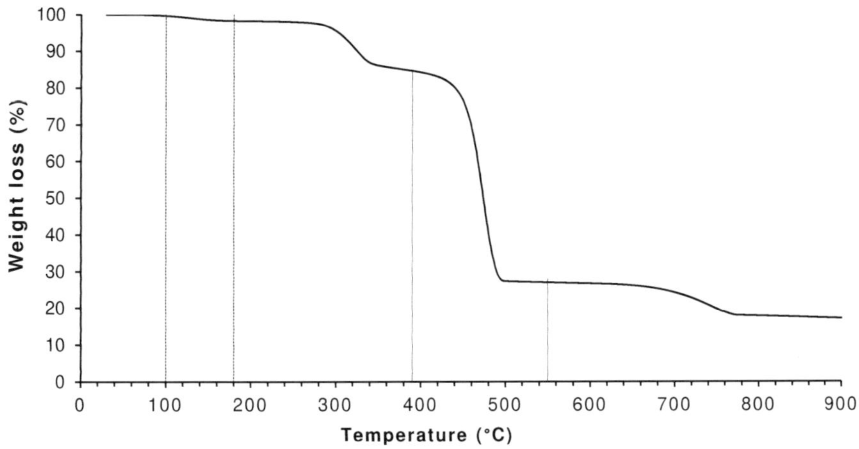

Figure 2. TGA curve of body section.

3.2. *Crystallinity analysis*

It was assumed that the PP and TPS components could recrystallized during storage because the storage temperature was above T_g of the PP component (–10°C) [20]. The actual T_g of the components could not be measured because of the limitation on operational temperature range of DSC. In DSC results, it was projected that the H_f would increase [5]. The single melting peak could also separate into individual peaks of TPS and PP components as each component recrystallized in the blend [11]. In XRD results, the increase in peak intensity and an increase of crystalline size could be observed [6]. New crystalline peaks could also be formed as observed in the study of van Soest *et al* [7].

3.2.1. Overall crystallinity analysis. The heat of fusion (H_f) was used to estimate the overall crystallinity of the TPS-PP composite [6,14]. Initially, there was only one endothermic peak with H_f of 45.9 ± 4.1 J/g. After 4 months, DSC results exhibited only one endotherm in both storage conditions (figure 1). The melting point range was 164–166°C. Figure 3 shows the effect of storage on the H_f. After 4 months, the H_f of the samples were 47.3 ± 5.0 J/g for samples stored at 20°C, 80% RH and 49.3 ± 0.1 J/g for samples stored at 40°C, 40% RH. The H_f did not increase significantly for both storage conditions. The DSC results suggested that the overall crystallinity did not increase significantly with storage.

3.2.2. Crystalline structure analysis. The crystalline pattern of body section at initial stage was shown in figure 4. The characteristic peaks are at 2θ = 9.8, 14.4, 17.3, 19.3, 29.0, 29.8, 31.4 and 31.9°. The crystalline peaks at 2θ = 14.4 and 29.0° were identified to be PP [17,21]. The peak at 2θ = 29.0° was sharp and intense. This peak was usually of low intensity in crystallinity studies of PP materials [6,17]. The peak at 2θ = 29.0 and 29.8° could also be processed induced crystallinity during the manufacturing process of the cup.

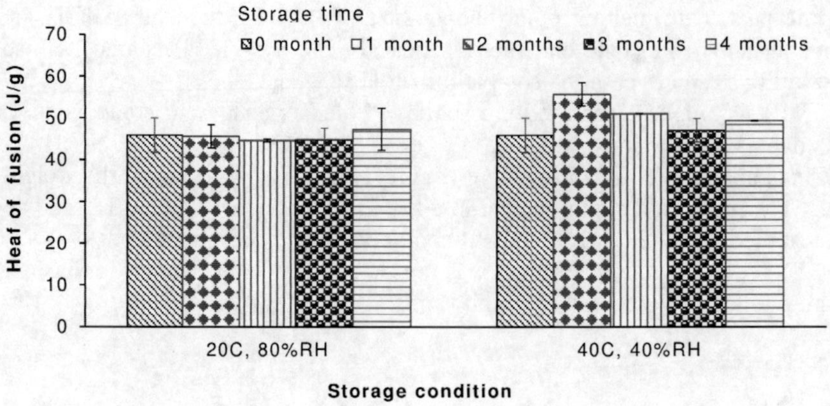

Figure 3. Effect of storage on the heat of fusion (body section).

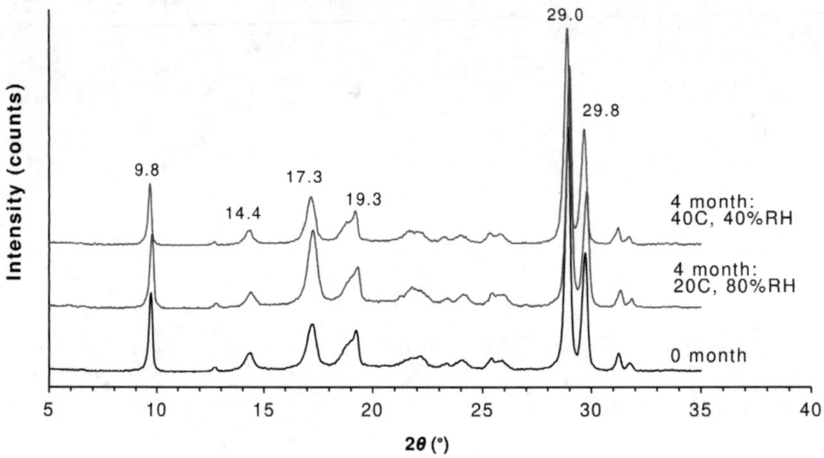

Figure 4. X-ray diffraction patterns of body section at initial stage and after 4 months of storage.

Table 1. XRD results of FWHM intensity of body section at initial stage and after 4 months storage.

2θ(°)	Full width at half maximum (FWHM) 0 month	after 4 months of storage 20°C, 80% RH	40°C, 40% RH
9.8	0.24 ± 0.03	0.22 ± 0.03	0.24 ± 0.01
14.4	0.45 ± 0.03	0.44 ± 0.03	0.43 ± 0.01
17.3	0.46 ± 0.00	0.45 ± 0.03	0.46 ± 0.00
19.3	0.54 ± 0.05	0.52 ± 0.04	0.49 ± 0.01
29.0	0.32 ± 0.05	0.31 ± 0.02	0.31 ± 0.01
29.8	0.32 ± 0.05	0.31 ± 0.02	0.31 ± 0.02
31.4	0.36 ± 0.12	0.31 ± 0.02	0.34 ± 0.05
31.9	0.43 ± 0.14	0.37 ± 0.01	0.33 ± 0.04

The crystalline peak at 2θ = 17.3° could be an overlapped of B-type of the starch component (2θ = 16.8°) [7] and PP component (2θ = 16.8°) [17]. The wide peak at 2θ = 19.3° could also be an overlapped of V_h-type of the starch component (2θ = 19.5°) [7] and PP component (2θ = 18.5°) [17]. The crystalline peak at 2θ = 9.8° was not a characteristic peak of starch and PP component. This could also be processed induced crystallinity during the manufacturing process of the cup.

After 4 months of storage, there were no developments of new crystalline peaks. The FWHM of the sharp peaks did not decrease (table 1) which would imply increase in crystallite size. For the PP component, the FWHM was expected to increase as the sample was stored above the T_g of PP. The FWHM of PP peaks at $2\theta = 14.4$, 29.0 and 29.8° did not decrease. Thus, an increase of crystalline size was not observed [7] as effect of storage. The XRD results agreed with the DSC results.

3.3. Compressive load

The compressive load of the starch-based biodegradable cup was 100 ± 31 N. After 4 months of storage, the compressive load was 95 ± 36 N for samples stored at 20°C, 80% RH whereas 94 ± 21 N for samples stored at 40°C, 40% RH. It was evident that the standard deviation was large (≈30 N) in all samplings. This could be due to manufacturing method of the cup.

The lower limit of compressive load was set to be 69 N based on 0 month and the percentage of samples below 69 N was ≈27%. After 4 months of storage (table 2), the percentage of samples below 69 N was ≈40% when stored at 20°C, 80% RH while ≈27% when stored at 40°C, 40% RH. The compressive load of samples stored at 20°C, 80% RH decreased during retention while the samples stored at 40°C, 40% RH remained the same.

Table 2. Percentage of samples below the lower limit of compressive load during storage.

Storage time (month)	Percentage (%) of samples below 69N* 20°C, 80% RH	40°C, 40% RH
0	27	27
1	7	0
2	40	13
3	40	27
4	33	13

* Lower limit of compressive load based on 0 month

3.4. Moisture content

After 4 months of storage, the moisture content of the samples stored at 20°C, 80% RH increased from 2 to 3.1%. The high humidity environment of 80%RH increased the moisture content of the samples. The TPS component absorbed water, thus increasing its plasticizer content [3,10]. The increase in moisture content decreased the compressive load when stored at 20°C, 80% RH.

When samples were stored at 40°C, 40%RH, moisture content decreased from 2 to 1.6%. The loss of moisture was due to the evaporation of water at low humidity environment of 40% RH. The loss of moisture of the TPS component could equate to decrease in plasticizer content and it could lead to embrittlement [2]. The compressive load of samples stored at 40°C, 40% RH was not affected by the decrease in moisture content.

4. Conclusion

The body section of the starch-based cup was identified as the area that more likely to be deformed during excessive compression. Different characterization techniques were used to understand its mode of failure as a takeaway packaging. The recrystallization phenomenon was not observed based on DSC and XRD results after 4 months of storage. When stored at high relative humidity of 80%, the starch component absorbed moisture which would become its additional plasticizer. The material became more flexible which leaded to the decrease of compressive load during storage. It is recommended to measure T_g for further studies on retrogradation of TPS blend composites.

Acknowledgments

This work was financially supported by Industrial Technology Development Institute, Department of

Science and Technology, Philippines. The authors were grateful to Advanced Device and Materials Testing Laboratory (ADMATEL) for TGA analysis.

References

[1] Carvalho A 2013 Starch: major sources, properties and applications as thermoplastic materials *Handbook of Biopolymers and Biodegradable Plastics* S Ebnesajjad ed (USA: Elsevier Inc) chapter 7 pp 129-522
[2] Nafchi A, Moradpour M, Saeidi M and Alias A 2013 Starch/Starke **65** 61-72
[3] Ottenhoff M A and Farhat I A 2004 *Biotechnol. Genet. Eng.* **21** 215-28
[4] Zhang Y and Rempel C 2012 Retrogradation and antiplasticization of thermoplastic starch *Thermoplastic Elastomers* A El-Sonbati ed (IntechOpen) pp 117-34 DOI: 10.5772/35848 Available from: https://www.intechopen.com/books/thermoplastic-elastomers/retrogradation-and-antiplasticization-of-thermoplastic-starch
[5] Kohyama K, Matsuki J, Yasui T and Sasaki T *Carbohyd. Polym.* **58** 71-7
[6] Milicevic D, Micic M, Stamboliev G, Leskovac A, Mitric M and Suljovrujic E 2012 *Fiber Polym* **13** 466-70
[7] van Soest J and Knooren N 1997 *J. Appl. Polym. Sci.* **64** 1411-22
[8] Fama L, Goyanes S and Gerscehnson L 2007 *Carbohyd. Polym.* **70** 265-73
[9] Pushpadass H A and Hanna M A 2009 *Ind. Eng. Chem. Res.* **48** 8457-63
[10] Suppakul P, Chalernsook B, Ratisuthawat B, Prapasitthi S and Munchukangwan N 2013 *LWT - Food Sci. Technol.* **50** 290-7
[11] Garalde R A, Orendain E and Urbona J 2019 *Heliyon* **5** 1-20
[12] Dang K M and Yoksan R 2015 *Carbohyd. Polym.* **115** 575-81
[13] ASTM 2003 *D3418-03 Standard Test Method for Transition Temperatures and Enthalpies of Fusion and Crystallization of Polymers by Differential Scanning Calorimetery* (USA: ASTM International) pp 1-8
[14] Roy S B, Ramaraj B, Shit SC and Nayak S K. 2011 *J. Appl. Polym. Sci.* **120** 3078-86
[15] ASTM 2015 *D695 – 15 Standard Test Method for Compressive Properties of Rigid Plastics* (USA: ASTM International) pp 1-8
[16] Coates J 2006 Interpretation of infrared spectra, a practical approach *Encyclopedia of Analytical Chemistry: Applications, Theory and Instrumentation* R Meyers ed (John Wiley & Sons) pp 1-23 DOI: 10.1002/9780470027318.a5606
[17] Parparita E, Darie R N, Popescu C M, Uddin M A and Vasile C 2014 *Mater. Des.* **56** 763-72
[18] Martins A B and Santana R M C 2016 *Carbohyd. Polym.* **135** 79-85
[19] Raee E and Kaffashi B 2018 *J. Appl. Polym. Sci.* **135** 1-12
[20] Giles H, Wagner J and Mount E 2005 Single screw extrusion *Extrusion: The Definitive Processing Guide and Handbook* (USA: William Andrew, Inc.) chapter 4 pp 141-50
[21] Favaro M M, Branciforti M and Bertas R 2009 *Mater. Res.* **12** 455-64

IOP Conf. Series: Materials Science and Engineering **634** (2019) 012027 doi:10.1088/1757-899X/634/1/012027

Piezoelectric characteristics of inorganic (Na, K, Li) (N_b, S_b, T_a)O_3 – (Bi, Na)(Sr)ZrO_3 ceramics with sintering time

J H Yoo[1, 2] and J H Lee[1]

[1]Department of Electrical Engineering, Semyung University, Jecheon 390-711, Korea

E-mail: juhyun57@semyung.ac.kr

Abstract. In recent days, the Rhombohedral-Tetragonal (R-T) region composition (Na,K)NbO_3 ceramics have been actively studied because of their excellent piezoelectric properties. In this study, in order to develop the ceramics capable of being sintered at low temperature for piezoelectric actuator, $0.97(Na_{0.52}K_{0.443}Li_{0.037})(Nb_{0.923}Sb_{0.04}Ta_{0.037})O_3 - 0.03(Bi_{0.5}Na_{0.5})_{0.9}(Sr)_{0.1}ZrO_3$ R-T region ceramics were manufactured using the sintering aids of Bi_2O_3-CuO-Fe_2O_3 according to sintering time by the conventional solid state method and their piezoelectric and dielectric properties were systematically investigated. At the 9 hour sintering specimens, the excellent physical properties of d_{33}=244pC/N ε_r=1475, k_p=0.44, and Qm=240 were obtained for piezoelectric actuator application.

1. Introduction

Recently, Pb (Zr,Ti)O_3 (PZT) materials were utilized in various kind of devices such as piezoelectric speakers, AE sensors for detecting the rear and side of automobiles, and ultrasonic motor owing to their best piezoelectric and dielectric characteristics [1-2] . But, PbO is composed of more than 60% of the PZT ceramic compositions. Generally, when the ceramics is sintered at above 1200 ° C, the PbO causes environmental pollution and is fatal to peoples because it rapidly volatilizes from the temperature of 1000°C [3-8].

Accordingly, in order to replace Pb–based ceramics, it is required to develop the environmentally friendly ceramics that is comparable to the excellent characteristics of PZT [9].

(Na, K) NbO_3 (NKN) based ceramics reported by Satio et al.in 2004 have excellent piezoelectric properties d_{33} = 416 [pC / N] and high Tc (T_C = 460 ℃). Therefore, the ceramics have been attracting attention as a substitute for PZT system ceramics. However, it is very difficult to manufacture excellent ceramics owing to the volatilization and high deliquescence of alkaline elements of K and Na during sintering at high temperatures [10]. Accordingly, by adding Fe_2O_3, Bi_2O_3, CuO, etc. with low melting point, the liquid phase sintering of the ceramics can be performed at low temperature through prohibiting the volatilization of K and Na, thereby promoting its densification.

In recent years, piezoelectric speakers have been adopted in LG smartphones in korea, which employ laminated piezoelectric actuators that require low-temperature sintering to use the costly Ag-Pd internal electrodes. And also, the d_{33} and the kp should be large on order to enhance the strain and the characteristics of the piezoelectric actuators, respectively. The excellent physical characteristics of the Pb (Zr,Ti)O_3 system ceramics show in the tetragonal-rhombodedral (R-T) regions [11]. So, the R-T regions of the ceramics have been utilized for manufacturing the piezoelectric devices by many researchers.

Accordingly, in this study, (Na,K,Li)(Nb,Sb,Ta)O_3 – (Bi,Na)(Sr)ZrO_3 R-T region ceramics were fabricated and their piezoelectric characteristics were investigated.

2. Experimental procedure

In this experiment, the specimens were prepared using a conventional mixed oxide process. The following composition formulas were used.

(Na, K, Li) (Nb, Sb, Ta) O_3 – (Bi, Na) (Sr)ZrO_3 + 0.5wt% Bi_2O_3 + 0.4wt% Fe_2O_3 + 0.3wt% CuO

The pure reagents of Na_2CO_3, K_2CO_3, Sb_2O_5, Nb_2O_5, Li_2CO_3, Ta_2O_5, Bi_2O_3, Sr_2CO_3, ZrO_2 Fe_2O_3, and CuO with more than 99.0% were weighed to 10^{-4} g, and the dispersion medium was mixed with acetone in a 3Φ zirconia ball for 24 hours. After drying, they were calcined at 850 ° C for 5 hours. The additive Bi_2O_3, CuO and Fe_2O_3 were added to the calcined power, followed by secondary mixing and drying.

A polyvinyl alcohol (5%PVA) was added to the dried powders and the powders were molded under a pressure of 15 MPa with a mold having a diameter of 17Φ. The molded powders were sintered at 1020 ℃ for 5, 7, 9, and 11 hours. The sintered specimen was polished to a thickness of 1 mm, and the Ag electrode was coated by a screen printing method and then heat-treated at 600 ° C for 10 min. A 3 kV / mm electric field was applied to the specimen in a silicon oil bath at room temperature for polarization for 30 minutes. fr and fa of the polarized specimens were measured using an impedance analyzer (Agilent 4294), and also the its dielectric properties were measured at a frequency of 1 kHz using an LCR meter (Instek LCR-819). Characteristics of the specimens were calculated using the resonance and antiresonance method according to the IEEE standard.

3. Results and discussion

Figure 1 shows the density of specimens with sintering time. The maximum value of the density was 4.403 [g / cm^3] in the specimen sintered for 9 hours. These results show that Bi_2O_3 (817℃), CuO (1060 ℃) and Fe_2O_3 (730℃), which have low melting point, added as sintering aids form a liquid phase during the sintering process. It was thought that the formed liquid phase induced the densification of the ceramics. And also, the formed liquid phase was volatilized by over sintering for 11 hours and the density decreased.

Figure 2 shows the piezoelectric constant (d_{33}) with sintering time. As the sintering time increased up to 9 hours, the piezoelectric constant increased and then decreased. The maximum value of 244 [pC / N] was obtained for the 9 hours sintered specimen. These results show that as the sintering time increases, the particles grow in the specimen.

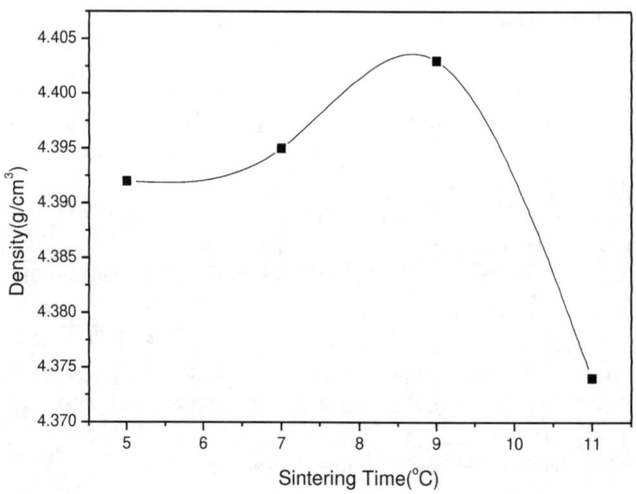

Figure 1. Density of specimens with sintering time.

IOP Conf. Series: Materials Science and Engineering **634** (2019) 012027 doi:10.1088/1757-899X/634/1/012027

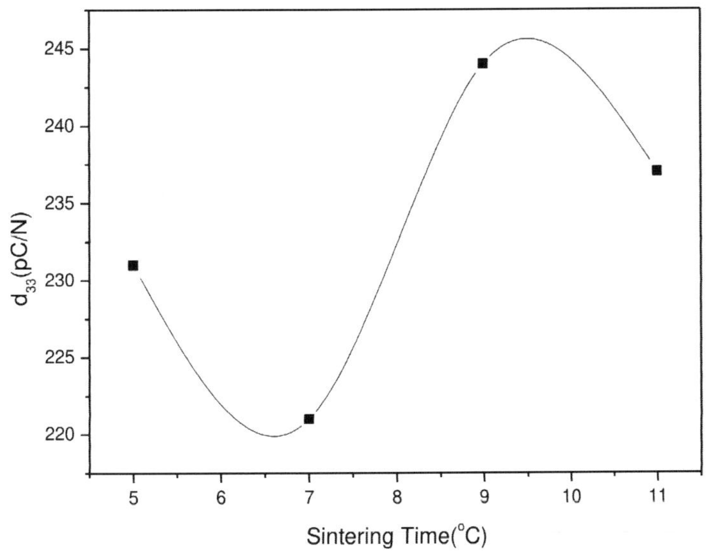

Figure 2. Piezoelectric constant (d_{33}) of specimens with sintering time.

It is considered that the piezoelectric constant (d_{33}) is increased because the domain switching of the ceramics is facilitated and the polarization efficiency is increased. The d_{33} value showed the same tendency as that of density.

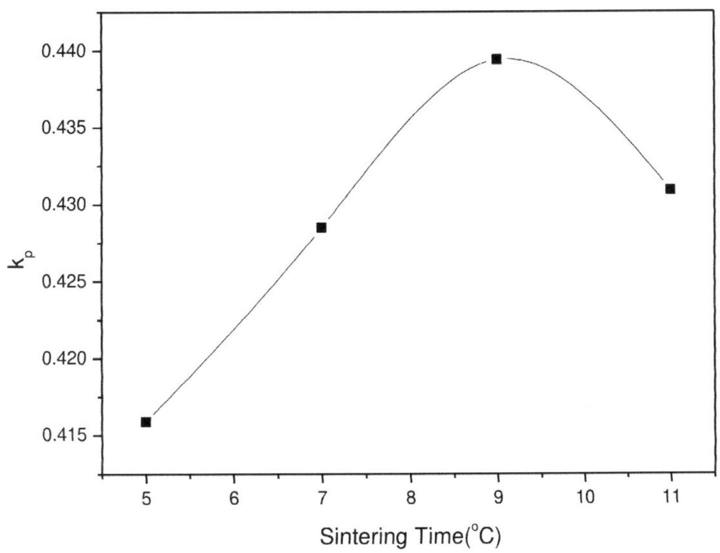

Figure 3. Electromechanical coupling factor(k_p) of specimens with sintering time.

Figure 3 shows the kp with sintering time. Like the piezoelectric constant, the kp tends to increase and decrease with increasing sintering time. The maximum value of 0.44 was obtained from the 9 hour sintered specimen. The results can be illustrated by the liquid phase sintering effect of the ceramic particles.

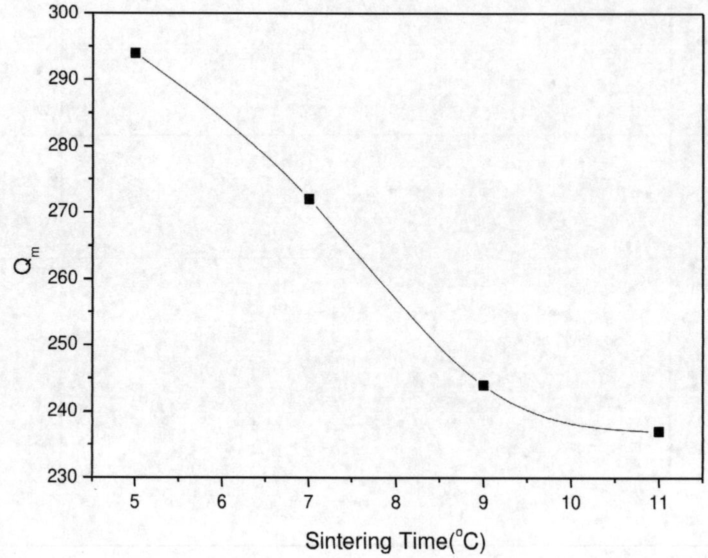

Figure 4. Mechanical quality factor(Q_m) of specimens with sintering time.

Figure 4 shows the Qm with sintering time. As the sintering time increases, the Qm tends to decrease. These results suggest that the increase of the sintering time induces the increase of the piezoelectric constant d_{33} and the electromechanical coupling factor (kp). As the results, the soft properties of the ceramics enhance, and then the Qm is decreased.

Figure 5 shows the εr with sintering time. The maximum value of 1475 was obtained for the 9 hour sintered specimen. As the sintering time increased, the dielectric constant also tended to increase because the soft material properties were promoted. Table 1 shows the physical properties of specimens as a function of sintering time.

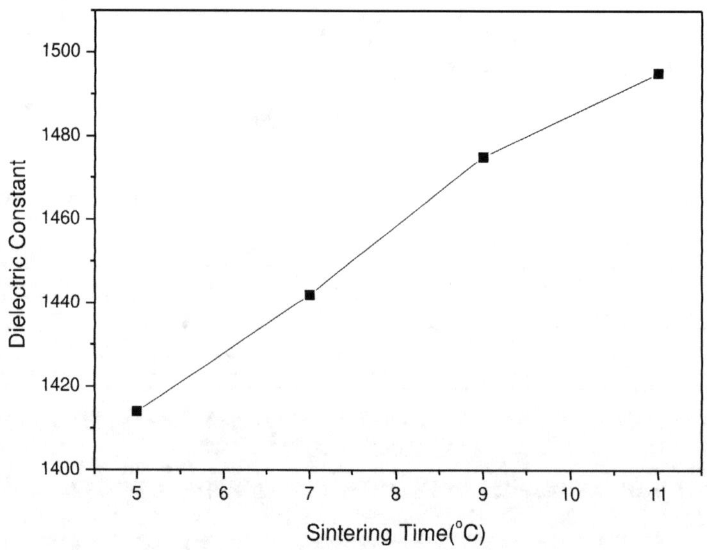

Figure 5. Dielectric constant(ε_r) of specimens with sintering time.

Table 1. Physical properties of specimens as a function of sintering time.

Sintering Time (°C)	Density [g/cm^3]	k_p	Dielectric constant	d_{33} [pC/N]	Q_m
5	4.392	0.415	1414	231	294
7	4.395	0.427	1442	221	272
9	4.403	0.440	1475	244	240
11	4.374	0.431	1495	237	235

4. Conclusions

In this study, (Na,K,Li)(Nb,Sb, Ta)O_3 – (Bi,Na,) (Sr) ZrO_3 R-T region ceramics were manufactured using the sintering aids of Bi_2O_3-CuO-Fe_2O_3 according to sintering time and their piezoelectric characteristics were systematically investigated.

1. The maximum value of the density was 4.403 [g / cm^3] in the specimen sintered for 9 hours

2. At 9 hours sintered specimens, the excellent physical properties of d_{33}=244pC/N ε_r=1475, k_p=0.44, and Qm=240 were obtained for piezoelectric actuator application.

Acknowledgements

This paper was studied by 2016 National Research Foundation of Korea.

References

[1] Azough F, Wegrzyn M, Freer R, Sharma S and Hall D 2011 *J. Eur. Ceram. Soc*, **31**, 569-76.
[2] Sunmin B and Juhyun Y 2011 *Ferroelectrics*, **425**, 98.
[3] Zuo R, Xu Z and Li L 2008 *J. Phys. Chem. Solids*, **69**, 1728-32.
[4] Yiping G, Ken-ichi K and Hitoshi O 2004 *Solid. State. Commun.*, **129**, 279.
[5] Juhyun Y 2017 *Ferroelectrics*, **507**, 12.
[6] Jungrae N and Juhyun Y 2013 *J. Electroceram*, **30**, 139.
[7] Sang-Hoon S, Juhyun Y and Sang-Don L, 2014, *Ferroelectrics Lett.*, **41**, 56.
[8] Coondoo N, Panwar R, Rai H and Amorín A L, 2013 *Phase Transit.*, **86**, 1130.
[9] Zhao Y, Zhao Y, Huang R, Liu R and Zhou H 2012, *Mater. Lett*, **75**, 146-8.
[10] Saito Y, Takao H, Tani T, Nonoama T, Takatori K, Homma T, Nagaya T and Nakamura M 2004 *Nature*, **432**, 84-87.
[11] Gou Q, Zhu J, Wu J, Li F, Jiang L and Xiao D 2018 *J. Alloys. Compd.*, **730**, 311.

IOP Conf. Series: Materials Science and Engineering **634** (2019) 012028 doi:10.1088/1757-899X/634/1/012028

The research about ultimate load of CFRP repaired pipes under long-term seawater immersion and bending moment

J H Xin[1], Y Zhang[1,2], Q Huang[1] and N Y Cheng[1]

[1]College of Safety and Ocean Engineering, China University of Petroleum, Beijing102249, China

E-mail: zhangyu@cup.edu.cn

Abstract. Corrosion is a common failure form for subsea pipes. Corroded pipes are increasingly repaired by the composite material because of the low cost and the simple and quick process. However, the repair will be affected by factors such as long time bending moment and seawater immersion. To study the law of the effect, the bending moment and seawater immersion factors are considered in this paper. For the experiment of CFRP repaired pipes, 14 groups of CFRP-pipe specimens were designed, and 30% bending moment was applied, the seawater concentration 5 times more than that recommended by ASTM D1141-98 (2013) was adopted to accelerate the experiment. The pipe specimens were immersed in seawater for different time period (200 hours, 400 hours, 600 hours and 800 hours). The experimental result showed that the ultimate bending load capacity of the repaired pipe decreases with the increasing of time. To study the reasons for the decrease of the ultimate load of the repaired pipes, CFRP-steel plate and CFRP durability tests were set up, 5 groups of CFRP and 5 groups of CFRP-steel plate specimens were designed. It was found that the decrease of the interfacial bonding strength of CFRP repaired pipe and decrease of the CFRP strength resulted in the decline of ultimate load of repaired pipe.

1. Introduction

With the development of offshore and deep-water oil and gas resources in world, the construction for subsea pipes known as the "lifeline" of offshore oil and gas fields is increasing. As an important part of offshore oil and gas field development, it plays a key role in the development, production and transportation of oil and gas. However, due to the harsh marine environment, subsea pipes are easy to be damaged. Corrosion is a common failure form for subsea pipes. Therefore, effective anti-corrosion measures must be taken for subsea pipes. Three technologies are commonly used for repair of subsea pipes: welding repair, fixture repair and composite material repair technology [1]. Composite material repair technology has been widely applied in the maintenance and repair of the oil and gas pipes because of its low cost, relatively simple process and short time consumption [2,3]. However, with the increase of service time, the repair effect will be affected by factors such as long time bending moment and seawater immersion, so the performance of the composite repair will degrade. If the degradation is great, it may even cause the failure of the repaired pipe, resulting in serious environmental pollution and huge economic losses. Therefore, it is of great significance to study the effect of bending moment, seawater immersion on pipes repaired by composites.

Fiber composite repair technology was initially used to reinforce bridge concrete structures. In the 1970s, the technology of repairing pressure pipes and vessels with fiber composite materials emerged in foreign countries. At present, a great deal of research has been done on composite material repair technology and the degradation of composite material. Borrie [4] studied the durability of steel structures strengthened with CFRP laminates and sheets under seawater environment with increasing temperature. The results showed that even in short-term exposure, protecting and maintaining the bonding interface of the repaired structure were essential to ensure the strength and elasticity of the reinforced steel. Kabir [5] studied the durability of CFRP structural members under long-term low temperature environment by means of experiments and finite element analysis. The results showed that the strength and stiffness of CFRP structural members would be reduced in 12 months under low temperature environment. Almosawe [6] studied the bond behavior between CFRP laminates and steel members under impact load. The results showed that there is a significant bonding strength enhancement under impact load, while high load rate has little effect on the effective bonding length. Nguyen [7] studied the bonding properties of CFRP-steel interface at different temperatures. It was proposed that the bonding failure time was the result of the triple action of temperature, time and load. Compared with the specimens subjected to constant temperature, 47% of the specimens subjected to cyclic temperature have been failed at the same load level. Peck [8] studied the fiberglass wound repair pipeline and studied the mechanical properties of the repaired pipeline by internal pressure test and simple four-point bending test, and verified by finite element analysis results. Abarilla M A [9] immersed the carbon fiber epoxy resin matrix composites at different temperatures, the results showed that the strength and elastic modulus of the composite decreased after immersing for a long time, and the degree of degradation for strength is great, but the elastic modulus is relatively small. Ramirez [10] studied the mechanical properties of fiberglass and carbon fiber composites after different immersion time in seawater. It was found that the properties of carbon fibers were better, and the strength of both fiberglass composites and carbon fiberglass composites decreased after immersion in seawater, among them, the strength of glass fiber decreases more obvious. Based on a theoretical method and the finite element method (FEM) with different crack lengths, Zhang [11] studied the stress distributions of adhesive in a cracked steel plate repaired with CFRP. The results showed that the maximum shear stress and peel stress in the adhesive were obtained from products of stress ratios and stresses for the case of a through-wall crack, and the adhesive failure or the maximum load can be simply predicted by using quadratic stress criteria. Kafodya [12] studied the durability of carbon fiber reinforced polymer (CFRP) plates in water and seawater under 30% and 50% ultimate bending strain. The results showed that the shear strength of carbon fiber composites decreases significantly when immersed in two kinds of media, and the effect of immersion on the tensile strength and modulus of carbon fiber composites is small. Chan [13] studied the mechanical properties of steel pipes repaired by fiber reinforced polymer composites under bending load by experimental method and finite element method. The result showed that use CFRP as repairing material, the bending stiffness of corroded pipes and the ability to bear bending load can be improved.

The durability of CFRP repaired pipe under long-term bending moment and seawater immersion is considered in this paper. For the experiment of CFRP repaired pipes, 14 groups of CFRP-pipe specimens were designed, 30% bending moment was applied to the specimens. By setting the used seawater concentration 5 times more than that recommended by ASTM D1141-98 [14] to accelerate the experiment, the pipe specimens were immersed in seawater for different time period (200 hours, 400 hours, 600 hours and 800 hours). Then the 5 groups of CFRP-steel plate and 5 groups of CFRP durability tests were set up to further explore the reasons for the decrease of ultimate load of repaired pipe.

2. Ultimate bending load of CFRP repaired pipe

2.1. Experimental condition

The CFRP cloth is made of 3K-200g high strength two-way 0/90 braided carbon fiber cloth, and the API 5L X60 seamless pipe is selected as the base material. The material parameters are shown in tables 1 and 2.

Table 1. Material parameters of CFRP cloth.

Material parameter	CFRP cloth
Thickness/mm	0.22
Density /g/cm^3	1.8
Elastic modulus/GPa	240
Tensile strength/MPa	1700
Poisson’s ratio	0.3
elongation	2.07%

Table 2. Material parameters of the API 5L X60 seamless pipe.

Material parameter	Pipe
External diameter/mm	21.3
Thickness/mm	1.6
Yield strength/MPa	355
Tensile strength/MPa	600
Poisson's ratio	0.3
Elastic modulus/GPa	210
Length of pipe/mm	300
Corrosion length/mm	10
Corrosion depth/mm	1.5
Corrosion angle/	15

The adopted adhesive was Araldite 2015, as shown in figure 1. The elastic modulus of the adhesive is 2 GPa, the tensile strength is 22.3 MPa and the ultimate elongation is 1.85%. Double bond DB406 metal corrosion resistant filler is used as repair filler, as shown in figure 2. The elastic modulus of the filler is 5.5 GPa, Poisson's ratio is 0.4 and the shear strength is 15 MPa.

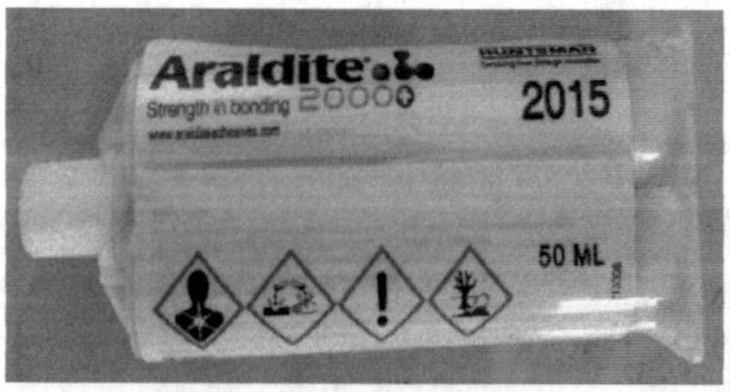

Figure 1. Araldite 2015.

Figure 2. Repair filler.

Two groups of experiments were set up in this paper. The experimental schemes are shown in tables 3 and 4.

In the first group, CFRP repaired pipes were subjected to seawater immersion and bending moment for a long time before the test to measure the ultimate load. The long-term bending the pipes were first subjected to 30% ultimate bending moment, which is equal to the ultimate bending moment of the original undamaged pipe. Then put into the 5 times seawater concentration at 17°C. The seawater immersion of the pre-loaded pipe specimens lasted for different time period (200 hours, 400 hours, 600 hours and 800 hours). To study the effect of a single factor on the repairing effect, another group of experiment was set up, the pipe specimens were only subjected to 30% ultimate bending moment.

The effects of bending moment and seawater immersing time on the ultimate load of the CFRP repaired pipe was studied by three-point bending experiment.

Table 3. Experimental scheme with seawater immersion and bending moment.

Type	Bending moment	Time/hours	Seawater concentration	Temperature /°C
CFRP repaired pipe	0%	0	5 times ASTM standard	17
		200		
		400		
		600		
		800		
	30%	0		
		200		
		400		
		600		
		800		

Table 4. Experimental scheme with bending moment.

Type	Bending moment	Time/h
CFRP repaired pipe	30%	0
		200
		400
		600
		800

14 groups of CFRP-pipe specimens were designed, According to the standard ASME PCC-2-2011 [15], the thickness and length of CFRP repair winding were calculated as 0.6mm and 50mm respectively. The Araldite 2015 structural adhesive was used for bonding between the pipe and the CFRP cloth and between the layers of the CFRP cloth during winding, as shown in figure 3. The two ends of the pipe specimens are roughly polished by a grinder to prevent slippage during loading.

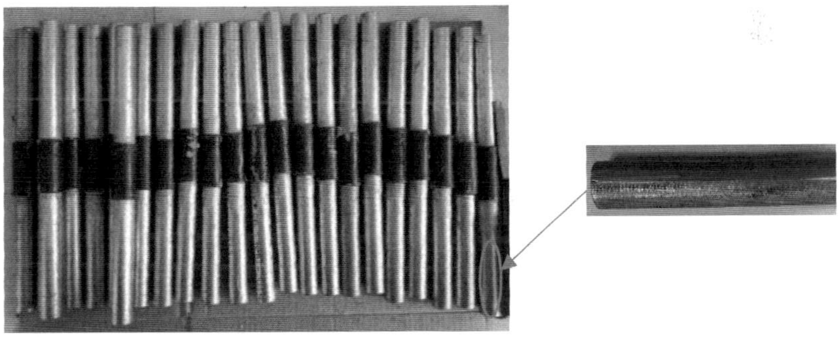

Figure 3. CFRP repaired pipe specimens.

The bending device of CFRP-pipes are shown in figure 4. By applying a certain torque to the nut, the bending moment was applied to the CFRP pipes, and the vertical bending distance of the pipe is measured at the bottom center of the repair region. The pre-loaded specimens were placed in the seawater, as shown in figure 5. The three-point bending test device for studying the ultimate bending load of the CFRP pipes is shown in figure 6. The span for installation of the pipe specimens is 210 mm, the load is controlled by the displacement control mode, and the load speed is 0.5mm/min.

Figure 4. Bending device of CFRP-pipe.

Figure 5. Seawater container with CFRP-pipe.

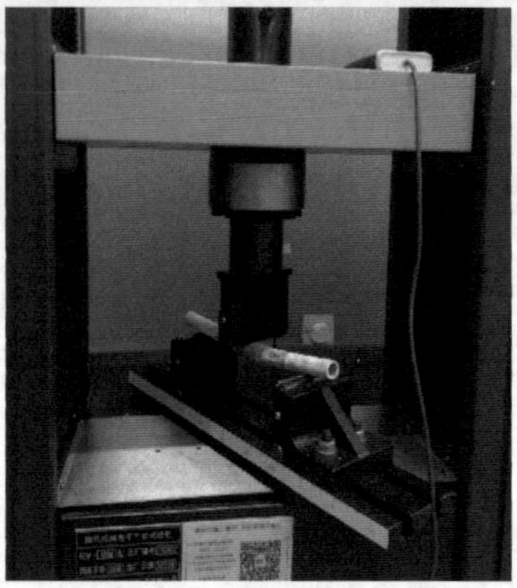

Figure 6. Three-point bending test device.

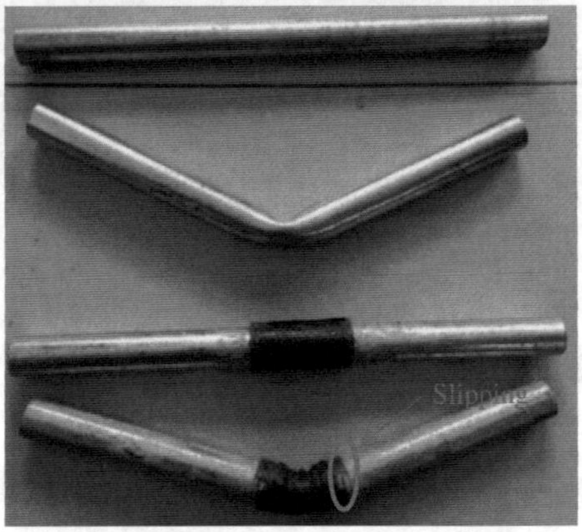

Figure 7. Shape diagram of the pipes before and after deformation.

2.2. Experimental result

The shape diagram of the corroded pipes and CFRP repaired pipes before and after deformation is shown in figure 7. Through the experiment, it was found that the deformation of the pipes mainly developed from circular to flat section. For the CFRP repaired pipe, under the continuous loading, partial interface slipping occurred between the carbon fiber cloth and the pipe.

To avoid the error caused by the experiment, each group of experiments were tested twice and then the average value was obtained. For example, when the CFRP repaired pipes were subjected to no immersion and bending moment, the ultimate loads are 6.8 KN and 6.9 KN, and the average load was 6.85 KN. The load-displacement curves of corroded pipe and CFRP repaired pipe is shown in figure 8. With the increase of load, the displacement increases rapidly and reaches the peak after a short hardening effect, and then the load drops with the increase of deformation. The whole curve has obvious characteristics of early elasticity, mid-term plastic strengthening and post-softening. The ultimate load of CFRP repaired pipe with no immersion and bending moment is 6.85 KN, and the ultimate load of corroded pipe is 5.35 KN, the experimental results showed that the pipe repaired with CFRP can improve the ultimate bending capacity.

The load-displacement relationship of CFRP repaired pipe under different conditions are shown in figure 9. With the increase of time, the ultimate load of CFRP repaired pipes under different conditions decrease gradually. For example, the ultimate loads are 6.7 KN, 6.4 KN, 6.2 KN and 5.9 KN under seawater immersion and bending moment for 200, 400, 600 and 800 hours, separately. When the time is 800 hours, the ultimate loads decrease by 15% compared to the repaired pipe with no immersion and bending moment.

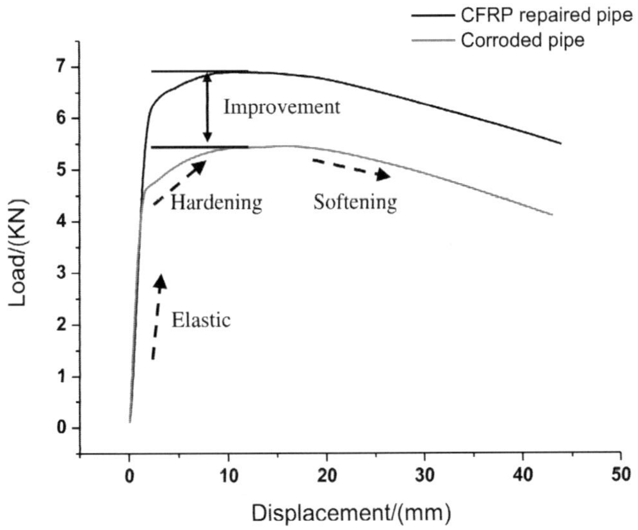

Figure 8. Load-displacement relationships of the corroded pipe and the repaired pipe.

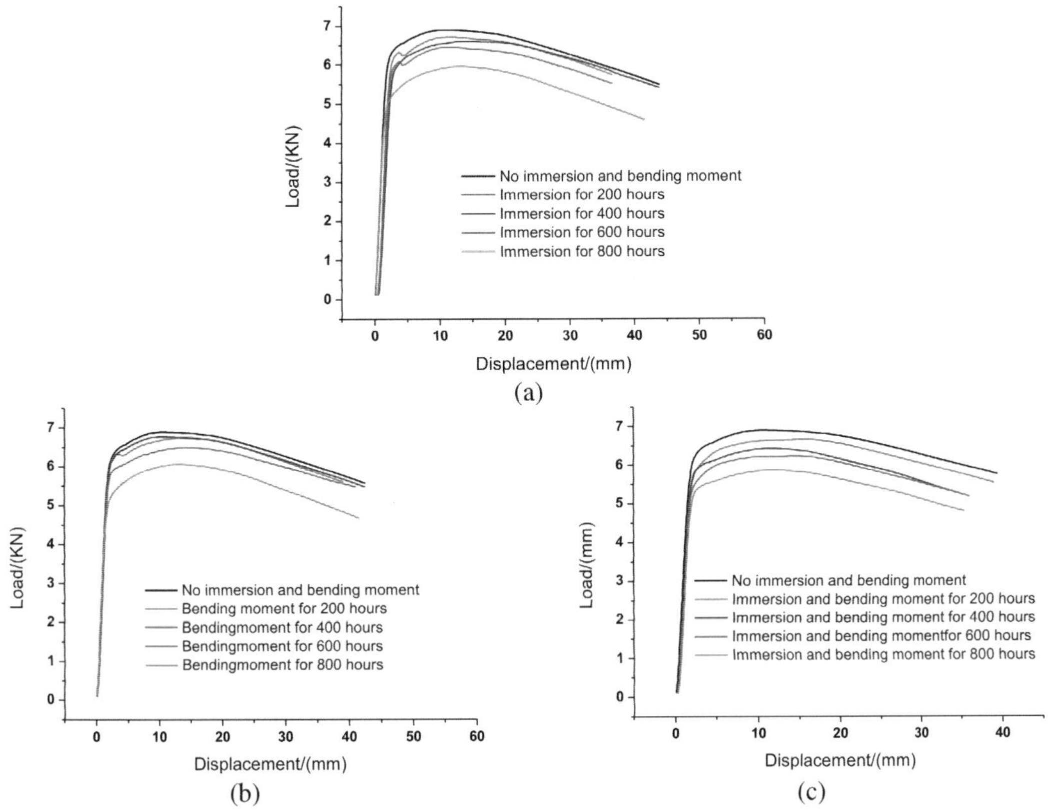

Figure 9. Relationship of load-displacement under different conditions. (a) Relationship of load-displacement under different immersion time, (b) Relationship of load-displacement under different bending moment time and (c) Relationship of load-displacement under different immersion and moment time.

The relationships between the ultimate load and time under different conditions are shown in figure

10. With the increase of time, the ultimate loads of CFRP repaired pipes decrease linearly. When the time was 800 hours, the decline of the ultimate load was the most serious. It was found that the effect of bending moment on the ultimate load of pipe was less than that of the seawater immersion, it decreases about 15% when both of the bending moment and the seawater immersion time acted. Compare to only consider bending moment or seawater immersion, the ultimate loads decrease seriously.

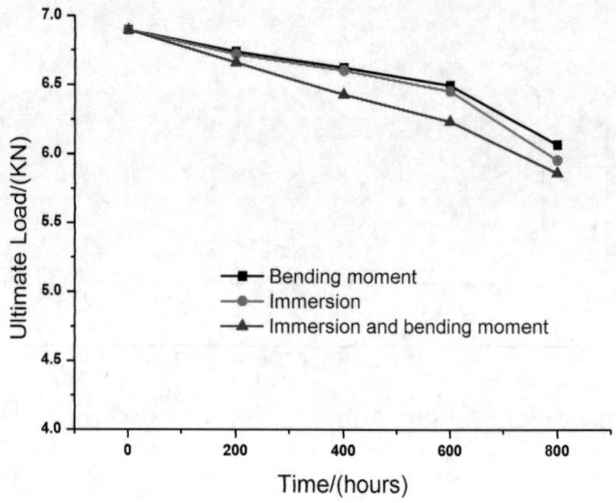

Figure 10. Relationship between time and load under different conditions.

From the above results, it can be found that ultimate load decreased gradually with time under different conditions. Through the experiment, it is found that the interface between the carbon fiber cloth layer and the pipe sliped partially under bending moments, which was considered to be one of reasons for the decrease of the ultimate load of the CFRP repaired pipe. Another reason was assumed that the mechanical properties of CFRP decreased with time under different conditions. To explain the reasons for the decrease of ultimate load of the CFRP repaired pipes, the durability of CFRP-steel plate and CFRP were studied in the followings.

3. Durability of CFRP strength and CFRP-steel plate interfacial bonding strength

3.1. Experimental condition

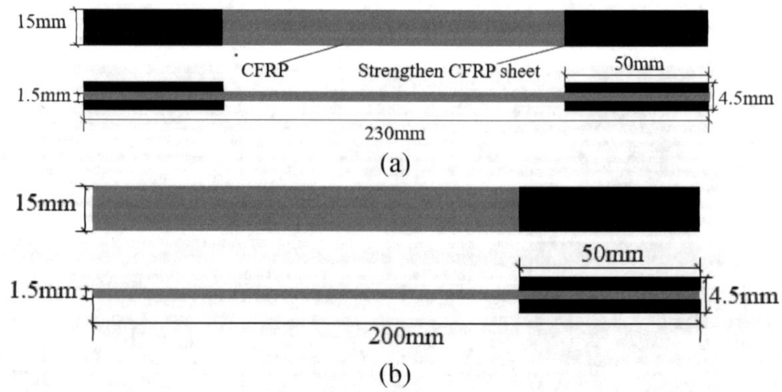

Figure 11. Sizes of CFRP plate. (a) CFRP plate and (b) CFRP for bonding with steel plate.

Two different sizes of CFRP plates were selected for the experiment. One group was used for CFRP durability test, and the other group was used for CFRP-steel plate interface strength test. Since the cutting edge of the fixture of universal testing machine is easy to destroy the surface of carbon fibers. Therefore, the strengthen sheet is pasted on the specimen to prevent the early failure of CFRP plate. The size of the test specimen is shown in figure 11, tensile strength and elastic modulus of CFRP material are 1700 MPa and 235 GPa. The material of steel plate is Q235, its material parameters are shown in table 5.

Table 5. Material parameters of steel plate.

Material parameter	Steel plate
Length/mm	250
Thickness/mm	5
Width/mm	30
Yield strength/MPa	258
Tensile strength/MPa	414
Elastic modulus/GPa	198

The adopted adhesive between CFRP and steel is also Araldite 2015. For the experiment of CFRP and CFRP-steel plate, 5 groups of CFRP-steel plate and 5 groups of CFRP were set up in this paper, as shown in table 6. The CFRP plates and CFRP-steel plates were put into the 5 times seawater concentration at 17°C for different time (200 hours, 400 hours, 600 hours and 800 hours).

Table 6. Experimental scheme.

Type	Time/hours	Seawater concentration	Temperature/°C
CFRP	0	5 times ASTM standard	17
CFRP-steel	200		
	400		
	600		
	800		

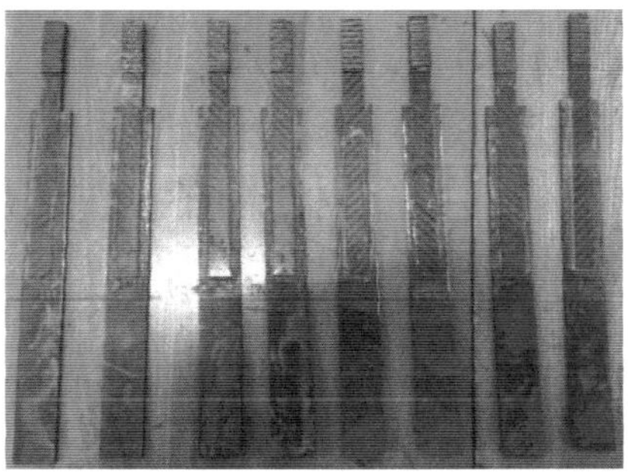

(a)

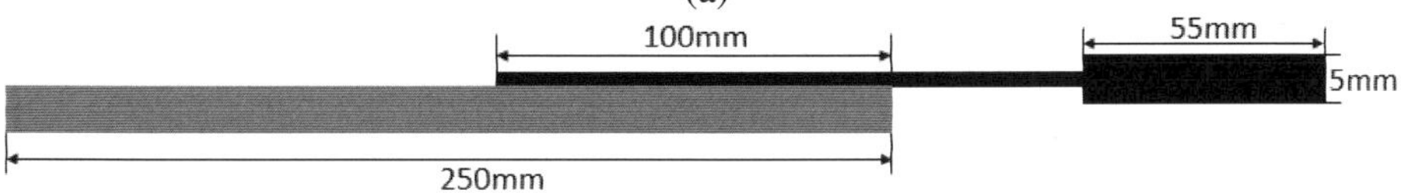

(b)

Figure 12. CFRP-steel plate specimens. (a) CFRP-steel plate from top view and (b) CFRP-steel plate from front view.

According to the experimental scheme, CFRP and CFRP-steel plate specimens are designed. The bonding length and width of CFRP-steel plate are 100 mm and 15 mm, as shown in figure 12. To obtain good bonding performance, before bonding CFRP and steel, the surface of steel plate was polished, and the surface of steel plate and CFRP was cleaned with acetone. The container for seawater is shown in figure 13, the test load device of CFRP and CFRP-steel plate is shown in figure 14.

Figure 13 The container for seawater.

Figure 14. Test device of CFRP and CFRP-steel.

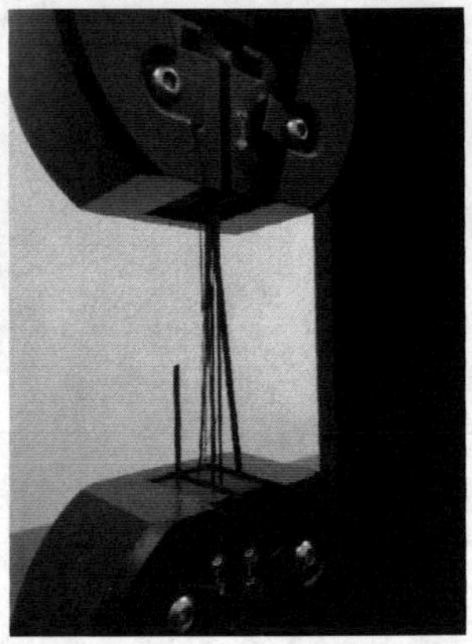

Fig. 15 Fracture of CFRP specimens.

3.2. Experimental result

When the CFRP sheet tensile test is carried out, the CFRP sheet will occasionally emit the splitting sound in the linear stage. As the load increases, the sound becomes more and more obvious. The tensile fracture of the CFRP sheet is shown in figure 15. The process of CFRP tension includes a long linear stage and a very short yield stage. When the yield stage is reached, the fiber breaks. The sound is dense when the fiber test piece is completely broken, and the break is accompanied by a lot of fiber spatter, while emitting a strong burnt smell.

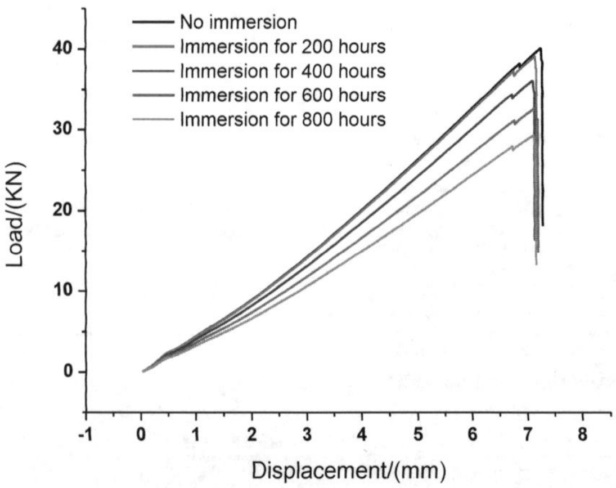

Figure 16. Relationship of load-displacement under different immersion time.

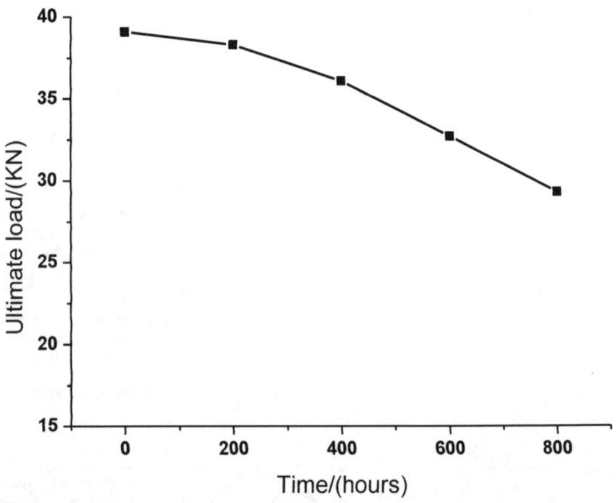

Figure 17. Relationship between ultimate load and time.

The experimental results of CFRP sheets are shown in figure 16. The ultimate loads are 38.3 KN, 36.1 KN, 32.7 KN and 29.3 KN under seawater immersion at 200, 400, 600 and 800 hours. With the increase of immersion time, the ultimate load decreases gradually. When the immersion time is longer, it decreases more quickly. The relationship between ultimate load and time is shown in figure 17.

For the CFRP-steel plate experiment, with the increase of load, the adhesive layer at the load end gradually destroys, and then the peeling of CFRP plate gradually spreads to the free end. When the peeling of CFRP plate approaches the free end or the peeling of CFRP layer occurs, the load drops rapidly and then the specimen fails. The failure modes of CFRP-steel plate under different immersion time are shown in figure 18. When there is no seawater immersion, the interfacial failure may happens from steel-adhesive, adhesive-CFRP or the fracture of CFRP (shown in figure 18(a)). When the seawater immersion time is 800 hours, it can be seen that the interface of CFRP and steel plate has been seriously eroded by seawater, resulting the debonding strength of CFRP-steel decreases greatly.

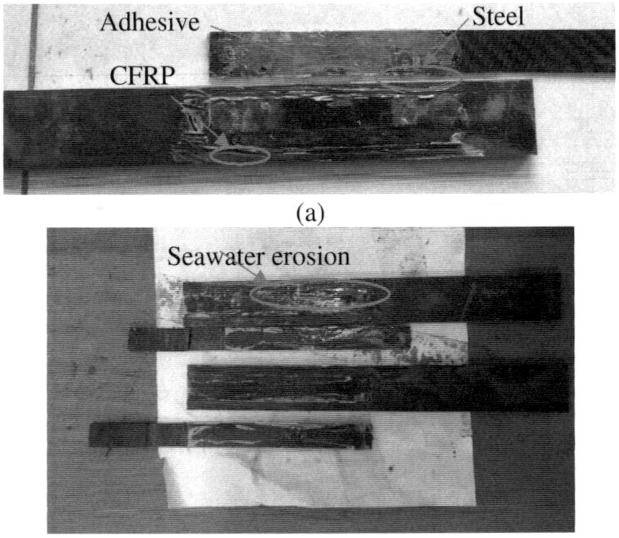

(b)

Figure 18. Failure modes of CFRP-steel plate. (a) No immersion and (b) Immersion for 800 hours.

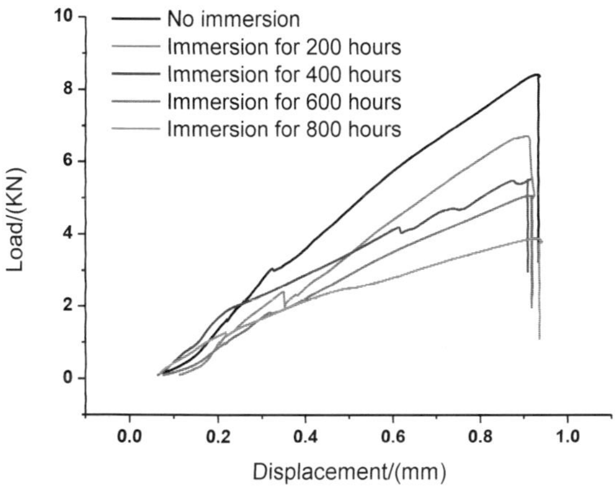

Figure 19. Relationship of load-displacement under different immersion time.

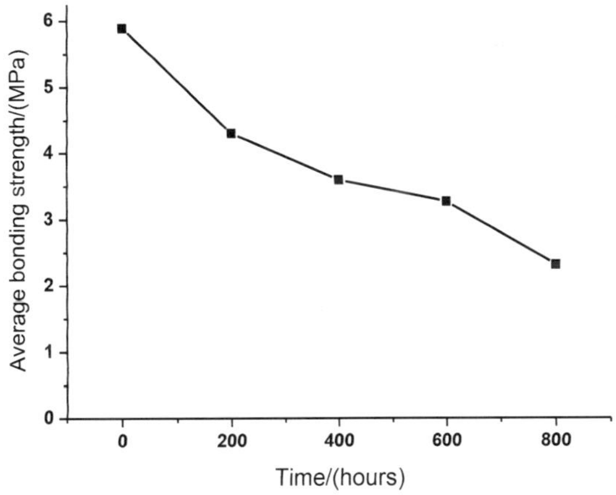

Figure 20. Relationship between average bonding strength and time.

As the load increases, the interface between CFRP plate and steel slips and gradually peels off. The load-displacement curve is basically linear before the specimen fails, and the load suddenly decreases when failure occurs. The load-displacement curves under different seawater immersion time is shown in figure 19. The ultimate loads are 6.5 KN, 5.4 KN, 4.9 KN and 3.5 KN under seawater immersion at

IOP Conf. Series: Materials Science and Engineering **634** (2019) 012028 doi:10.1088/1757-899X/634/1/012028

200, 400, 600 and 800 hours. With the increase of immersion time, the ultimate load decreases gradually. When the immersion time is 800 hours, the ultimate load decreases by 60% compared to the CFRP-steel plates with no seawater immersion, according to the analysis, it can be seen that the bonding interface is seriously eroded by seawater, and the interface is easily damaged.

The average bonding strength of CFRP and steel plate can be calculated by the ultimate load, and the relationship between time and average bonding strength is shown in figure 20. It can be found that the average bonding strength decreases with the increase of time.

From the above results, it can be found that the strength of CFRP and the bonding strength of CFRP-steel decreased as the immersion time increases, which leads to the decrease of ultimate load of repaired pipe.

4. Conclusion

In this paper, the durability of long-term bending moment and seawater immersion on CFRP repaired pipes were considered. The ultimate load of CFRP repaired pipe under different seawater immersion time and bending moments were tested by three-point bending test. Then the CFRP-steel plate and CFRP durability tests were set up to study the reasons for the decrease of ultimate load of repaired pipe. The following conclusions are obtained:

- With the increase of time, the ultimate load of repaired pipe decreases under long-term bending moment and seawater immersion. The effect of bending moment on the ultimate load of repaired pipe is less than that of immersion time. When both of bending moment and immersion time act, the ultimate load decreases seriously.
- The strength of CFRP and the bonding strength of CFRP-steel decreases with the increase of immersion time, as the immersion time increases, the speed of decline increases, when the immersion time is 800 hours, the decrease is the largest.
- Through the durability test of CFRP and CFRP-steel plate, it can be proved that the interfacial bonding strength of CFRP repaired pipe and CFRP ultimate strength decrease under long-term immersion, which lead to the decrease of ultimate load of the repaired pipe.

Acknowledgment

This research was supported by the National Natural Science Foundation of China (grant number 51779266) and the National Key Research and Development Plan (grant number 2016YFC0303705).

References

[1] Wang H Q 2014 Corrosion pipe repair technology *Oil and Gas Field Surface Engineering* **33** 109

[2] Lu M X, Chen Y F and Dong S H 2006 Pipe maintenance reinforcement technology and its development trend *City Gas* **24** 129-32

[3] Michael B J 2009 Mechanical damage in pipelines: Final report (Washington DC: Department of Transportation, PHMSA) pp 13-5

[4] Borrie D, Liu H B and Zhao X L 2015 Bond durability of fatigued CFRP-steel double-lap joints pre-exposed to marine environment *Composite Structures* **131** 799-809

[5] Kabir M H, Fawzia S and Chan T H T 2016 Durability of CFRP strengthened circular hollow steel members under cold weather: Experimental and numerical investigation *Construction & Building Materials* **123** 372-83

[6] Almosawe A, Almahaidi R and Zhao X L 2016 Bond behaviour between CFRP laminates and steel members under different loading rates *Composite Structures* **148** 236-51

[7] Nguyen T C, Bai Y and Zhao X L 2011 Mechanical characterization of steel/CFRP double strap joints at elevated temperatures *Composite Structures* **2011** 1604-12

[8] Peck J A, Li G and Pang S S 2004 Light intensity effect on UV cured FRP coupled composite pipe joints *Composite Structures* **64** 539-46

[9] Abanilla M A, Li Y and Karbhari V M 2006 Durability characterization of wet layup

graphite/epoxy composites used in external strengthening *Composites Part B Engineering* **37** 200-12

[10] Ramirez F A, Carlsson L A and Acha B A 2008Evaluation of water degradation of vinylester and epoxy matrix composites by single fiber and composite tests *Journal of Materials Science* **43** 5230-42

[11] Zhang Y, Zhang K and Zhao H K 2018 Stress analysis of adhesive in a cracked steel plate repaired with CFRP *Journal of Constructional Steel Research* **145** 210-7

[12] Kafodya I, Xian G and Li H 2015Durability study of pultruded CFRP plates immersed in water and seawater under sustained bending: Water uptake and effects on the mechanical properties *Composites Part B: Engineering* **70** 138-48

[13] Chan P H, Tshai K Y and Johnson M 2015 The flexural properties of composite repaired pipeline: Numerical simulation and experimental validation *Composite Structures* **133** 312-21

[14] Standard Practice for the Preparation of Substitute Ocean Water 2013 ASTM International: ASTM D1141-98 (2013) (West Conshohocken, PA)

[15] The American Society of Mechanical Engineers 2011 Repair and testing standard: ASME PCC-2-2011 (New York: ASME Publishing Services) pp 139-71

IOP Conf. Series: Materials Science and Engineering **634** (2019) 012029 doi:10.1088/1757-899X/634/1/012029

Preparation and application of a novel high temperature resistant filtration reducer in water-based drilling fluids

X F Chang[1], J S Sun[1,2] F Zhang[1] and K H Lv[1]

[1]School of Petroleum Engineering, China University of Petroleum (East China), Qingdao, Shandong 266580, China

E-mail: sunjinsheng@petrochina.com.cn

Abstract. 4-vinylpyridine (VP), N, N-dimethylacrylamide (DMAA), 2-Acrylamido-2-methylpropane sulfonic acid (AMPS) and sodium styrene sulfonate (SSS) were synthesized to a novel filtrate reducer in aqueous solution by free radical copolymerization using ammonium persulfate and sodium bisulfite as redox systems. The structure of copolymer was characterized by Fourier transform infrared spectroscopy (FT-IR). Thermogravimetric analysis (TGA) showed that the thermal decomposition temperature of the copolymer was higher than 310°C, which indicated that the copolymer had excellent high temperature resistance. At the same time, the rheological and filtration properties of water-based drilling fluids (WBDFs) were further evaluated by adding copolymer as a filtrate reducer. The results showed that the fluid loss (FL_{API}) of WBDFs was only 5.0 mL when the copolymer content is 2.0 wt%, Even after hot rolling at 260°C for 16 hours, the FL_{API} of WBDFs was 8.2 mL. The FL_{API} of fresh water mud was 25.2 mL, and that of aged fresh water mud was 48 mL. In addition, WBDFs with copolymer effectively resists salt and calcium intrusion before and after hot rolling at high temperature for 16 hours, which indicates that the copolymer can be used as an efficient filtration reducer in ultra-deep well drilling operations.

1. Introduction

In recent years, with the exploration and development of oil and gas resources to deep strata, drilling formation is more complex, and the requirement of temperature and salt resistance of drilling fluid is higher [1-3]. Filtration reducer is one of the key types of drilling fluid additives. The main direction is to develop filtrate reducer with high temperature, salt and calcium resistance [4,5]. In particular, the polymer filtrate reducer synthesized with AMPS as the main monomer is the most widely studied and the number of products is increasing year by year [6,7]. At present, the methods to improve the temperature resistance of such filtrate reducers are mainly focused on introducing strong adsorption groups (such as amino group, silicon hydroxyl group) into the molecular chain to improve the adsorption capacity at high temperature [8,9]. However, the structure of the main chain of C-C bond of these filtrate reducers has not changed greatly, but a new type of adsorption group has been introduced into the side chain. The effect of molecular rigidity on the temperature resistance of filtrate reducers has not been taken into account in terms of molecular flexibility, which limits the renewal speed of these products to a certain extent [10,11]. Increasing the rigidity of polymer molecular chains can reduce the thermal movement at high temperatures and slow down the desorption process of molecules at high temperatures [12-14]. DMAA, AMPS, SSS and VP were used for free radical copolymerization to synthesize a kind of Quaternary copolymer filtrate reducer with high temperature and salt-calcium corrosion resistance under redox system initiator, and added to water-based drilling fluid to further evaluate its rheological properties and filtration properties of water-based drilling fluid. It is expected to change a series of problems such as wellbore instability and collapse.

2. Materials and methods

2.1. Materials

Ammonium persulfate, (AR) sodium bisulfite (AR), DMAA (AR), AMPS, SSS and VP were purchased from Sinopharm Chemical Reagent Co., Ltd (Shanghai, China)

Bentonite-based mud was obtained by mixing 40 g bentonite in 1000 mL distilled water at high speed for 20 minutes, then keeping it in airtight condition at room temperature for 24 hours.

2.2. Synthesis of copolymer

AMPS was dissolved in 80 g distilled water, and then the pH was adjusted to a neutral environment with sodium hydroxide solution. DMAA, VP, SSS with molar ratio of 5:3:1:1 were added to the mixed solution under stirring conditions, protected by nitrogen for 30 minutes, and 0.5 wt% ammonium persulfate and sodium bisulfite were added as initiators to react at 60°C for 2 hours, then filtered in MD44 dialysis bag for 24 hours, dried and crushed to obtain the target product.

2.3. Characterization of copolymer

Fourier transform infrared spectroscopy (FTIR) of copolymer was carried out by spectrometer (IR Tracer-100 Shimadzu Corporation). Thermal degradation of copolymer was measured by thermal analyzer (TGA, Mettler Toledo Co., Switzerland).

2.4. Performance of copolymer in water-based drilling fluids

Thirty minutes of fluid loss (FL_{API}) was tested by medium pressure filter (SD3, Qingdao Haixin Optical Co., China) according to API standard which is as follows:

- Measure and record the initial temperature of the drilling fluid to the nearest degree Celsius (degree Fahrenheit).
- Inspect the cell and all components, and especially the screen, to ensure that all parts are clean and dry, and the gaskets are not cracked, distorted, or worn. Assemble the cell by placing a gasket in the base cap, place the screen on top of the gasket, place the filter paper on top of the screen, place the sealing gasket on top of the paper, then assemble the cell. Pour the drilling fluid sample into the cell to within 12 mm (0.5 in.) of the top and complete the assembly.
- Place a clean dry graduated cylinder under the drain tube to collect the filtrate. Close the relief valve and adjust the regulator so that a pressure of 690 kPa ±35 kPa (100 psi ±5 psi) is applied within 30 s or less. Start the 30 min-timer immediately at the time of pressure application.
- At the end of the 30 min-test period, measure and record the filtrate volume. Shut off the flow through the pressure regulator and open the relief valve carefully. The time interval, if other than 30 min, shall be reported.

The rheological properties of bentonite-based mud are measured using a flow rate rotating viscometer. The apparent and plastic viscosities are calculated by reading 300 and 600 rotating speeds of the viscometer.

$$\text{Apparent Viscosity(AV)}=\theta_{600}/2(\text{mPa}\cdot\text{s})$$

$$\text{Plastic Viscosity(PV)}=\theta_{600}-\theta_{300}(\text{mPa}\cdot\text{s})$$

$$\text{Yield Point(YP)}=(\theta_{300}-\text{PV})/2(\text{Pa})$$

3. Results and discussion

3.1. Characterization of copolymer

Figure 1 shows that the peak at 2944 cm^{-1} is attributed to the stretching vibration of the product methylene. The peak at 1633 cm^{-1} is attributed to the C=O bond v stretching vibration of tertiary amines in DMAA. The vibration absorption peaks at wavelength 1552 cm^{-1} and 775 cm^{-1} are caused by the vibration of pyridine group in VP molecular chain. The peak at 1490 cm^{-1} is attributed to the stretching vibration of C-N bond in DMAA. The peak absorption vibration at 1188 cm^{-1} and 1037 cm^{-1} is attributed to the stretching vibration of sulfonic groups on AMPS. The peak at 605 cm^{-1} is attributed to the stretching vibration of C-S bond. The peak at 680 cm^{-1} is the absorption peak of benzene ring. It can be seen from the infrared spectrum that the synthesized PDANS is the target product.

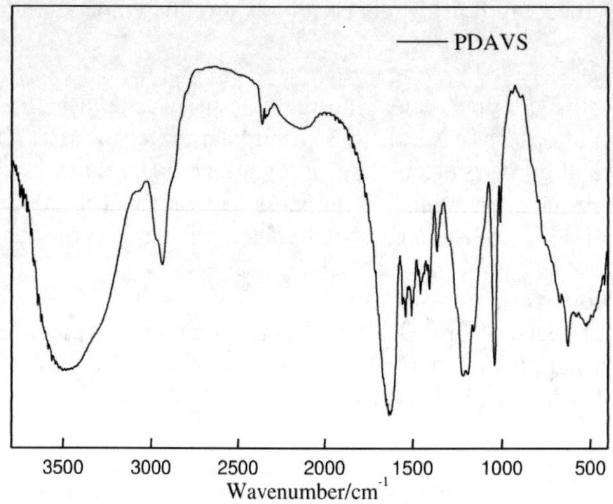

Figure 1. FT-IR of copolymer.

3.2. Thermal stability of copolymer

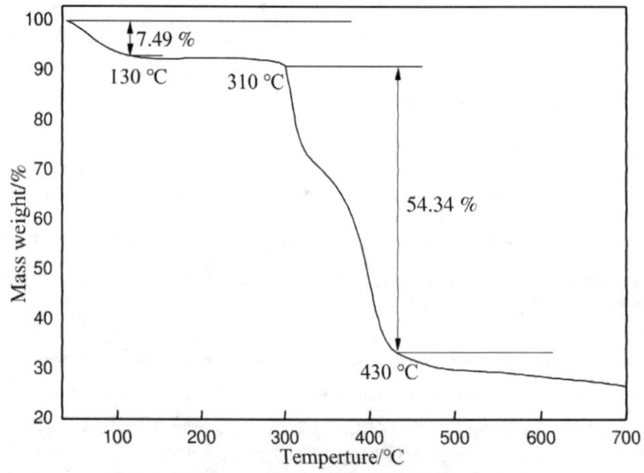

Figure 2. TGA of copolymer.

As shown in figure 2, the temperature of the copolymer decreases as the high temperature. The mass loss rate of the copolymer was 7.49% in the temperature range of 40 to 130°C. It is because the copolymer molecule contains adsorption groups, which can absorb moisture in the air. The temperature rise leads to the evaporation of free water, which leads to the mass loss of the copolymer in the temperature range. The mass loss rate of the copolymer was 54.34% in the temperature range of 310 to 430°C. It is because the side chains of amide and sulfonic acid groups in the copolymer molecular chains are thermally degraded and separated from the main chains with the further increase of temperature, which results in the mass loss of the copolymer at this stage. The mass loss of the copolymer was 38.17% after 430°C. In this stage of mass loss, the main chain of the copolymer is basically broken and decomposed at high temperature. Generally speaking, copolymers have excellent thermal stability.

3.3. Performance evaluation of copolymer in WBDFs

Figure 3 shows that the increase in viscosity of bentonite base mud is attributed to the addition of a large number of copolymers. When the concentration of copolymer is 2%, AV is 34 mPa·s, and it still reaches 27 mPa·s after high temperature aging, the FL_{API} was 5.0 mL and 8.2 mL after high temperature aging. It is due to the high molecular weight of copolymer, when it is added as drilling fluid additive, it will increase the viscosity of WBDFs. At the same time, due to the extension of copolymer macromolecular chains in water-based drilling fluids, the adsorbing groups on the side chains of molecular chains can effectively adsorb clay particles and prevent clay particles from coalescing, thus reducing the amount of fluid loss [15,16].

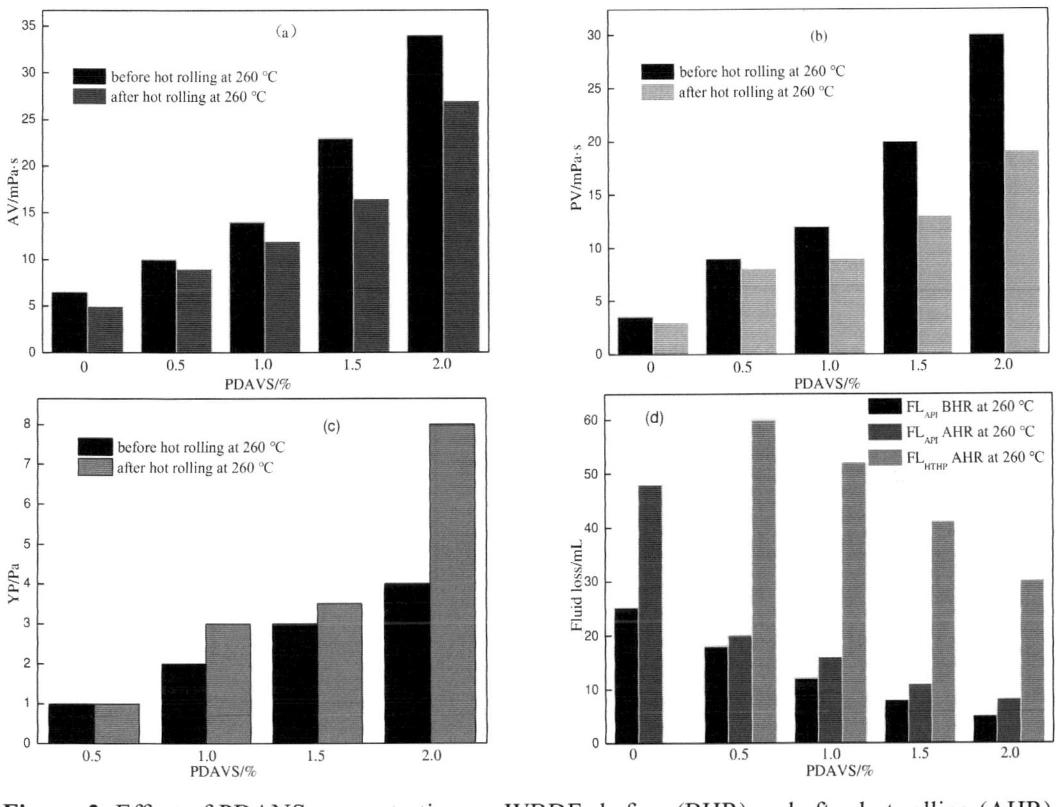

Figure 3. Effect of PDANS concentration on WBDFs before (BHR) and after hot rolling (AHR) at 260°C for 16 h.

Figure 4 shows that Viscosity of bentonite-based mud containing copolymer decreases at high temperature, while FL_{API} and FL_{HTHP} increase gradually. The FL_{API} is 8.2 mL when the temperature is 260°C. In high temperature environment, the degradation and hydrolysis of copolymers are the main reasons for the chemical degradation of copolymers [17-19]. This chemical degradation will result in the change of functional groups of side chains, the decrease of molecular weight and intrinsic viscosity of copolymers, and eventually lead to the failure of copolymers, resulting in the increase of fluid loss in WBDFs [20,21]. Compared with other products, the new product PDANV has the most obvious effect in reducing drilling fluid filtration.

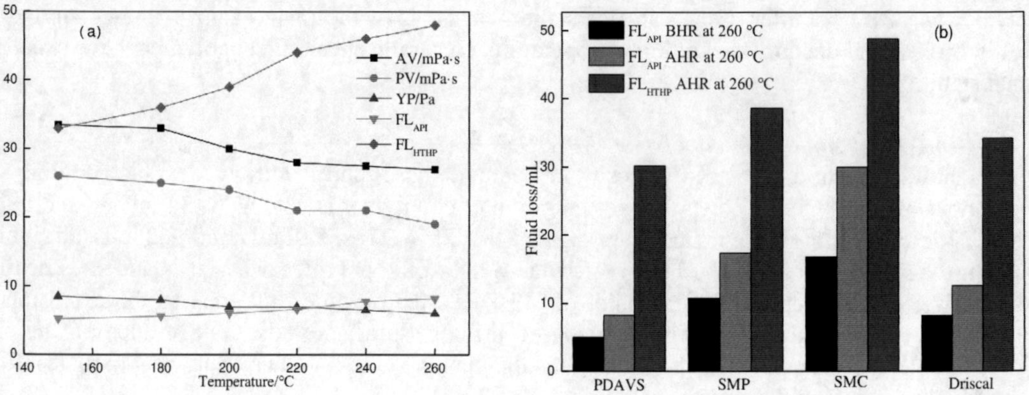

Figure 4. Effect (a) of temperature on performance of water-based drilling fluids with PDANS and comparisons with other products (b).

Table 1 shows that when a large number of salt ions are added, the viscosity of water-based drilling fluid decreases gradually before and after hot rolling at high temperature. When the salt content is 35%, the apparent viscosity is 28.5 mPa·s, and the AV can still reach 15 mPa·s after hot rolling at high temperature. At the same time, the FL_{API} is 11.2 mL and 14.5 mL after hot rolling, which shows that the product is insensitive and can effectively protect the stability of drilling fluid in salt-gypsum formation.

Table 1. Effects of NaCl concentration on WBDFs with PDANS.

NaCl/%	AV/mPa·s		PV/mPa·s		FL_{API}/mL		YP/Pa	
	BHR	AHR	BHR	AHR	BHR	AHR	BHR	AHR
0	34.0	27	30	19	5.0	8.2	4.0	8
5	33.0	20	21	18	6.6	7.4	12.0	2
10	32.0	18	21	17	6.8	8.9	11.0	1
20	30.0	17	20	16	8.5	9.6	10.0	1
35	28.5	15	19	14	11.2	14.5	9.5	1

Table 2. Effects of $CaCl_2$ concentration on WBDFs with PDANS.

$CaCl_2$/%	AV/mPa·s		PV/mPa·s		FL_{API}/mL		YP/Pa	
	BHR	AHR	BHR	AHR	BHR	AHR	BHR	AHR
0	34.0	27	25	19	5.0	8.2	4.0	8
0.5	32.5	25	26	20	5.8	10.5	7.5	5
1.0	30.0	22	24	18	7.6	14.0	4.0	4
2.0	27.0	20	14	14	9.7	16.4	3.0	3

Table 2 shows that with the increase of calcium content, the viscosity of water-based drilling fluid decreases gradually before and after hot rolling at high temperature. When the salt content is 2%, the AV is 27 mPa·s, and the AV can still reach 20 mPa·s after hot rolling at high temperature. At the same time, the FL_{API} is 9.7 mL and 16.4 mL after hot rolling, which indicates that the copolymer is insensitive to calcium intrusion.

The toxicity and biodegradability of PDAVS were determined by modern water Delta tox II, BOD rapid determinator and COD determinator (LEICI-571) respectively. It can be seen from table 3 that PDAVS is an environmentally friendly filter loss reducer.

Table 3. Environmental friendly evaluation of PDAVS.

sample	EC_{50} (mg/L)	toxicity	BOD_5/COD (%)	Biodegradability
PDAVS	30500	Non-toxic	26.4	Easily degraded

4. Conclusions

4-vinylpyridine (VP), N, N-dimethylacrylamide (DMAA), 2-Acrylamido-2-methylpropane sulfonic acid (AMPS) and sodium styrene sulfonate (SSS) were successfully synthesized to a novel filtrate reducer in aqueous solution by free radical copolymerization using ammonium persulfate and sodium bisulfite as redox systems. FT-IR spectra showed that the target product has been successfully synthesized and had excellent thermal stability. Meanwhile, the properties of WBDFs were further revealed by adding copolymer as a filtrate reducer. The results revealed that the FL_{API} of WBDFs was only 5.0 mL when the copolymer content is 2.0 wt%, Even after hot rolling at 260°C for 16 h, the FL_{API} of water-based drilling fluids was 8.2 mL. In addition, water-based drilling fluid containing copolymer effectively resists salt and calcium intrusion. Therefore, copolymers are recommended as effective filtrate reducers for ultra-deep well drilling operations with high salinity.

References

[1] Xu C C, Zou W H, Yang Y M *et al* 2017 Status and prospects of exploration and exploitation of the deep oil & gas resources onshore China *Nat. Gas Geosci* **28** 1139-53
[2] Zhao Z, Shujuan Xu, Jiang X *et al* 2016 Deep strata geologic structure and tight sandy conglomerate gas exploration in Songliao Basin, East China *Pet. Explor. Dev* **43** 13-25
[3] Dong D, Wang Y, Li X *et al* 2016 Breakthrough and prospect of shale gas exploration and development in China *Nat. Gas Ind B* **3** 12-26
[4] Mao H, Qiu Z, Shen Z *et al* 2014 Synthesis and mechanism of hydrophobic associated polymer based nano-silica filtrate reducer *Acta Pet. Sin* **35** 771-8
[5] Cheng M A, Zhen J, Wang Z *et al* 2014 Synthesis and evaluation of filtrate reducer SPH-F for calcium chloride clay-free drilling fluids *Drill. Fluid Completion Fluid* **2014** 95-102
[6] Tang X, Yuan B, Yang Y *et al* 2016 Preparation and performance of AMPS/AA/DMAA/SA copolymer as a filtrate reducer for oil well cementing *J. Appl. Polym. Sci* **133** 43824 (1-9)
[7] Zhang Y M, Zhu H, Wang F H *et al* 2008 Preparation and characterization of CTAB/polymer composite modification of montmorillonite as filtrate reducer *J. Funct. mater* **12** 2028-2031
[8] Zhao X, Xie J and Zhang L 2012 Synthesis and properties of high temperature-resistant phenyl cationic starch filtrate reducer *Petrochem. Technol* **41** 801-805
[9] Yu W U, Sun D, Tao W U *et al* 2001 Study on mud properties of AMPS/AM/IA terpolymer as temperature resistant and salt tolerant filtrate reducer *Oilfield Chem* **18** 101-104
[10] Sun J, Ning F, Wu N *et al* 2015 The effect of drilling mud properties on shallow lateral resistivity logging of gas hydrate bearing sediments *J. Pet. Sci. Eng* **127** 259-69
[11] Xin J, Xi F, Lv D *et al* 2011 The effect of the chitosan membrane properties on the enzyme adsorption and performance for the construction of horseradish peroxidase biosensors *Carbohydr. Polym* **85** 786-91
[12] Palmeri J, Manghi M and Destainville N 2008 Thermal denaturation of fluctuating finite DNA chains: the role of bending rigidity in bubble nucleation *Phys. Rev. E: Stat., Nonlinear, Soft Matter Phys* **77** 011913
[13] Tendero R, Miguel P, Jimenez *et al* 2006 Viscoelastic and thermal characterization of crosslinked PVC *Eur Polym J* **42** 961-9
[14] Fenouillot F, Cassagnau P and Majesté J C 2009 Uneven distribution of nanoparticles in immiscible fluids: Morphology development in polymer blends *Polym* **50** 1333-50
[15] Mao H, Qiu Z, Shen Z, et al 2015 Hydrophobic associated polymer based silica nanoparticles

composite with core–shell structure as a filtrate reducer for drilling fluid at utra-high temperature *J. Pet. Sci. Eng* **129** 1-14

[16] Ritter H and Spee A G V 1994 Bacterial production of polyesters bearing phenoxy groups in the side chains, 1. Poly (3-hydroxy-5-phenoxypentanoate-co-3-hydroxy-9-phenoxynonanoate) from Pseudomonas oleovorans *Macromol. Chem. Phys* **195** 1665-72

[17] Zaikov G E and Livshitz V S 1987 The mechanism of chemical degradation of polymers: Part III—The anomaly in the hydrolysis of glycolide copolymers *Polym. Degrad. Stab* **17** 65-9

[18] Siracusa V, Rocculi P, Romani S, et al 2008 Biodegradable polymers for food packaging: a review *Trends Food Sci. Technol* **19** 634-643

[19] Middleton J C, Tipton A J 2000 Synthetic biodegradable polymers as orthopedic devices *Biomater* **21**(23) 2335-2346

[20] Peron J, Shi Z and Holdcroft S 2011 Hydrocarbon proton conducting polymers for fuel cell catalyst layers *Energy Environ. Sci* **4** 1575-91

[21] Chang X F *et al* 2019 Synthesis of a novel environment-friendly filtration reducer and its application in water-based drilling fluids *Colloids Surf., A* **568** 284-93

Prediction of parameters of microscale coating-metal interface phase based on finite element method

Z K Jia[1], Y Zhang[1,2], J H Xin[1] and Q Cheng[1]

[1]College of Safety and Ocean Engineering, China University of Petroleum, Beijing102249, China

E-mail: zhangyu@cup.edu.cn

Abstract. The coating-metal transition interface can form a stable gradient interface phase composite. In this paper, a finite element model of gradient interface phase is established based on Tyson Polygon Method (TPM), and the microstructure equivalent parameters of interface material of the element are obtained by means of average method. By analyzing the functional relationship between the equivalent elastic parameters of the gradient interface element and the metal volume ratio, the functional expression of the microstructure parameter distribution of the interface phase is obtained. The results show that the equivalent elastic modulus and equivalent shear modulus of the interface satisfy the distribution of exponential function, linear function and power function in the local scope. The change rule of equivalent Poisson's ratio with metal phase volume fraction in local scope can be expressed by logistic function. It is found that the global parameter distribution of interface phase is different from that of local material. The equivalent elastic modulus and equivalent shear modulus of the global interface satisfy the double exponential function distribution, while the distribution law of the equivalent Poisson's ratio can be piecewise expressed by the logistic function and the exponential function.

1. Introduction

Corrosion is one of the most important failure modes of metal structure in ships and marine engineering etc, and anti-corrosion coating is a widely used method to preventing corrosion. When spraying anti-corrosive coating, interfacial phase materials with different properties of metal and coating materials are often formed [1-3], as shown in figure 1(b) [4]. Although the thickness of interface phase is very small, its structure and properties largely determine the interface strength.

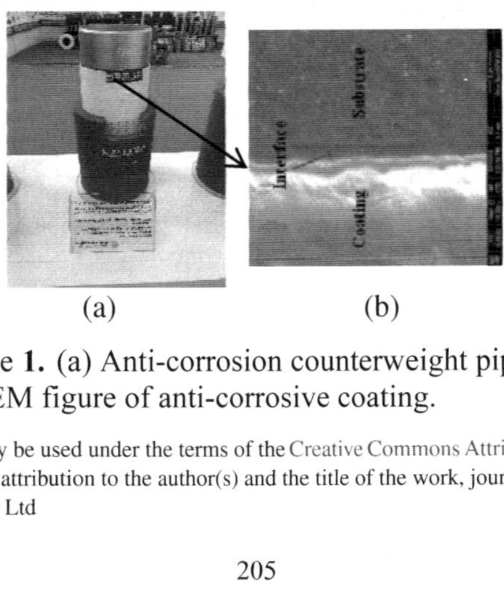

(a) (b)

Figure 1. (a) Anti-corrosion counterweight pipe and (b) SEM figure of anti-corrosive coating.

Structure and performance parameters of the interface phase have been widely concerned. The third-phase material between coating and metal has regular characteristics, and many studies about microscle parameters have been conducted. Sun *et al* [5] studied the influence of interfacial phase parameters on the uniaxial tensile behavior of ceramic matrix composites, and discussed the effects of interfacial phase component volume fraction, elastic modulus and Poisson's ratio on tensile behavior. However, the microscale elastic constant of the interface phase is heavily dependent on the experimental value. Li *et al* [6] established the composite material microscopic representative cellular model, four kinds of Voronoi cells with mathematical and statistical significance were established by using TPM. Leslie [7] calculates effective elastic material constants of composite materials by standard mechanics method and homogenization method, and discusses the influence of reinforcement arrangement on material performance parameters. Wakashima [8] proposed to establish the stress-strain relationship between the macroscopic gradient interface and the microscopic cellular model by applying the classical laminated plate theory. Li *et al* [9] studied the load-deformation behavior of the typical sub-laminated plate and derived the effective elastic constant from the properties of its constituent layers. The explicit expression of effective elastic constant of thick plate is given. Some scholars assume that the parameters change according to specific functions, such as exponential function [10,11] or power function [12,13]. At the same time, in order to consider the arbitrary variation form of material parameters, Wang put forward the laminated plate model [14], and Gao put forward a new layered model [15] based on the fact that any continuous curve can be approximated by a series of broken lines. Experiments have confirmed that the mechanical properties of the interface change regularly with the increase of the volume fraction of a phase [16]. In addition, the above studies based on theoretical analysis often assume that the mechanical parameters of the interface conform to a certain function distribution without further study. In this paper, the functional relations of equivalent parameters of gradient interface are predicted by studying the microscale parameters properties of interface materials by the numerical fitting method.

2. Interface layer model and finite element realization

2.1. Interface layer model

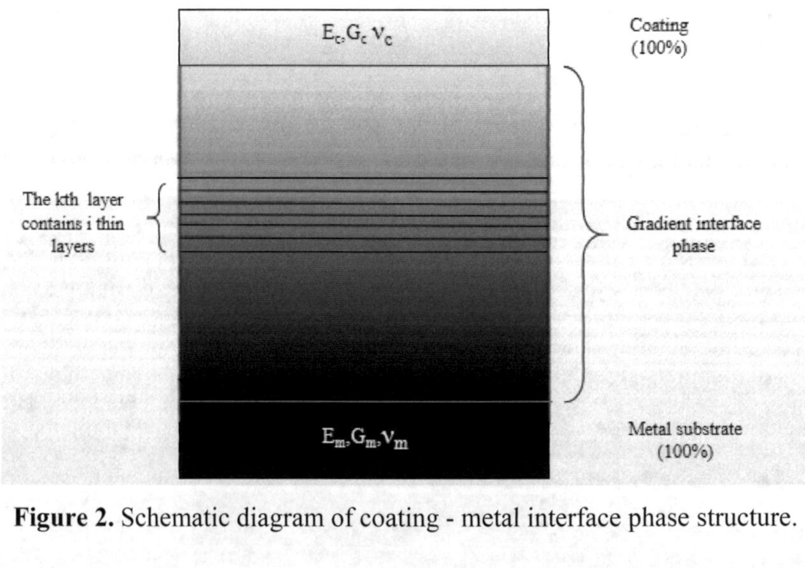

Figure 2. Schematic diagram of coating - metal interface phase structure.

In the coating - metal gradient interface phase, it is assumed that the volume fraction of the metal phase changes uniformly along the thickness direction, as shown in figure 2. In this paper, the *k*th thick layer of interface phase material is selected randomly to predict the changes of local interface phase material parameters. Since the model size established in this paper is very small, it is assumed that the thin layer i within the thick layer is an ideal single layer without thickness, and the stress and displacement are continuous at and on the upper and lower surfaces of the thin layer. In this paper, based on the homogenization theory, by calculating the equivalent material parameters of each thin layer, the relationship between the interface phase equivalent parameters and the volume fraction of the metal phase is established. On this basis, the mechanical properties of the global interface phase are analyzed.

2.2. Thin layer microscopic model
The geometrical structure distribution of coating metal interface materials is random and discrete. The representative volume element (RVE) is established by using TPM [17] and the basic idea of this method is to construct a Tyson polygon containing only one discrete point data by connecting the vertical bisecting lines of the two adjacent points. The points in the Tyson polygon are closest to the corresponding discrete points, and the points on the side of the Tyson polygon are equal to the discrete points on both sides. The unit cell established by this method can fully explain the micro structure and profit strain field distribution of composite materials.

Metal particles can be virtually of any shape. However, in order to simplify the calculation, we choose the cube shape inclusion as the particle shape and the coating was used as the matrix to seamlessly fill the cube particles. Suppose the metal particles are arranged in a simple cubic way as one of the spatial arrangements of metal crystals. By using the Tyson polygon method, as shown in figure 3(a), combining the translation and symmetry properties of the unicellular geometric structure, we establish the 1/8 interface layer single-layer unit as shown in figure 3(b).

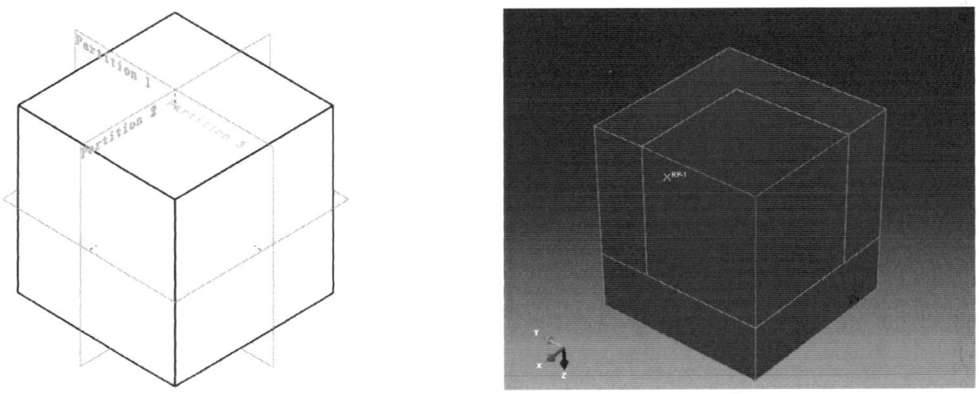

(a) (b)

Figure 3. (a) Schematic diagram of division unit by Tyson polygonal method and (b) Schematic diagram of single-layer cell of 1/8 interface phase.

2.3. Periodic boundary conditions and its finite element realization

2.3.1. Periodic boundary conditions. In the direction perpendicular to the interface layer, the material properties vary with the different components of the two-phase mixture, and show a certain regularity. However, in each single layer, the micro structure presents a good periodicity. In the finite element calculation, the stress and strain distribution at the material microscopic level can be obtained by establishing single layer cell model of the interface and applying certain boundary conditions. Reasonable boundary conditions are the key to ensure accurate calculation. Xia *et al* [18] proposed uniform periodic boundary conditions for a cellular model with relatively parallel boundary surfaces. For materials with periodic microscopic cell structure, the expression of periodic displacement field is as follows:

IOP Conf. Series: Materials Science and Engineering **634** (2019) 012030 doi:10.1088/1757-899X/634/1/012030

$$u_i = \bar{\varepsilon}_{ik} x_k + u_i^* \tag{1}$$

Where: $\bar{\varepsilon}_{ik}$ is the average strain of the unit cell; x_k is the coordinates of any point in the unit cell; u_i^* is the correction of periodic displacement.

Xia *et al* [18] used the symmetry property of cells to establish the relative displacement relationship between parallel planes, and obtained the expression of general periodic boundary conditions:

$$u_i^{j+} - u_i^{j-} = \bar{\varepsilon}_{ik}(x_k^{j+} - x_k^{j-}) = \bar{\varepsilon}_{ik} \Delta x_k^j \tag{2}$$

Obviously, the above equation satisfies the periodicity and continuity of the displacement on the boundary, and the displacement difference of the corresponding node on the parallel periodic boundary remains constant. Therefore, the above equation is adopted to apply boundary conditions to the unit cell in this paper.

2.3.2. Finite element realization of periodic boundary conditions. In the micro-mechanical analysis of composite materials, RVE can be a complete unit cell or 1/8 unit cell. However, periodic boundary conditions are closely related to the selection of unit cell. According to the different unit cell model, Li gives the simplified model considering mirror symmetry boundary conditions and not to consider a complete model of mirror symmetry boundary condition equation [6,19]. Considering model simplification, 1/8 cell was established as the interface layer microscopic mechanics finite element model analysis, the corresponding periodic boundary conditions are as follows [20]:

The boundary conditions for the case under ε_x^0 are in equation (3)

$$\begin{aligned} u|_{x=0} &= 0 \text{ and } u|_{x=bx} = b_x \varepsilon_x^0 \\ v|_{y=0} &= 0 \text{ and } v|_{y=by} = b_y \varepsilon_y^0 \\ w_{z=0} &= 0 \text{ and } w|_{z=bz} = b_z \varepsilon_z^0 \end{aligned} \tag{3}$$

With similar considerations as given above, the boundary conditions for the unit cell under, ε_y^0 and ε_z^0 are identical to those in equation (3). However, the boundary conditions under the shear load with mirror symmetry are completely different from equation (3).

The boundary conditions for the case under γ_{yz}^0 are in equation (4).

$$\begin{aligned} & u|_{x=0} = 0 \;;\; u|_{x=bx} = 0 \\ & u|_{y=0} = w|_{y=0} = 0 \;;\; u|_{y=by} = w|_{y=by} = 0 \\ & u|_{z=0} = v|_{z=0} = 0 \;;\; u_{z=bz} = v_{z=bz} = b_z \gamma_{yz}^0 \end{aligned} \tag{4}$$

With similar considerations as given above, the boundary conditions for the unit cell under, γ_{xy}^0 and γ_{xz}^0 are identical to those in equation (4).

The above formula is the boundary condition of the constraint surface, and different edge and corner nodes have different boundary conditions.

3. Prediction of effective elastic parameters of interface layer

3.1. Calculation method of material parameters in RVE

The unit cell of interface phase can be regarded as homogeneous anisotropy, and its equivalent constitutive relation vector can be expressed as

$$\bar{\varepsilon}_i = S_{ij}\bar{\sigma}_j, (i, j = 1,2,3,4,5,6) \tag{5}$$

Where S_{ij} is the flexibility matrix, $\bar{\varepsilon}_i, \bar{\sigma}_j$ are the mean stress and mean strain of the unit cell respectively. Where the relation between the flexibility coefficient matrix and the elastic constant is:

$$S_{ij} = \begin{vmatrix} \frac{1}{E_1} & \frac{\nu_{12}}{E_2} & \frac{\nu_{31}}{E_3} & 0 & 0 & 0 \\ & \frac{1}{E_2} & \frac{\nu_{32}}{E_3} & 0 & 0 & 0 \\ & & \frac{1}{E_3} & 0 & 0 & 0 \\ & & & \frac{1}{G_{12}} & 0 & 0 \\ & symm & & & \frac{1}{G_{23}} & 0 \\ & & & & & \frac{1}{G_{13}} \end{vmatrix} \tag{6}$$

Where E_x, E_y, E_z are elastic modulus of interface layer element material in *x, y* and *z* directions respectively; G_{xy}, G_{yz}, G_{xz} are shear modulus in plane *xy, yz* and *xz* respectively; $\mu_{xy}, \mu_{yz}, \mu_{xz}, \mu_{yx}, \mu_{zy}, \mu_{zx}$ are the Poisson's ratio in plane *xy, yz, xz, yx, zy* and *zx* respectively.

In this paper, six sets of linearly independent boundary conditions of strain load equation are applied to 1/8 cellular model respectively.

For cubic cells, in this paper, unit cell strain ε_i is already know as the displacement boundary condition, and σ_j can be expressed as:

$$\bar{\sigma}_j = \frac{(P_i)_j}{S_j} \tag{7}$$

Where $(P_i)_j$ is the sum of the constrained counter-forces in the direction of node *i* on plane *j*; S_j is the *j* plane of the unit cell.

The elastic constant of a unit cell can be calculated from equation (8).

$$\begin{aligned} E_i &= \frac{\bar{\sigma}_i}{\bar{\varepsilon}_i}(i, j = 1,2,3;) \\ \mu_i &= -\frac{\bar{\varepsilon}_j}{\bar{\varepsilon}_i}, (i, j = 1,2,3,4,5,6;) \\ G_i &= \frac{\bar{\sigma}_i}{\bar{\varepsilon}_i}, (i, j = 4,5,6;) \end{aligned} \tag{8}$$

3.2. Finite element calculation of element material parameters

Finite element method is widely used to solve material characteristic parameters. In this paper, finite element large commercial software ABAQUS is used to calculate the material parameters of unit cell

IOP Conf. Series: Materials Science and Engineering **634** (2019) 012030 doi:10.1088/1757-899X/634/1/012030

of interface phase. The unit cell size established in this paper is 2×10^{-2} um. In order to make the mesh division smooth, we used the swept central axis algorithm to divide the mesh, and selected C3D8R linear hexahedral element. The metal phase and the matrix phase in the model are bound together. In this paper, the epoxy coating commonly used in underwater anti-corrosion equipment and the material properties of steel pipe are selected, as shown in table 1. Boundary conditions are applied as described above. It should be noted that when there are common edge nodes on both sides, boundary constraints must be imposed on edge nodes alone to prevent excessive constraints, as well as common angle nodes.

Table 1. Properties of metal particles and epoxy materials.

Material	E (GPa)	ν
Epoxy matrix	3.43	0.38
Metal particles	207	0.3

In this paper, the global size of grid is 0.46 and 0.30 respectively to study the size of this effect. As shown in table 2, material parameters of cells with metal phase volume fraction of 30%, 40% and 50% are calculated in both grid sizes, and the parameter corresponding to the 0.30 grid ruler in brackets. It can be seen that grid has little influence on the calculation results. In this paper, the global size of the grid is selected as 0.46, and the divided cell model is shown in figure 4.

Table 2. The calculation difference of effective material parameters with global grid size between 0.46 and 0.30.

volume ratio of Metal (%)	$E_1=E_2=E_3$(GPa)	$G_{12}=G_{23}=G_{13}$(GPa)	$v_{12}=v_{13}=v_{23}=v_{21}=v_{32}=v_{31}$
30	8.45318 (8.4642)	2.19234 (2.19862)	0.315176 (0.31489)
40	11.79322 (11.82504)	2.6409 (2.6495)	0.284402 (0.285554)
50	16.50422 (16.63406)	3.22862 (3.25118)	0.241836 (0.252542)

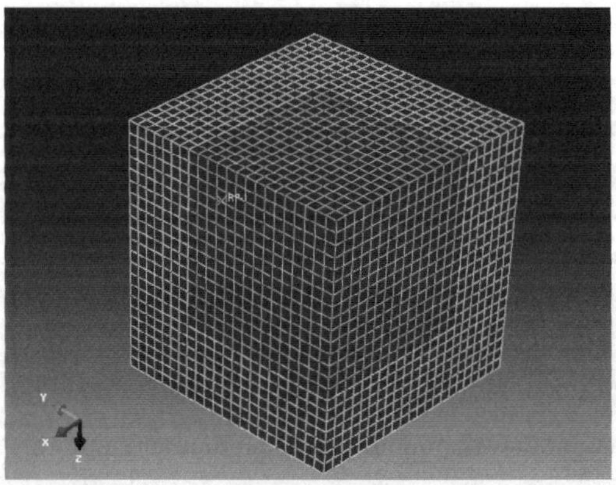

Figure 4. Mesh of the unit cell for interface layer single layer.

When the y-plane of the interface unit is subjected to a displacement load of 0.005×10^{-3} um, the stress and displacement fields of the unit cell are obtained through finite element calculation, as shown in figures 5(b) and 5(d). As the material unit is symmetric in the direction of the coordinate system spindle, the finite element calculation results show that the symmetry still exists in the stress, strain, and displacement fields of the unit cell.

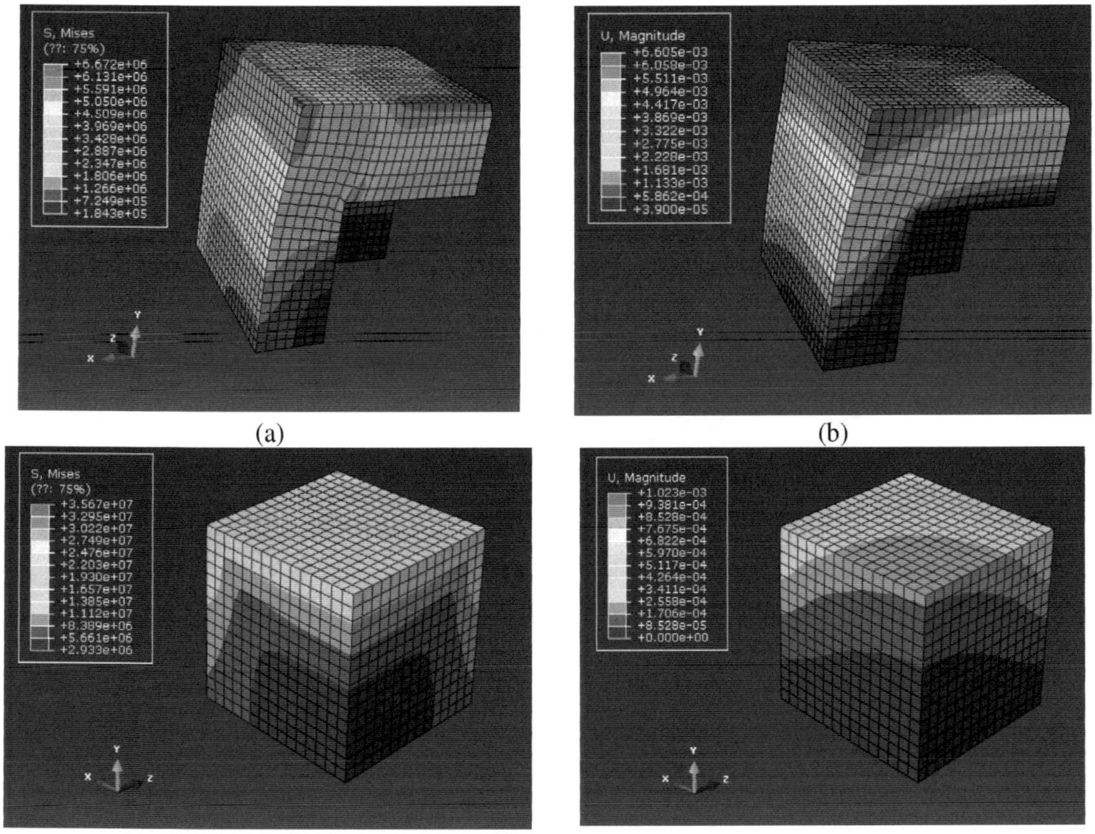

Figure 5. Unit cell finite element calculation results with 30% volume fraction of metal phase: (a) contours of von Mises stress without coating matrix. (b) contours of equivalent displacement without coating matrix. (c) contours of von Mises stress with metal particles. (d) contours of equivalent displacement with metal particles.

4. Results

The relationship between the volume fraction of the metal phase and the material parameters of the local interface layer and the whole interface phase is discussed below.

4.1. Parameter properties of local interface layer

In the finite element calculation results of the model, the stress, strain, and displacement fields of the analysis model are given priority to, and the material microscopic parameters are obtained through the calculation of equations (5)-(8). The finite element model established in this paper, due to the symmetrical nature of the geometric structure, presents a high degree of orthogonal isotropy. Therefore, in the subsequent result analysis, plane y is selected to apply the load for solution. The calculated parameters of the model are shown in table 3.

The functional relationship between material parameters of local interface phase and volume fraction of metal phase can be obtained by interpolation fitting. As shown in figure 6(a), when the volume fraction of the metallic phase changes from 30% to 40%, the linear characteristics of the equivalent elastic modulus of the thin layer material of the interface phase are significant. At the same time, it can be found that the change relation of equivalent elastic modulus of local interface layer materials can also be well expressed by exponential function and power function. Considering the simplicity and accuracy of the function form, it is recommended to use the linear function as the change rule of the equivalent elastic modulus of the local interface phase.

Table 3. Effective material properties of interface unit cell for the volume ratio of metal from 30% to 40%.

volume ratio of metal (%)	E_1=E_2=E_3(GPa)	G_{12}=G_{23}=G_{13}(GPa)	$\nu_{12}=\nu_{13}=\nu_{23}=\nu_{21}=\nu_{32}=\nu_{31}$
30	8.4532	2.1923	0.3152
31	8.7366	2.2327	0.3101
32	9.0304	2.2739	0.3094
33	9.3349	2.3162	0.3093
34	9.6507	2.3586	0.3089
35	9.9778	2.4040	0.3059
36	10.3170	2.4499	0.3005
37	10.6608	2.4926	0.2935
38	11.0244	2.5407	0.2871
39	11.4014	2.5901	0.2849
40	11.7932	2.6409	0.2844

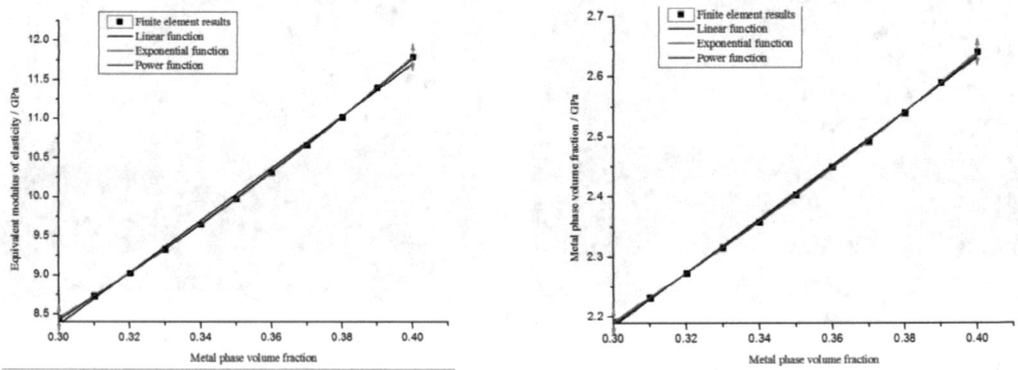

(a) (b)

Figure 6. (a) Diagram of the relationship between the equivalent elastic modulus of the local material in the interface phase and the volume fraction of the metal phase; (b) Diagram of the relationship between the local equivalent shear modulus of the interface phase and the volume fraction of the metallic phase.

Figure 6(b) is an interpolation diagram of the equivalent shear modulus of the interface phase material and the volume fraction of the metal phase. The change relation of the equivalent shear modulus of the local interface layer material with the increase of the metal phase is similar to that of the equivalent elastic modulus, which can also be expressed in three functional forms.

As shown in figure 7, the linear or exponential relationship can only be established between very small layers. In other words, the material parameter only satisfies the linear distribution or exponential distribution within the small change range of the volume fraction of the metal phase. when the local interface layer contains more thin layers, the change trend of material equivalent Poisson's ratio and metal phase volume fraction is closer to the logistic growth model, and the form of logistic function is shown in equation (9), where the first term represents the Poisson's ratio component of the surface in the cell affected by the lateral deformation of the metal-phase restricted cell, and the second term represents the Poisson's ratio component of the surface in the cell not affected by the lateral deformation of the metal-phase restricted cell.

$$v = \frac{A_1 - A_2}{1+(\mathrm{v}/\mathrm{v}_0)^p} + A_2 \tag{9}$$

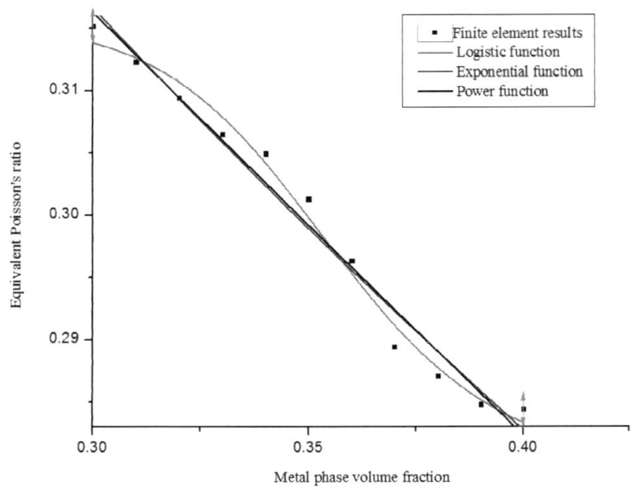

Figure 7. Diagram of the relationship between local material equivalent Poisson's ratio of interface phase and volume fraction of metal phase.

When the metal phase volume fraction is between 30% and 60%, the equivalent Poisson's ratio of the interface phase material changes greatly. When the volume fraction of the metal phase exceeds 36%, the equivalent Poisson's ratio of the composite interface is smaller than the Poisson's ratio of the two single-phase materials. This is because the reaction between the metal particles and the substrate contact surface in the model greatly limits the lateral deformation of the composite system, thereby reducing the equivalent lateral strain of the unit cells.

4.2. Parameter characteristics of global interface phase

Generally, researchers believe that the variation law of the global material parameters of the gradient interface phase is the same as that of the local interface phase. In this paper, thin material parameters with volume fraction of 0, 10%, 20%, 30%, 40%, 50%, 60%, 70%, 80%, 90% and 100% were selected to fit the relationship between the global material parameters of interface phase and the volume ratio of metal phase. The selected thin material parameters are shown in table 4.

Table 4. Effective material properties of interface monolayer model.

volume ratio of Metal (%)	E_1=E_2=E_3 (GPa)	G_{12}=G_{23}=G_{13} (GPa)	v_{12}=v_{13}=v_{23}= v_{21}=v_{32}=v_{31}
0	3.4300	1.2428	0.3800
0.1	4.5322	1.5284	0.3649
0.2	6.1268	1.8351	0.3422
0.3	8.4532	2.1923	0.3152
0.4	11.7932	2.6409	0.2844
0.5	16.5042	3.2286	0.2418
0.6	23.5278	4.1029	0.2222
0.7	33.9836	5.4632	0.1962
0.8	52.4520	8.0405	0.1840
0.9	89.7366	14.7388	0.1854
1	207	79.6154	0.3000

Exponential function, double exponential function and power function were used to fit the volume fraction of the metal phase and the material parameters of the global interface phase. The fitting effect of the exponential function and the power function is poor. The numerical results show that the relationship between the global equivalent elastic modulus of the interface phase material and the metal phase gradient can be well expressed by the double exponential function, as shown in figure 7(a). The form of the double exponential function is shown in equation (10) ,which reflects the contribution of the two components of the interface phase to the equivalent elastic modulus.

$$E = E_0 + A_1 e^{A_2 v} + A_3 e^{A_4 v} \tag{10}$$

When the volume fraction of the metallic phase is lower than 80%, the equivalent elastic modulus of the gradient interface phase material changes slowly with the volume. In combination with the research of section 4.1, it is found that when the volume fraction of the meta phase is lower than 80%, the local interface phase material parameter changes have the characteristics of linear distribution. When the volume fraction of the metal phase exceeds 80%, the equivalent elastic modulus of the interface phase increases rapidly in a small range, which is consistent with the change characteristics of double-exponential function. The equivalent shear modulus of the global interface phase has similar laws, as shown in figure 8(b).

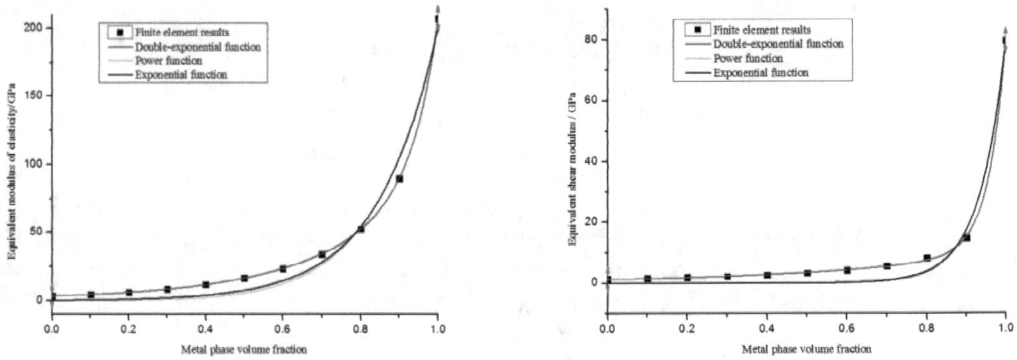

(a) (b)

Figure 8. (a)Diagram of the relationship between the equivalent elastic modulus of the interface phase integral material and the volume fraction of the metallic phase; (b) Diagram of the relationship between the equivalent shear modulus of the interface phase integral material and the volume fraction of the metallic phase.

According to the characteristics of the data, the equivalent Poisson's ratio of the interface phase material is segmented and fitted, as shown in figure 9. When the metal phase volume fraction is less than 80%, the logistic growth model better reflects the relationship between the equivalent material Poisson's ratio of the interface phase and the metal phase volume fraction. When the metal phase volume fraction is greater than 80%, the equivalent Poisson's ratio increases rapidly in a small range, and the exponential function can well express the relationship of the equivalent Poisson's ratio in this range.

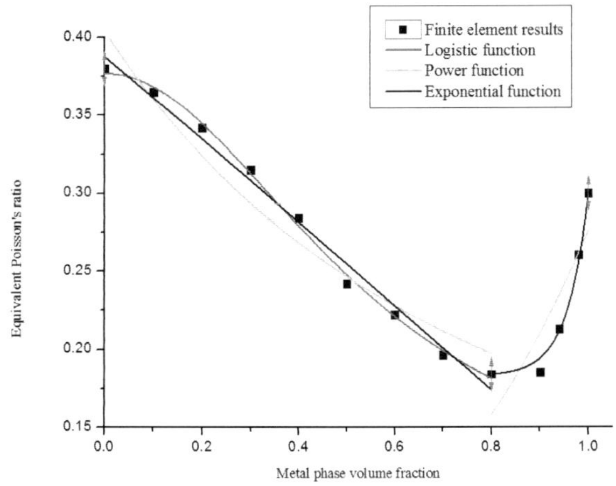

Figure 9. Diagram of the relationship between equivalent Poisson's ratio of interface phase integral material and volume fraction of metallic phase.

4.3. Prediction of interface phase equivalent material parameters

Before the interface phase equivalent material parameters are predicted, the following assumptions are proposed in this paper:

- when uniaxial load is applied to the thin layer of interface phase material, only corresponding internal stress is generated in different thin layers, and other internal stress components are zero.
- The internal stresses in different thin layers are equal under axial action.

According to sections 4.1 and 4.2, the equivalent material parameters of interface phase are distributed continuously and regularly. Combined with the above assumptions, the mixing rate model can be used to predict the value of interface phase equivalent material parameters.

$$E_i = \sum_{k=1}^{N} \mathrm{E}_\mathrm{i}^k \mathrm{v}_k \approx \int_0^1 E(\mathrm{v})d(\mathrm{v}) \tag{11}$$

$$G_i = \sum_{k=1}^{N} \mathrm{G}_\mathrm{i}^k \mathrm{v}_k \approx \int_0^1 G(\mathrm{v})d(\mathrm{v}) \tag{12}$$

$$v_i = \sum_{k=1}^{N} v_\mathrm{i}^k \mathrm{v}_k \approx \int_0^1 v(\mathrm{v})d(\mathrm{v}) \tag{13}$$

Combined with equations (11)-(13), the equivalent elastic modulus, equivalent shear modulus and equivalent Poisson's ratio of interface phase can be calculated to be 33.5075 Gpa, 7.1857 Gpa and 0.1412 respectively.

The relationship between the material parameters of the interface phase materials in the volume fraction of different metal phases was studied, and the functional form of the local interface layer and the global interface phase material along with the metal volume fraction was obtained in this paper. It can provide a basis for further study of interface phase materials

5. Conclusions

Based on the homogenization method, the finite element cell model of composite interface layer material was established in this paper. It is necessary to pay attention to the division of scale in the mechanical analysis and research of coating - metal gradient interface phase. The elastic parameters of the materials vary with the volume fraction of the two-phase components at different scales. By calculating and analyzing the parameters of the cell material, the following conclusions are obtained:

- The elastic modulus and shear modulus in interface layer have linear relations with volume fraction of metal in local interface layers, and the equivalent Poisson's ratio has a linear relation with volume fraction in thinner interface layers, but in local layers it can be expressed by the logistic function.
- The relationship between the overall interface phase material parameters and the volume fraction of the metal phase is different from the local material parameter distribution law. The equivalent elastic modulus and equivalent shear modulus of the global interface phase material satisfy the double exponential function distribution. The change rule of equivalent Poisson's ratio is significantly affected by the volume fraction of metal phase, which can be expressed piecewise by logistic function and exponential function .

Acknowledgment
This study was funded by The National Key Research and Development Program (Grant Number: 2016YFC0303705).

References

[1] Du Z Q, Zhang W F *et al* 2011 Application of fused epoxy powder coating technology in submarine pipelines *Steel Pipe* **40**
[2] Kwei T K and Kumins C A 1964 Polymer- filler interaction: Vapor absorption study *Appl Polym Sci*. **8** 1483-90
[3] Fan J Z, Yao Z K *et al* 1997 Research progress on interface of aluminum matrix composites reinforced with silicon carbide *Rare Metal*. **2** 55-9
[4] Xia Y F, Jiang W G and Wu Z K 2015 Analysis of residual stresses and failure of aluminum honeycomb with Al2O3 coating under thermal loads *Transactions of Materials and Heat Treatment* **37** 193-7
[5] Sun Z G, Li L *et al* 2010 Effect of interface layer parameters on uniaxial tensile behavior of ceramic matrix composites *Journal of Aerodynamics* **25** 597-602
[6] Li S G and Wongsto A 2004 Unit cells for micromechanical analyses of particle-reinforced composites *Mechanics of Materials* **36** 543-72
[7] Leslie B S, Leiderman V and Fang D 1997 On the effect of particle shape and orientation on elastic properties of metal matrix composites *Composites Part B* **28B** 465-81
[8] Wakashima K, Hirano T and Nino M 1990 Space application of advance structure materials *ESP* **303** 97-100
[9] Delate F and Erodgan F 1988 On the mechanica lmodeling of the interface region in bonded half- planes *ASME J Appl Mech*. **55** 317-24
[10] Choih J 1997 A periodic array of cracks in a functionally graded nonhom ogeneousm edium loaded under in-plane normal and shear *Int J Fract*. **88** 107-28
[11] Wang X Y, Zou Z Z and Wang D 1997 On the penny-shaped c rack in a nonhom ogeneous interlayer of adjoining two different elastic materials *Int J Solids Struct*. **34** 3911-21
[12] Li C Y and Weng G J 2001 Dynamic stress intensity factors of a cylindrical interface crack with a functionally graded interlayer *Mech Mater*. **33** 325-33
[13] Wang B L, Hang J C and Du S Y 1999 Dynamic response for functionally graded materials with penny-shaped cracks *Acta Mech Solida Sin* **12** 106-13
[14] Wang Y S and Ross D G 2000 Analysis o f a crack in a functionally gradient interface layer under static and dynamic loading *Key Engn Mat*. **183-187** 331-6
[15] Gao T K 2005 A series of boundary value problems of partial differential equations in plane fracture of functionally gradient materials and composites *Journal of Shanxi Normal University* **19** 52-6

[16] Delale F and Erdogan F 1988 On the mechanical modeling of the interface region in bonded half-planes *Journal of Applied Mechanics* **55** 317-24
[17] Li T and Yao M F 2017 Natural fractals--application of Voronoityson polygons in architectural design *Chinese and Foreign Construction* **7** 139-44
[18] Xia Z H, Zhang Y F and Ellyin F 2003 A unified periodical boundary conditions for representative volume elements of composites and applications *International Journal of Solid sand Structures* **40** 1907-21
[19] Li S 2000 General unit cells for micromechanical analyses of unidirectional composites *Composites: Part A* **32** 815-26
[20] Li S G 2008 Boundary conditions for unit cells from periodic microstructures and their implications *Composites Science and Technology* **68** 1962-74

Research and application of key technology of submersible oil reciprocating pumping unit

Y A Sun[1,2]

[1]Petrochina Daqing Oil Field Production Engineering Research Institute, Daqing, 163453, China

E-mail: qk13836956215@163.com

Abstract. The submersible oil reciprocating pumping unit is a new generation of rod-less oil recovery equipment for low permeability reservoirs. The optimization and upgrading of key technologies is conducive to its large-scale application and cost reduction and efficiency enhancement. This paper proposes a series of optimization and upgrading methods for the problems existing in oilfield production, including the research of closed-loop control technology based on power carrier communication, the research on harmonic suppression technology of inverter output, and the research and development of wireless IOT remote oil recovery equipment management system. Research on new low-cost clear wax-proof process and intelligent composite material coiled tubing supporting technology. At the same time, indoor and on-site test evaluation is carried out. The effect shows that the new process of optimization and upgrading can improve the reliability, adaptability and economy of the submersible reciprocating pumping unit, laying the foundation for its intelligent, digital and large-scale development. It has certain practical significance.

1. Introduction

In the development of oilfields, the artificial lift production wells mainly consist of two pumping and screw pumps, which account for about 90% of the total number of oil wells, but there are rods in the application of low permeability and heavy oil reservoirs. Serious partial wear and high energy consumption [1]. In foreign countries, the application of linear motors to underground oil pumping units has not been reported. It has formed small-scale applications in China, and the prospects are good. However, the test wells of submersible oil reciprocating pumping units generally have open-loop control methods that cannot achieve operating parameters. Automatic optimization and fault warning, low system efficiency, serious frequency harmonic pollution, high cost of clearing anti-wax, low level of intelligence, etc., resulting in a short cycle of overall process inspection [2], urgently needed to carry out in-depth research and optimization of key technologies .Based on this, this paper creatively proposes a series of optimization and upgrading methods: research on closed-loop control technology based on power carrier communication, research on harmonic suppression technology at inverter output, research and development of wireless IOT remote oil recovery equipment management system, and development of new low-cost cleaning wax Research on technology and intelligent composite material coiled tubing supporting technology. At the same time, indoor and on-site test evaluation is carried out. The effect shows that the new process of optimization and upgrading can improve the reliability, adaptability and economy of the submersible reciprocating pumping unit, laying the foundation for its intelligent, digital and large-scale development. It has certain practical significance.

2. Research on closed-loop control technology

2.1. Overall system design

As shown in figure 1, The original control system of the submersible reciprocating pumping unit is an open-loop control mode, which is a constant rigid drive. There are problems such as instantaneous hard start and serious commutation impact. It is impossible to realize automatic optimization of operation parameters and fault warning [3,4]. A closed-loop control technology based on power carrier communication, clear design scheme, and indoor test, indicating that the closed-loop control system works normally, laying a foundation for the intelligent and digital development of the submersible oil reciprocating pumping unit.

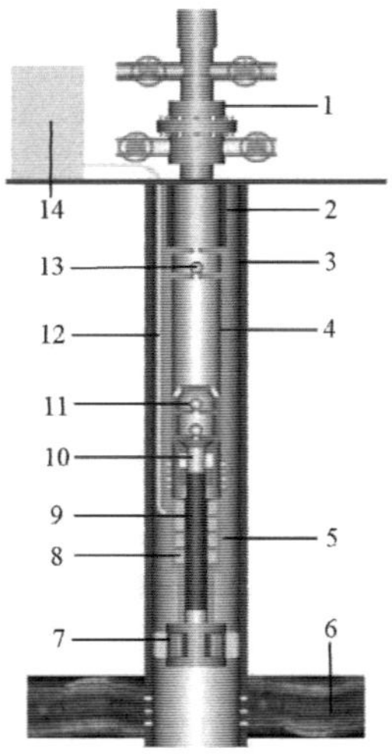

1.Wellhead;
2.Tubing;
3.Casing;
4.Plunger pump;
5.Submersible linear motor;
6.Oil layer;
7.Centralizer;
8.Stator;
9.Mover;
10.Plunger;
11.Swimming valve;
12.Cable connector;
13.Fixed valve;
14.Control system

Figure 1. Submersible oil reciprocating pumping unit.

The closed-loop control system consists of ground control equipment and downhole monitoring equipment. The system uses submersible cable and cable suede as transmission channels to supply power to the underground monitoring system through star point equipotential method, and uses the suede as the ground line to transmit monitoring signals. Real-time monitoring and automatic adjustment of the operating state of the submersible reciprocating pumping unit to optimize equipment operation and prolong pumping cycle.

2.2. Development of underground monitoring equipment

Equipment composition: multi-parameter sensors and various types of circuits, downhole single-phase reactors, cable traversers, drain seals, stainless steel capillaries, anti-vibrators, mechanical structural parts and seals.

Design requirements: design a cylindrical structure with an outer diameter of Φ114 mm, connected to the tail of the submersible linear motor, determine the type of monitoring of the working conditions, technical indicators, various circuit designs and low-power performance and reliability of the device, and then mechanical structure The parts and seals form a high pressure seal chamber that houses all components in the sealed chamber and is isolated from the external well fluid.

Process requirements: All joints are quenched and tempered, which can be reliably installed in the lower part of the downhole motor or pump, requiring sealing, oil immersion, vibration resistance, pressure resistance and temperature resistance. Performance indicators are as shown in table 1:

Table 1. Monitoring parameter types and performance index requirements.

1. Use environment: well temperature is not higher than 125°C, downhole pressure is not higher than 40Mpa			
2. Technical parameters: (Select the parameter type according to actual needs)			
Acquisition parameter	**Measuring range**	**Precision**	**Resolution**
Inlet pressure	0 ~ 6000Psi	0.1%	1psi
Outlet pressure	0 ~ 6000Psi	0.1%	1psi
Ambient temperature	0 ~ 150°C	0.1%	0.1°C
Motor temperature	0 ~ 250°C	0.1%	0.1°C
Export flow	0 ~ 30000bbi/day	5%	1bbi/day
Vibration (x and z)	0 ~ 10G	1.0%	0.01G
Leakage current	0 ~ 25mA	0.05%	1μA
Supply voltage	DC,0 ~120V	0.2%	0.1V
3. Insulation index: 25°C/85% humidity/2500vdc/, insulation resistance ≥2000mω			
4. Sealing performance: pressure bearing capacity ≥ 40MPa			

2.3. Research on underground communication method
Downhole communication methods include two aspects of downhole power supply and signal transmission [5,6]. The power supply mode of the underground monitoring equipment mainly includes cable power supply, battery power supply, power input from the motor and star point DC power supply. Among them, the cable power supply method needs to be put into a special power supply cable, which has high cost and is limited by the working space under the well; battery-powered mode, because the battery itself has limited battery life, regular replacement of the battery requires the oil well to stop production and construction work; The way of taking power at the input end will affect the balance of motor power supply, damage the sealing and insulation performance of the motor, and the implementation process is complicated. The star-point DC power supply mode can meet the power requirements of the downhole sensor without affecting the normal operation of the AC-driven motor. The implementation method is simple. Therefore, the star-point DC power supply method is the most ideal and most suitable power supply mode for the underground oil reciprocating pumping unit downhole monitoring system.

Transmission methods of downhole monitoring signals mainly include dedicated signal cable transmission, sound wave transmission, and power carrier communication. Among them, the use of dedicated signal cable transmission is not susceptible to external interference, easy to achieve high transmission accuracy and speed, but the dedicated signal cable needs to work in high temperature, high pressure, strong corrosion downhole complex environment, insulation performance and reliability is difficult to guarantee, and The cost is higher; based on acoustic wave transmission, the sound wave is transmitted by the oil pipe to transmit the specific frequency, but the noise of the submersible reciprocating pumping unit is large, the frequency is difficult to extract, and the communication based on the power carrier is the use of the submersible cable. The physical entity medium selects a reasonable signal transmission form according to the channel characteristics, does not need to develop a special signal cable, can realize power supply and signal transmission at the same time, has strong reliability and anti-interference ability, and is a submerged oil reciprocating pumping unit downhole signal. Ideal for transmission, this article uses the submersible cable to transport the signal as a ground wire.

2.4. Development of ground control equipment
Equipment composition and requirements: The ground control equipment and the frequency conversion control system are integrated into a cabinet structure, which mainly includes a three-phase reactor and a secondary instrument (a power module, a data storage module, a data wireless transmission module, a data display module, a control unit, Data demodulator, etc.), to realize data

demodulation processing, real-time communication of monitoring data and frequency conversion control system, and remote data transmission. The design requirements should have certain moisture, dust, rain, snow and oil proof functions, and the process is safe and reliable.

3. Research on negative problems of variable frequency drive

The variable frequency control system drives the submersible linear motor. There is "harmonic pollution" at the output end of the inverter. When it is used for long-distance submarine cable (800 m—3000 m), there are a series of negative problems [7-9]: (1) The severe harmonic current and harmonic voltage in the cable will generate overvoltage and overcurrent, which will cause the cable to heat up continuously, forming skin effect and proximity effect, generating harmonic reactive power and reducing power factor; (2) Harmonics The submersible motor coil is continuously heated, generating high-frequency damped oscillation, and operating malfunction; (3) Harmonics seriously affect the life and accuracy of electronic and electrical equipment in the downhole monitoring instrument and control system. Eventually, the insulation pressure of the rod-less oil production equipment is intensified, resulting in insulation aging, breakdown or burning of the submersible cable and the motor. The overall process vibration is severe, and as the cable length increases and the harmonic frequency increases, the harmonic hazard is continuously amplified.

The harmonic suppression at the output of the inverter is mainly from three angles: First, the inverter control system is optimized and upgraded, including the optimization of the inverter control strategy and the applicable filtering device. This method is convenient for finding economical and practical suppression measures and improving the motor. Efficiency; Second, the optimization and upgrading of submersible cables, including improving the cable structure, reducing the cable distribution capacitance, improving the cable insulation level, thermal stability, and overvoltage and current resistance, but the cost is large. At the same time, the connection mode of the submersible cable connector is optimized and upgraded, and the flat and flat cable joints are integrated, no manual connection is required at the site, and the connector between the small flat and the motor lead is plugged to improve the insulation performance and reliability of the joint; third, the submersible motor Optimize and upgrade, add electrostatic shielding layer between the motor stator and mover and increase the insulation performance of the motor winding.

In order to suppress the higher harmonics generated at the output of the variable frequency control system of the submersible reciprocating pumping unit, to realize the protection of the submersible cable and the linear motor, reduce the failure rate of the overall process, and prolong the inspection cycle, this paper designs a sine wave. The filter is a kind of rlc low-pass filter, which is mainly composed of output inductor, bypass capacitor and damping resistor. The installation position and topology are shown in figure 2.

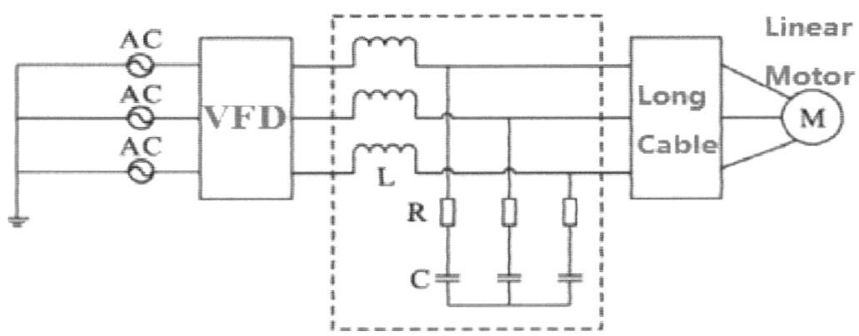

Figure 2. sine wave filter structure.

Considering the rated electrical parameters of the submersible reciprocating pumping unit, the length of the submersible cable, the fundamental frequency, the switching frequency, the output voltage drop, the power factor, the harmonic THD value and other factors, the sine wave filter is set. The cutoff frequency f is 800 Hz, the filter inductance Lf is 15.25 mH, the filter capacitor Cf is 4.2 μF, and the damping resistor Rf is 0.02 Ω. Carry out the indoor test, the voltage waveform of the output

IOP Conf. Series: Materials Science and Engineering **634** (2019) 012031 doi:10.1088/1757-899X/634/1/012031

end of the inverter is shown in figure 3. After installing the sine wave filter, the output voltage waveform is shown in figure 4. It can be seen that the sine wave filter is designed by adding the design to the output end of the inverter. After that, the harmonic components are significantly attenuated and the output waveform is significantly improved.

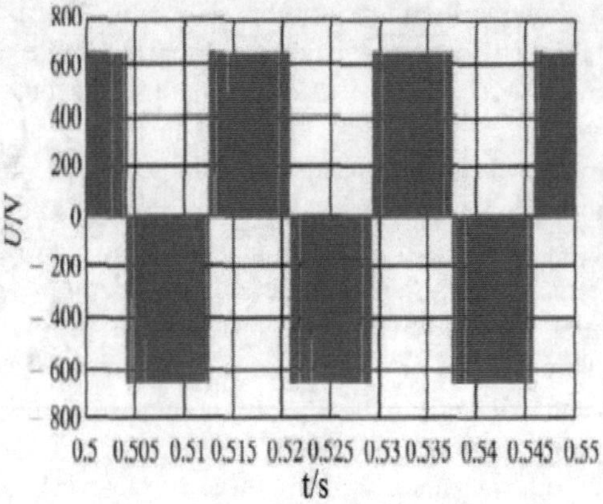

Figure 3. Voltage waveform at the output of the inverter.

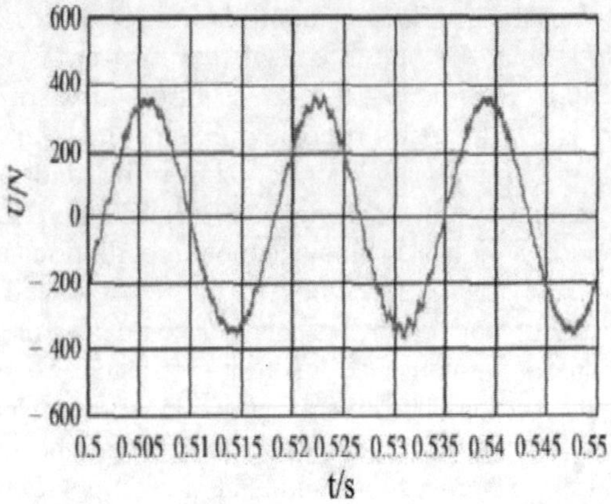

Figure 4. Output voltage waveform of the sine wave filter.

4. Main supporting technologies

4.1. Wireless IOT remote Oil production equipment management system

The application of the wireless IOT remote device management system realizes the digital management of the process, which is mainly composed of management software and hardware. The software part is composed of office remote management control system and mobile remote management monitoring system; the hardware part consists of data management server, management office computer, mobile data management equipment, wellhead remote terminal control system (rtu), Internet of Things data transmission equipment (dtu), Protocol converter, intelligent linear motor control system, pressure monitoring system, temperature monitoring system, electric metering equipment.

The management system continuously records the real working conditions through the continuous collection and analysis of data, and quickly responds to abnormal working conditions. The manager can realize data storage, management analysis, production parameter setting and office in the office or any other place. Operation status monitoring, real-time monitoring of the operation status of each production well by using a computer or an authorized Android system connected mobile terminal, including parameters such as real-time electricity, current curve, downhole pressure, downhole temperature, etc., and coordinate analysis by protocol converter to set a straight line Motor rotor running frequency, stroke, stroke and other parameters to ensure that the submersible reciprocating pumping unit operates stably under the optimal conditions. Dtu can transmit encrypted work data through mobile or China Unicom network. At present, 86 wells are tested in the field, and the application effect is good.

4.2. New low-cost anti-wax process

The submersible oil reciprocating pumping unit belongs to the rod-less oil recovery process, which is prone to tube and wax formation. The application of the electric heating rod clearing and anti-wax process basically meets the production demand, but the cost is high and the production and operation cost is high. Based on this, a new type of low is developed. Cost of solid chemical anti-wax process research. Analyze the physical properties of oil in the oilfield block, and use the content of wax, colloid and asphaltene as the reference basis for the formulation of the targeted formula. After repeated trials, the best ratio can be obtained. At the same time, in order to solve the problems of short effective cycle of solid anti-wax technology, high cost of periodic dosing operation and complicated operation, a solid chemical anti-wax device capable of fishing is developed, and a solid anti-wax agent wellhead dosing method is proposed. There is no need to use a moving column, which greatly facilitates the replenishment of solid anti-wax agents. The medicament can be taken into the wellbore with the disposable wax-proof device, and the dosage is determined according to the production condition of the wellbore; the oil well does not need to be periodically heated, the heat washing cost is saved, and the oil layer is protected from water damage; Low technical advantages of low fishing process.

The anti-wax device is composed of a salvage dekaler, a solid anti-wax device and a well lower limiter. According to the single well production and the well fluid flow rate, the solid anti-wax agent charge and the sustained release rate of the chemical are determined, and the production liquid is prevented. The hollow portion of the wax device flows upward, and the flow process slowly dissolves the anti-wax agent. At the time of delivery, the salvage stripper is docked on the ground with the solid wax-proof device. After the wire rope winch is lowered into the wellbore to reserve the depth, the tail of the solid wax-proof central tube enters the center hole of the stopper to achieve the fixed position. Then, the card unloader and the fishing head are disengaged, and the wire rope can be lifted to complete the device placement. The anti-wax agent is continuously released during the production process. When the predetermined period of use is reached, the salvage dragger is lowered from the wellhead, docked with the anti-wax head, and replaced with a new wax-proofing device as needed. At present, 224 oil wells are tested in the field, and the application results are shown in figures 5 and 6.

Figure 5. The liquid wax crystals were extracted before the measures.

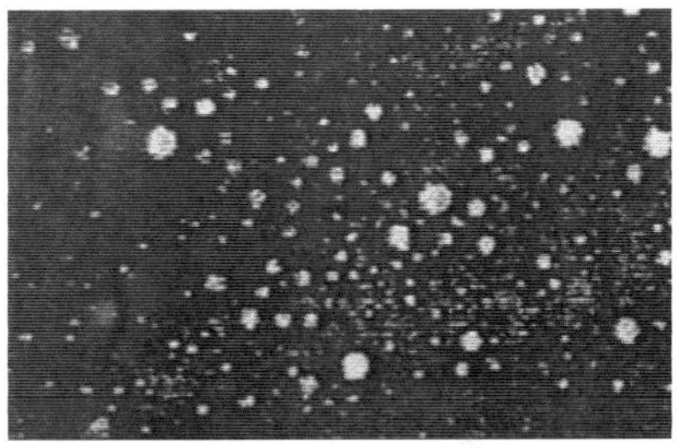

Figure 6. The liquid wax crystals were extracted after the measures.

It can be seen from the comparison of the wax crystals that the anti-wax agent makes the volume of the wax crystals in the oil flow significantly smaller, showing a dispersed state, and the crystallization rate of the crystal nucleus is reduced, thereby slowing the waxing speed of the wellbore, and the effectiveness of the medicament is proved.

4.3. Research on the matching technology of intelligent composite material coiled tubing

At present, the rod-less oil recovery of the submersible oil reciprocating pumping unit is continued by metal fuel pipe, and then the armored submersible cable is bundled on the outside of the metal pipe to achieve oil lift, and there are pipe wall scaling, waxing, corrosion, etc. The problem is that the external cable is frequently damaged during the construction operation and operation, the repetition rate is low, and the operation process is complicated. Based on this, this paper proposes an intelligent composite material coiled tubing supporting submersible linear motor technology. The coiled tubing consists of intelligent inner lining layer, fiber reinforced layer and outer protective layer. Its technical features are power cable, signal transmission cable and heating cable. Three kinds of cables are built in the intelligent inner liner to solve the above practical problems. At the same time, the heating cable is used to clear the wax; the structural layer is composed of fibers impregnated with thermoplastic resin, and is subjected to internal pressure, external pressure and axial tensile force; the outer protective layer can be It has a protective function in transportation, operation and production. Six wells were tested in the field, and the operation was normal, and the technical advantages were outstanding.

5. Conclusion

The submersible reciprocating pumping unit is a new generation of rodless oil recovery equipment. The in-depth research and large-scale application of key technologies is an inevitable trend of its development. This paper creatively proposes the optimization method of the process: research on closed-loop control technology based on power carrier communication, research on harmonic suppression technology at inverter output, research and development of wireless IOT remote oil recovery equipment management system, and development of new low-cost clear wax prevention process, Research on supporting technology of intelligent composite coiled tubing. The indoor and field test results show that the new process of optimization and upgrading can further improve the reliability of the submersible reciprocating pumping unit and its supporting technology, and broaden the technical and economic adaptability. At the same time, this paper points out the future research direction of the process, and provides technical support for the development of intelligent, digital, high-efficiency, energy-saving, safety and environmental protection, and lays a foundation for further research and development of the process to adapt to deep wells.

References

[1] Qian K, Sun Y A and Du W S 2018 Research and application of key technology of electric submersible plunger pump *Materials Science and Engineering* **372** 012024

[2] Liu Y H and Yu Z W 2016 LLE for submersible plunger pump fault diagnosis via joint wavelet and SVD approach *Neurocomputing* **185**

[3] Shao K Y, Qian K *et al* 2014 Research and application of fault diagnosis of electric submersible piston pump based on indirect dynamometer measurement *Chemical Automation and Instrumentation* **41** 1023-7

[4] Shao K Y, Qian K, Zhang F W *et al* 2014 Design and application of fault monitoring and diagnosis system for numerically controlled reciprocating submersible electric pump *Chemical Automation and Instrumentation* **41** 1155-7

[5] Du P 2013 Research on multi-state parameter monitoring system in underground (Liaoning, China: Shenyang University of Technology)

[6] Zhang M X, Dong Z G and Zhang B L 2009 Study on state monitoring technology of submersible electric pump unit and downhole sensor *Journal of Electric Machines and Control* **13** 28-33

[7] Ma W Z, Li Y H and Kong L 2006 Research on the negative problem of variable frequency drive of submersible electric pump *Acta Petrolei Sinica* **27** 133-6

[8] Zhang W, Dai P, Zong W L *et al* 2013 Study on the effect of long cable transmission pulse wave on motor terminal overvoltage *Electric Drive* **43** 36-9

[9] Li X H and Wang W D 2006 Influence and suppression of higher harmonics in frequency converter lead cables *Microcomputer Information* **22** 161-3

IOP Conf. Series: Materials Science and Engineering **634** (2019) 012032 doi:10.1088/1757-899X/634/1/012032

Photoelectric emission and work function of silver-zinc oxide Ag-ZnO (92/8) pseudo-alloy electrical contacts

M Akbi[1,3] and M Cherif M'ziane[2]

[1]Université M'hammed Bouguerra de Boumerdes, Avenue de l'indépendance, Boumerdes, DZ-35000, Algeria
[2]Fundamental and applied Physics Laboratory, (FUNDAPL), Université Saad Dahleb Blida 1, PV 08, Route de Soumaa, B.P. N° 270, DZ-9000, Blida, Algeria

E-mail: akbim656@gmail.com; m.akbi@univ-boumerdes.dz

Abstract. Arc roots models require a good knowledge of the physical constants characterizing the contact material used in various kinds of switchgears. Electron work function (EWF) is one of the most important parameters of electron emission for metals and has not yet been studied in detail for pseudo-alloy contact materials. For pure metals, the values of these constants are well known and can easily be found in tables. This is not the case for the new silver-based industrial materials recently available on the market, which are more efficient without the manufacturers knowing the cause. The main goal of this paper is to understand better the photoelectron emission occurring in silver-metal oxide Ag-ZnO (92/8) electrical contacts, before and after several heat treatments from room temperature up to 813 K, under UHV conditions. The photoelectric work function of Ag-ZnO (92/8) pseudo-alloy electrical contacts has been measured by photooelectric effect, using both Fowler's methods, i.e. isothermal curves method and linearized plots method. It was demonstrated that after several heating treatments in vacuum, the EWF of the silver–zinc oxide, Ag–ZnO (92/8), determined at T = 300 K, is kept within the following interval limits $]\phi(Ag), \phi(ZnO)[=]4.26$ eV, 4.68 eV[. This is due to the evaporation and/or diffusion phenomena which influences the surface composition of the pseudo-alloy electrical contact. Moreover, all the measurements carried out in this study give clearly a demonstrable proof that the EWF of the Ag-ZnO pseudo-alloy should be between the EWF values of each of the two components 4.26 eV for Ag and 4.64 eV for ZnO. Observations and surface analyses carried out under the scanning electron microscope illustrated the development of the surface composition of the Ag-ZnO contact material after several heating treatments in vacuum. Thus, surface analysis of two Ag–ZnO electrical contacts performed before and after VH treatments showed a significant increase of weight proportion for ZnO nanoparticles on the contact surface.

1. Introduction

Silver metal oxides Ag-MeO have found many applications in electrical and electronics industries. In particular, Ag-ZnO electrical contact materials are widely used in various kinds of switchgears, such us relays, contactors, engine starters and electrical switchers [1, 2]. The Ag-ZnO material is used for switching contacts in low-voltage (between 24 and 600 V) contactors (up to 1000 A) [1, 2]. In addition, high capacity and low-voltage electrical switchgears are widely used in industrial automatisms, in transport such as cars and railways, in lifts and forklift trucks, and in control and regulating devices, such as thermostats [3].

The Ag-CdO material has virtually dominated medium to high current switching applications (between 50 and 5000 amperes) for several decades after its debut in 1950 [4-5]. However, due to the toxicity of cadmium, environmentally friendly materials, such as Ag-SnO_2 [6, 7] and Ag-ZnO [8], have been developed as an alternative to Ag-CdO [9]. Ag-SnO_2 materials have progressively replaced Ag-CdO materials over the last four decades, as they present superior erosion resistance and anti-welding characteristics [10]. However, silver tin oxide shows a high temperature rise and its treatment is more delicate [11]. Due to these undesirable effects, the properties of silver and zinc oxide (Ag-ZnO) enabled it to effectively replace the contact material (Ag-SnO_2) [12].

To replace Ag-CdO contact material, Ag-ZnO was developed in the 1970s. Ag-ZnO contact material exhibits, comparatively with other contact materials, higher electrical and thermal conductivities, higher thermal stability, higher hardness and strength, higher welding resistance and low arc erosion. This new material is highly recommended when contacts are subjected to intense closing and opening cycles during which an electrical arcing occurs. Another favorable quality is that it is harmless and does not pollute the environment.

Ag-ZnO (92/8) innovative contact materials having multi-layered architecture were supplied by Metalor (France) and fabricated by powder metallurgy [3]. It should be emphasized that the performances achieved by Ag-ZnO contact materials have not yet been understood by the manufacturers. Therefore, the purpose of this study is to understand better the electron emission of the pseudo-alloy Ag-ZnO by examining the modifications of the photoelectron work function changes under various UHV heat treatments. The electron work function (EWF) of silver zinc oxide contacts was determined by photoelectric effect [13−20] using both Fowler's methods: isothermal curves and linearized plots [21, 22].

In addition, to study the influence of vacuum heat treatments on the composition of Ag-ZnO cathode surfaces, on the surface distribution of ZnO nanoparticles and consequently on the electron work function of such electrical contacts, the microstructure of the Ag-ZnO contact surface has been observed by SEM and then analyzed by EDS.

2. Experimental

2.1. Materials

The electrical contacts carried out in the present study are manufactured by Metalor (Courville-sur-Eure, France). They have a cylindrical shape, a diameter of 8 mm and a height of 3.5 mm, with a radius of curvature of 16 mm. In this paper, a set of new experiments carried out with silver-zinc oxide Ag-ZnO (92/8) is highlighted. Numbers in parentheses represent the mass proportion of pure silver particles and granular ZnO particles, respectively. The grain size of ZnO is between 1 and 5 μm. It should be pointed out that the size of oxide particles distributed in the silver matrix plays a key role in diffusion processes which occur during vacuum heat treatment and in electron emission properties of the Ag-ZnO composite since it has been established that electron work function depends upon surface composition and microstructure [10] of silver-zinc oxide (Ag-ZnO) electrical contact. The conditioning processes of contacts have been given in much detail elsewhere [13−16]. However, the main process of cleaning the surface has been achieved by repeated heating cycles at 700 K under very high vacuum conditions. EWF measurements were carried out at low pressure (10^{-9} mbar), after repeated annealing cycles at 700 K, and for short periods (less than thirty minutes for one test). The experimental set-up has been reported in details in previous papers [13−20].

2.2. Experimental method

EWF data were obtained using Fowler's theory of photocurrent. Fowler developed a theory of electron energy distribution, founded on the hypothesis that the free electrons of a metal satisfy Fermi-Dirac's statistics [23]. In accordance with this theory, the emitted photoelectric current is:

$$I = \gamma \frac{4\pi e\, m(kT)^2}{h^3} f(x) = \gamma A_0 T^2 f\left(\frac{h(\nu - \nu_0)}{kT}\right) \quad (1)$$

where:

$$f(x) = e^x - \frac{e^{2x}}{2^2} + \frac{e^{3x}}{3^2} - \text{L} \quad (x<0)$$

$$= \frac{\pi^2}{6} + \frac{1}{2}x^2 - \left[e^{-x} - \frac{e^{-2x}}{2^2} + \frac{e^{-3x}}{3^2} - \text{L}\right] \quad (x>0)$$

and

$A = \gamma \frac{4\pi e\, mk^2}{h^3} = 1.2 \cdot 10^6\ A \cdot m^2 \cdot K^{-2}$ is Dushman thermionic constant.

Equation (1) is used to determine the photoelectron work function of the sample and the constant $A = \gamma A_0$ (γ is the probability of photo-ionization of the cathode surface) from experimental data since this equation can be written in the following form:

$$Log \frac{I}{T^2} = B + F(x); \qquad x = \frac{h(\nu - \nu_0)}{kT} \quad (2)$$

where B is a constant independent of the frequency ν of the incident radiation and T the absolute temperature of the emitting contact surface. The constant B represents the photoelectric surface yield; it depends on the optical absorption of the sample, the photoelectric efficiency of the emitting surface and the probability of escape of photoelectrons.

$F(x) = \log f(x) = \log f\left(\frac{h(\nu - \nu_0)}{kT}\right)$ is a universal function of x; it is important to note that its graph has the same shape for all metals and alloys as well as for all temperatures.

More precisely,

$$F(x) = \ln\left\{\frac{\pi^2}{6} + \frac{1}{2}x^2 - \left[\exp(-x) - \frac{\exp(-2x)}{2^2} + \frac{\exp(-3x)}{2^3} - \text{L}\right]\right\} \quad (3)$$

To assess the results of the measurements, the Fowler isothermal curve method can be used. In this method, photocurrents I are measured for different frequencies ν at a given temperature T and as a function of the same number of photons of absorbed light. This technique requires a photon source of calibrated spectral intensity and variable energy in the range around the threshold frequency for emission ν_0. The ultraviolet radiation device has been described in much detail elsewhere [13–16]. Photocurrent measurements (of order of 10^{-13}A) were performed using a picoammeter (Keithley's model 466). The uncertainties in the EWF measures were ± 0.03 eV, ± 0.07 eV and ± 0.08 eV at T = 300 K, T = 500 K and T = 800 K, respectively.

By plotting $\ln\left(\frac{I}{T^2}\right)$ versus $\frac{h\nu}{kT}$, we can compare the theoretical Fowler curve $F(x)$ versus x to the experimental Fowler curve. A vertical offset of the data gives B which is nevertheless unimportant and a horizontal one gives the EWF ϕ. This is one of the most accurate methods for determining the photoelectron work function of metallic and allied materials, since in most cases both theoretical and experimental curves fit very well.

In addition, the evaluation of the EWF may as well as be founded on the simplified theory of Fowler [21, 22] which establishes a linear relationship between the square root of the photoelectric current, $\sqrt{I}$, and the incident photons energy $h\nu$ in accordance with the relationship [14, 24]

$$\sqrt{I} \approx (h\nu - \phi), \qquad h\nu > \phi \quad (4)$$

where ϕ is the electron work function (EWF) of the sample surface.

The interception of $\sqrt{I}$ versus $h\nu$ is used to determine $\phi = h\nu_0$. We have assumed that the absorptivity of the surface is almost constant in the region where the photon energy, $h\nu$, and the quantum yield, $\sqrt{I}$ follow the Fowler curve.

3. Results and discussion

The reliability and service life of the components have been improved for recent electromechanical equipments containing electrical contacts with alternative materials such as silver-zinc oxide (Ag-ZnO) and silver-tin oxide ($Ag\text{-}SnO_2$). However, the physics of this phenomenon is still poorly known [4]. In addition, it should be stressed that there is no rational scientific explanation for the improved performance of new contact materials; It should also be noted that the new switching devices currently available on the market perform thousands of operations under load without any malfunction.

Notwithstanding that a large number of contact materials surveys and their application criteria have been published over the past decades, there is relatively little information on the properties of the innovative contact materials, principally in terms of electron emission mechanisms. Moreover, the theoretical models of the electric arc, mainly those of cathodic phenomena, considered today do not reflect the evolution of electron emissivity with the conditions on the arc feet and the nature of the material on which it originates. Theoretical studies conducted in recent decades have established that the lack of knowledge of the EWF of the emitted electrode causes a significant error on the physical quantities of the arc root (electric field, temperature, cathode spot radius...). For example, a variation of 0.1 eV of EWF causes a variation of 100 K in the cathode temperature for an electric arc conducting a 1000 ampere current.

For these reasons, this investigation tries to describe the work function measurements of these pseudo alloy contact materials using the photoelectric effect. Therefore, the objective of this work is to describe, quantify and explicate the phenomena of electron emission from both theoretical and experimental angles, in particular the effects of the annealing temperature on the EWF for silver-zinc oxide materials, Ag-ZnO (92/8), pure silver is the material reference. On the other hand, the effects of heating treatment in vacuum on the variations of EWF have also been extensively studied.

There are a variety of effective techniques, such as powder metallurgy and internal oxidation, to prepare the silver-zinc oxide materials used in circuit breakers. However, silver-zinc oxide contact materials containing mainly from 6 to 10% by weight of oxides including other small metal oxides, are produced exclusively by powder metallurgy. Figure 1 shows the microstructure of Ag-ZnO (92/8) before annealing. The higher magnification (x 8000) corresponds to the lower micrograph, while the higher one has a magnification (x 2000).

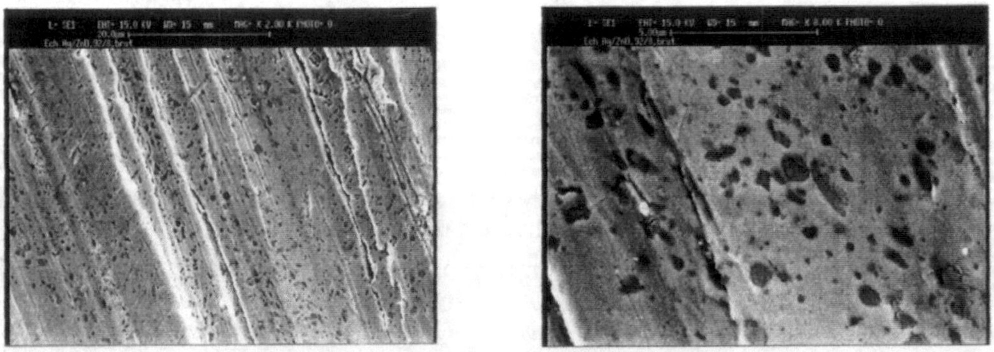

(a) (b)

Figure 1. Microstructure of Ag-ZnO (92/8).

Notes: (a) SEM magnification x2000, (b) SEM magnification x8000.

IOP Conf. Series: Materials Science and Engineering **634** (2019) 012032 doi:10.1088/1757-899X/634/1/012032

3.1. EWF measurements of Ag–ZnO (92/8) electrical contacts

EWF is one of the most important physical parameters that characterizes the surface condition of metallic and allied contact materials, and checks their electron emission. It is defined as the minimum amount of energy required for an electron to release from a metal surface. Moreover, it depends upon both bulk and surface of the metal. The contribution of the contact volume can be described as the difference in energy between the Fermi level and the vacuum level. On the other hand, the contribution of the surface is due to the electrostatic dipole barrier which depends strongly on the surface conditions. Therefore, it can be said that EWF is a surface characteristic that is highly dependent on the microscopic details of the surface. Therefore, for the same metal, it is different from one microscopic surface to another.

The present investigation deals with the surface state modifications of the Ag-ZnO (92/8) contact pastille with annealing temperature under ultra high vacuum conditions. It is worth noticing that all the changes in surface electronic structures of a material are related with the movement of electrons which depends on Fermi level and electron work function (EWF) [27]. We demonstrated that annealing temperature in UHV environment affects the surface characteristics of Ag-ZnO contact materials, as for example microstructure, roughness and in particular photoelectron work function.

In order to determine the electron work function, the generated photocurrent was determined as a function of the energy of the incoming photons. The EWFs of the unpolished and virgin contact Ag-ZnO (92/8), that have been measured after 10 cleaning cycles by heating treatments and 8 days of outgassing, under UHV conditions at different temperatures, are summarized in Table 1.

Table 1. Electron work function of the virgin contact Ag-ZnO (92/8).

Cycle	Surface Treatments	Heating Time, [hour]	Cleaning Time, [day]	Maximal Temperature, [K]	Temperature Measurement,	Electron Work Function
1	Without heating	–	1	299	299	4.62 ± 0.03
2	Heating N°1	3	2	408	408	4.57 ± 0.07
3	Heating N°2	4	3	470	470	4.47 ± 0.07
4	Heating N°3	16	4	533	533	4.52 ± 0.07
5	Heating N°4	4	4	713	713	4.38 ± 0.08
6	Heating N°5	3	5	823	823	4.32 ± 0.08
7	Without heating	-	5	300	300	4.48 ± 0.03
8	Heating N°6	18	8	753	296	4.34 ± 0.03
9	Heating N°7	4	9	533	533	4.46 ± 0.07
10	Heating N°8	4	9	813	813	4.59 ± 0.08

IOP Conf. Series: Materials Science and Engineering **634** (2019) 012032 doi:10.1088/1757-899X/634/1/012032

Studies on the influence of the heating temperature on the photoelectron work function for silver-zinc oxide contacts were carried out. For this purpose, photocurrent measurements have been extended to higher temperatures. As the temperature increases, the EWF measurements become rather difficult because the thermionic current is superimposed on the photoelectric one. As a result, the total current obtained fluctuates considerably, so that it was not possible to carry out precise photoelectric measurements. A method based on averaging was used, in which the total current was measured every 10 seconds at a given high temperature. Thus, twenty values were registered by a Keithley picoameter (model 486) for each wavelength of incident radiation. The photoemission was then interrupted by cutting off the power supply of the deuterium lamp and similar measurements were made for the thermionic current alone. The photoelectric current can be then deduced by subtracting the thermionic current from the total current.

When photocurrent readings were attempted at higher temperatures, large current fluctuations started to appear around 700 K and became more pronounced as the temperature was rising. Regrettably, these fluctuations have made it very difficult to measure the photocurrent.

Typical photoelectron spectral distribution curves for silver-zinc oxide contacts at T = 297 K, T = 533 K and T = 823 K, and under ultrahigh vacuum, are shown in Figure 2.

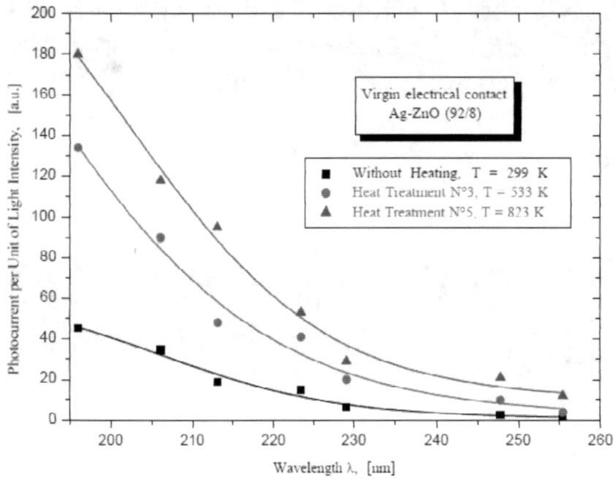

Figure 2. Spectral distribution curves.

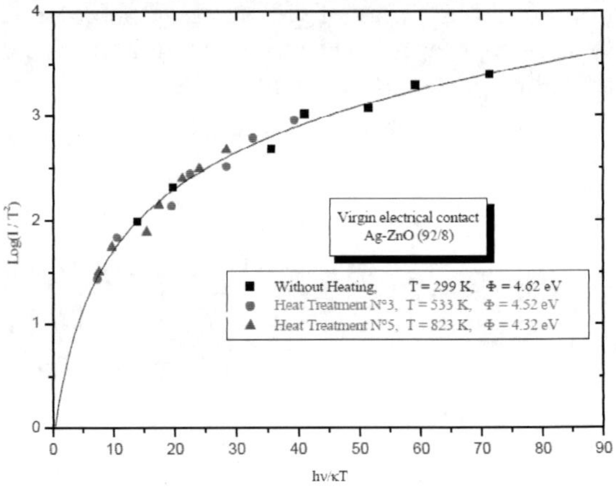

Figure 3. Typical Fowler isothermal curves (Equation (1)).

Figure 3 compares the Fowler theoretical curve with the the experimental one which corresponds to the above measurements for a virgin Ag-ZnO electrical contact (92/8).

In addition, the evaluation of the EWF may as well as be based on the the simplified theory of Fowler [21, 22]. Consequently, the Fowler curve allows to determine the EWF $\phi = h\nu_0$ directly from an adequate extrapolation of the straight lines drawn in Figure 4; the intersections with the abscissa axis give I = 0 and consequently the electron work function $\phi = h\nu_0$.

3.2. Influence of heating treatment in vacuum

The EWF of the raw and virgin silver-zinc oxide contact, Ag-ZnO (92/8), before the UHV heating treatment of the sample and before the baking of the experimental chamber is equal to (4.62 ± 0.03) eV. The work function was taken to be 5.3 eV for the ZnO materials from T. Minami et al [25]. These authors have measured the work function of ZnO films by ultraviolet photoelectron spectroscopy operated in air. In our measurements, the EWF electrical contact of Ag-ZnO (92/8) began to decrease as soon as the annealing temperature increased more.

After five successive vacuum heating operations at 408 K, 473 K, 533 K, 713 K and 823 K for 3 h, 4 h, 16 h, 4 h and 3 h respectively on the fifth day of the measurements, a surface structure modification of the Ag-ZnO cathode surface occurs which explains the EWF variations in the contact surface. Thus, after these vacuum heat treatments, and in particular at the seventh cycle at 300 K, the EWF takes an intermediate value of 4.48 eV which indicates a surface covered with more nanoparticles of zinc oxide than silver atoms. In addition, after the sixth surface heat treatment, during a prolonged heating period of 18 h at 753 K, the EWF is 4.34 eV, which is quite close to that of pure silver. The surface of the contact should then be almost covered with silver atoms.

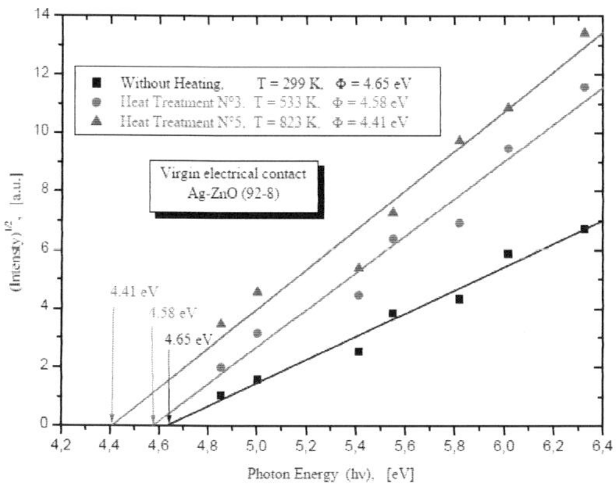

Figure 4. Typical linearized Fowler plots (Equation (4)).

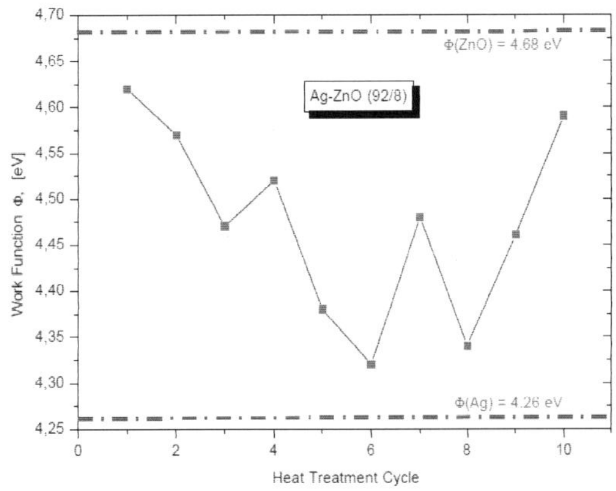

Figure 5. Vacuum heat treatment effect on the EWF of an outgassed virgin Ag-ZnO (92/8) contact.

Obviously, it seems that the sample cleaned by heating treatments in vacuum has become exclusively covered with silver atoms since surface impurities have been removed by thermal desorption. Long-term vacuum heating has moved the EWF values of the pseudo-alloys silver-zinc closer to those of silver. This could be explained by the diffusion of silver atoms on the surface, which will then overwhelm the crystalline nanoparticles of zinc oxide. The latter hypothesis is attributable to the lower melting point of silver (961.8°C) compared to zinc oxide (1975°C), so that under the effect of temperature, the diffusion rate of Ag atoms is higher than that of ZnO nanoparticles. As a result, the EWF value of the Ag-ZnO contact is reduced, and becomes close to that of Ag.

This behaviour has already been observed and demonstrated in our previous work [15, 19], for Ag-Me multilayer contact materials, i.e. for silver alloys such as silver-nickel and silver-tungsten. However, this phenomenon is noticed for the first time for Ag-MeO pseudo-alloys, especially for Ag-ZnO (92/8) electrical contacts. Further investigations should be carried out subsequently for silver-pseudoalloys.

As the temperature of the Ag-ZnO (92/8) cathode increases with silver and zinc oxide, a gradual evaporation of the silver atoms of the surface layers is likely to occur. At the same time, the zinc oxide nanoparticles slowly and uniformly diffuse from the underlying layers to the contact surface. All experiments conducted in this study indicate that the EWF of the Ag-ZnO pseudo-alloy is confined between the EWF values of each of the two components, i. e. (4.26 ± 0.03) eV for Ag and 4.68 eV for ZnO.

During heating under ultra high vacuum, the two compounds present at different percentages on the surface of the contact evaporate therefrom at different speeds, which allow the zinc oxide nanoparticles to degass more slowly than the silver atoms, and the composition of the surface layer therefore varies considerably for the two complementary components. As shown in Figure 5, it is worth noting that the phenomena of diffusion and/or evaporation during the vacuum heating treatment, which affect the surface composition of the pseudo-alloy electrical contact, in a way limit the EWF of Ag-ZnO, measured at room temperature in the range] Ag , ZnO [=] 4.26 eV, 4.68 eV [.

The multilayer appearance of the Ag-ZnO contact material is at the origin of these outcomes: the two compounds present in the electrical contact do not mix as in a solution and therefore occupy successive layers [3].

4. Discussion

To study the influence of UHV heat treatments on Ag-ZnO pseudo-alloys, observations by SEM and surface analyses by EDS were carried out on three samples before and after the surface treatments (polishing and annealing). EDS line scans should be reported in a future article.

After several periods of 30h heat treatment at different temperatures within the range [299 K, 823 K], it was found that the EWF of the Ag-ZnO contact (92/8) was equal to (4.32 ± 0.08) eV at 823 K. It is very likely that this would be due to a significant increase in the atomic proportion of silver in the total composition of the contact surface of the Ag-ZnO contacts after many UHV heating procedures (see Table 1).

Other series of vacuum heat treatments, with a total duration of 26 hours, significantly increased the value of of the work function (i.e. 4.59 eV) of the Ag-ZnO electrical contact (92/8) measured at 813 K. It should be noted that the EWF value after these different heat treatments is quite similar to that of ZnO (4.68 eV) [26]. This substantial variation in the chemical composition of the contact surface that has undergone prolonged heating under vacuum is probably due to the diffusion of zinc oxide from the contact core towards its contact surface. This also implies that, during heating, the silver atoms evaporate from the surface layers of the electrical contact to the vacuum vessel, so that the surface of the sample becomes covered with a majority of zinc oxide nanoparticles.

After a heat treatment of 30 min at 623 K, electron microscopy observations show that the zinc oxide blue islets are much larger than before heating (Figure 6 (a) and Figure 6 (b)), while the silver grey areas decrease with increasing temperature.

With regard to the multilayer structure of silver-zinc oxide pseudo-alloys, the results obtained must therefore be well interpreted. In addition, the EWF of pure metals decreases with increasing temperature [13, 16], while an opposite result has been found for Ag-MeO pseudo-alloys, i.e. Ag-ZnO (92/8), and previously for Ag-Me alloys [15, 19].

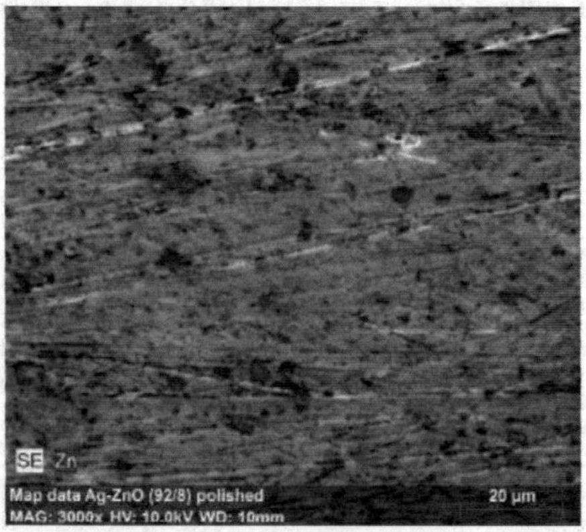

(a)

(b)

Figure 6. EDS elemental mapping of (a) a polished Ag-ZnO (92/8) contact, (b) an annealed and polished Ag-ZnO (92/8) contact. Microscope magnification x 3000.

5. Conclusion

The EWF of a pseudo-alloy contact material is highly dependent on the condition of its surface. The main results obtained in this survey are as follows:

1. The multi-layer nature of the Ag-ZnO electrical contact (92/8) has been demonstrated. By heating a pseudo-alloy contact under ultra-high vacuum (UHV), significant variations in the EWF were observed, resulting from the vaporization of the material components by sheets.

2. The EWF of pure metals decreases with increasing temperature while an opposite result was found for Ag-MeO pseudo-alloys, and previously for Ag-Me alloys. These opposite results are significant, because depending on the use of contacts (low voltage, high voltage, small arc duration...), we can predict the nature of the contact material which must be either a pure metal or an alloy or pseudo-alloy.

In conclusion, it could be inferred from this study that further research is needed to better elucidate these complex phenomena for which physics is still poorly known. Additional investigations will focus on the influence of vacuum heat treatments on the morphology, microstructure and EWF of silver-tin oxide Ag-SnO2 (88/12) and silver cadmium oxide Ag-CdO (88/12) electrical contacts.

Acknowledgement
The first author gratefully thanks Dr. C. Bourda (Metalor, France) for providing Ag-ZnO (92/8) pseudo-alloy samples.

References

[1] Braunovic M, Konchits V V and Myskin N K 2006 *Electrical Contacts: Fundamentals, Applications and Technology* (CRC Press, London) 672

[2] Schoepf T J, Behrens V, Honig T and Kraus A, 2002 Development of silver Zinc Oxide for General-Purpose Relays *IEEE Trans. Components Packaging Technologies* **25** 656–62

[3] http://WWW.Metalor.com/en/electrotechnics /products/ silver-alloys, and pseudo-alloys

[4] Schröder K H 1987 Silver-Metal Oxides as Contact Materials *IEEE Trans. Components Hybrids Manufacturing Technology* **10** 127–87

[5] Slade P G 2014 *Electrical contacts: Principales and Applications* (Second edition, CRC Press, Boca Raton, Florida) 1311

[6] Meng F B, Lu J G, Zhu Y Q, Chen Z C and Huang Q T 2006 Mass material transfer of Ag-SnO_2 contact material in DC Level *Proceedings of the 52nd IEEE Holm Conference on Electrical Contacts* (Montreal) 11–4

[7] Liu X M, Wu S L, Chu P K, Chung C Y, Zheng J and Li S L 2006 Effects of coating process on the characteristics of Ag-SnO_2 contact materials *Materials Chemistry and Physics* **98** 477–80

[8] Wu C P, Yi D Q, Li J, Xiao L R, Wang B and Zheng F 2008 Investigation on microstructure and performance of Ag-ZnO contact material Journal of Alloys and compounds **457** 565–70

[9] Nilsson O, Hauner F and Jeannot D 2004 Replacement of AgCdO by $AgSnO_2$ in DC Contactors *Proceedings of the 50th IEEE Holm Conference on Electrical Contacts and 22nd International Conference on Electrical Contacts* (Seattle) 70–4

[10] Sabairi K, Tsuji H and Tsuchiya K 1982 Manufacturing process and characteristics of Ag-SnO_2 for electrical contact material *Proceedings of the 28th Annual Meeting of the Holm Conference on Electrical Contacts* (Illinois) 77–81

[11] Michal R and Saeger K E 1989 Metallurgical aspects of silver-based contact materials for air-break switching devices for power engineering *IEEE Trans. Components, Hybrids, and Manufacturing Technology* **12** 74–81

[12] Pandey A, Verma P and Pandey O P 2008 Comparison of properties of Silver-tin oxide electrical contact materials through different processing routes *Indian Journal of Engineering of Materials Sciences* **15** 236–40

[13] Akbi M and Lefort A 1998 Work function measurements of contact materials for industrial use *J. Phys. D, Appl. Phys.* **31** 1301–8

[14] Akbi M, Bouchou A and Ferhat-Taleb M 2014 Effects of surface treatments on photoelectric Work function of silver-nickel alloys Vacuum **101** 257–66

[15] Akbi M, Bouchou A and Zouache N 2014 Effects of vacuum heat treatment on the photoelectric work function and surface morphology of multilayered silver-metal electrical contacts *Appl. Surf. Sci.* **303** 131–9
[16] Akbi M 2014 A method for measuring the photoelectric work function of contact materials versus temperature *IEEE Trans. Compon. Packag. Manuf. Technol.* **4** 1293–302
[17] Lefort A, Akbi M and Parizet M J 1993 Experimental determination of work function about silver alloys *Proc 2nd Int. Conf. Elect. Contacts, Arcs, A. Appl.* (Beijing) 363–7
[18] Lefort A, Akbi M and Parizet M J 1994 Electron emission of silver alloys used for breakers *Proc. SPIE, 16th Int. Symp. Discharges Elect. Insul. Vac.* (Moscow) 486–90
[19] Akbi M 2014 Effect of heat surface treatment on photoelectric work function of silver-metal alloys *European conference on heat treatment and 21st IFHTSE Congress* (Munich, Germany) 489–96
[20] Akbi M 2015 Effect of arcing in air on the photoelectric work function of silver-based contacts *IEEE Trans. Plasm. Sci.* **43** 637–42
[21] Wooten F and Stuart R N 1969 Fowler's hypothesis and the determination of photoemission thresholds *Phys. Rev. B* **186** 592–3
[22] Berglund C N and Spicer W E 1964 Photoemission on studies of copper and silver: experiment *Phys. Rev* **136** A1044–64
[23] Fowler R H 1931 The analysis of photoelectric sensitivity curves for clean metals at various temperatures *Phys. Rev.* **38** 45–56
[24] Eastman D E 1970 Photoelectric work function of transition, rare-earth, and noble metals *Phys. Rev. B* **2** 1–2
[25] Moormann H, Kohl D and Heiland G 1979 Work function and band bending on clean cleaved zinc oxide surfaces *Surface Science* **80** 261–4
[26] Swank R K 1967 Surface properties of II-VI compounds *Phys. Rev.* **153** 844–9
[27] Wu H, Xue M, Ou J, Wang F and Li W 2013 Effect of annealing temperature on surface morphology and work function of ZnO nanorod arrays *Journal of Alloys and Compounds* **565** 85–9
[28] Septier A, Sabary F, Dudek J and Boumiz A 1992 Une photocathode à réserve alcaline (Al-Li)-Ag-O-Li = An alkaline dispenser photocathode (Al-Li)-Ag-O-Li *Comptes Rendus de l'Académie des Sciences* **314** 569–74

Biodegradable, spun nanocomposite polymeric fibrous dressings loaded with bioactive biomolecules for an effective wound healing: A review

H P Felgueiras[1,2], T D Tavares[1] and M T P Amorim[1]

[1]Centre for Textile Science and Technology, University of Minho, Campus de Azurém, 4800-058 Guimarães, Portugal

E-mail: helena.felgueiras@2c2t.uminho.pt

Abstract. In recent years, spinning techniques have been highlighted for their capability of producing complex, fibrous and reticulated constructs with great potential for biomedicine. Indeed, biodegradable constructs based on polymeric fibers are desirable in wound healing due to their large surface area, interconnectivity, open pore structure, and controlled mechanical strength. Modern dressings made of spun biopolymers can incorporate active compounds within their matrix, which can be delivered topically, very beneficial to the healing process. In the present work, we explore the spinning techniques used to process polymers in the form of fibers, assess the properties of some of the most common biodegradable polymers used in wound dressing production, and highlight the potentialities of important biomolecules used in the production of new generation of bioactive wound dressings.

1. Introduction

For many years, the use of artificial fibrous structures was restricted to applications in clothing and decoration. In the past century though, fiber constructs made breakthroughs in engineering with applications in filtration, composite fabrication, energy systems, microfluids and even medicine [1]. In fact, fibers manufactured as mono- or multi-filaments quickly became very useful in medicine regenerative, particularly in skin management and treatment.

Biotextiles based on natural and synthetic fibers are highly desirable in tissue engineering. They are defined as ''structures composed of textile fibers and designed for uses in a specific biological environment, where their performance depends on their interactions with cells and biological fluids as measured in terms of biocompatibility and biostability'' [2]. Depending on their production and arrangement processes, fibers can be knitted, braided, woven and nonwoven, in aligned or random way, to produce complex structures with tuned flexibility, surface area, porosity and mechanical performance [3]. Wound dressing design and production follows these same principles. However, aside from protecting the wound, an effective and functional wound dressing should also take into consideration the wound type and the healing time, so the highest healing rate and the best aesthetic repair is attained in the shortest period possible [4]. In recent years, known spinning techniques have been highly sought to produce fibrous dressings with large surface area for wound healing. Indeed, polymeric constructs in the form of films, mats or even scaffolds have been spun and functionalized with active biomolecules, giving rise to a new generation of wound dressings, the bioactive [5,6].

In the present work, we highlight the most important spinning techniques and polymers (natural and synthetic) applied to the production of wound dressings and offer some examples of the many potentialities of these constructs. Surface functionalization methods and the most desirable biomolecules in wound healing are also analyzed. This work focuses on the use of biodegradable polymer nanocomposites. Contrary to non-biodegradable polymers, which are formed of long chains of C and H atoms that form inter-atomic bonds very difficult to break, biodegradable polymers get decomposed under aerobic or anaerobic conditions, in result of microorganism and enzymes actions.

2. Production of fibrous dressings via spinning techniques

For many years, man-made fibers have been part of our quotidian. Yet, it was only in the last 50 years that efforts have been devoted to the understanding of the fundamental theories behind fiber production, required for the development of complex fibrous systems for biomedical uses. Between the many spinning techniques used to produce polymeric fibers, the following four can be highlighted. Melt spinning is perhaps the preferred method due to the simplicity of polymer processing, as it does not require any solvents. Yet, the electrospinning technique has gained more attention in wound healing due to the structural similarities between the electrospun mats and the extracellular matrix [7].

2.1. Melt spinning

This spinning technique allows the production of customized fibrous constructions, including mono- and multi-filaments, by melting polymer granules at high temperatures and then by drawing the extruded polymer through the spinneret, followed by fiber solidification via rapid quenching or gradual cooling. Mel spinning can only be applied to polymers that resist thermal degradation, i.e. polyethylene terephtalate (PET), starch–polycaprolactone (PCL), poly (L-lactic acid) (PLLA) and poly(lactic acid) (PLA), for instance. Melt spinning is an efficient method to combine two or more polymers to produce fibers without concerns with polymer solubility or solvent miscibility. The polymers' molecules interact via collision during movement and establish interactions via secondary forces such as hydrogen bonding, dipole, ionic or van-der Waals forces. By varying the processing parameters, it is possible to fabricate fibers in the range of few micrometers with tailored morphology, diameter, cross-section and texture. Also, the mechanical properties can be improved at the spin-line by heat treatment or at the drawing and winding station [8-11].

2.2. Wet spinning

Wet spinning is based on the principle of precipitation, in which mass transfer of both solvent and polymer must be considered. It produces fibers of diameters between 30 and 600 μm by extruding the polymeric solution through the spinneret directly into a liquid bath of a poor or non-solvent. This triggers polymerization while drawing off to form long fibers. The coagulation bath removes the solvent as the filaments solidify (from the surface to the core). A diffusional interchange between the newly formed filaments and the coagulation bath causes the bath components to diffuse into the polymeric filament, rendering it insoluble by chemical reaction, and the solvent to diffuse out of the fiber. Alginate, collagen, chitosan, PCL and many of their composites have been processed in the form of fibers using wet spinning, for a variety of tissue engineering uses. Because of the fibers relatively larger diameters (in the micrometer range), fibrous constructs are fashioned with intrinsically high porosity (larger pore size) and interconnected open pore structure, which are desirable for cell penetration, adhesion and proliferation [12-14].

2.3. Dry spinning

In dry spinning, a bi-component system of polymer and solvent is employed. Here, production of filaments is function of both gas temperature and polymer concentration and is coupled to the mass transfer mechanisms of solvent evaporation. The solution containing the polymer (the also known spin dope) is processed and filtered to achieve the right consistency and viscosity, and then is forced at a controlled rate through a fine jet (spinneret). The polymeric solution solidifies as the highly volatile solvent evaporates. As a binary system, mass transport must be considered inside the filament, as a molecular diffusion process, and at the boundary, as the interphase mass transfer of solvent vapor.

Conventional and less conventional polymers like cellulose acetate, polymers and co-polymers of vinyl chloride, PLLA, and dibutyrylchitin have been processed by extrusion into a hot gas stream. Dry spinning is a common alternative for polymers vulnerable to thermal degradation [15-17].

2.4. Electrospinning

Electrospinning is extensively used for fiber production of thinner diameters, 2 nm to several micrometers, and large surface area. In electrospinning, fibers are produced from a polymeric solution pumped at constant rate through a syringe or capillary tube that is subjected to an electrical field, which once reached a critical value surpasses the surface tension and extrudes the solution in the direction of the collecting plate. As the fibers are extruded, they assume a conical shape, known as the Taylor cone. It elongates, subdividing into nanofiber jets that are collected at an optimal distance, while the volatile organic solvent evaporates. The electrospinning technique allows for fibers to be formed in the submicron range with extremely high surface-to-volume ratio, tunable porosity and malleability, interconnected open pore structure, and desirable composition. This is perhaps the most important spinning technique in wound dressing production. Its reticulated network of fibers allows for a breathable dressing to be formed, with a structure that resembles closely the extracellular matrix, this way facilitating tissue regeneration [5,6].

3. Biodegradable polymers

For many years now, biodegradable polymers (hydrolytically and enzymatically degraded) have been used for biomedical applications. In wound healing, biodegradable polymers are required to exhibit specific properties: do not instigate or induce a toxic response upon use; present an acceptable shelf life; degrade at a rate that matches the regeneration process; possess appropriated mechanical properties that vary with degradation in a proportion compatible with regeneration; do not produce toxic degradation by-products; and are permeable and easily processed [18].

Table 1. Biodegradable synthetic polymers: physical and mechanical properties [20-23].

Polymer	Melting Point (°C)	Glass Transition Temperature (°C)	Tensile Modulus (GPa)	Elongation (%)	Degradation time (months)
Polycaprolactone (PCL)	58 - 63	(-65) - (-60)	0.2–0.4	300–1000	> 24
Poly(glycolic acid) (PGA)	220 - 233	35 - 40	6.0 - 7.0	1.5 - 20	6 - 12
Poly(lactic-co-glycolic acid) (PLGA)	Amorphous	45 - 55	1.4 - 2.8	3 - 10	1 - 12 (adjustable)
Poly(lactic acid) (PLA)	150 - 162	45 - 60	0.4 - 3.5	2.5 - 6	> 24
Poly (L/D-lactide) (PLLA or PDLA)	170 - 200	55 - 65	2.7 - 4.1	3 - 10	> 24
Poly (DL-lactide) (PDLLA)	Amorphous	50 - 60	1 - 3.5	2 - 10	12 - 16

Degradation of synthetic and natural polymers requires cleavage of bonds sensitive to hydrolytic or enzymatic action. Synthetic polymers are characterized by predictable properties and a certain uniformity in site-to-site and patient-to-patient outcomes. They can be processed with specific properties that respond to specific demands or application requirements and are most of the time hydrolytically degraded. Compared to polymers susceptible to enzymatic degradation, those hydrolytically degraded are preferred due to the predictability in body response. Table 1 provides a list of some of the most common synthetic biodegradable polymers used in wound dressing production and their inherent properties, including average degradation rates (time to complete resorption) [19].

PCL is a hydrophobic, semi-crystalline, linear resorbable aliphatic polyester, obtained by either ring-opening polymerization of -caprolactone or via free radical ring-opening polymerization of 2-methylene-1-3-dioxepane. Its biodegradation is associated with its aliphatic ester linkage susceptibility to hydrolysis. Poly(glycolic acid) (PGA) is a highly crystalline, biocompatible polyester with good mechanical features and degradation profile, and low solubility in organic solvents. PGA has been used for a variety of purposes including bone fixation devices, drug carriers and scaffolds. PLA is a biodegradable, aliphatic polyester derived from lactic acid. It is a versatile polymer made of renewable materials, i.e. corn starch or sugar cane, that can be fermented into lactic acid and prepared via cyclic dilactone, lactide, ring-opening polymerization. By varying the ratios between its copolymers PGA and PLA, the poly (lactic-co-glycolic acid) (PLGA) co-polymer offers a wide range of degradation rates. Its degradation kinetics is governed by both the hydrophobic/hydrophilic balance and crystallinity, which makes PLGA particularly desirable for medicine regenerative. PLA exists in two stereoisomeric forms giving rise to PDLA and PLLA, two stereoregular polymers, and PDLLA, a racemic polymer obtained from the mixture of D- and L-lactic acid. PDLLA is an amorphous polymer commonly used for drug delivery due to its ability to disperse homogeneously the active species within a monophasic matrix [20-23].

Regarding natural polymers, the majority degrades via enzymatic degradation. The rate of degradation dependents on the application site, accessibility to and concentration of enzymes, and possible chemical alterations made to their structure. They are also susceptible to cell-triggered proteolytic degradation. Natural polymers are classified as polysaccharides, such as alginate, cellulose and chitosan, and as polypeptides, that include collagen and gelatin. Alginate is a polyanion, typically obtained from brown seaweed, that possesses solubility in water, great biocompatibility, low toxicity and is of relatively low cost. Its physical and mechanical properties are dependent on the length, molecular weight and proportion of the guluronate block within the polymeric chain. Cellulose is an abundant polysaccharide based on glucose, that is present in plants, bacteria, fungi, algae and animals. Cellulose can also be biosynthesized by bacteria in the form of bacterial cellulose. It exhibits a unique nanostructure, remarkable physical-chemical properties and biocompatibility and is resistant to hydrolysis, strong alkali and oxidizing agents. Chitosan, derived from partially deacetylated chitin, is composed of D-glucosamine and N-acetylglucosamine repeat units, forming the only pseudonatural cationic polymer. By varying the degree of deacetylation, viscosity and molecular weight, a series of chitosan polymers may be generated. Chitosan is known for its biocompatibility, biodegradability, antimicrobial activity, and wound healing abilities. Collagen is the most abundant mammalian protein, accounting for around 30% of all body proteins, and a major component of ligaments, tendons, skin and bone. It is a good surface-active agent and is capable of penetrating within lipid-free interfaces. It is desirable in many biomedical applications because of its biodegradability, weak antigenicity and superior biocompatibility compared with other natural polymers, i.e. gelatin. Gelatin is a natural polymer derived from the controlled structural and chemical degradation of collagen. Gelatin contains many functional groups and cell binding sites in its structure, increasing its cell binging ability. It is very common in the production of biocompatible and biodegradable drug delivery systems [24-28].

4. Functionalized biomolecules and their effect in wound healing

Skin is the largest organ in the body, formed mainly of two layers, the epidermis and the dermis, and primarily serving as a protective barrier against the environment. Skin wounds normally heal in the predictable amount of time by forming an epithelialized scar tissue. However, whenever burns, trauma, irreversible damaged skin or chronic wounds occur, the need for substitutes or more efficient protective barriers is raised. Engineered skin tissue and high-performance wound dressings are an excellent solution. Although allografts and autografts have been the most recurrent choice in the past, temporary 3D constructs mimicking the skin architecture and loaded with fibroblasts, keratinocytes and endothelial cells or bioactive biomolecules have gained more attention [5,6].

Biodegradable fiber-based structures have been proposed for the healing of dermal and epidermal injuries. In fact, PLGA fibrous dressings have been already successfully used in clinical patients diagnosed with diabetic foot ulcers (i.e. Dermagraft and Dermagraft-TC) [3]. Thin biodegradable hybrid meshes of PLGA and collagen have also been produced for culture of human skin fibroblasts. Results indicated the web-like collagen formations distributed along the PLGA knitted fibers increased the fibroblasts seeding and distribution and facilitated rapid formation of dermal tissue with uniform thickness [29]. Recently, Norouzi *et al* produced nanofibers of PLGA and gelatin via electrospinning and demonstrated desirable bioactivity and hemostasis of the fibrous scaffolds with controlled release of the protein [30]. Electrospun PLGA nanofibers have also been encapsulated with epidermal growth factors (EGF) resulting in desirable biocompatible and bioactive scaffolds with EGF controlled release [31]. Collagen-based scaffolds have been the most popular for skin regeneration both alone or in combination with other polymeric matrices or molecules (growth factors, proteins, etc.). For instance, dermal substitutes composed of type I collagen and elastin hydrolysate, applied in combination with split-skin mesh grafts contributed to the healing of full-thickness wounds [32] and improved dermal regeneration [33]. It has also been shown that preparations of collagen type I and PCL result in optimal degradation kinetics and mechanical performance while supporting dermal fibroblasts attachment and proliferation. Developing a porous structured expedited the healing process, assisted in re-epithelialization and follicle regeneration, and promoted the formation of dermal tissue with a matrix architecture resembling normal, unwounded skin [34]. Additionally, they revealed these microporous electrospun constructs pre-seeded with fibroblasts to promote greater wound healing than acellular scaffolds [35].

Synthetic biodegradable polymers have been used as matrices for skin regeneration both individually and in combination with other natural origin polymers, and more recently have served as base substrates for the incorporation of bioactive molecules. Kenawy *et al* were the first to report the incorporation of antibiotics within nanofiber meshes produced via electrospinning [36]. Since then, more reports on the antibacterial performance of immobilized antibiotics onto PLA, PCL or PLGA nanofibers have been published [37,38]. As the incorporation of bioactive molecules affects the polymer spinnability, more complex systems have been developed; for instance, two stream electrospinning setups have been engineered so the influence of the antibiotic was only seen in one layer of the composite mat and, hence, the elasticity and tensile strength were maintained while the antibacterial performance was enhanced [39]. Biocides, such as QACs, triclosan and chlorhexidine, have also been applied to conventional fibers and textiles for antibacterial purposes. Even though they are not recommended for skin care, they are quite useful in hospital environments [40]. Metal oxide nanoparticles (NPs) have long been known as effective antibacterial agents for medical applications. Silver (Ag), zinc oxide (ZnO) and titanium dioxide (TiO_2) NPs are perhaps the most common and widely studied. AgNPs have attracted interest in wound healing for their large surface area and strong and broad-spectrum of antimicrobial action. They have been loaded onto PCL, PVA, PLGA, chitosan or gelatin for topical and systemic administration with successful results against both Gram-positive and Gram-negative bacteria [41-43]. However, with the rising of antibiotic resistant pathogens and the increasing concern over the impact of NPs in the environment, new alternatives have been researched.

Antimicrobial peptides (AMPs) form an integral part of the innate immune system, working as a defense mechanism in multicellular organisms to control microbial proliferation and modulate the host's response. AMPs are low molecular weight molecules, often cationic, composed of 5-100 amino. Because of their amino acids-based composition, they can be easily surface immobilized and structurally modified, being possible to obtain new AMPs with improved stability and greater targets range. AMPs have also been seen to display synergistic effects with antibiotics, increasing their overall antimicrobial action. Contrary to antibiotics which act selectively against bacteria, AMPs act at multiple sites within microbial cells reducing their resistance [5,6,44]. They play a key role as endogenous mediators of wound healing making their inclusion in active wound dressings, with the goal of both fighting infections and aiding with the skin regeneration, highly demanded. Incorporation of AMPs by simple co-electrospinning or physical binding has been recently explored. The release kinetics of fluorescein labelled inverse-Crabrolin AMP incorporated into PCL nanofibers was examined against *Escherichia coli* and *Bacillus subtilis* with very promising results [45]. In another study, the AMP motif Cys-KR12 originated from the human cathelicidin peptide LL37 was immobilized onto electrospun silk fibroin nanofiber membranes via click chemistry. Data revealed great antimicrobial action against four pathogenic bacterial strains (*Staphylococcus aureus*, *Staphylococcus epidermidis*, *Escherichia coli*, and *Pseudomonas aeruginosa*) without biofilm formation, and functionalized mats were seen to facilitate the proliferation of keratinocytes and fibroblasts and to promote the differentiation of keratinocytes with enhanced cell-cell attachment [46]. Lysozyme and nisin AMPs were functionalized onto poly(acrylic acid) (PAA) and PVA electrospun mats and their antibacterial activity assessed against *Staphylococcus aureus*, the most commonly isolated bacterium in infected wounds. Results showed the ability of the functionalized mats to eliminate bacteria colonies, being able to free the affected area of microbial growth within 14 days [47]. There is still much to explore about the incorporation of AMPs within fibrous constructs; however, as research evolves quickly, we expect progress in this area to arrive very soon.

5. Conclusions and future perspectives

Since the early 20th century that great advances have been introduced and achieve in the area of fibrous constructs, namely bioactive wound dressings. As a result, it is now possible to produce uniform nanofibrous systems with the most complex architectures and features mimicking the extracellular matrix. Natural and synthetic polymeric matrices functionalized with small but highly effective biomolecules have revolutionize and continue to revolutionize the way we treat wounds and protect our skin. Even though significant advances have been made, several issues still need to be further addressed, particularly the biomolecules functionalization process. The inefficient immobilization of metal nanoparticles has caused significant impact in the environment. The activity of many AMPs cannot reach optimal values since, in many instances, is lost when in contact with physiological media and instead of being topically delivered is systemically, putting our immune system at risk. Additionally, the slow and controlled release of the antibacterial agents *in vitro* is not entirely mimicked *in vivo*, with the initial antimicrobial burst being less easy to predict.

In conclusion, great advances have been achieved in wound healing with the use of spinning techniques, particularly those capable of producing fibers at the nanoscale. Nevertheless, more research is required to fully realize the potential of this new generation of bioactive dressings.

Acknowledgments

Authors acknowledge the Portuguese Foundation for Science and Technology (FCT), FEDER funds by means of Portugal 2020 Competitive Factors Operational Program (POCI) and the Portuguese Government (OE) for funding the project PEPTEX with reference POCI-01-0145-FEDER-028074. Authors also acknowledge project UID/CTM/00264/2019 of 2C2T funded by FCT/MCTES.

References

[1] Tamayol A, Akbari M, Annabi N, Paul A, Khademhosseini A and Juncker D 2013 *Biotechnol. Adv.* **31** 669-87
[2] King M W 1991 *Can. Text. J.* **108** 24-30
[3] Tuzlakoglu K and Reis R L 2008 *Tissue Eng. Part B Rev.* **15** 17-25
[4] Zahedi P, Rezaeiana I, Ranei-Siadatb S-O, Jafaria S-H and Supapholc P 2010 *Polym. Adv. Technol.* **21** 77-95
[5] Felgueiras H P and Amorim M T P 2017 *Colloids Surf. B Biointerfaces* **156** 133-48
[6] Felgueiras H P and Amorim M T P 2017 *IOP Conf. Ser. Mater. Sci. Eng.* **254** 062004 1-6
[7] Sumanasinghe R D and King M W 2003 *JTATM*, **3** 1-13
[8] Mezghani K and Spruiell J 1998 *J. Polym. Sci.* B **36** 1005-12
[9] Hufenus R, Reifler F A, Maniura-weber K, Spierings A and Zinn M 2012 *Macromol. Mater. Eng.* **297** 75-84
[10] Liang J Z, Zhong L and Wang K 2012 *J. Appl. Polym. Sci.* **125** 2202-6
[11] Lowe A, Deng W, Smith D W and Balkus K J 2012 *Macromolecules* **45** 5894-900

[12] Lin H-Y and Wang H-W 2012 *Biomatter* **2** 321-8
[13] Rickman J, Tronci G, Liang H and Russell S J 2019 *J. Mater. Sci.* **54** 10529-47
[14] He Y, Zhang N, Gong Q, Qiu H, Wang W, Liu Y and Gao J 2012 *Carbohydr. Polym.* **88** 1100-8
[15] Ohzawa Y, Nagano Y and Matsuo T 1969 *J. Appl. Polym. Sci.* **13** 257-83
[16] Gou Z and McHugh A 2001 *Int. Polym. Process.* **19** 244-53
[17] Abdel-Mohsen A M, Jancar J, Massoud D, Fohlerova Z, Elhadidy H, Spotz Z and Hebeish A 2016 *Int. J. Pharm.* **510** 86-99
[18] Asghari F, Samiei M, Adibkia K, Akbarzadeh A and Davaran S 2017 *Artif. Cells Nanomed. Biotechnol.* **45** 185-92
[19] Patel H, Bonde M and Srinivasan G 2011 *Trends Biomater. Artif. Organs* **25** 20-29
[20] Middleton J C and Tipton A J 2000 *Biomaterials* **21** 2335-46
[21] Farah S, Anderson D G and Langer R 2016 *Adv. Drug Deliv. Rev.* **107** 367-92
[22] Rezwan K, Chen Q Z, Blaker J J and Boccaccini A R 2006 *Biomaterials* **27** 3413-31
[23] Maitz M F 2015 *Biosurf. Biotribol.* **1** 161-76
[24] Mele E 2016 *J. Mater. Chem. B* **4** 4801-12
[25] Rogina A 2014 *Appl. Surf. Sci.* **296** 221-30
[26] Mogoşanu G D and Grumezescu A M 2014 *Int. J. Pharm.* **463** 127-36
[27] Sell S A, Wolfe P S, Garg K, McCool J M, Rodriguez I A and Bowlin G L 2010 *Polymers* **2** 522-53
[28] Silva I O, Ladchumananandasivam R, Nascimento J H O, Silva K K O, Oliveira F R, Souto A P, Felgueiras H P and Zille A 2019 *Nanomaterials* **9** 1064-85
[29] Chen G, Sato T, Ohgushi H, Ushida T, Tateishi T and Tanaka J 2005 *Biomaterials* **26** 2559-66
[30] Norouzi M, Shabani I, Ahvaz H H and Soleimani M 2015 *J. Biomed. Mater. Res.* A **103** 2225-35
[31] Norouzi M, Shabani I, Atyabi F and Soleimani M 2015 *Fiber. Polym.* **16** 782-7
[32] de Vries H J, Middelkoop E, Mekkes J R, Dutrieux R P, Wildevuur C H and Westerhof H 1994 *Wound Repair Regen.* **2** 37-47
[33] de Vries H J, Zeegelaar J E, Middelkoop E, Gijsbers G, Van Marle J, Wildevuur C H and Westerhof W 1995 *Br. J. Dermatol.* **132** 690-7
[34] Bonvallet P P, Culpepper B K, Bain J L, Schultz M J, Thomas S J and Bellis S L 2014 *Tissue Eng. A* **20** 2434-45
[35] Bonvallet P P, Schultz M J, Mitchell E H, Bain J L, Culpepper B K, Thomas S J and Bellis S L 2015 *PLOS One* **10** e0122359 1-17
[36] Kenawy el-R, Bowlin G L, Mansfield K, Layman J, Simpson D G, Sanders E H and Wnek G E 2002 *J. Control Release* **81** 57-64
[37] Zahedi P, Karami Z, Rezaeian I, Jafari S-H, Mahdaviani P, Abdolghaffari A H and Abdollahi M J 2012 *Appl. Polym. Sci.* **124** 4174-83
[38] Kim K, Luu Y K, Chang C, Fang D F, Hsiao B S, Chu B and Hadjiargyrou M J 2004 *Control. Release* **98** 47-56
[39] Hong Y, Fujimoto K, Hashizume R, Guan J, Stankus J J, Tobita K and Wagner W R 2008 *Biomacromolecules* **9** 1200-7
[40] Gao Y, Truong Y B, Zhu Y and Kyratzis I L 2014 *J. Appl. Polym. Sci.* **131** 40797-810
[41] Abdelgawad A M, Hudson S M and Rojas O J 2014 *Carbohydr. Polym.* **100** 166-78
[42] Zhuang X, Cheng B, Kang W and Xu X 2010 *Carbohydr. Polym.* **82** 524-27
[43] Coelho D, Sampaio A, Silva C J S M, Felgueiras H P, Amorim M T P and Zille A 2017 *ACS Appl. Mater. Interfaces* **9** 33107-18
[44] Querido M M, Felgueiras H P, Rai A, Costa F, Monteiro C, Borges I, Oliveira D, Ferreira L and Martins M C L 2018 *Adv. Mater. Interfaces* **5** 1801390-400
[45] Eriksen T H B, Skovsen E and Fojan P 2013 *J. Biomed. Nanotechnol.* **9** 492-8
[46] Song D W, Kim S H, Kim H H, Lee K H, Ki C S and Park Y H 2016 *Acta Biomater.* **39** 146-55
[47] Amariei G, Kokol V, Boltes K, Letón P and Rosal R 2018 *RSC Adv.* **8** 28013-23

IOP Conf. Series: Materials Science and Engineering **634** (2019) 012034 doi:10.1088/1757-899X/634/1/012034

A study on the nonlinear elastic behavior of ballistic gelatin using a rotational rheometer

L Liu[1], C Ding[2, 3], Q Q Lu[1] and X F Zhang[1]

[1]School of Engineering, Zhejiang University City College, Hangzhou 310015, China
[2]Key Laboratory of Special Purpose Equipment and Advanced Manufacturing Technology, Ministry of Education & Zhejiang Province, Zhejiang University of Technology, Hangzhou 310013, China

E-mail: chuanding@zjut.edu.cn

Abstract. The present work was focused on the study of large elastic behavior and constitutive modeling of ballistic gelatin. The simple shear tests were conducted using a rotational rheometer. It is observed that the mechanical parameters of the material should be analyzed directly by the data of rotational angle and torque measured by the rheometer but not the apparent shear stress and strain data provided by the rheometer directly. A hyperelastic constitutive model, a 2-term reduced polynomial strain energy potential, was adopted to describe the nonlinear stress-strain relationship. The material parameters were identified and the dependency of the parameters on the temperature was analyzed. It's found that the modulus at small deformation decreases with increasing temperature following an approximately linear relationship; while, the modulus at large deformation showed a process of first increasing and then decreasing with increasing temperature.

1. Introduction

Ballisti Gelatin (BG) is a tissue simulant whose properties can be easily modified to match various human organ tissue properties by simply varying water to powder ratio. Due to its super mimic ability, gelatin has become a popular model material or surrogate in many bioengineering or biomedical studies [1-3]. There are several reports on the uniaxial compression or elongation experiments of gelatin [4-7]; and found that gelatin can bear very large elastic deformation. When the elastic limit is exceeded, the material fails due to fracture. To evaluate the mechanical stiffness of BG, the elastic or hyperelastic constitutive models have been developed to describe the stress-strain relationship, in which the stiffness parameters were calibrated for individual strain rates. Cronin and Falzon [5] used the Neo-Hookean constitutive model to describe the compressional stress-strain relationship for strain rates from 0.01/s to 1/s. Amborn *et al*. [8] adopted three forms of hyperelastic models (Marlow form, Neo-Hookean form and Ogden form), and obtained the values of the material parameters using the compression data at a specific strain rate of 300/s, which was considered to be consistent with the application in ballistic penetration. Ravikumar *et al*. [9] employed an Ogden type strain energy density function for the nonlinear elastic response and a single Prony exponential term for the rate-dependent response of BG for multiple strain rates. The calibration of the material parameters was based on uniaxial compression tests in which the gelatin samples were compressed to the maximum engineering strain of 0.25 and the nominal compression strain rates in the range of 0.001/s~0.1/s. In our previous work [10], the Mooney-Rivlin model was adopted to describe the hyper-elastic behavior of BG and the material parameters were obtained as functions of the strain rate from the quasi-static compression tests (~0.208/s) and SHPB tests of Salisbury and Cronin (2009) (~1550/s). The drawbacks of this method are that the fitting errors in calibrating the material parameters are quite large

The shearing test has great advantage in studying the mechanical properties of soft materials[11]. The volume of the samples can be very small (in the present study, the diameter of the samples is 25 mm and thickness 2 mm), making the temperature control not as difficult as that in the compression or elongation test. The deformation rate can be easily controlled to be a constant value and the experimental results are much more repetitive. In the present study, a rotational rheometer with parallel plates (see Figure 1a) was used to study the non-linear elastic behaviors of gelatin. The parallel plates geometry was sketched in Figure 1b. With one plate stationary and the other rotating at angular velocity of ω, neglecting inertial forces, the shearing strain rate and shear strain in the sample must be,

$$\dot{\gamma}(r)=\frac{r\omega}{h},\quad \gamma(r)=\frac{r\theta}{h} \tag{1}$$

In which θ is the rotating angle of the upper plate which is measured by the rheometer to calculate the shear strain or strain rate. Therefore, the shear strain and stress in the sample is nonhomogeneous and depends on position. From a torque balance, the load torque of the rotating shaft is,

$$M=2\pi\int_0^R r\tau(r)rdr \tag{2}$$

In which $\tau(r)$ is the shear stress. The value of M can be measured by the rheometer to calculate the shear stress at the edge of the sample, approximately,

$$\tau(R)=\frac{2M}{\pi R^3} \tag{3}$$

This approximation is valid for a wide range of liquids[11]. The rheometer provides the representative stress and strain at the point r/R=3/4 to study the viscosity of the sample,

$$\gamma_a=\frac{3R}{4h}\theta,\ \tau_a=\frac{3M}{2\pi R^3} \tag{4}$$

However, for BG under 29℃, the sample is a viscoelastic solid and the approximation in Eq. (3) is not necessarily valid and the shear stress can not be calculated directly. In the present study, it's supposed that BG is incompressible and its large elastic behavior can be described by the following hyperelastic constitutive model,

$$U=C_{10}(I_1-3)+C_{20}(I_1-3)^2 \tag{5}$$

In which, U is the strain energy per unit of reference volume; C_{10} and C_{10} are material parameters; I_1 is the first deviatoric strain invariant. Then, the shear stress and strain can be related by,

$$\tau=c_1\gamma+c_2\gamma^2\ (c_1=2C_{10}, c_2=4C_{20}) \tag{6}$$

Then the torque of the rotating shaft can be calculated by substituting Equation (6) to equation (2),

$$M=2\pi\int_0^R\left[c_1\frac{\theta}{h}r^2+c_2\left(\frac{\theta}{h}\right)^3r^3\right]r^2dr=\frac{2\pi R^3}{3}c_1\gamma_a+\left(\frac{32}{27}\right)\frac{2\pi R^3}{3}c_2\gamma_a^{\ 3} \tag{7}$$

In the present work, the elastic modulii of the gelatin material, c_1 and c_2, were estimated both by the apparent shear stress and strain using equation (6) (method 1) and by the torque *vs*. apparent strain using equation (7) (method 2) and the two methods were compared to study the error of calculating the shear stress with equation (3). Further, the mechanical parameters of BG and their dependency on sample concentration was analyzed.

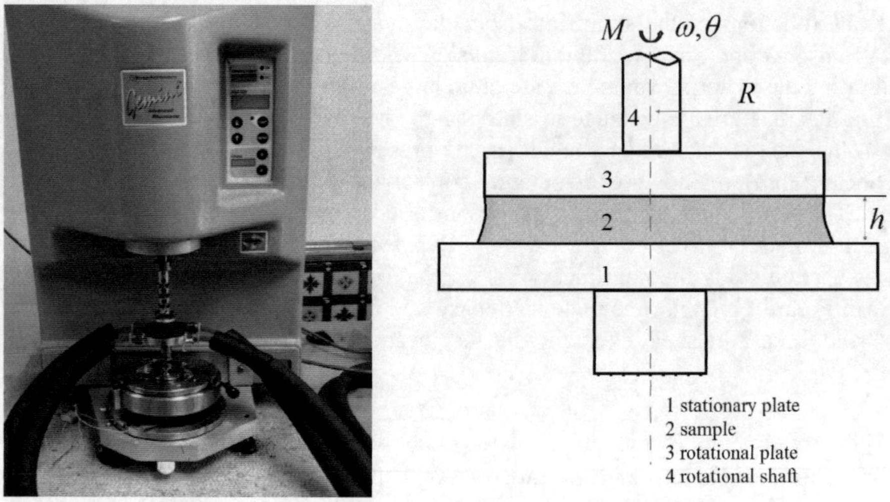

(a) (b)

Figure 1. Rheomter used in the present study (a) and the sketched configuration of the shear test (b).

2. Materials and methods

2.1. Gelatin gel preparation

The Type-A gelatin powder with a Bloom number of 250, provided by QingHai Gelatin Factory (China), was used in the present study. Preparation of the gelatin solution was described in a previous publication [12]. The solution was left in a incubator chamber at 40°C for hydration for 24 hours. After that, the solution was poured gently onto the stationary plate of the rheometer. After the rotational plate descending to contact with the solution and the gap between the two plates achieving the predetermined value (2mm), the superfluous solution was scraped off and the water-cooling system connected to the stationary plate was turned on to keep the sample at 4-19℃. After 24 hours' aging, the gelatin gel was ready to be tested.

2.2. Shearing test

The selected shearing rates are 0.0005/s, 0.01/s, and 0.1/s. The angle of rotation and torque of the rotational shaft were recorded, and the shearing stress and strain were provided using Eq. (4). Each test was replicated at least three times. It's observed from the test that BG can bear very large elastic deformation before fracture. Figure 2a and figure 2b show a typical set of test results displayed with τ_a *vs.* γ_a and M *vs.* γ_a, respectively.

The results showed that the material behaves nonlinearly at large deformation and the stiffness of the material is dependent on strain rate. The elastic modulii, c_1 and c_2, can be estimated both by the apparent shear stress and strain in figure 2a using equation (6) (method 1) and by the torque *vs* apparent strain in figure 2b using equation (7) (method 2). The comparison of the results obtained by the two methods was presented in table 1.

The errors of the parameters obtained by method 1 were close to 20% compared that with method 2. This observation confirmed that it is not proper to use the apparent shear stress and strain directly provided by the rheometer with parallel plates to study the mechanical parameters of BG. Otherwise, the torque of the rotating shaft should be used for analysis indirectly, e.g. equation (7). It should be pointed out the specific relationship between the M *vs.* γ_a is dependent on the constitutive model chosen as in equation (5).

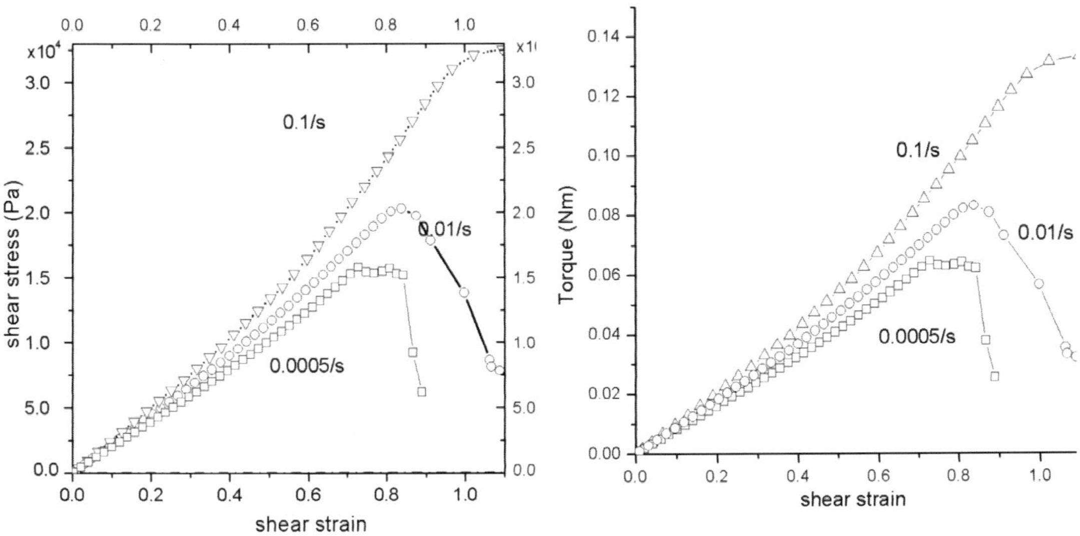

(a) (b)

Figure 2. Typical sets of shear test results: (a) apparent shear stress *vs.* shear strain and (b) torque *vs.* apparent shear strain.

Table 1. material parameters, c_1 and c_2, obtained by two methods.

Strain rate	Method 1		Method 2	
	c_1	c_2	c_1	c_2
0.0005/s	19300	4500	16300	3800
0.01/s	22000	4600	18500	3900
0.1/s	24800	8300	21300	7000

2.3. Temperature dependence of the mechanical parameters

The shearing test under different temperatures were carried out with a selected shearing rate, 0.01/s, to study the temperature dependency of the mechanical parameters of BG. The torque *vs.* apparent strain were used to estimate the elastic modulii, c_1 and c_2, using equation (7). The curve fitting results were shown in figure 3a and the elastic modulii variation with temperature in figure 3b.

Figure 3b indicated that the modulus c_1 changes monotonically with the temperature. As anticipated, decreasing the temperature increased the material stiffness. Nevertheless, the value of c_2 seemed to reache a maximal value between the temperature of 4℃ and 14℃. This phenomenon needs further study.

3. Discussions and conclusions

The mechanical behaviors of ballistic gelatin can be obtained using a rheometer with parallel plates. However, for ballistic gelatin, direct using the apparent shear stress and strain reported by the rheometer to estimate the mechanical parameters produces relatively large errors. Instead, the shear stress-strain relationship can be pre-assumed, e.g., equation (6), and the torque of the rotating plate can be calculated, equation (7) and used to fit the test results. It is observed that the errors of the elastic modulii of 10% ballistic gelatin at 4℃ obtained by the traditional method (method 1) were close to 20% compared that with the new method (method 2). And the large errors exist for different loading rates in the range of 0.0005~0.1/s.

The temperature dependence of the mechanical parameters of ballistic gelatin was also studied by using the shearing test under different temperatures with a selected shearing rate, 0.01/s. The elastic

IOP Conf. Series: Materials Science and Engineering **634** (2019) 012034 doi:10.1088/1757-899X/634/1/012034

modulus, c_1, which is the shear modulus (under small deformation) of the material, decreases monotonically with the increasing temperature. But the variation of c_2 with the temperature seems more complicated, which reaches a maximal value approximately at 9℃ and further tests are needed to verify this conclusion.

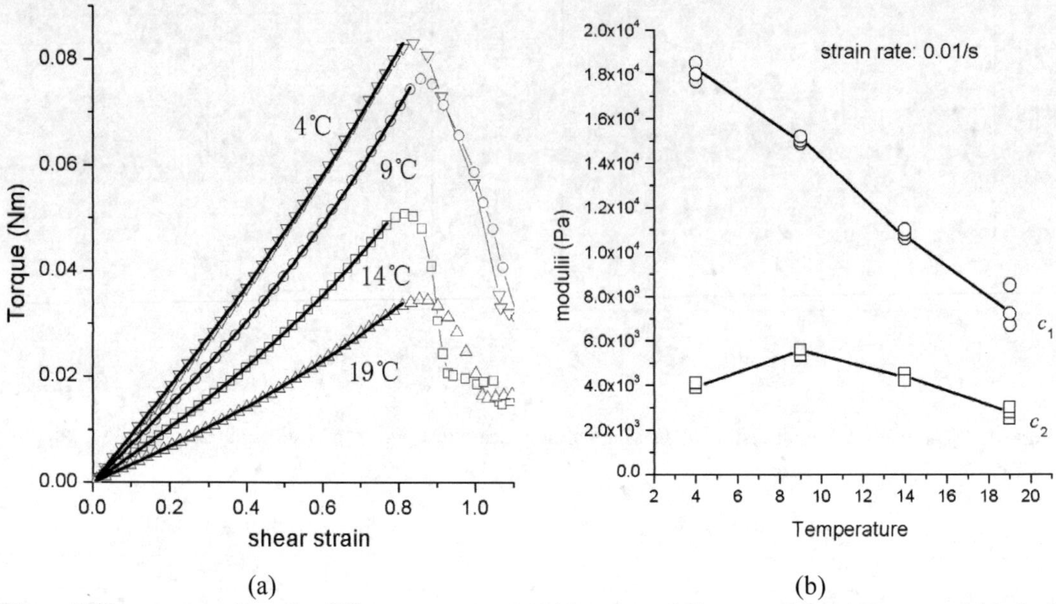

Figure 3. Shear test results under different temperature: (a) test data of *M vs.* γ_a and fitted curves, and (b) elastic modulii variations with temperature.

Acknowledgements
This work is supported by the Zhejiang Provincial Natural Science Foundation of China (Q19A020012)

References
[1] Pirlot M D G B 2001 Soap and Gelatine for simulating human body tissue: an experimental and numerical evaluation *19th International Symposium of Ballistics* Interlaken Switzerland 1011-7
[2] Jussila J 2004 Preparing ballistic gelatine-review and proposal for a standard method *Forensic Science Int* **141** 91-8
[3] Grover C N, Cameron R E and Best S M 2012 Investigating the morphological, mechanical and degradation properties of scaffolds comprising collagen, gelatin and elastin for use in soft tissue engineering *J Mech Behav Biomed* **10** 62-74
[4] Moy P, Gunnarsson C A and Weerasooriya T 2009 Tensile deformation and fracture of ballistic gelatin as a function of loading rate *SEM Conf Proceedings* (New Mexico, USA)
[5] Cronin D S and Falzon C 2011 Characterization of 10% ballistic gelatin to evaluate temperature, aging and strain rate effects *Exp Mech* **51** 1197-206
[6] Koene L, Barou J L and Viot P 2011 Ballistic gelatine behaviour under quasi-static and dynamic loadings *26th International Symposium on Ballistics* (Miami, USA)
[7] Czerner M, Fasce L A, Martucci J F, Ruseckaite R and Frontini P M 2016 Deformation and fracture behavior of physical gelatin gel systems *Food Hydrocolloids* **60** 299-307
[8] Amborn E K, Mucikùchler K H and Hinz B J 2012 Experimental and numerical study of soft tissue surrogate behavior under ballistic loading *IMECE2012-85724*
[9] Ravikumar N, Noble C, Cramphorn E and Taylor Z A 2015 A constitutive model for ballistic gelatin at surgical strain rates *J Mech Behav Biomed* **47** 87-94

[10] Liu L, Jia Z, Ma X L, Fan Y R, Li W and Liu H W 2014 A spherical cavity expansion model of large elastic deformation and its application to ballistic gelatin penetration problems *Int J Impact Eng* **71** 106-17
[11] Macosko C W 1994 Rheology: principles, measurements, and applications *New York: WILEY-VCH*
[12] Liu L, Jia Z, Ma X L and Fan Y R 2013 Analytical and experimental studies on the strain rate effects in penetration of 10wt % ballistic gelatin *J Phys: Conf Ser* **451** 012035

IOP Conf. Series: Materials Science and Engineering **634** (2019) 012035 doi:10.1088/1757-899X/634/1/012035

Computational design of dummy molecularly imprinted polymers via hydrogen bonding investigation for oxytetracycline determination

N B D Reyes[1,2] and B B Pajarito[1]

[1]Polymer Research Laboratory, Department of Chemical Engineering, University of the Philippines, Diliman, Quezon City 1101, Philippines

E-mail: nbdelosreyes@gmail.com

Abstract. Oxytetracycline (OTC), a banned broad-spectrum antibiotic, currently requires highly selective and specific determination methods to measure its concentration below 200 ppb in foods of animal origin. Molecular imprinting technology could be utilized to construct a highly selective and cost-effective synthetic receptor for OTC. In this work, we investigated different monomers in designing a dummy MIP for OTC detection using TC as the dummy template. Template-monomer complexes of pre-polymerization mixtures were modeled using density functional theory for geometry optimization, intermolecular hydrogen-bonding situation, and interaction energies. O-phenylenediamine (OPD) at TC:OPD molar ratio = 1:7 was shown to be the optimum monomer, forming 11 stable intermolecular hydrogen bonds with TC and having the lowest interaction energy among the complexes. We also presented indole, pyrrole, and carbazole to be plausible monomers for imprinting TC; however, they are energetically less-favored than OPD. This study provides aid in dummy MIP design for OTC concentration measurement using the molecular level interaction of different monomers with the template TC.

1. Introduction

Oxytetracycline (OTC) is a broad-spectrum antibiotic used against gram-positive bacteria, gram-negative bacteria and atypical organisms including mycoplasmas, protozoan parasites, and chlamydiae. However, it is currently an agricultural contaminant, which could lead to serious multi-drug resistance of bacteria, environmental pollution, and chronic toxicity [1]. As a countermeasure, the Codex Alimentarius Commission implemented a maximum residue limit of 200 ug kg^{-1} for OTC determination in food of animal origin [2]. Traditional method for determining trace amount of OTC utilizes high performance liquid chromatography (HPLC) coupled with either mass spectroscopy or UV/Vis spectroscopy. Unfortunately, traditional assays suffer from expensiveness, laborious sample pretreatment, and complex extraction procedures. In effect, researchers are developing cost-effective and simpler assays, which include immunological methods, electrochemical methods, colorimetric methods, and molecularly imprinted polymer (MIP)-based sensing [3].

MIP-based techniques develop a synthetic receptor for a particular analyte through self-assembled complexes between functional monomers and the target analyte. The active imprint created in the polymer acts as a selective recognition site in the MIP, which the target analyte rebinds to via non-covalent interactions such as hydrogen bonding, electrostatic attraction, and dispersion forces [4].

However, some recognition sites have non-extracted template due to imperfect template removal via solvent extraction. The limitation of solvent extraction results to template bleeding during the rebinding process and the problem extends to quantification inaccuracies. To address template bleeding, researchers develop a dummy MIP, which uses a dummy template with an analogous structure to the analyte of interest, as reported in cyproheptadine, polybrominated diphenyl ethers, and nevirapine determination [5-7]. OTC has eight structural analog molecules, closest of which is tetracycline (TC) differing only with a single hydroxyl moiety, as shown in figure 1.

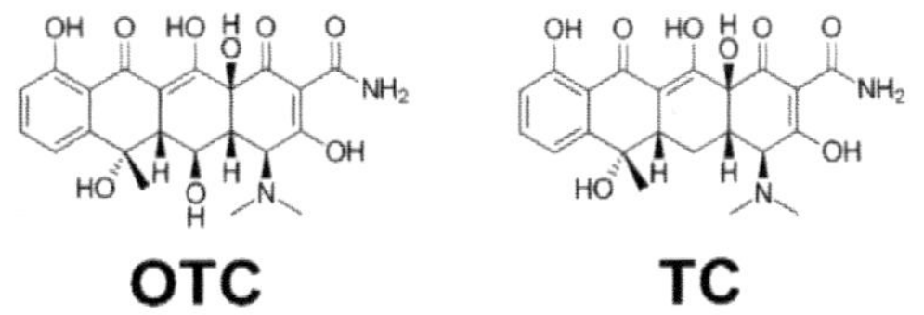

Figure 1. Chemical structures of OTC and TC.

Moreover, development of dummy MIP utilizes computational approach through *ab initio* calculations such as density functional theory (DFT). Recent advances in dummy MIPs used the DFT approach for antibiotic detection such as clenbuterol and chloramphenicol [8,9].

DFT methods provide efficient theoretical description of the template molecules, the functional monomers, and the intermolecular hydrogen bonds present in the pre-polymerization complex. Calculated theoretical descriptions include prediction of equilibrium geometry, interaction energy ΔE, and sites of action between molecules [10]. DFT approach selects optimum functional monomers with the lowest predicted ΔE value of the resulting template-monomer complexes. Currently, researchers attribute low ΔE values to the presence of intermolecular hydrogen bonding between the template and the functional monomers [11]. Thus, to develop an optimum dummy MIP, we explored the resulting intermolecular hydrogen bonds between TC and functional monomers through their sites of action, equilibrium geometry, and interaction energy using DFT calculations. This paper presents the optimum functional monomer for dummy MIP-based sensing of OTC as determined from DFT calculations, using TC as dummy template.

2. Experimental

2.1. Instrumentation and computational methods

DFT calculations were performed using Spartan Student Edition Version 7.2.7 (Wavefunction Inc.) on a personal computer with an Intel Core i5 processor, 3.20 GHz CPU, 8 GB memory, and a 1 TB hard disk running on Windows 8.1 operating system [12]. B3LYP was chosen as gradient corrected functional to provide accurate structural and energetic description of hydrogen bond systems in the template-monomer complex. 6-31G* was used as the basis set to provide accurate results with the minimum computational time [13]. Moreover, basis set superposition error (BSSE) was reported negligible for systems with large clusters of hydrogen bonding and was not considered in this study [10].

2.2. Analysis of reaction sites

During the imprinting and rebinding process, the template molecule and functional monomer interact at specific sites of action via non-covalent bonding [4]. The sites of interaction of the template were analyzed through chemical functional description (CFD) and electrostatic potential (ESP) diagrams. CFD shows proton donors and acceptors of a molecule, aiding in possible hydrogen bonding sites between the template and the functional monomer. ESP diagram shows the concentration of electronegative and electropositive regions of a molecule, assisting in the determination of electrostatic attraction sites between the template and functional monomer.

2.3. Construction of template-monomer complexes

To produce optimum template-functional monomer complex during the imprinting process, the template-monomer complex ratio, was modeled in terms of energy. This work proceeds on the idea that generating the most possible number of stable hydrogen bonds would yield the most stable imprinted complexes [14]. The number of hydrogen bonds, bond length, bond angles, and bond types was further analyzed. Hydrogen bonding strength based on geometry was discussed using Jeffrey's classification of hydrogen bonding geometry shown in table 1 [13].

Table 1. Hydrogen bonding categories following the classification of Jeffrey [15].

Type	Strong	Moderate	Weak
interaction type	strongly covalent	mostly electrostatic	electrostatic/dispersion
bond length [Å]	1.2 – 1.5	1.5 – 2.2	2.2 – 3.2
bond angle [°]	175 – 180	130 – 180	90 – 150

2.4. Selection of functional monomer

Introduction of conducting functional monomers lead MIP-based sensing to improved MIP processability, stability, and cost effectiveness. This route of MIP production proceeds with electropolymerization, removing the need for the crosslinking agent and initiator [16]. Four conducting functional monomers were screened: o-phenyldiamine (OPD), carbazole (CAR), indole (IND), and pyrrole (PYR). The molecular structures of the functional monomers are shown in figure 2. Each conformation of functional monomer and template was optimized separately to obtain their respective energies. Using the magnitude of ΔE, the optimum functional monomer and template:monomer molar ratio was selected.

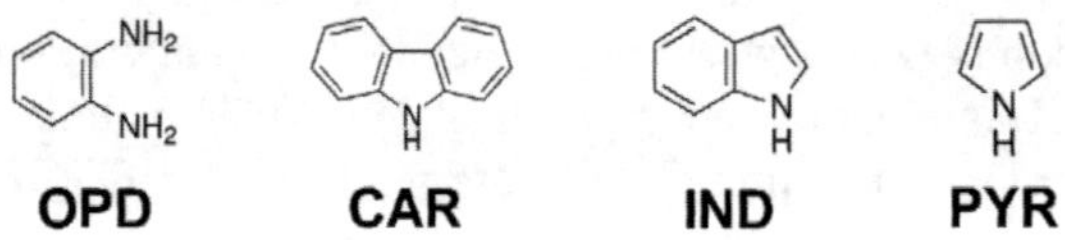

Figure 2. Chemical structures of OPD, CAR, IND, and PYR.

ΔE was computed using equation (1) [16]:

$$\Delta E = E_C - E_T - nE_M \quad (1)$$

where E_C is the equilibrium monomer-template complex energy, E_T is the equilibrium template energy, E_M is the equilibrium monomer energy, and n is the number of monomers present.

3. Results and discussion

3.1. Determination of template molecule's site of action

With the goal of forming stable complex configuration, molecular self-assembly was simulated with the assistance of CFD and ESP to investigate all possible sites of action. Figure 3 shows the CFD and ESP diagrams of TC. There are 10 hydrogen bonding sites in the TC molecule: hydroxyl groups - O1, O2, O3, O5, and O6; carbonyl groups – O4, O7, and O8; and amine groups – N1 and N2. All these functional sites were proton acceptors, while only 7 of these were also proton donor sites for functional monomer adduction (O1, O2, O3, O5, O6, N1, and N2). The ESP diagram shows the electroactive regions of TC. TC has 9 electronegative regions: O1, O2, O3, O4, O5, O6, O7, O8, and N1; and has 2 electropositive regions: H5(N2) and H7(O2). This investigation proceeded to use the 10 sites of action of TC as proton acceptor, and each functional monomer as a single proton donor. No functional monomer was paired with H5(N2) and H7(O2), to avoid steric crowding of functional monomers on O2 and N2, respectively [18,19]. In effect, the selected sites of action for the template molecule were O1, O2, O3, O4, O5, O6, O7, O8, N1, and N2.

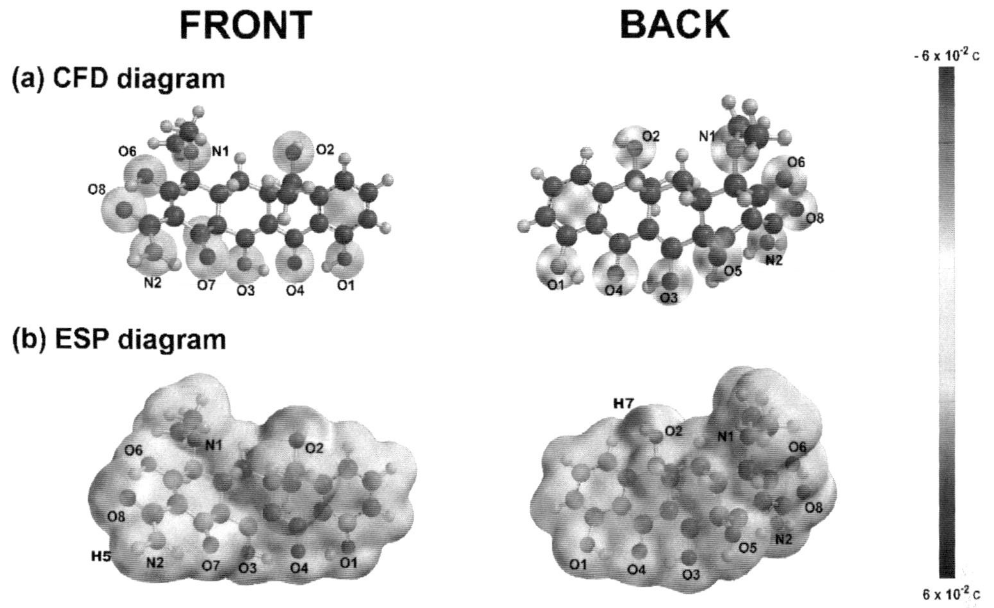

Figure 3. CFD (a) and ESP (b) diagrams of TC.

3.2. Formation of the template-monomer complexes

The template-monomer molar ratio must be optimized for high selectivity of the imprinted polymer. In this study, each imprinting system was simulated to start at template-monomer ratio = 1:10. Each type of functional monomer was initially placed at a hydrogen bond distance of 1.6 Å and angle of 180° with each site of action of the template molecule to simulate the strongest intermolecular hydrogen bond according to Jeffrey [15]. Conformational search was implemented in each constructed complex using Merck Molecular Force Field calculation. Resulting non-hydrogen bonded functional monomers from each run were removed from the complex, using a 3.2 Å and 120° cut-off hydrogen bond distance and angle, until all functional monomers have stable hydrogen bond with the template molecule [12]. After which, hydrogen bonds were verified using equilibrium geometry calculation via DFT at B3LYP level, with a basis set of 6-31G*. This top-down procedure eliminates the ambiguity of comparing ΔE values for 1:1 template-monomer molar ratio for template molecules with multiple sites of action.

Figure 4 shows the final configuration of each template-monomer complex. The optimum conditions were as follows: TC:OPD = 1:7, TC:CAR = 4, TC:IND = 5, and TC:PYR = 5. Differences in the optimum molar ratio were accounted to steric crowding between the functional monomers. Bulky, tricyclic functional monomer CAR has its maximum monomer limit at 4, compared to smaller molecules: bicyclic functional monomer IND, and single ring functional monomers, PYR and OPD. The number of electroactive sites of functional monomers also affect the results. IND, PYR, and CAR have only 1 active amine group to act as a proton donor and proton acceptor, compared to OPD with 2 active amine groups, providing four proton donors and two proton acceptors.

Table 2 illustrates a quantitative analysis of the formed hydrogen bonding networks. Considering the 4 equilibrium complexes, all formed hydrogen bond lengths ranged from 1.9 - 2.5 Å and hydrogen bond angles are within 121.5° - 173.5°. The mean hydrogen bond length of each complex was 2.1, 2.1, 2.2, and 2.0 Å for TC + 7OPD, TC + 4CAR, TC + 5IND, and TC + 5PYR, respectively. The mean hydrogen bond angles of each complex was 152.0°, 158.2°, 145.4°, 154.8° for TC + 7OPD, TC + 4CAR, TC + 5IND, and TC + 5PYR, respectively. The mean hydrogen bond lengths and angles of the formed complexes indicate moderate hydrogen bond strength, where intermolecular hydrogen bonds were mostly electrostatic in nature, as shown in table 1 [13].

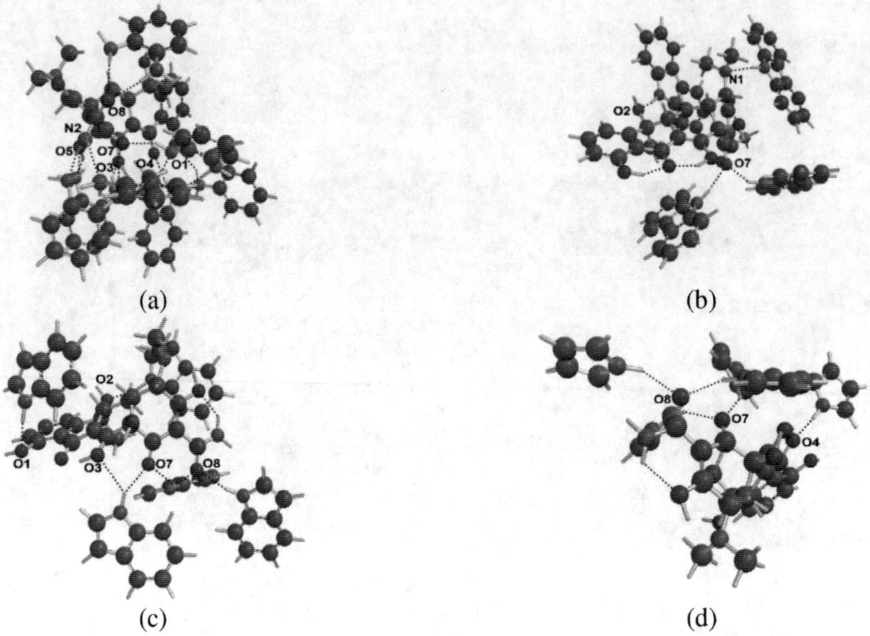

Figure 4. Final conformation of template-monomer complexes after DFT optimization: TC + 7OPD (a), TC + 4CAR (b), TC + 5IND (c), and TC + 5PYR (d).

Table 2. Geometrical parameters of intermolecular hydrogen bonding in each template-functional monomer complex. Double hydrogen bonds observed in specific sites have superscripts [a] and [b].

Complex	Type	Length [Å]	Angle [°]	Complex	Type	Length [Å]	Angle [°]
TC + 7OPD	N2[a]	2.3	144.9	TC+ 5IND	N1	2.2	173.5
	N2[b]	2.1	167.3		O2	2.0	160.1
	O1	2.2	142.6		O3	2.5	130.0
	O3	2.0	153.2		O7[a]	2.5	121.5
	O4	1.9	152.6		O7[b]	2.0	149.3
	O5[a]	2.2	139.3		O8[a]	1.9	166.4
	O5[b]	2.0	152.8		O8[b]	2.1	153.4
	O7[a]	2.1	142.6				
	O7[b]	2.1	148.3	TC + 5PYR			
	O8[a]	2.0	169.3		O4	2.0	169.5
	O8[b]	2.2	158.6		O7[a]	2.0	146.9
					O7[b]	2.1	145.0
TC + 4CAR	N1	2.2	173.5		O8[a]	2.0	158.3
	O2	2.1	160.2		O8[b]	2.0	154.1
	O7[a]	2.0	159.3				
	O7[b]	2.1	139.8				

Double hydrogen bonds were present in each template-monomer complex at the following sites of action: carbonyl groups O7 and O8, hydroxyl group O5, and amine group N2 for TC + 7OPD; carbonyl group O7 for TC + 4CAR; carbonyl groups O7 and O8 for both TC + 5IND and TC + 5PYR. Double hydrogen bonding promotes the stability of the template-monomer complex and increase the affinity of the recognition site during the rebinding process [20].

3.3. Theoretical screening of functional monomer

Functional monomers interact uniquely with a template molecule and in each site of action during the imprinting process. Using DFT optimization, equilibrium geometry and interaction energy were calculated for each complex. All functional monomers have negative ΔE values, indicating their feasibility during the imprinting process as shown in table 3. Among the functional monomers, OPD has shown the most stability with the most negative ΔE value. The significant ΔE value of OPD was attributed to its high template-monomer molar ratio and to the presence of double hydrogen bonds in N2, O5, O7, and O8 [17]. Contrary to the prior hypothesis, the number of stable hydrogen bonds does not necessary determine the electronic stability of a template-functional monomer complex. The results suggest that the molecular size and available functional groups in a functional monomer are key factors in the resulting magnitude of electronic stability of a template-functional monomer complex.

Table 3. Interaction energies of complexes formed with the description of the hydrogen bonding situation.

Complex	ΔE [kJ/mol]	Hydrogen bonds formed	Double hydrogen bonds formed	Mean Length [Å]	Mean Angle [°]
TC + 7OPD	-357.102	11	4	2.087	151.95
TC + 4CAR	-84.131	5	1	2.096	158.19
TC + 5IND	-70.573	7	2	2.213	145.37
TC + 5PYR	-62.593	4	2	2.040	154.76

4. Conclusion

In this study, geometry optimization, bonding situation, and binding energies of template-monomer complexes were presented using TC as the dummy template molecule. Through DFT methods, OPD was presented as the best functional monomer with a template-monomer molar ratio = 1:7. The complex TC + 7OPD formed produced 11 stable intermolecular hydrogen bonds and 4 double hydrogen bonds in distinct sites of action. This work provides the theoretical basis for the development of dummy MIP for OTC sensing.

Acknowledgments

This work is fully supported by the Engineering Research and Development for Technology (ERDT) program of the Department of Science and Technology (DOST).

References

[1] Harnisz M, Korzeniewska E and Gołaś I 2015 The impact of a freshwater fish farm on the community of tetracycline-resistant bacteria and the structure of tetracycline resistance genes in river water *Chemosphere* **128** 134-41

[2] Codex Alimentarius 2017 Maximum Residue Limits (MRLs) and Risk Management Recommendations (RMRS) for Residues of Veterinary Drugs in Foods CAC/MRL 2-2017 19, 25, 35, 38 y 40

[3] Gao F, Feng S, Chen Z, Li-Chan E C Y, Grant E and Lu X 2014 Detection and quantification of chloramphenicol in milk and honey using molecularly imprinted polymers: Canadian penny-based SERS nano-biosensor *J. Food Sci.* **79** N2542-9

[4] Kowalska A, Stobiecka A and Wysocki S 2009 A computational investigation of the interactions between harmane and the functional monomers commonly used in molecular imprinting *J. Mol. Struct. Theochem* **901** 88-95

[5] Feás X, Seijas J A, Vázquez-Tato M P, Regal P, Cepeda A and Fente C 2009 Syntheses of molecularly imprinted polymers: Molecular recognition of cyproheptadine using original print molecules and azatadine as dummy templates *Anal. Chim. Acta* **631** 237–44

[6] Marć M, Panuszko A, Namieśnik J and Wieczorek P P 2018 Preparation and characterization of dummy-template molecularly imprinted polymers as potential sorbents for the recognition of selected polybrominated diphenyl ethers *Anal. Chim. Acta* **1030** 77–95
[7] Kareuhanon W, Lee V S, Nimmanpipug P, Tayapiwatana C and Pattarawarapan M 2009 Synthesis of Molecularly Imprinted Polymers for Nevirapine by Dummy Template Imprinting Approach *Chromatographia* **70** 1531–7
[8] Zhang B, Fan X and Zhao D 2018 Computer-aided design of molecularly imprinted polymers for simultaneous detection of clenbuterol and its metabolites *Polymers (Basel).* **11** 17
[9] Liu J B, Wang G Y, Tang S S, Gao Q, Liang D D and Jin R F 2019 Theoretical and experimental research on self-assembly system of molecularly imprinted polymers formed via chloramphenicol and methacrylic acid *J. Sep. Sci.* **42** 769–77
[10] Del Sole R Del, Lazzoi M R, Arnone M, Sala F Della, Cannoletta D and Vasapollo G 2009 Experimental and computational studies on non-covalent imprinted microspheres as recognition system for nicotinamide molecules *Molecules* **14** 2632–49
[11] Piacham T, Nantasenamat C, Isarankura-Na-Ayudhya C and Prachayasittikul V 2013 Synthesis and computational investigation of molecularly imprinted nanospheres for selective recognition of alpha-tocopherol succinate *EXCLI J.* **12** 701–18
[12] Spartan'07, Wavefunction, Inc., Irvine, CA
[13] Zhu W L, Jiang H L, Puah C M, Tan X J, Chen K X, Cao Y and Ji R Y 1999 Density functional theory (DFT) study on the interaction of ammonium (NH4+) and aromatic nitrogen heterocyclics *J. Chem. Soc. Perkin Trans. 2* 2615–22
[14] Zhang B, Fan X and Zhao D 2018 Computer-aided design of molecularly imprinted polymers for simultaneous detection of clenbuterol and its metabolites *Polymers (Basel)* **11** 17
[15] Sánchez-García E 2006 Computational Study of Weakly Interacting Complexes, www-brs.ub.ruhr-uni-bochum.de/netahtml/HSS/Diss/SanchezGarciaElsa/.
[16] Dakshayini B S, Raghava K, Mishra A, Shetti N P, Malode S J, Basu S, Naveen S and Raghu A V 2019 Role of conducting polymer and metal oxide-based hybrids for applications in ampereometric sensors and biosensors *Microchem. J.* **147** 7-24
[17] Riahi S, Edris-Tabrizi F, Javanbakht M, Ganjali M R and Norouzi P 2009 A computational approach to studying monomer selectivity towards the template in an imprinted polymer *J. Mol. Model.* **15** 829–36
[18] Sharma P S, Iskierko Z, Pietrzyk-Le A, D'Souza F and Kutner W 2015 Bioinspired intelligent molecularly imprinted polymers for chemosensing: A mini review *Electrochem. Commun.* **50** 81–7
[19] Mojica E R E, Autschbach J, Bright F V and Aga D S 2011 Tetracycline speciation during molecular imprinting in xerogels results in class-selective binding *Analyst* **136** 749-55
[20] Wang Y, Liu J, Tang S and Jin R-F 2015 Preparation of melamine molecularly imprinted polymer by computer-aided design *J. Sep. Sci.* **38** 2647-54F

IOP Conf. Series: Materials Science and Engineering **634** (2019) 012036 doi:10.1088/1757-899X/634/1/012036

Feasibility of oxidized soybean oil for rubber devulcanization

C C Pegollo[1,2] and B B Pajarito[1]

[1] Polymer Research Laboratory, Department of Chemical Engineering, University of the Philippines Diliman, Quezon City 1101 Philippines

E-mail: ccpegollo@gmail.com

Abstract. This study investigates the feasibility of using oxidized soybean oil (Ox-SO) for mechano-chemical devulcanization of sulfur-cured natural rubber. A 2^2 full factorial design of experiment was used to determine the effect of devulcanization time (20, 40 min) and amount of Ox-SO (6, 10 phr) in the crosslink density and sol-gel content of devulcanized VR (DVR). After conductive heating and stirring of soybean oil for 60 h, a peroxide value of 316.3 meq O_2 kg^{-1} oil was achieved which corresponds to approximately 2500 ppm of lipid hydroperoxides. Decreasing the devulcanization time (from 40 to 20 min) and amount of Ox-SO (from 10 to 6 phr) decreased the crosslink density by 71.1% and increased the sol content by 228.2%. The optimum conditions are 20 min of devulcanization time and 6 phr of Ox-SO where the crosslink density of VR is reduced to 6.2×10^{-5} mol cm^{-3} and the sol content is 10.2% (control VR crosslink density: 2.1×10^{-4} mol cm^{-3}, sol content: 3.1%). In summary, Ox-SO was found to be a feasible agent for rubber devulcanization. DVR blends with virgin rubber are required to assess Ox-SO's commercial readiness.

1. Introduction

The inherent physical and chemical stability of waste vulcanized rubber (VR) makes it a significant contributor to solid waste in the Philippines. According to the status report of the National Solid Waste Management Commission of the Philippines status report, rubber waste comprises at most 0.37% which is currently equivalent to 58.7 thousand tons and 62.6 thousand tons in 2020. To mitigate this problem, waste VR finds use in energy recovery and filler applications. However, with the continued rise in rubber production, rubber recycling through devulcanization remains the most sustainable method to decrease the accumulation of waste rubber [1].

Mechanical devulcanization uses mechanical shear along with chemical modifiers for devulcanization. The mechanical shear is to be supplied during the milling of the waste rubber in which the crosslinks are subject to thermal and mechanical stresses which result to breakage of bonds. This type of devulcanization is the most common owing to its direct applicability to existing equipment of rubber processing facilities. This would mean higher ease of introducing the technology to the current process with less capital and operating costs. Some of the chemical agents that have been used in combination with mechanical shearing are disulfides, mixtures of rubber accelerators and activators collectively known as De Link reactant, a mixture of quinone group and benzoic acid which acts as a counter cure reagent, the modifying composition used in the Lev Gum process, benzoyl peroxide and various organic compounds such as dienophiles and acids incorporated with high pressure high temperature sintering [1,2].

Soybean oil (SO) is characterized by its high degree of unsaturation, with it having one of the highest amounts of polyunsaturated fatty acids (PUFAs) among edible oils [3]. Because of this, SO is prone to lipid peroxidation upon subjection to various conditions. Lipid peroxidation is the transformation of

PUFAs into lipid hydroperoxides in a degradation process initiated by either light, heat or oxygen [4]. Compared with other vegetable oils such as from corn, rapeseed and sunflower, SO appeared to have the highest oxidation rate when heated in a microwave oven reaching a peroxide value of ~30 meq O_2 kg^{-1} after only 12 min of microwave exposure [5,6].

This work explores the possibility of the use of oxidized soybean oil (Ox-SO) as a devulcanizing agent for mechanochemical devulcanization of sulfur - cured VR compounds. The effect of devulcanization time and amount of Ox-SO were determined by analyzing the crosslink density and sol content of the devulcanized VR (DVR). Ox-SO has the potential to be an abundant and low-cost waste resource for recycling rubber, eliminating 2 sources of waste in one process that can benefit the community environmentally and economically.

2. Experimental

2.1. Materials

Pure and refined SO made from American soybeans were purchased from a local supermarket. PUFAs, monounsaturated fatty acids, and saturated fatty acids content are approximately 50, 14, and 19 g per 100 ml, respectively as stated in the product label (Simply Pure Soybean Oil). Glacial acetic acid (ACS grade, RCI Labscan), chloroform (ACS grade, RCI Labscan), potassium iodide KI (AR grade, Ajax Finechem), sodium thiosulfate pentahydrate (ACS grade, Duksan Reagents) and pure potato starch (Jandy Goods Corp.) were used for the peroxide value (PV) determination of Ox-SO. All rubber compounding ingredients – natural rubber SPR10, sulfur, N-cyclohexyl-2- benzothiazole sulfenamide (CBS), zinc oxide, and stearic acid are industrial grade and purchased from Philippine Belt Manufacturing Corporation (Quezon City, Philippines). Lastly, Acetone (AR grade, RTC), benzene (ACS grade, Merck) and isopropanol (ACS grade, Scharlau) were used for the characterization of DVR.

2.2. Preparation and characterization of Ox-SO

SO underwent induced lipid peroxidation through direct conduction heating using a hotplate stirrer (IKA C-MAG HS 10 digital) at 80°C. Samples were taken every 2 h and stored in a freezer at -20°C to halt peroxidation. The open-air oxidation of PUFAs below 100°C yields conjugated hydroperoxides which produces free radicals [7]. Storage of Ox-SO at low temperatures and without the presence of metals stabilize the primary oxidation products [4].

Iodometric titration was performed for each sample to measure PV. A small amount of oil is mixed with 50 ml of a 3:2 glacial acetic acid:chloroform solvent. 0.5 ml of saturated KI solution was then added and left for 2 min with gentle swirling every 30 s. 30 ml H_2O was added immediately followed by titration with 0.01 N sodium thiosulfate solution until equivalence point indicated by addition of starch indicator. SO and Ox-SO samples also underwent Fourier transform infrared spectroscopy - attenuated total reflection (FTIR-ATR) analysis (Thermo Scientific Nicolet TM 7000). The samples were scanned at a resolution of 4 cm^{-1} in the mid-infrared range of 1000-1700 cm^{-1} and the band intensities were expressed in absorbance.

2.3. Mechanochemical devulcanization of VR

Compounded rubber with the formulation – NR SPR10(100 phr) , zinc oxide(5 phr), stearic acid(2 phr), sulfur(1.4 phr) and CBS(1.2 phr) was processed in the laboratory using a twin roll mill (Bharaj BM-06) and then vulcanized at 160°C for 15 min using a hydraulic lab press model (Chemplast). The resulting rubber sheets were then ground using an industrial cracker mill (Rhodeco Rubber Processing Services, Inc.) at ambient conditions. Ground vulcanized rubber (GVR) (<2 mm) were then stored in air-tight plastic bags and in cool dark cabinets before devulcanization.

DVR was produced by loading GVR into an open twin roll mill with 0.5-mm gap at ambient temperature (see figure 1). After 2 min of passing the GVR between the rolls, Ox-SO was added, and the process was continued to reach desired devulcanization. A 2^2 full factorial design of experiment (DOE) was employed with the factors and levels – sulfur ratio (high,medium,low), devulcanization time

(20, 40 min) and amount of Ox-SO (6,10 phr). The operational matrix is shown in table 1. Main effects and interaction effects factorial plots were generated using Minitab® 17.1.0.

Table 1. Operational DOE matrix used in the study.

Devulcanization Conditions	Control	DVR6-20	DVR10-40	DVR6-40	DVR10-20
Amount of SO (phr)	6	-	-	-	-
Amount of Ox-SO (phr)	-	6	10	6	10
Devulcanization time (min)	20	20	40	40	20

Figure 1. Mechanochemical devulcanization of VR.

2.4. Sol-gel analysis and crosslink density determination of DVR

To determine the crosslink density, DVR samples were swollen in 50 ml of benzene for 72 h at ambient temperature. The swollen samples were dried in a forced convection oven (Binder FED 56) for 24 h at 70°C. The dried samples were then weighed. Density of the dried rubber ρ_{dr} was measured using a 10 ml pycnometer (Blaubrand®) and isopropanol at 25°C. The mass of the pycnometer with sample and alcohol A and the mass of pycnometer with alcohol only B were also measured. Equation (1) was used to calculate ρ_{dr}.

$$\rho_{dr} = \frac{0.9971 \times m_{dr}}{m_{dr} - (A - B)} \, x \, \rho_{isop} \quad (1)$$

Where m_{dr} is the weight of the dried rubber and ρ_{isop} is the density of isopropanol at 25°C. The Flory-Rehner equations (2) and (3) is used for calculation of crosslink density v_e.

$$V_r = \frac{\frac{m_{dr}}{\rho_{dr}}}{\frac{m_{dr}}{\rho_{dr}} + \frac{(m_s - m_{dr})}{\rho_{benzene}}} \quad (2)$$

$$v_e = \frac{-[\ln(1 - V_r) + V_r + \chi_1 V_r^2}{[V_1(V_r^{1/3} - V_r)/2]} \quad (3)$$

Where m_s is the weight of the rubber at the end of swelling and before drying, χ_1 is the polymer-solvent interaction parameter at 25 °C and V_1 is the molecular volume of benzene [8].For sol-gel analysis, the samples first underwent Soxhlet extraction with hot acetone for 8 h to eliminate acetone-solubles. After drying, weighed samples were each subjected to Soxhlet extraction with benzene at 80 °C for 6 h. The DVR samples were then dried in a forced convection oven before being weighed. Sol content was calculated using equation (4) [9].

$$\text{Sol content} = \frac{D - E}{D} \times 100 \quad (4)$$

Where D is the mass of the sample after acetone extraction and drying and E is the mass of the sample after benzene extraction and drying.

3. Results and discussion

3.1 Synthesis of Ox-SO through lipid peroxidation

Figure 2 shows the increase in PV value in the SO as it is being oxidized through direct conductive heating with stirring. The oil sample starts to become rancid after the first 24 h where it reaches a value of 50 meq O_2 kg^{-1} oil. After 60 h of heating, the PV value reached 316.3 meq O_2 kg^{-1} oil which corresponds to about 2500 ppm of hydroperoxides. Heating was stopped after such time because it was observedthat further heating results to the polymerization of the oil. The polymerized oil is solidified and is unsuitable for use in rubber processing.

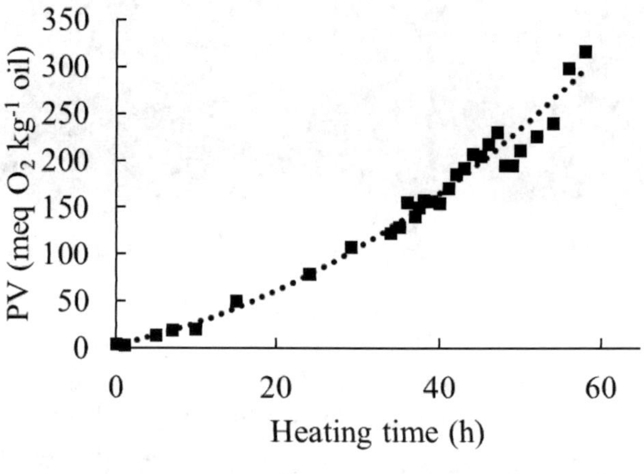

Figure 2. PV of Ox-SO as a function of heating time.

Figure 3 shows the FTIR-ATR spectra of Ox-SO and SO. The major changes that are observed are the increase in band intensity of peak "a" and the disappearance of peak "b", which are results of the oxidative deterioration that SO was subjected to during heating. The increase of peak "a" at 1160 cm^{-1} (C-O ester groups) is caused by the hydrolysis of triglycerides which produces fatty acids and glycerols. Moreover, the disappearance of peak "b" at 1637 cm^{-1}(C=C stretching vibration) is caused by the loss of double bonds upon oxidative degradation [10].

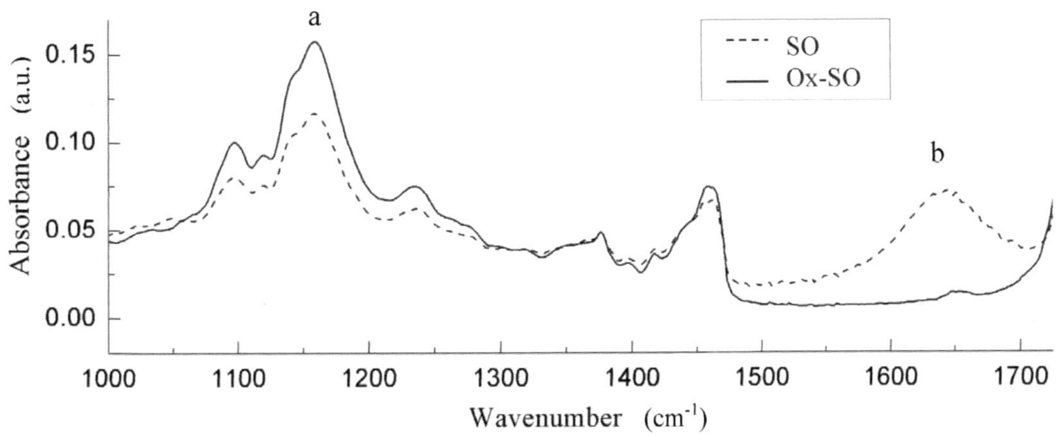

Figure 3. FTIR-ATR spectra of Ox-SO and SO.

3.2 Extent of devulcanization by Ox-SO

Figure 4 shows the effect of lipid peroxidation process of SO on sol content and crosslink density of DVR. Devulcanization is achieved when the crosslink density is decreased, and the sol content is increased due to the cleaving of sulfur-sulfur and carbon-sulfur bonds in the rubber matrix. Upon devulcanization of VR with SO, the sol content increased by 35.7% and the crosslink density decreased by 18.1% only. This can be attributed to the mechanical and thermal stresses present during devulcanization wherein SO acted as a plasticizer to make the process more efficient [11]. On the other hand, devulcanization of VR with Ox-SO increased the sol content by 94.2% and the decreased the crosslink density by 71.1%. This shows the clear potential of Ox-SO in the devulcanization of VR.

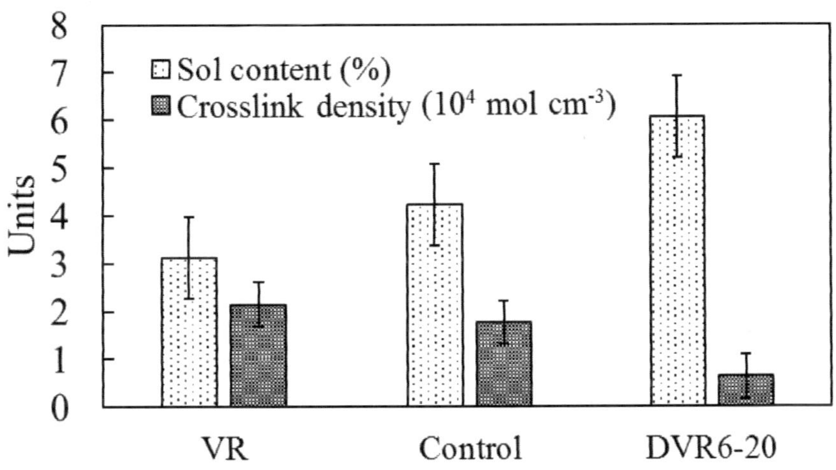

Figure 4. Sol content and crosslink density data of VR, Control DVR (with SO) and DVR6-20 (with Ox-SO) devulcanized for 20 min with 6 phr of agent added.

3.3 Main factor and interaction effects on extent of devulcanization

In this section, the main factor effects - figure 5 (a) and (b), and interaction effects – figure 5 (c) and (d), are determined from factorial plots generated from the 2^2 full factorial DOE that was used. In figure 5 (a) and (b), a higher extent of devulcanization (lower crosslink density and higher sol content) is consistently observed at 40 min of devulcanization time and 6 phr of added Ox-SO. A longer processing time allows for a better dispersion of the Ox-SO into the rubber matrix, making the process more efficient. This also allows for more interaction with mechanical and thermal stresses from the milling equipment, which helps with the devulcanization. The preference for a lower amount of Ox-SO is counter-intuitive, knowing that this also means there is a lower amount of lipid hydroperoxides present to devulcanize the rubber. This is probably due to the excessive plasticizing effect of the Ox-SO at the higher levels (10 phr) wherein the rubber matrix flows easily during milling, decreasing the friction felt, and subsequently lowering the efficiency of devulcanization.

This explanation is further supported by figure 5 (c) and (d) where the factor interactions are shown. The optimal conditions for a higher extent of devulcanization are 40 min of devulcanization time and 6 ml of Ox-SO added. This combination resulted in the 71.1% decrease in crosslink density and 228.2% increase in sol content (control VR crosslink density: 2.1×10^{-4} mol cm^{-3}, sol content: 3.1%).

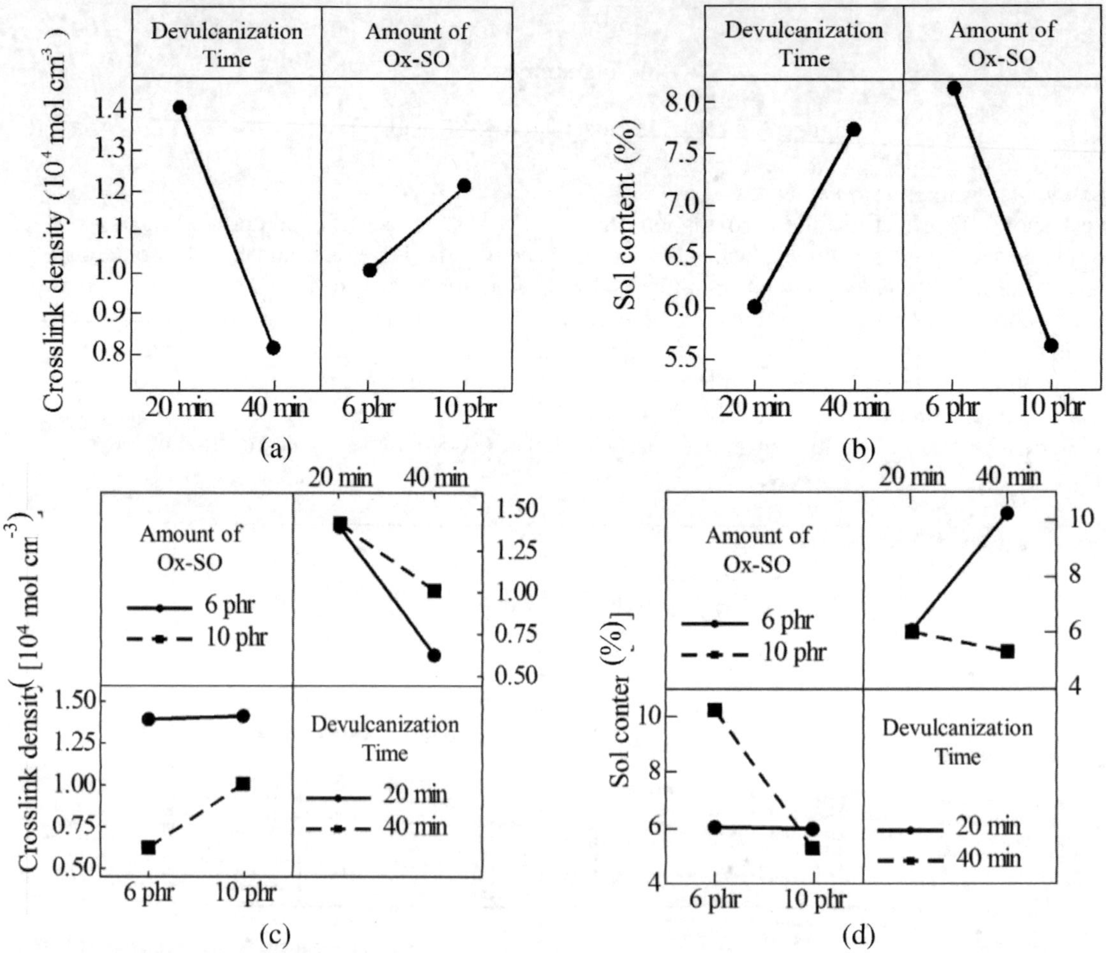

Figure 5. Main factor effects factorial plots for (a) crosslink density, (b) sol content and interaction effects factorial plots for (c) crosslink density and (d) sol content from 2^2 full factorial DOE.

4. Conclusions

Ox-SO was successfully prepared from SO using direct conductive heating via a hotplate stirrer at 80°C. The PV of the resulting Ox-SO is 316.3 meq O_2 kg^{-1} oil which is approximately 2500 ppm of lipid hydroperoxides. The Ox-SO was used as a chemical agent for mechanochemical devulcanization of VR. At 10 min devulcanization time and 6 ml, Ox-SO was found to be a feasible devulcanization agent for VR, as it lowers the crosslink density and increases the sol content of VR by 71.1% and 228.2%, respectively. Revulcanization experiments using blends of DVR with virgin rubber are currently in progress to assess applicability for commercial use

Acknowledgements
The researchers would like to thank the Department of Science and Technology Science Education Institute, the Engineering Research and Development for Technology, the Philippine Council for Industry, Energy and Emerging Technology Research and Development and the Advanced Device and Materials Testing Laboratory for funding this study and the University of the Philippines Diliman Materials Research and Development Consulting Facility for the use of its twin roll mill.

References
[1] Anu Mary J, Benny G, Madhusoodanan K N and Rosamma A 2016 The current status of sulphur vulcanization and devulcanization chemistry: devulcanization *Rubber Sci.* **29** 62–100
[2] Rooj S, Basak G C, Maji P K and Bhowmick A K 2011 New route for devulcanization of natural rubber and the properties of devulcanized rubber *J. Polym. Environ.* **19** 382–90
[3] Javidipour I, Erinç H, Baştürk A and Tekin A 2017 Oxidative changes in hazelnut, olive, soybean, and sunflower oils during microwave heating *Int. J. Food Prop.* **20** 1582-92
[4] Choe E and Min D B 2006 Mechanisms and factors for edible oil oxidation *Compr. Rev. Food Sci. Food Saf.* **5** 169–86
[5] Lukešová D, Dostálová J, El-Moneim Mahmoud E and Svárovská M 2009 Oxidation changes of vegetable oils during microwave heating *Czech J. Food Sci.* **27** 178-81
[6] Vieira, Thais M. F. S., & Regitanoo-D'Arce, Marisa A. B.. 1998 Stability of oils heated by microwave: UV spectrophotometric evaluation *Food Science and Technology.* **18** 433-7
[7] Morris B A 2017 Effect of processing on quality *The Science and Technology of Flexible Packaging Elsevier* 465–513
[8] ASTM Standard D6814, 2002 (2013) Standard test method for determination of percent devulcanization of crumb rubber based on crosslink density *ASTM International*, West Conshohocken, PA, 2006, DOI: 10.1520/D6814-02R13.2, www.astm.org
[9] Garcia P S, Sousa F D B De, Lima J A De, Cruz S A and Scuracchio C H 2015 Devulcanization of ground tire rubber : physical and chemical changes after different microwave exposure times *EXPRESS Polym. Lett.* **9** 1015–26
[10] Sim S F, Lee T Z E, Mohd Irwan Lu N A L and Samling B 2014 Synchronized analysis of FTIR spectra and GCMS chromatograms for evaluation of the thermally degraded vegetable oils *J. Anal. Methods Chem.* **2014** Article ID 271970
[11] Petrović Z S, Ionescu M, Milić J and Halladay J R 2013 Soybean oil plasticizers as replacement of petroleum oil in rubber *Rubber Chem. Technol.* **86** 233–49

Computational screening of functional monomers for bitertanol sensing using molecularly imprinted polymer

C A L Cayabyab[1,2] and B B Pajarito[1]

[1]Polymer Research Laboratory, Department of Chemical Engineering, University of the Philippines Diliman, Diliman, 1101 Quezon City, Philippines

E-mail: clcayabyab@up.edu.ph

Abstract. Pesticide residue monitoring in agricultural products is required by governments and organizations to minimize its toxic effects of pesticides in humans. Bitertanol is a fungicide used in various crops to control plant diseases but also poses harmful consequences in human health when misused. Molecularly imprinted polymers (MIPs) are artificial materials that can be used to selectively isolate and recover this substance from contaminated crops. In this work, we determined the best monomer to imprint bitertanol by evaluating the interaction between different functional monomers and bitertanol in the pre-polymer complex. Density functional theory used to optimize the structure of the complex and evaluate the interaction. The computational results showed that hydroquinone produces the most stable complex at 1:3 template-monomer ratio. The hydroxyl functionalities of hydroquinone can effectively form hydrogen bonds with the triazole ring and the hydroxyl group of bitertanol. Furthermore, parameters such as Mulliken atomic charges, bond lengths, and frontier molecular orbitals were also evaluated to confirm the formation of stable hydrogen bonds. This study can be used as a theoretical foundation for the preparation of MIPs for bitertanol using.

1. Introduction

Molecularly imprinted polymers (MIPs) are synthetic materials that contain cavities that are specifically designed to bind a target analyte. Similar to biological enzymes, these materials are capable of selective binding due to the highly specific structure and functionality of the active site cavities. The superior selectivity of MIP is owed to the interaction of the functional monomer and analyte formed in the pre-polymer complex. Non-covalent interactions such as hydrogen bonding, electrostatic attraction, and van der Waals forces are desired over covalent interactions for ease of analyte removal in the MIPs. Non-covalent interactions should be maximized in the pre-polymer complex in order to produce MIPs with high affinity over the target analyte [1]. Nowadays, computational chemistry has been exploited in the design of MIPs by performing molecular simulations to evaluate the nature and energetics of the interactions present between the functional monomer and analyte in the pre-polymer complex. Density functional theory (DFT), an *ab-initio* quantum chemical calculation, is applied to screen functional monomers to determine the best monomer for the design of highly selective MIPs. This type of calculation predicts the formation of favorable interactions between monomer and analyte, which is dictated by the energy associated with them. Several studies have used DFT to design MIP systems for a wide range of analytes such as disulfoton, amoxicillin, and tinidazole [2-4].

Bitertanol (BIT) is a triazole fungicide (structure shown in figure 1) that inhibits the production of ergosterol, a hormone responsible for fungal growth, that prevents fungi from proliferating. This pesticide is used in banana, apple, tea, and wheat. Studies have shown that large amounts of BIT residue in crops can result in different human health diseases such as reproductive disorder, cancer and endocrine dysfunction [5]. Hence, different government organizations established a maximum residue limit (MRL) on BIT. The European Food Safety Authority only allows 0.01 mg of BIT per kg of sample for plant and animal commodities while the European Union established an MRL of 0.05 mg BIT per kg of sample [6]. The traditional way of monitoring BIT residue in crops is performed via a chromatographic technique using high-performance liquid chromatography (HPLC) and gas chromatography (GC). This technique can provide accurate measurements and is considered the golden standard for analysis but the tedious and expensive operation of this process is inconvenient when monitoring a large volume of samples. On the other hand, MIPs have been prevalently used as a sensing material for pesticides like disulfoton, endosulfan, and methyl parathion [2,7,8]. In this study, we screened different monomers using DFT to design an MIP for BIT sensing. The monomers were allowed to interact with the template BIT in vacuum until the optimized geometry, the structure with the lowest energy, was achieved.

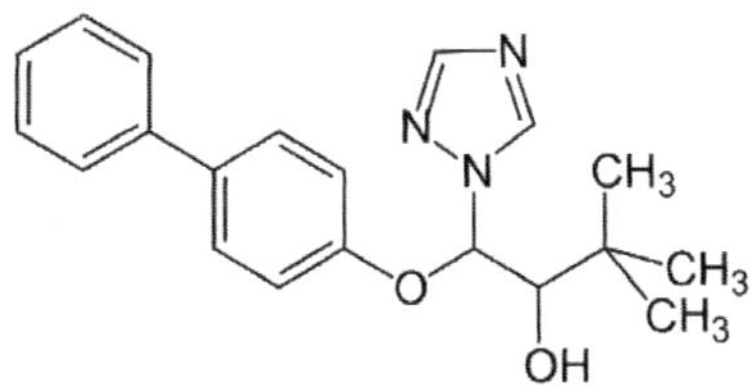

Figure 1. Chemical structure of bitertanol.

2. Methodology

2.1. Instrumentation
All computer simulations were performed on a workstation equipped with Intel Core i5 CPU with a clock speed of 3.20 GHz and 8 GB RAM.

2.2. Molecular modelling
Equilibrium geometry optimization and property calculation of each monomer (structures shown in figure 2): aniline (ANI), diaminonapthalene (DAN), hydroquinone (HYQ), indole (IND), pyrrole (PYR), and vinyl pyrrole (VP); the template BIT, and template-monomer complexes at different template-monomer ratios (1:1, 1:2, and 1:3 for the top three complexes of 1:2) were performed using *Spartan Student Edition* Version 7.2.7 (Wavefunction Inc.) at the quantum mechanical level of theory. At first, 3D molecular models of the compounds involved in this study were built in the graphical user interface of the software. An initial global minimum conformational search was performed by implementing a Monte Carlo Search using Merck Molecular Force Field (MMFF). Pre-optimization of equilibrium geometry of all the conformations found using MMFF was done using the PM3 Semi-Empirical (SE) calculation using the Pople type basis set 6-31G* in the vacuum-phase. The three lowest energy conformations were further optimized using DFT with the same basis set used in the pre-optimization stage. The conformations with the lowest energy for each monomer (E_M), template (E_T), and each monomer-template complex (E_C), defined as BIT-MONOMER$_{ratio}$ in the text, at different template-monomer ratios (n) were used to calculate the energy of interaction (ΔE) of the pre-polymer complex using equation (1).

$$\Delta E = E_C - [E_T + nE_M] \quad (1)$$

Hydrogen bond distances and angles were also optimized using DFT. Other properties such as HOMO-LUMO energies and Mulliken atomic charges were also determined using DFT.

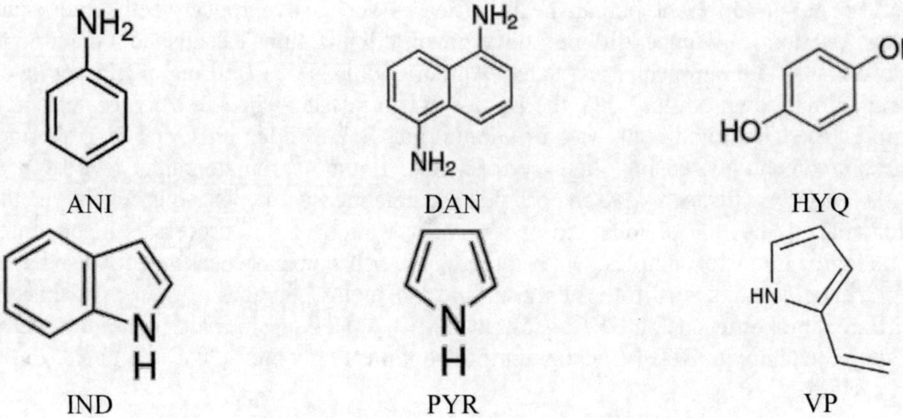

Figure 2. Chemical structures of: ANI, DAN, HYQ, IND, PYR, VP.

3. Results and discussion

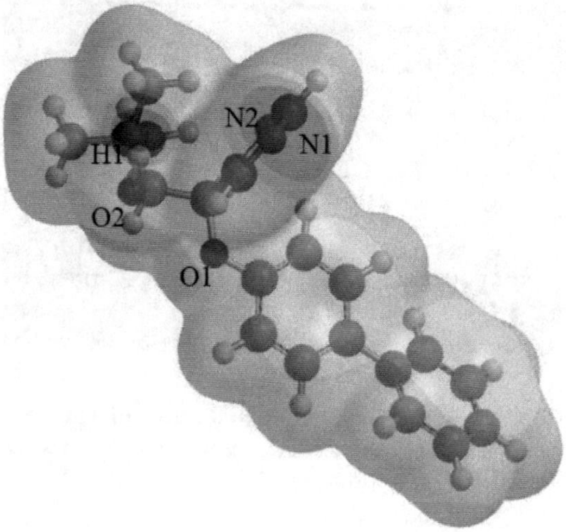

Figure 3. Electrostatic potential map of BIT calculated using DFT.

The design of highly selective MIP for BIT relies heavily on the interaction between the functional monomer and BIT. By observing the electrostatic potential map of BIT molecule (see figure 3), possible interaction sites will occur at the regions in red (proton acceptor) and blue (proton donor) which are in the space of the atoms O1, O2, N1, N2, and H1. In this study, we only considered monomers that contain functionalities such as 1° ($-NH_2$) amine, 2° ($-NH$) amine, or hydroxyl ($-OH$) group that can both act as proton donors and acceptors. These monomers are also electropolymerizable which allows the formation of thin polymeric films that is advantageous when extracting the template from the MIP. The geometry of the complexes formed by BIT and functional monomers were optimized in order to determine the nature of its interaction and ΔE. Based on the calculated values of interaction energies in vacuum reported in table 1, all complexes at different template-monomer ratios formed stable complexes (negative ΔE) which imply the ability of the functional monomers to effectively interact with BIT. At 1:1 ratio, HYQ obtained the most negative ΔE value of -47.1 kJ mol^{-1} followed by IND and DAN with -34.9 kJ mol^{-1} and -30.6 kJ mol^{-1} respectively. Complexes formed at the ratio of 1:2 were observed to form a more stable complex than at 1:1 ratio. The order of ΔE in increasing magnitude is as follow: ΔE (BIT-HYQ_2) > ΔE (BIT-IND_2) > ΔE (BIT-VP_2) > ΔE (BIT-PYR_2) > ΔE(BIT-DAN_2) > ΔE(BIT-ANI_2).

Hydrogen bonds play an important role in the formation of stable complexes. Systems containing a good number of short hydrogen bonds are found to have stronger interactions [9]. Furthermore, increasing the number of monomers per molecule of template in the pre-polymer complex increases the chance of more hydrogen bonds being formed. For this reason, we explored the possibility of improving the ΔE value by increasing the ratio of the three monomers that formed the most stable complexes at the ratio of 1:2. The top three monomers were only considered for further analysis to save computation time. Upon observation, there was indeed a significant improvement in the ΔE value of the monomer-template complex as the ratio is increased to 1:3. HYQ retained to form the most stable complex with ΔE of -145.6 kJ mol^{-1} which indicates that BIT will strongly interact with this functional monomer. Figure 4 shows the optimized structures of BIT-HYQ complexes at the three different ratios. The geometries show that the hydroxyl groups of HYQ function as a proton donor to form hydrogen bonds with the proton acceptors of BIT (N2, N3, and O2). The location of these hydrogen bonding interaction sites will serve as the binding sites of the MIP.

Table 1. Interaction energy and number of hydrogen bonds formed of template-monomer complexes at gas phase.

Complexes	ΔE (kJ mol^{-1})	H-bonds
BIT-ANI	-21.8	1
BIT-DAN	-30.6	1
BIT-HYQ	-47.1	1
BIT-IND	-34.9	1
BIT-PYR	-30.4	1
BIT-VP	-21.5	1
BIT-ANI$_2$	-46.6	1
BIT-DAN$_2$	-52.1	2
BIT-HYQ$_2$	-86.9	2
BIT-IND$_2$	-60.9	1
BIT-PYR$_2$	-54.8	2
BIT-VP$_2$	-58.6	2
BIT-HYQ$_3$	-145.6	3
BIT-IND$_3$	-90.9	2
BIT-VP$_3$	-94.6	2

Hydrogen bonding parameters such as X-H bond length, hydrogen bond length, and angle are presented in table 2. The formation of hydrogen bond was observed to elongate and weaken the O-H bond (from 0.969 Å) of HYQ due to the outflow of electron density from the H atom to O atom. This resulted in the strengthening of the C-O bond, evidenced by bond shortening, connected to O-center to protect its octet structure [10]. Furthermore, these also led to changes in Mulliken atomic charges as a consequence of the redistribution of electron density. At 1:3 ratio, the Mulliken atomic charges before complexation are: -0.650 au on O in HYQ, 0.405 au on H available for hydrogen bonding in HYQ, -0.274 au on N42 in BIT, -0.434 au on N47 in BIT, and -0.629 au on O26 in BIT. After complexation, these atoms changed to -0.733 au on O59, 0.468 au on H60, -0.695 au in O26 for BIT. The following changes in the atomic charge suggest that there are hydrogen bonding interactions present between HYQ monomer and BIT molecule [11,12]. This charge transfer interplay is further supported by the location of the frontier molecular orbitals, the highest occupied molecular orbital (HOMO) and the lowest unoccupied molecular orbital (LUMO). These orbitals point the direction of charge transfer (from HOMO to LUMO) in molecular systems. We observed from the simulation of BIT-$(HYQ)_3$ complex that the HOMO is located at HYQ monomers while the LUMO resides at BIT molecule which indicates that the charge transfer occurs from the functional monomers to the template molecule [13].

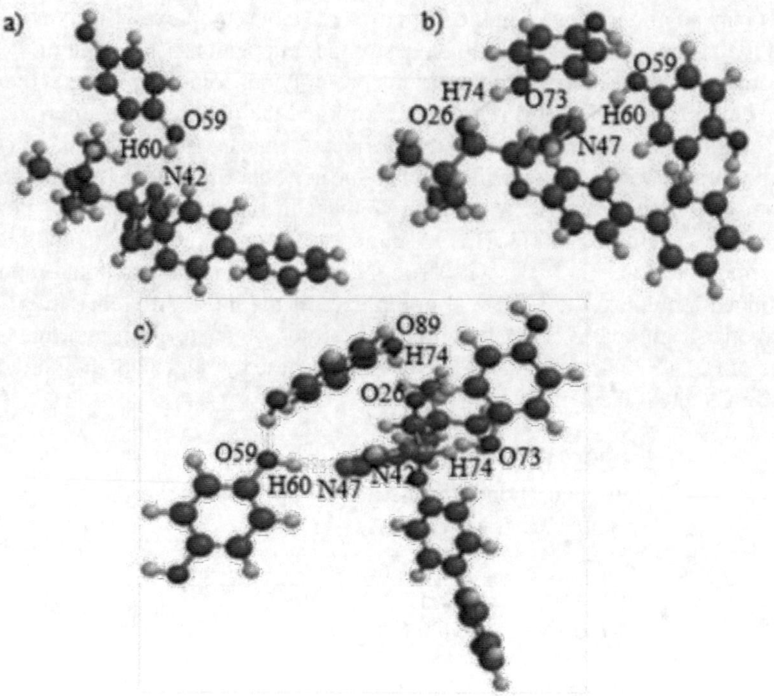

Figure 4. Optimized structures of the most stable complexes at different ratios between BIT and functional monomer in the gas phase: (a) BIT-HYQ, (b) BIT-HYQ_2, and (c) BIT-HYQ_3.

Table 2. Hydrogen bonding parameters of BIT-HYQ complexes at different template-monomer ratios.

Complex	Hydrogen bonding position	X-H bond length (Å)	H - bond length (Å)	Angle (°)
BIT-HYQ	O59-H60-N42	0.987	1.87	161.8
BIT-HYQ_2	O73-H74-O26	0.979	1.95	153.9
	O59-H60-N47	0.990	1.90	162.9
BIT-HYQ_3	073-H74-N42	0.983	1.93	158.1
	059-H60-N47	0.989	1.88	169.0
	O89-H90-O26	0.976	1.97	145.3

4. Conclusion

The result of computational screening shows that HYQ is the best monomer to imprint BIT among the monomers considered in this study at all different template-monomer ratios. This monomer has the best interaction with the template molecule at 1:3 ratio forming multiple hydrogen bonds (at O2, N2, and N3) based on the interaction energy and optimized structure of the corresponding complex calculated using DFT. These hydrogen bonding sites will serve as the adsorption site during template rebinding. The computational results from this study can be useful in the actual preparation of highly selective MIP for BIT.

Acknowledgments

This work is fully supported by Engineering Research and Development for Technology (ERDT) program of the Department of Science and Technology (DOST) of the Republic of the Philippines. We also extend our gratitude to the Advance Device and Materials Testing Laboratory (ADMATEL) under the DOST.

References

[1] Yan M and Ramström O 2005 *Molecularly Imprinted Materials: Science and Technology* (New York: Marcel Dekker)
[2] Qader B, Baron M, Hussain I, Sevilla J M, Johnson R P and Gonzales-Rodriguez J 2019 *Electrochim.* **295** 333-9
[3] Ayankojo A G, Reut J, Boroznjak R, Öpik A and Syritski V 2018 *Sensors Actuators B. Chem* **258** 766-74
[4] Wang H, Qian D, Xiao X, Deng C, Liao L, Deng J and Lin Y 2018 *Biochemistry* **121** 115-24
[5] Ghisari M, Long M, Tabbo A and Bonefeld-Jorgensen E C 2015 *Toxicol. Appl. Pharmacol.* **284** 292-303
[6] Blasco C, Font G and Pico Y 2006 *Food Control* **17** 841-6
[7] Bow Y, Sutriyono E, Nasir S and Iskandar I 2017 *International Journal on Advance Science, Engineering and Information Technology* **7** 662
[8] Lin X H, Ling Aik S X, Angkasa J, Le Q, Chooi K S and Li S F Y 2018 *Sensors Actuators, B Chem.* **258** 228-37
[9] Salajegheh M, Ansari M and Mehdi M 2019 *J. Pharm. Biomed. Anal* **162** 215-24
[10] Grabowski S 2018 *J. Mol.* **24**
[11] Huang Y and Zhu Q 2015 *J. Chem* **2015** 1-9
[12] Pardeshi S, Patrikar R, Dhodapkar R and Kumar A 2012 *J. Mol. Model* **18** 4797-810
[13] Karmakar A and Singh B 2019 *J. Mol. Struct.* **1180** 698-707

IOP Conf. Series: Materials Science and Engineering **634** (2019) 012038 doi:10.1088/1757-899X/634/1/012038

Optimization of reaction parameters of esterification on the synthesis of palm oil-based alkyds using response surface methods

M D H Beg[1,3], M R Islam[2,3], B Ahmed[1], R M Yunus[1] and A Abdullah[1]

[1]Faculty of Chemical and Natural Resources Engineering, Universiti Malaysia Pahang, Gambang, Pahang, Malaysia
[2]Malaysian Institute of Chemical and Bioengineering Technology, University of Kuala Lumpur, Alor Gajah, Melaka, Malaysia

E-mail: dhbeg@yahoo.com (M D H Beg)/muhammad.remanul@unikl.edu.my (M R Islam)

Abstract. In this research, polyalkyd resins were prepared from palm oil by polycondensation reaction. A two-step method viz., alcoholysis and esterification were conducted to synthesize polyalkyd resins. Response surface methodology was used to optimize the reaction parameters for esterification like reaction temperature, reaction time, catalyst concentration, acid anhydride to mono-glyceride ratio and agitation speed. The optimum condition was used to prepare various types of polyalkyds using different types of acid anhydrides such as, maleic anhydride, succinic anhydride, phthalic anhydride, 3,4,5,6-tetrahydrophthalic anhydride and cis-1,2-cyclohexanedicarboxylic anhydride. Result analyses through RSM revealed a desirability of 0.985 for the reaction time of 88.64 min. Moreover, 91.5% fractional conversion was achieved actually, which is close to the predicted value. The analysis revealed that 3,4,5,6-tetrahydophthalic anhydride-based resin showed improved resinous property than others due to the high degree of crosslinking.

1. Introduction

There are a huge number of vegetable oils available in the world. Nearly, 4000 botanical species have been identified in the planet, from which oil can be extracted [1]. These vegetable oils can be utilized to produce polymers and polymeric materials [2]. Many useful polymers like, polyesters, polyurethanes, polyamides, etc. have been prepared from different vegetable oils and studied in previous investigations [3]. The polymerization techniques and methods with different features of the prepared polymers are also highlighted. Moreover, palm oil along with other vegetable oils has been studied for the preparation of different kinds of polymers, especially polyalkyds.

The common process to prepare polyesters is the two-steps method includes alcoholysis followed by esterification [4,5]. In alcoholysis process, the oils or triglycerides are subjected to alcoholysis to produce mono-glycerides (as partial glycerides), catalysed by various acid or base catalysts such as, PbO, NaOH, CaO, $Ca(OH)_2$, $CaCO_3$, para-toluene sulphonic acid (PTSA), etc. After this, the mono-glycerides are subjected to polymerization process with different acid anhydrides such as, aliphatic acid anhydride (glutaric anhydride, maleic anhydride, succinic anhydride, etc.) and aromatic acid anhydrides (phthalic anhydride) to produce polyalkyd based resin [6,7].

The aim of this study was to optimize the reaction parameters for esterification process using palm oil. In the previously published article, the esterification was performed using different kind of vegetable oils, but optimization of the reaction parameters, especially for pal oil-based polyalkyd preparation yet not documented. Moreover, a solution based on the analysis was suggested which may help to predicts the reaction parameters for other kinds of vegetable oil to be used for the same purpose.

2. Experimental

2.1. Materials

Palm oil was provided by Malaysian Palm Oil Board (MPOB). Glycerol and calcium oxide were procured from Aladdin Chemistry Co. Ltd., Shanghai, China. Methyl ethyl ketone peroxide and 3,4,5,6-tetrahydrophthalic anhydride (Molecular weight: 152.16 g/mol; chemical formula: $C_8H_8O_3$) was purchased from Aladdin chemical corporation, Shanghai China. Sulphuric acid was purchased from Fisher scientific, USA. Maleic anhydride (MA) was purchased from ACROS, USA. Cobalt napthenate and styrene were purchased from Sigma Aldrich, USA

2.2. Optimisation of esterification process using RSM

The solutions for the desired response and the corresponding set of optimum experimental conditions for the esterification process were suggested by response surface methodology (RSM). The minimum numbers of experimental runs were predicted by central composite design. The total numbers of experiments were based on five (5) different factors at various distinct settings, which suggest 50 runs including 32-full factorial distinct points, ten (10) runs at unique axial points and eight (8) constant runs at center points. The primary analysis with crude oil helped to find out the range of the parameters. The experimental factors and pre-assumed range of values for polyesterification reaction are presented in table 1. The factors are selected from the previous study based on polyalkyd preparation [8]. Table 2 represents the fifty (50) experimental runs as suggested by central composite design.

Table 1. Experimental factors and pre-assumed range of values for esterification.

Factors	Units	Low	High	-α	α
(A) Temperature	Celsius	220	260	192.432	287.568
(B) Time	Min	30	180	-73.3811	283.381
(C) Agitation	rpm	300	1000	-182.445	1482.44
(D) Molar ratio	Unit less	0.1	0.5	-0.175683	0.775683
(E) Catalyst concentration	wt.%	0.02	0.08	-0.0213524	0.121352

Table 2. Analysis of variance for response surface quadratic model.

Source	Sum of squares	df	Mean Square	F value	p-value Prob>F
Model	15594.52	20	779.73	31.82	<0.0001
A-Temperature	2691.36	1	2691.36	109.83	<0.0001
B-Time	312.03	1	312.03	12.73	0.0013
C-Agitation	196.32	1	196.32	8.01	0.0084
D-Molar ratio	246.78	1	246.78	10.07	0.0036
E-Catalyst concentration	34.20	1	34.20	1.40	0.2470
Residual	710.62	29	24.50		
Lack of Fit	710.43	22	32.29	1159.21	<0.0001
Pure Error	0.19	7	0.028		
Cor Total	16305.14	49			
	Std. Dev.		4.93	R-Squared	0.9389
	Mean		51.24	Adj R-Squared	0.9270
	C.V. %		9.62	Pred R-Square	0.8982
	PRESS		1660.42	Adeq Precisior	27.790

The optimized data, obtained from RSM analysis, was used further in the preparation of different types of polyalkyds by using SA, MA, PA, TPA and CDA through esterification process. The name tags for the formulated resins using SA, MA, PA, TPA and CDA are SAR, MAR, PAR, TPAR and CDAR, respectively. Further, esterification was also carried out with TPA and MA at different ratios to find the effects of aromatic and aliphatic acid anhydride content on the resinous properties. The name tags for the samples containing 0, 50, 75 and 100% MA are MA0, MA50, MA75 and MA100, respectively.

3. Results and discussion

3.1. Factor analysis

Preparation of polyalkyds involves alcoholysis and esterification process. The alcoholysis process is usually completed within one or two hours after the reaction temperature is reached [9]. The screening and selection of the range of reaction parameters were performed by using crude oil for esterification process. The esterification process was optimized with the help of design expert software (DES). The completion of esterification process was determined by the acid value of the products. The reduction in acid value with respect to time is useful to calculate the fractional conversion of the alkyd resins. The variations of response were found to be significant due to the changes of reaction conditions, which is satisfying the usefulness of RSM application as well as indicating the importance of different factors. The minimum number of experimental runs was predicted by using central composite design. A full factorial central composite design matrix was used for five factors in against for a single response, acid value. The response for the individual run was the input in the system. Eventually, a solution was suggested by RSM, containing a set of reaction parameters. These parameters were economic, and the reaction based on that condition was the optimum with the highest desirability.

Analysis of the effects of different variables showed the significance of their contributions to the reaction. The analysis of variance (ANOVA) was generated after the computational tasks. A quadratic model of reduced order was formulated using the Equation (1) to find the values for different coefficients for the characteristics of the reaction. The process variables such as, temperature, time, catalyst concentration, molar ratio and agitation speed in the predictive model are designated as A, B, C, D and E, respectively. The results obtained from the predictive model indicate that the aforementioned factors are responsible for the main linear effects. The "Model F-value" of 31.82 implies the model is significant and there is only a 0.01% chance that a "Model F-Value" could occur due to noise. Additionally, the values of "Prob > F" less than 0.0500 indicates model terms are also significant. In this case A, B, C, D, BC, A^2, B^2, D^2 are significant model terms. Table 2 shows the statistical analysis of the variables. From these data a modified model was proposed by excluding the insignificant terms from the general predictive equation as mentioned below. From the analysis, it was found that the esterification process can be elucidated by the equation of the empirical model, with more than 93% overall system variability, as indicated by the R-Squared value of 0.9389. It was performed by the multivariate regression analyses. Thus, the general equation becomes as follows (equation (1)):

$$Y = 28.38 + 8.90A + 3.03B + 2.40C - 2.69D + 5.08BC + 20.27A^2 + 6.12\,B^2 + 7.22D^2 \tag{1}$$

Where, Y is the fractional conversion. The main linear effects of the different factors such as, temperature, time, agitation and molar ratio are denoted, respectively as A, B, C and D. The linear interaction effect for time and agitation are indicated by BC, whereas quadratic effects can be denoted as A^2, B^2, D^2, which corresponds to respective factors or variables. The "Pred R-Squared" of 0.8982 is in reasonable agreement with the "Adj R-Squared" of 0.9270. Usually, signal-to-noise ratio is desirable more than 4. The signal to noise ratio was found to be 27.790, which signifies adequate signal.

3.2. Process analysis

Different solutions for the esterification reaction were obtained by the predictive equation as stated earlier (equation (1)). RSM was used conveniently for this process, as it can give different choices to fix the response at different reaction conditions. Different interactions between different factors for the esterification process were evaluated by solving the equation. The conversation of alkyd resin at any of the points of interactions of two variables was calculated assuming the other parameters to be set at their mean value, zero. To understand the effects of different parameters on esterification process, the interactions of two individual factors and their 3-D responses were examined by the graphs generated with the help of DES.

Figure 1 represents the 3-D plots of agitation vs. temperature (figure 1(a)), agitation vs. time (figure 1(b)) and agitation vs. catalyst concentration (figure 1(c)), respectively. It is clear from the figure that, the interaction between agitation and temperature has much effect than that of others. The acid value changes sharply due to high temperature and high agitation. As the temperature and agition varied the fatty acids in the palm oil reacted with the glycerol and presence of acid also varied the product. Therefore, the acid valuse chages. However, the interaction for agitation vs. time and agitation vs. catalyst concentration was found to be almost a flat trend, indicatiing less affecting contribution in terms of variation of responses. Figure 1 represents the 3-D responses of catalyst concentration vs. molar ratio (figure 1(d)), catalyst concentration vs. temperature (figure 1(e)) and catalyst concentration vs. time (figure 1(f)). Sharp changes were observed in the curves of catalyst concentration and temperature, indicating good interaction. The higher catalyst concentration and the higher temperature was found to be responsible for a higher value of the response, whereas the molar ratio and time showed less affection to the reaction process, as observed from the nearly flat trend during the reaction. This is because of hgiher percentage of catalyst at higher temperature can stimulate more reactions to get affected on the response. The significant and quick changes observed in the case of molar ratio and temperature. This response was sharp and steady due to changes in temperature and molar ratio. This is happened so because, the reaction is mode dependent on the molar ratio of fatty acid and glycerol percentage or ratio. Therefore, if the molar ratio of the reactant changes, the development of the product also changes. The other responses were insignificant as predicted due to the quadratic effect of molar ratio. Finally, the time and temperature, which is also important and an indication for the sharp changes of the acid value. For a batch reaction system, the extent of reaction or conversion is higher with time at low reaction temperature [10] and seems to be a first order reaction kinetics. As the temperature increases the response was found to be increasing up to temperature around 235°C and further increasing of temperature affects the response slowly and behaves like a second order reaction kinetics [11]. The reversible nature of esterification reaction is probably the reason for slow responses at a higher temperature.

Among the process variables, the molar ratio of PA to monoglyceride and catalyst concentration is very important. These parameters need to be controlled to achieve a higher yield and conversation of oil to alkyd with high purity. It was found that the maximum conversion can be achieved by increasing the catalyst concentration up to a certain level for a molar ratio. It was also found that, at the lower molar ratio with higher catalyst concentration the conversion remains still high. However, the industrial importance in the manufacturing process is always depreciated to use of the high amount of catalyst. Thus, optimization of this process found to be a significant analysis and can be used for the manufacturing process at a larger scale.

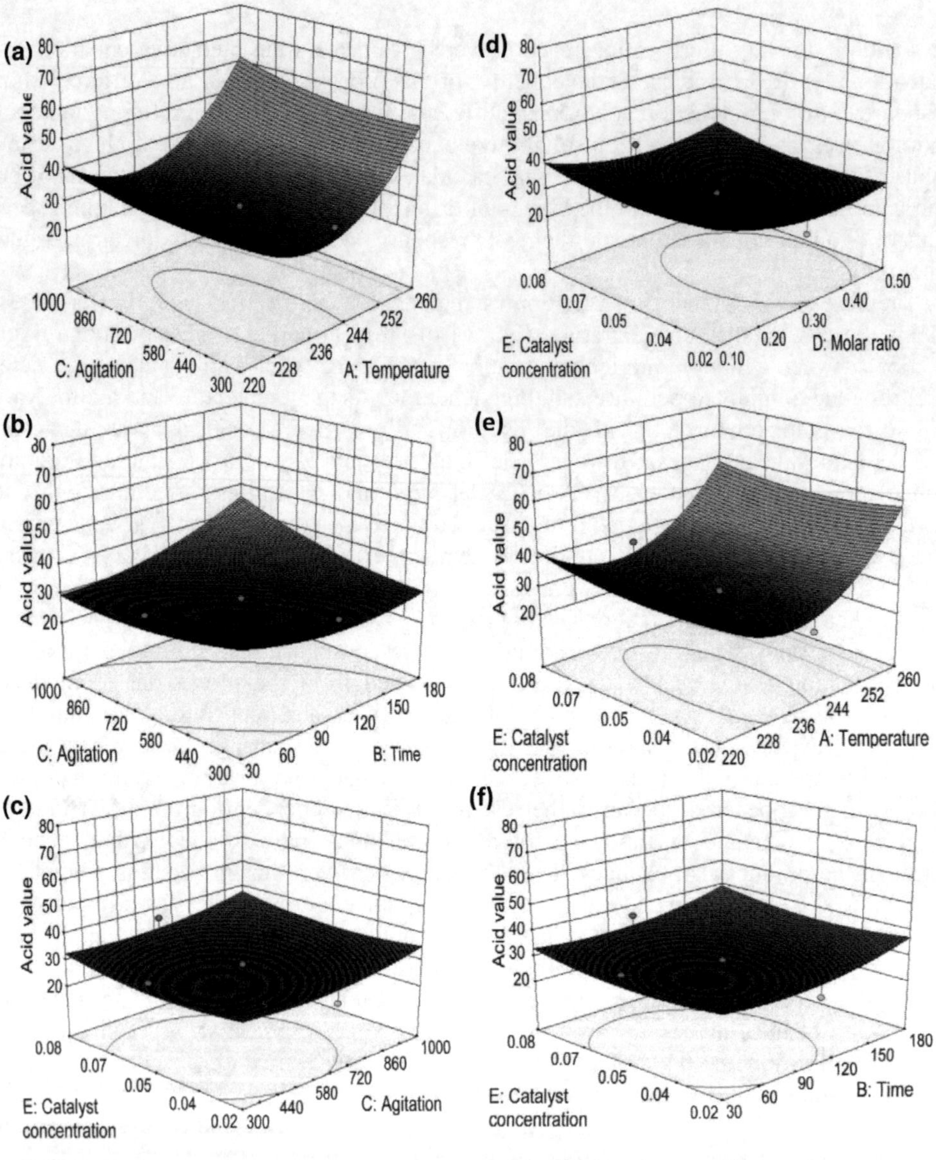

Figure 1. 3-D curves of agitation vs. temperature (a), agitation vs. time (b) and agitation vs catalyst concentration (c), catalyst concentration vs. molar ratio (d), catalyst concentration vs. temperature (e) and catalyst concentration vs time (f).

3.3. Process optimization

The optimization of the process is required not only to ensure or detect the optimum reaction parameters for a higher yield or desired response, but also to confirm high quality products with faster drying time, acceptable colour, low manufacturing cost, etc. The drying time is important for product quality. Additionally, it is always considered to reduce the product cost by reducing the reaction time and temperature. The optimization of the process was carried out by the optimization tool of the design expert software, while the criteria function helped to input the desired value to achieve a targeted response.

The best solutions were obtained by solving equation 1 and it was targeted to achieve the desired responses. There were a number of solutions available as suggested by the design expert software. The best one was considered based on the economic value of the process by evaluating different factors, individually and their effects on the response. Nearly, 91.5% of fractional conversion was achieved which is indicating a good fit of the model with the experimental statistical analysis. A comparison with the previous work showed an improvement in the reaction time and almost like the conversion rate, although a very few works related to palm oil-based alkyds and esterification were performed [8].

Several solutions were proposed by the software, although the best one was considered for further experiments. However, the solutions of optimization process for different sets of reaction parameters are presented in table 3, whereas the best solution was chosen as follows: reaction temperature = 240°C, reaction time = 88.64 min, agitation speed = 584.20, molar ratio = 0.35:1, catalyst concentration = 0.04 at response 27.12 of acid value with the highest desirability of 0.985. The optimum parameters are listed in table 4. The values for the other solutions were found to be very close to each other. Finally, to confirm the optimization process, the optimum experimental run was carried out, which was found similar to the values as suggested earlier by the software. In the previous investigation, the optimization of the esterification process was carried out with four factors [8]. The optimization process was considered by the dehydration process by $NaHSO_4$ in the presence of NaOH and PbO as the catalysts for alcoholysis. In a different work, castor oil-based esterification process was analyzed and nearly 97% of the reaction was completed in 2 h and 30 min in the range of temperature of 230 to 240°C [11]. The major differences from these works to the current study was the dehydration process and uses of different catalyst (CaO), which reduced the reaction time almost 50% with higher desirability and lower acid value. Moreover, agitation was found to be influential regarding the properties of the polyalkyds. Several research works were carried out for the transesterification of different vegetable oils for the production of biodiesel, and RSM was used [12,13] for the optimization process. The reported results were found different from this current work due to the high activation energy required for the establishment of the equilibrium reaction between PA and glyceride oil for polyalkyd preparation [8].

Table 3. The solutions suggested by RSM for the esterification process.

Number	Temperature	Time	Agitation	Molar ratio	Catalyst concentration	Acid value	Desirability
1	240.00	88.64	584.20	0.35	0.04	27.12	0.985
2	240.00	88.55	586.81	0.35	0.04	27.12	0.985
4	240.00	88.10	584.83	0.35	0.04	27.12	0.985
4	240.00	88.40	587.40	0.35	0.04	27.12	0.985

Table 4. The optimum solution suggested by RSM.

Parameters	**Optimum values**
Fractional conversion (%)	91.5
Reaction temperature (°C)	240°C
Reaction time (min)	88.64
Agitation speed (rpm)	584.20
Molar ratio of PA: MG	0.35
Catalyst concentration (wt.%)	0.04
Desirability (%)	0.985
Response (Acid value)	27.12

4. Conclusion

Palm oil was used to prepare polyalkyds through alcoholysis and esterification process. The crude oil was subjected for the dehydration process to increase the unsaturation. It was found that the usage of H_2SO_4 improves the unsaturation of oil as maximum as 165%. After performing a preliminary screening of the reaction parameters, the esterification process was optimized through RSM. Result analyses revealed very high desirability of 0.985 through RSM analyses. Moreover, 91.5% fractional conversion was achieved, which is close to the predicted value. The optimum reaction conditions were used to formulate different types of polyalkyds using different acid anhydrides, viz., SA, MA, PA,

IOP Conf. Series: Materials Science and Engineering **634** (2019) 012038 doi:10.1088/1757-899X/634/1/012038

TPA and CDA. The properties of the resins were found to be dependent on the types of acid anhydrides. Among the resins, the TPA-based resin was found to be the best in terms of different resinous properties.

Acknowledgment
The authors would like to thank the Ministry of Higher Education Malaysia for supporting this project through FRGS FRGS/1/2016/TK05/UMP/01/1.

References

[1] Santori G, Nicola G D, Moglie M and Polonara F 2012 A review analyzing the industrial biodiesel production practice starting from vegetable oil refining *Appl. Energy* **92** 109-32
[2] Islam M R, Beg M D H and Jamari S S 2016 Characterization of multi wall carbon nanotubes filled, palm oil based polyalkyds: Effects of loading and in situ reaction *J Appl Polym Sci* **133** 5
[3] Islam M R, Beg M D H and Jamari S S 2016 Dispersion of montmorillonite nanoclays and their effects on the thermomechanical, structural and drying properties of palm oil based coating *Progr Organic Coatings* **91** 17-24
[4] Islam M R, Beg M D H and Jamari S S 2015 The effects of five different types of acid anhydrides and incorporation of montmorillonite nanoclays on thermosetting resins *Polym Bulletin* **72** 3007-30
[5] Islam M R, Beg M D H and Jamari S S 2015 The effects of increasing unsaturation in palm oil and incorporation of carbon nanotubes on resinous properties *International Journal of Chemical, Nuclear, Materials and Metallurgical Engineering* **9** 475-80
[6] Islam M R, Beg M D H and Jamari S S 2014 Development of vegetable oil-based polymers *Journal of Applied Polymer Science* **131** 787-740,799
[7] Islam M R, Beg M D H and Jamari S S 2014 Alkyd based resin from non-drying oil *Procedia Engineering* **90** 78-88
[8] Uzoh C F, Onukwuli O D, Odera R S and Ofochebe S 2013 Optimization of polyesterification process for production of palm oil modified alkyd resin using response surface methodology *J. Environ. Che. Eng.* **1** 777-85
[9] Ikhuoria E, Aigbodion U and Okieimen A I F E 2004 Enhancing the quality of alkyd resins using methyl ester of rubber seed oil *Tropical J. Pharm. Res.* **3** 311-7
[10] Gunstone F D, Alander J, Erhan S Z, Sharma B K, Mc Keon T A and Lin J T 2007 Nonfood uses of oils and fats *The Lipid Handbook* 3rd ed. F D Gunstone, J L Harwood and A J Dijkstra eds (New York, USA: CRC Press) pp 591-635
[11] Hlaing N N and Mya O 2008 Manufacture of alkyd resin from castor oil *World Academy of Science, Engineering and Technology* **24** 115-61
[12] Encinar J M, Gonzalez J F, Rodriguez J J and Tejedox A 2002. Biodiesel fuels from vegetable oil: transesterification of Cynara Ardunculus L. oils with ethanol *J. Energy Fuels.* **16** 443-50
[13] Freedman B, Butterfield R O and Pryde E H 1984 Variables affecting the yields of fatty esters from transesterified vegetable oils *J. Am. Oil Che. Soc.* **61** 450-503

Potential anti-corrosion additives derived from waste plastic sachets

F D T Eugenio[1,2,3] **and B B Pajarito**[2]

[1] Environmental Engineering Program, College of Engineering, University of the Philippines, Diliman, Quezon City 1101 Philippines
[2] Polymer Research Laboratory, Department of Chemical Engineering, College of Engineering, University of the Philippines, Diliman, Quezon City 1101 Philippines

E-mail: fteugenio@up.edu.ph

Abstract. Corrosion is a perennial problem that has a significant burden on the economy and environment. Recent studies have suggested that organic fillers can serve as corrosion inhibitors. In this work, the use of comminuted waste plastic sachets (WPS) as an anti-corrosion additive to commercial coating was explored. The effect of WPS lamination and loading on the viscosity, film thickness, and corrosion behavior of a red oxide primer paint was investigated. Results show that the viscosity and film thickness of the coating were only affected by the addition of WPS at high loading. At low loading, comminuted WPS reduced the corrosion rate of the coating by 99%. The presence of metal lamination in WPS contributes to the corrosion resistance of the coating. In conclusion, comminuted WPS demonstrates its potential in enhancing the anti-corrosion properties of coatings.

1. Introduction

Metal corrosion is a serious problem that has grave economic and environmental consequences. Corrosion weakens or destroys metal to the point that in the long run its economic value is rendered useless [1]. Corrosion is caused by an electrochemical reaction between the metal and an electrolyte, or by the oxidation of the metal atoms. While corrosion may be inevitable, many processes have been devised to slow down corrosion. Coatings are one of the generally-accepted methods of providing corrosion resistance to metals as they provide a barrier between the metal and the environment, preventing electrochemical reactions that cause corrosion from occurring. Many coatings also include anti-corrosion additives usually made of inorganic metal compounds (e.g. lead oxide). However, in recent years, the use of these compounds has been discouraged as studies have shown that the use of such metals has detrimental health effects not only to humans but to the environment [2,3].

Recent studies show that another method of supplying corrosion resistance to coatings involves the use of organic fillers and composites, usually alone or combined with other organic polymers such as epoxy, acrylic and polyurethane [4,5]. These anti-corrosion additives can also act as pigments [6]. As mechanisms and applications of such anti-corrosion additives are varied, various research involving effective anti-corrosion additives continue [7]. Waste plastic sachets (WPS) is primarily composed of polymers, and as an additive may impart sufficient barrier properties to increase the corrosion resistance of coatings. The use of such materials is primarily motivated by the fact that the use of synthetic polymers, in general, is hounded by recycling issues and that developing countries usually rely on sachet-packaged products to fill their basic needs [8,9].

To confirm this assumption, this work studied the effects of adding comminuted WPS on the viscosity and corrosion resistance of commercially available primer coating.

2. Experimental

2.1. Materials
WPS was acquired from the industrial rejects of a manufacturing company (Doxo Ingredients Inc., Philippines). Grinding wheel (Qingdao Sisa Abrasives Co., Ltd, China) with a grit number of 100 was used for comminution. High purity ethanol (99.5 wt%, Scharlab S.L., Spain) was used for optical microscopy. Silicon carbide sandpaper with 220 and 600 grit numbers, and acetone (99.5 wt%, RTC Laboratory Services & Supply House, Philippines) were used for the preparation of metal substrates before coating. Red oxide primer paint (Pacific Paint (Boysen) Philippines, Inc.) was purchased from a local hardware store. A 5.7 cm platinum counter electrode (ALS Co., Ltd, Japan) was used along with a 6 mm Ag/AgCl reference electrode (ALS Co., Ltd, Japan) containing 3M NaCl solution (potential $E°$= 0.209 V vs normal hydrogen electrode/NHE at 25°C).

2.2. Preparation and characterization of waste plastic sachets
WPS were segregated according to presence of metal lamination before washing to remove trace amounts of dirt, grease, and residues. Comminution was done using a 4460 rpm bench grinder (John Benzen Power Tools, Philippines). These were collected and subsequently sieved (US Standard Sieve Mesh 325, particle size <45 µm) yielding the additive.

The additives were subjected to optical microscopy (Novex B-Range Optical Microscope, Euromex Microscopes, Netherlands) in order to observe the size, shape, and features of comminute WPS. Samples were diluted with ethanol prior to application via drop casting on pre-cleaned glass slides. This was done at the 200× magnification.

2.3. Viscosity measurement
Viscosity was measured using a model PDVdi-120 falling needle viscometer (Stony Brook Scientific, Ltd., USA). Methodology used is similar to ASTM D5478. A 25 ml-sample was prepared for each coating and was placed at the provided syringe. The density of the sample ρ_f was measured. The needle density ρ_s was calculated using equation (1). In the aforementioned equation, m_{needle} refers to the needle mass.

$$\rho_s = \frac{28.9673 + m_{needle}}{9.8988} \quad (1)$$

A stainless steel extension bar having etched markings was attached to a needle, gently placed on the exposed layer of the fluid and allowed to fall vertically. A stopwatch was used to measure the time t required by the second needle mark to be fully submerged in the solution. Fluid viscosity μ was calculated using equation (2) for fluids having a density not equal to 1 g cm^{-3}:

$$\mu = 9.1463\, t(\rho_s - \rho_f) \quad (2)$$

2.4. Preparation of coated substrates
Carbon steel A36 coupons having a dimension of 1 × 3 cm were prepared by polishing with sandpaper, and degreasing via sonication in acetone for 15 min using an ultrasonic water bath (WUC-D03H, DAIHAN Scientific, South Korea). Prior to use, metal coupons were dried and kept in vacuum using a vacuum oven (LVO-2030, LabTech Instruments, Inc., USA). Coupon thickness was measured using a digital micrometer (MDC-1 SX, Mitutoyo Corp., Japan). WPS was added to the commercial red oxide primer paint at the following loadings: 0, 1, 2, 4, and 10 wt%. The blended coatings were applied onto

the metal coupon substrates via spin coating (SCK-200 P, INTRAS Scientific, USA) at 1200 rpm for 10 s. The coated metal samples were air dried at least 24 h prior to testing.

2.5. Potentiodynamic polarization tests

Potentiodynamic polarization scan was performed following ASTM G102 with the coated metal coupons acting as the working electrode. A plate material evaluating cell (ALS Co., Ltd, Japan) made of Teflon was used. The platinum counter electrode and reference electrode were prepared and attached to the evaluating cell as shown in figure 1.

To simulate the corrosive environment, approximately 1 ml of 5 wt% NaCl solution was dropped in the 7.8-mm diameter hole of the evaluating cell. All electrodes were connected to a potentiostat (CS315, Wuhan Corrtest Instruments Corp. Ltd., China) which scanned the coupon from -0.5 to +0.5 V with scan rate 1 mV s^{-1} and frequency of 1 Hz. Corrosion potential E_{corr}, and corrosion current density i_{corr} were determined from Tafel extrapolation plots [10]. This was done in triplicate, with results being averaged prior to treatment.

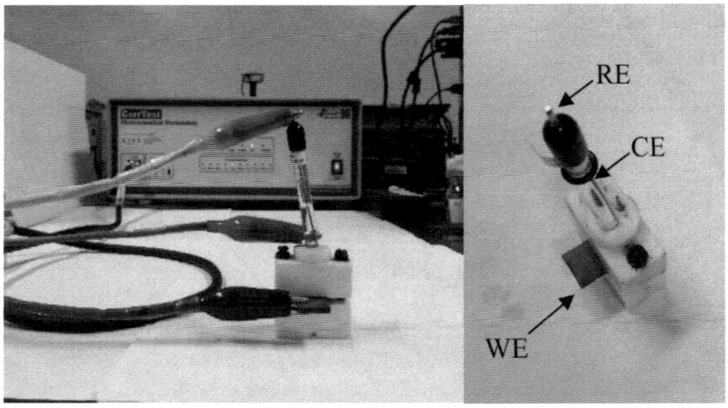

Figure 1. The potentiodynamic polarization scan set-up showing the plate material evaluating cell along with the attached counter electrode (CE), reference electrode (RE), and working electrode (WE).

3. Results

3.1. Morphology

Figure 2 shows the results of optical microscopy. It can be seen from the images that the comminuted WPS exhibits an irregular shape due to the abrasive nature of the grinding process. The contrasting focus within a particle in the images also show that the surface of each particle exhibits some irregular lumping. This is also an expected result of the abrasive grinding process. Based on the images, there are no apparent differences between the morphologies of laminated and non-laminated WPS.

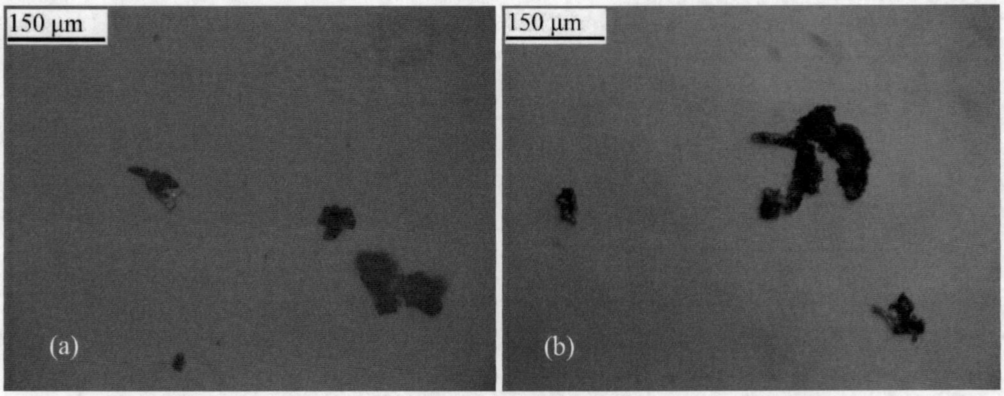

Figure 2. Morphology of (a) laminated WPS (b) non-laminated WPS.

3.2. Effect of additive on viscosity

Viscosity was measured primarily to investigate the ease of application of the coating mixture onto the metal substrate. Figure 3 shows the viscosities of coating mixtures having different loadings of WPS measured at 27 ± 1°C.

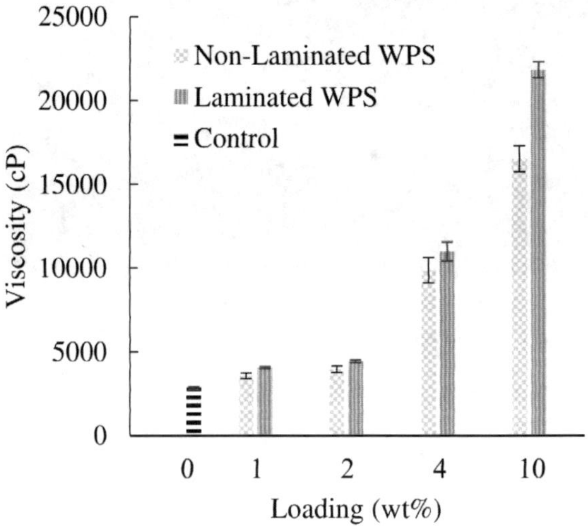

Figure 3. Viscosities of WPS-infused coating mixtures.

The data show that a higher WPS loading resulted in a higher dynamic viscosity. As solid particles, comminuted WPS contributes to the thickening and flow resistance of the coating. At 10 wt%, laminated WPS imparted a higher viscosity to the coating than non-laminated WPS. The metal film present in the laminated WPS offers more resistance to the flow of the coating. Viscosities of commonly available paints may vary depending on type, application, and preparation, as many paints are non-Newtonian fluids. For primer coatings, viscosities of commercially available primer paints are varied, many of them having viscosities ranging from 500 to 2000 cP [11]. The viscosity of paints, however, can be lowered by adding an appropriate solvent such as paint thinner.

3.3. Effect of additive on corrosion resistance

Figure 4 shows the representative Tafel plots of primer coatings blended with laminated and non-laminated WPS.

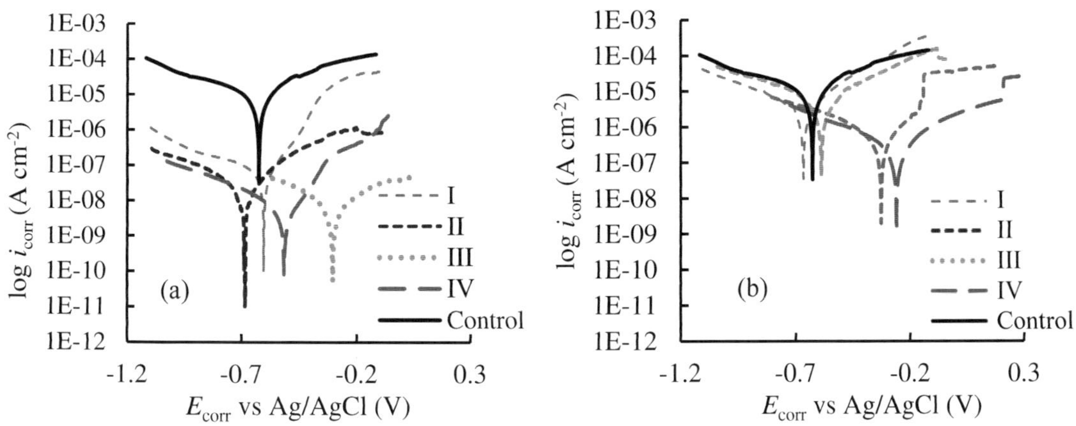

Figure 4. Representative Tafel plots of coatings infused with (a) laminated WPS and (b) non-laminated WPS at 0 (Control), 1 (I), 2 (II), 4 (III), and 10 wt% (IV).

Table 1 lists the properties of the WPS-infused coated substrates along with their respective coating thickness, E_{corr}, and i_{corr} values. Coating thickness was only affected by the addition of WPS at high loading. WPS shifted the E_{corr} of coatings to the positive (nobler) direction. Non-laminated WPS resulted in greater shifts in E_{corr} compared to laminated WPS. This denotes that the metal in the laminated WPS increases the overall tendency of the coating to lose electrons during exposure to corrosive medium, resulting to lower E_{corr} shift than non-laminated WPS. WPS lowered the i_{corr} of coatings by as much as 99%. As polymeric films, WPS improves the barrier properties of the coating by hindering the bulk diffusion of the corrosive medium into the coating [12,13]. The i_{corr} of coatings filled with laminated WPS are lower than with non-laminated WPS. The metal layer in the laminated WPS contributes better barrier properties to the coating. The same metal layer could also get passivated during exposure to corrosive medium which further enhances its contribution to the barrier characteristics of the coating. It is also possible that the metal layer lost electrons during corrosion and reacted with the dissolved oxygen to form a passivating oxide layer having stronger barrier properties [14,15]. On the other hand, higher loadings result in a lower drop of i_{corr} values as the dispersion of the filler interferes with the wet adhesion of the coating on the substrate [16].

Table 1. Properties of WPS-infused coatings.

Additive	Loading (wt%)	Coating thickness (mm)	E_{corr} (V)	$i_{corr} \times 10^7$ (A cm^{-2})	Change of i_{corr} relative to control (%)
None (control)	0	0.104	-0.7247	103.5	0.0
Laminated	1	0.100	-0.7232	0.610	99.4
Laminated	2	0.107	-0.5851	0.589	99.4
Laminated	4	0.123	-0.4457	0.223	99.7
Laminated	10	0.128	-0.6330	1.400	98.6
Non-Laminated	1	0.097	-0.5841	1.006	99.0
Non-Laminated	2	0.099	-0.3505	3.640	96.4
Non-Laminated	4	0.097	-0.5757	2.267	97.8
Non-Laminated	10	0.116	-0.3546	2.924	97.1

4. Conclusion

The addition of comminuted WPS to a red oxide primer paint improves the corrosion resistance of the coating. At low loading of laminated WPS, e.g. 1 and 2 wt%, the corrosion rate lowered by 99% with negligible change in coating thickness and viscosity. This work demonstrates for the first time the application of WPS in improving the anti-corrosion properties of coatings.

Acknowledgements

The authors would like to acknowledge the technical and financial support given by the Philippines' Department of Science and Technology – Engineering Research and Development for Technology (DOST-ERDT) program. The authors also express their gratitude to Doxo Ingredients, Inc., a private food company based in the Philippines, for providing waste plastic sachets.

References

[1] Wicks Jr. Z, Jones F N, Pappas S P and Wicks D A 2007 *Organic coatings science and technology* vol 13 (New Jersey: John Wiley & Sons, Inc.)
[2] Ravikumar H R, Rao S S and Karigar C S 2012 Biodegradation of paints: A current status *Indian J. Sci. Technol.* **5** 1977–87
[3] Turner A, Kearl E R and Solman K R 2016 Lead and other toxic metals in playground paints from South West England *Sci. Total Environ.* **544** 460–6
[4] Cai G, Wang H, Jiang D and Dong Z 2018 Degradation of fluorinated polyurethane coating under UVA and salt spray. Part I: Corrosion resistance and morphology *Prog. Org. Coatings* **123** 337–49
[5] Sanaei Z, Bahlakeh G and Ramezanzadeh B 2017 Active corrosion protection of mild steel by an epoxy ester coating reinforced with hybrid organic/inorganic green inhibitive pigment *J. Alloys Compd.* **728** 1289–304
[6] Armelin E, Oliver R, Liesa F, Iribarren J I, Estrany F and Alemán C 2007 Marine paint fomulations: Conducting polymers as anticorrosive additives *Prog. Org. Coatings* **59** 46–52
[7] Wu Y, Yu J, Zhao W, Wang C, Wu B and Lu G 2019 Investigating the anti-corrosion behaviors of the waterborne epoxy composite coatings with barrier and inhibition roles on mild steel *Prog. Org. Coatings* **133** 8–18
[8] Milios L, Holm Christensen L, McKinnon D, Christensen C, Rasch M K and Hallstrøm Eriksen M 2018 Plastic recycling in the Nordics: A value chain market analysis *Waste Manag.* **76** 180–9
[9] Sy-Changco J A, Pornpitakpan C, Singh R and Bonilla C M 2011 Managerial insights into sachet marketing strategies and popularity in the Philippines *Asia Pacific J. Mark. Logist.* **23** 755–72
[10] Papavinasam S 2014 *Corrosion Control in the Oil and Gas Industry* (London: Elsevier)
[11] NPCS Board of Consultants & Engineers 2009 *Handbook on Paints and Enamels* (Delhi: NIIR Project Consultancy Services)
[12] Moggridge G D, Lape N K, Yang C and Cussler E L 2003 Barrier films using flakes and reactive additives *Prog. Org. Coatings* **46** 231–40
[13] Pajarito B B, Caguntas A J F, Felices N B, Tubalinal H O S and Leuterio G L D 2018 Corrosion, wettability, and adhesion of acrylic coatings containing silane-treated mineral fillers on carbon steel *Mater. Sci. Forum* **917** 252–6
[14] Armelin E, Pla R, Liesa F, Ramis X, Iribarren J I and Alemán C 2008 Corrosion protection with polyaniline and polypyrrole as anticorrosive additives for epoxy paint *Corros. Sci.* **50** 721–8
[15] Mayne J E O 1952 The mechanism of the inhibition of the corrosion of iron and steel by means of paint *Off. Dig.* **24** 127
[16] Sangaj N S and Malshe V C 2004 Permeability of polymers in protective organic coatings *Prog. Org. Coatings* **50** 28–39

PVA/CA based electrospun nanofibers: Influence of processing parameters in the fiber diameter

M A Teixeira[1,2], M T P Amorim[1] and H P Felgueiras[1]

[1]Centre for Textile Science and Technology, University of Minho, Campus de Azurém, 4800-058 Guimarães, Portugal

E-mail: martaalbertinateixeira@gmail.com

Abstract. Recently, the electrospinning technique has been explored as a natural and synthetic polymer processing tool due to its versatility and potential to generate complex structures at a nanoscale. In this work, non-woven nanofibrous mats were electrospun, with a structure resembling the extracellular matrix, for prospective biomedical uses. Poly (vinyl alcohol) (PVA) and cellulose acetate (CA) based electrospun nanofibrous meshes were prepared at different ratios 100/0, 90/10, 80/20 and 70/30 and characterized in terms of fiber diameter. The process was kept as green as possible by resorting to a combination of acetic acid and distilled water as solvents. Optimal conditions for PVA/CA processing were established at 29 kV, feeding rate of 0.8 mL/h and distance between needle and collector of 17 cm. These allowed for the most uniform fibers with the smallest diameters to be produced.

1. Introduction

Since the late 20th century, the electrospinning has been garnering increasing attention in the scientific community, as well as in the industry [1] due its relatively simple setup, cost-effectiveness, ease of process and versatility [2,3]. Electrospun nanofibrous mats are characterized by their large surface area, controllable fibre diameters and structures (i.e. dense, hollow, and porous) and unique surface morphologies, which can be obtained by varying for instance the molecular weight of the polymers and the polymeric solution properties (viscosity, conductivity, dielectric constant, and surface tension). Processing parameters such as voltage power supply, feeding rate, distance between needle and collecting plate, as well as the use of coaxial or triaxial needles for hollow, core–shell or multi-sheathed structures, and even the temperature and humidity during processing, are also of extreme importance during mats production, exerting much influence in the final product [4,5]. Electrospinning requires the application of a high electric field to generate nanofibers from a charged polymer solution. Random, aligned and core-shell fibers can be obtained from natural or synthetic polymers [6,7].

Synthetic polymers have great flexibility during synthesis. They exhibit excellent mechanical properties, thermal stability and an appropriated degradation profile, being desirable for many biomedical applications that require stable, reticulated, nanofibrous constructs [8]. Poly (vinyl alcohol) (PVA), for instance, is known for its high hydrophilic, nontoxic and biocompatible nature, possesses good physicochemical properties that account for their mechanical stability, flexibility and slow degradation kinetics, well above nanofibrous made of natural polymers, and presents high capacity of spinnability [8,9]. However, synthetic polymers, including PVA, lack cell affinity or cell recognition sites. To overcome this limitation, synthetic polymers have been conjugated with natural polymers, which exhibit superior biocompatibility and low immunogenicity, and some even display intrinsic antibacterial properties. The acetate ester of cellulose, cellulose acetate (CA), has attracted wide interest in the production of nanofibers due to its biodegradability, chemical persistence, biocompatibility, and thermal constancy [10]. Regarding the CA low tensile strength, composites with synthetic polymers have been proposed. Here, the affinity of PVA/CA blends was explored in light of the electrospun processing parameters and their impact in the fiber's properties.

Most studies show the viscosity as a determinant factor for fiber diameter and morphology. However, the solution viscosity is not an independent variable and, as such, is intimately related to the concentration and molecular weight of the polymer, meaning that if one of these parameters increases the viscosity will also increase and, hence, larger diameter fibers are obtained. Still, for electrospinning processing, there is a critical value for each polymeric solution. If the viscosity is high, surpassing the polymeric solution critical value, the flow may be hindered and the droplet dries at the tip, blocking it. On the other hand, if the viscosity is low, the applied electric field and surface tension may cause the entangled polymer chains to break into fragments before reaching the collector, forming beaded nanofibers [11,12]. Regarding the solution conductivity, it not only affects the Taylor cone formation but also helps in controlling the diameter of the nanofibers. The increase of electrical conductivity of the solution, leads to a significant decrease in the diameter, whereas at low conductivities the jet elongation is not sufficient to produce uniform fibers. The conductivity is influenced by solvent type, polymer concentration and temperature [13].

In the processing parameters, the minimum voltage required to produce nanofibers increases with the solution's surface tension but this behavior is not always linear, its critical value varies from polymer to polymer [14]. High voltages lead to the formation of smaller fiber diameters attributed to the stretching of the polymer solution in correlation with the charge repulsion within the polymer jet [15]. Feeding rates and distances between the needle tip and the collector should be adjusted depending to the selected solvent. The solution feeding rate influences jet velocity and transfer rate. For proper solvent evaporation and to acquire well defined, solid nanofibers lower feeding rates are desirable. Ideally, the feeding rate must match the solution removing rate from the tip. Very low feeding rates may block the needle tip and prevent the polymer exit, while high feeding rates may result in beaded, large diameter fibers due to the short period allowed for solvent evaporation prior to fibers reaching the collector [16]. Large diameter nanofibers are also formed when the distance between needle and collector is kept small, since solvent evaporation is inefficient, whereas the diameter of the nanofibers decreases as the distance increases [17,18]. Environment parameters, such as relative humidity and temperature, also affect the diameter and morphology of the nanofibers. Humidity causes changes in the nanofibers diameter by controlling the solidification process of the charged jet. This phenomenon is, however, dependent on the chemical nature of polymer. It has been reported that the increase in the humidity may lead to a reduction in fibers diameter [19]. Temperature causes two opposing effects in the fibers diameter: (i) increases the rate of evaporation of solvent and (ii) decreases the viscosity of the solution. However, both have been shown to lead to the reduction of the nanofibers mean diameter [20,21].

In the present work, the previous parameters were altered and adapted in function of the concentration and polymeric ratios of PVA and CA. The goal was to determine their intimate relation and, in particular, their influence in the nanofiber's diameter.

2. Materials and methods

2.1. Materials

PVA with Mw 78,000 and 88% hydrolyzed was purchased from Polysciences, Warrington, USA. CA with Mw 30,000 was purchased from Sigma Aldrich, USA. The acetic acid (glacial) 100% was purchased from Merck, Darmstadt, Germany.

2.2. Fabrication of nanofibrous meshes

Polymeric solutions of PVA/CA were prepared at 11% and 9% (w/v) concentrations in acetic acid/distilled water at 60/40 v/v. PVA and CA were combined at varying ratios: 100/0 (or 100 PVA),

90/10, 80/20 and 70/30 (v/v). These solutions were continuously stirred at 80°C for 3 h and were electrospun (figure 1) at specified processing conditions of applied voltage, feed rate, collection distance and needle inner diameter, which was established at 18 gauge (1.02 mm). After the electrospinning process, films were placed in a drying oven for 12 h at 40°C, for a complete solvent evaporation. Four squared samples of 1x1 cm^2 were cut from each mesh and mounted on glass slides for microscopy observation. The microscope Leica DM750, with an integrated high definition digital camera, was used to verify the production of fibers. Five pictures were taken from each mesh and the respective fiber diameters were measured using the program Image J. Ten fibers were measured per picture and the average diameters reported.

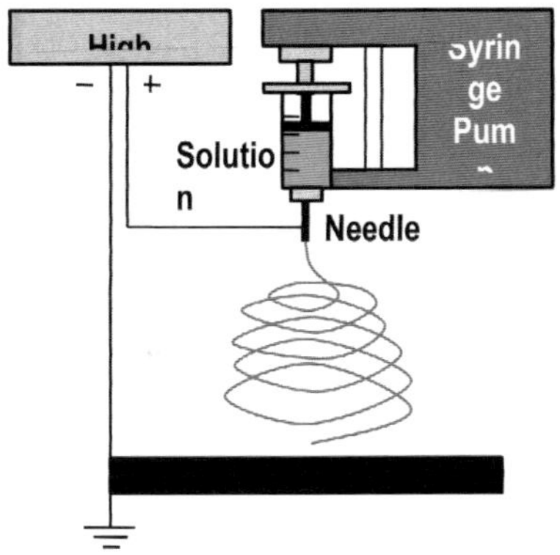

Figure 1. Schematic representation of the electrospinning setup.

3. Results and discussion

The main goal of this work was to produce nanofibrous meshes via electrospinning using green and sustainable methodologies. The polymers, PVA and CA, were selected based on their ease of process, biocompatibility and biodegradability. The solvents ratio of 60/40 v/v of acetic acid/distilled water was established as the most suitable to prevent polymer precipitation. Electrospun processing parameters, particularly voltages and feeding rates, which were found the most influential in the fibers diameter of this particular polymeric blend (PVA/CA), were altered to improve the fibers' features.

Table 1 reports the variations in processing parameters considered, namely electric potential, feed rate and distance to collector as well as polymer concentration and polymeric blend ratio, and the respective average diameter obtained. Initial tests were conducted with a 100% solution of PVA. Various concentrations were tested, being the 9% and 11% (w/v) the most successful. Upon close analysis, larger fiber diameters were evidenced at 11% (w/v) concentration and as such all subsequent polymeric solutions (including blends) were prepared at 9%.

Table 1. Average fiber diameters in response to polymer concentration, blend ratio and electrospun processing parameters [average ± standard deviation (SD), with n=10].

PVA/CA Meshes	Processing Parameters Voltage (kV)	Feed Rate (mL/h)	Distance (cm)	Fiber Diameter ± SD (nm)
11 % (w/v)				
100/0	28.0	1.2	17.0	705 ± 179
	29.0	0.8	17.0	851 ± 229
9 % (w/v)				
100/0	20.0	0.7	18.0	838 ± 246
	22.0	1.3	14.0	921 ± 155
	24.0	1.3	15.5	790 ± 188

	25.0	0.7	18.0	781 ± 190
	29.0	0.8	17.0	**665 ± 151**
	29.0	1.3	17.0	877 ± 227
	29.0	1.5	15.5	682 ± 188
	29.0	1.7	15.6	795 ± 170
90/10	20.0	0.8	17.0	788 ± 216
	22.0	0.8	17.0	772 ± 164
	29.0	0.8	17.0	**632 ± 117**
80/20	20.0	1.2	17.0	626 ± 149
	25.0	1.2	17.0	516 ± 128
	25.0	0.8	17.0	742 ± 181
	29.0	0.8	17.0	**665 ± 177**
70/30	29.0	0.8	17.0	**731 ± 156**

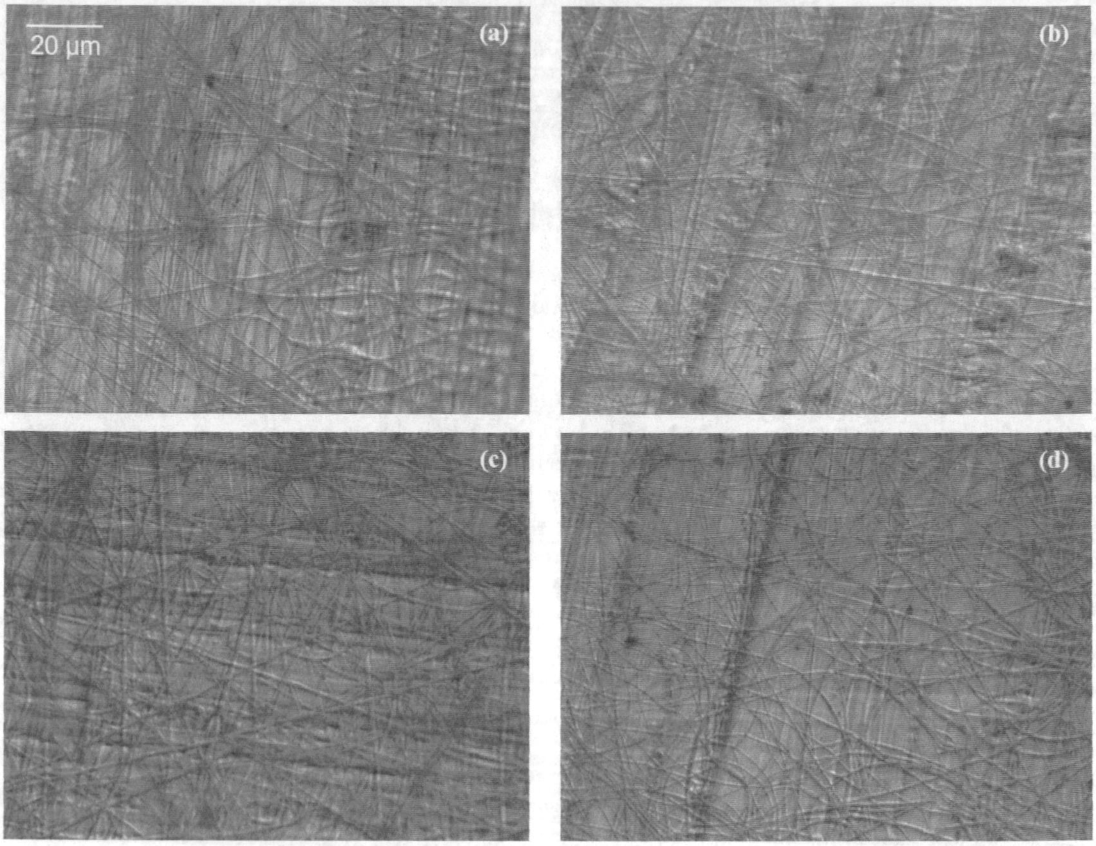

Figure 2. Optic micrographs (bright field) of (a) 100/0, (b) 90/10, (c) 80/20 and (d) 70/30 PVA/CA nanofibrous electrospun meshes, processed at 29 kV, 0.8 mL/h and 17 cm distance. All imagens were collected with the objective magnification of 100x and ocular magnification of 10x (scale bar of 20 μm).

Optimal processing conditions were selected based on gradual alterations in voltage. Extremes were defined between 20 kV (lowest) and 29 kV (highest). According to the literature, higher voltages are more likely to give rise to smaller diameters [13]. Data from 100 PVA showed that by increasing the voltage, the mean fibers diameter decreased. Changes in the feeding rate were also decisive to obtain diameters close to those required for biomedical applications (mimicking the extracellular matrix). It is clear that by increasing the feeding rate the fiber diameters also increased since the solvent did not have enough time to evaporate, becoming entrapped within the polymeric matrix. As such high voltages and low feeding rates were found more appropriated to process the PVA/CA meshes. Higher distances were also observed to promote the formation of continuous, beadless fibers with small diameters. To confirm these conclusions random processing parameters were applied both to 90/10 and 80/20 PVA/CA blends. In general, data collected confirmed our latest premise. A reduction of almost 200 nm in average diameter was registered when passing from a low voltage of 20 kV (838 nm) to a high voltage of 29 kV (665 nm), for 100 PVA, with similar observations being made for the combination 90/10. In the 80/20 ratio this linearity was not completely verified; in fact, the most successful combination of parameters required the lowest potential and the highest feeding rate. In light of previous reports, it is likely this to be an anomaly and a result of environmental conditions interference, such as temperature and humidity; this specific test is carried out in a different month than the rest. The blend 70/30 was only tested with the parameters defined as ideal.

In this research, it is proven the stability of the polymeric blend and the homogeneity of the solution, since optimal processing parameters were found for all mixtures at 29 kV, 0.8 mL/h and 17 cm distance, with resulting average diameters in the same ranges (figure 2).

4. Conclusion

Our main goal was to generate polymeric mats from PVA/CA with the minimal diameter possible using the most environmentally friendly and sustainable processes and materials. After several attempts, PVA and CA solvents were established at 60/40 acetic acid/distilled water, and the electrospinning processing conditions at 29 kV, 0.8 mL/h and 17 cm. These were the most successful in giving rise to continuous, beadles, small diameter nanofibers for all polymeric combinations. Further tests will be conducted to ensure the reproducibility of these results in different environment conditions. Still, preliminary data are very promising, revealing the potential of this polymeric combination, PVA/CA, to promote the formation of intricate and complex nanofibrous meshes for prospective biomedical uses.

Acknowledgments

Authors acknowledge the Portuguese Foundation for Science and Technology (FCT), FEDER funds by means of Portugal 2020 Competitive Factors Operational Program (POCI) and the Portuguese Government (OE) for funding the project PEPTEX with reference POCI-01-0145-FEDER-028074. Authors also acknowledge project UID/CTM/00264/2019 of Centre for Textile Science and Technology (2C2T), funded by national funds through FCT/MCTES.

References

[1] Haider A, Haider S and Kang I K 2015 A comprehensive review summarizing the effect of electrospinning parameters and potential applications of nanofibers in biomedical and biotechnology *Arab. J .Chem.* **11** 1165-88

[2] SalehHudin H S, Mohamad E N, Mahadi W N L and Afifi A 2018 Multiple-jet electrospinning methods for nanofiber processing: A review *Mater. Manuf. Process* **33** 479-98

[3] Felgueiras H P and Amorim M T P 2017 Electrospun polymeric dressings functionalized with antimicrobial peptides and collagen type I for enhanced wound healing *IOP Conf. Ser. Mater. Sci. Eng.* **254** 062004

[4] Agarwal S, Wendorff J H and Greiner A 2008 Use of electrospinning technique for biomedical applications *Polym.* **49** 5603-21

[5] Wang X and Hsiao B S 2016 Electrospun nanofiber membranes *Curr. Opin. Chem. Eng.* **12** 62-81

[6] Saadatmand M M, Yazdanshenas M E, Khajavi R, Mighani F and Toliya T 2019 Patterning the surface roughness of a nano fibrous scaffold for transdermal drug release *Int. J. Nano. Dimens.* **10** 78-88

[7] Coelho D, Sampaio A, Silva Carla J S M, Felgueiras H P, Amorim M T P and Zille A 2017 Antibacterial electrospun PVA/enzymatic synthesized poly(catechol) nanofibrous mid-layer membrane for ultrafiltration *ACS Appl. Mater. Interfaces* **9** 33107

[8] Miguel S P, Figueira D R, Simões D, Ribeiro M P, Coutinho P and Ferreira P 2018 Electrospun polymeric nanofibres as wound dressings: A review *J. Colloid Surf. B Biointerfaces* **169** 60-71

[9] Kasai D, Chougale R, Masti S, Chalannavar R, Malabadi R B and Gani R 2018 Influence of Syzygium cumini leaves extract on morphological, thermal, mechanical, and antimicrobial properties of PVA and PVA/chitosan blend films *J. Appl. Polym. Sci.* **135** 1-17

[10] Kanimozhi K, Khaleel Basha S and Sugantha Kumari V 2016 Processing and characterization of chitosan/PVA and methylcellulose porous scaffolds for tissue engineering *Mater. Sci. Eng. C* **61** 484-91

[11] Khoshnevisan K, Maleki H, Samadian H, Shahsavari S, Sarrafzadeh M H and Larijani B 2018 Cellulose acetate electrospun nanofibers for drug delivery systems: Applications and recent advances *Carbohydr. Polym.* **198** 131-41

[12] Okutan N, Terzi P and Altay F 2014 Affecting parameters on electrospinning process and characterization of electrospun gelatin nanofibers *Food Hydrocoll* **39** 19-26

[13] Salehhudin H S, Mohamad E N, Nor W and Mahadi L 2018 Multiple-jet electrospinning methods for nanofiber processing: A review *Mater. Manuf. Process*. 33 479-98

[14] Bhardwaj N and Kundu S C 2010 Electrospinning: A fascinating fiber fabrication technique *Biotechnol. Adv*. **28** 325-47

[15] Mercante L A, Scagion V P, Migliorini F L, Mattoso L H C and Correa D S 2017 Electrospinning-based (bio) sensors for food and agricultural applications: A review *Trends Anal. Chem.* **91** 91-103

[16] Kai D, Shy S and Jun X 2014 Biodegradable polymers for electrospinning: Towards biomedical applications *Mater. Sci. Eng. C* **4**5 659-70

[17] Mirjalili M and Zohoori S 2016 Review for applica.tion of electrospinning and electrospun nanofibers technology in textile industry *J. Nanost. Chem.* **6** 207-13

[18] Ghorani B and Tucker N 2015 Fundamentals of electrospinning as a novel delivery vehicle for bioactive compounds in food nanotechnology *Food Hydrocoll.* **51** 227-40

[19] Pelipenko J, Kristl J, Jankovi B and Kocbek P 2013 The impact of relative humidity during electrospinning on the morphology and mechanical properties of nanofibers **456** 125-34

[20] Fashandi H and Karimi M 2012 Pore formation in polystyrene fi ber by superimposing temperature and relative humidity of electrospinning atmosphere *Polym.* **53** 5832-49

[21] Felgueiras H P and Amorim M T P 2017 Functionalization of electrospun polymeric wound dressings with antimicrobial peptides *Colloids Surf. B* **156** 133-48

IOP Conf. Series: Materials Science and Engineering **634** (2019) 012041 doi:10.1088/1757-899X/634/1/012041

Carbonization performance of pre-oxidized PAN fibers prepared by microwave heating

T H H Elagib[1], E A M Hassan[1], B H Liu[1], M H Yu[1] and K Q Han[1,2]

[1]State Key Laboratory for Modification of Chemical Fibers and Polymer Materials, College of Materials Science and Engineering, Donghua University, Shanghai 201620, China

E-mail: hankeqing@dhu.edu.cn

Abstract. We have carbonized the modified PAN fibers which coated with a hybrid microwave absorbent and pre-oxidized by microwave heating for 50 min. The carbonization process of pre-oxidized CB/S-PAN fibers process was performed under nitrogen atmosphere at different temperatures. The carbonized fibers were characterized by mean of Raman spectroscopy, elemental analyzer, X-ray diffraction (XRD), thermogravimetric analyzer (TGA), mechanical tester, and scanning electron microscopy (SEM). The results indicated that more ordered crystal structure and highly graphitic content was obtained at 950°C, arise from the formation of aromatic structures. The fibers' surface appeared roughly without defects suggesting a strong graphitic structure. This rough nature which results from the deposition of nanoparticles at an earlier stage before the pre-oxidation and the microwave penetration can strongly improve the interlocking between the fibers and matrix in the carbon fibers' composites.

1. Introduction

Carbon fibers (CF) have attracted considerable attention due to their high mechanical properties, light weight, good chemical and thermal resistance, and excellent electrical and thermal conductivity [1-6]. Carbon fibers can be classified based on the precursor types into polyacrylonitrile (PAN)-based CFs, pitch-based CFs, and rayon-based CFs. The commonly used precursor is PAN due to its higher melting point [7], relatively lower cost, and greater carbon yield [8]. The conversion process of PAN precursor into CFs involved two main stages: pre-oxidation where the linear structure is transformed into conjugated ladder structure [9] and carbonization where a graphite layer structure is formed [10].

Many researches concentrated on studying the thermal pre-oxidation and subsequent carbonization [11] of CFs. A few studies about the optimization of PAN conversion process have been done [12]. Kim *et al* [13] introduced a new microwave plasma system for the carbonization process which resulted in higher surface roughness. In our previous study, hybrid nanoparticles (Carbon black/SiC) were used as microwave absorbent and coated onto the surface of the PAN fibers in order to improve their microwave absorbability. Typical new microwave system was designed and used to induce the pre-oxidation process for the modified CB/S-PAN fibers. The results showed that the cyclization reaction was accelerated and cyclization degree was optimized [14]. In this work, structural evolvements in the conversion of microwave pre-oxidized CB/S-PAN fibers into CFs as well as the influences of fibers modification on the carbonization performance were investigated.

2. Experimental

2.1. Material

The CB/S-PAN fibers used in this study were modified in our laboratory. The modified PAN fibers were pre-oxidized at low temperature (110-170ºC) by microwave heating. In the process of carbonization, the pre-oxidized fibers were passed through a tubular furnace with a ceramic tube (length: 1 m). The pre-oxidized fibers were carbonized at different temperature (550ºC, 750ºC, and 950ºC) in nitrogen (N_2) with a flow rate of 3-5 L/min.

2.2. Characterization

The surface structure was analyzed by Raman spectra (inVia-Reflex, Japan). The elements' content was measured by Vario EL III elemental analyzer (Germany). The phase structures were detected by X-ray diffraction (XRD, D/max 2550 PC, Japan). Gravimetric thermal analysis (TGA) was carried out using Discovery TGA (Q5000IR) instrument. A XQ-1 computerized mechanical tester was used for mechanical analysis. The morphology was studied by field emission scanning electron microscopy (FESEM, JSM-5600LV, Japan). Results and discussion

3. Results and discussions

3.1. The evolution of chemical structure

Raman patterns of carbonized CB/S-PAN fibers were presented in figure 1. Two characteristic peaks were found, Raman-allowed band around 1580 cm^{-1} assigned to the graphitic structure (G-mode) and disorder-induced band around 1360 cm^{-1} (D-mode) which represented disorder carbon structure corresponded to the increase in the amount of unorganized carbon and the decrease in crystal size [15].

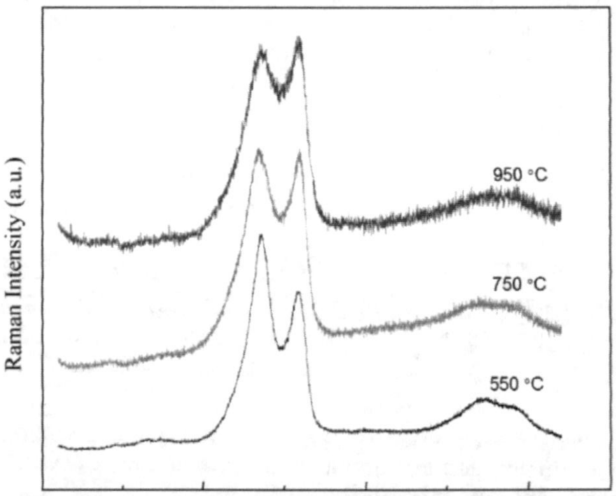

Figure 1. Raman spectra of carbonized CB/S-PAN fibers.

The graphitization degree R which is sensitive to the average distance between defects [16] was characterized by the ratio of the integrated intensity of disorder-induced band to Raman-allowed band (R=ID/IG). The Raman shifts of the G and D peaks and the R values of all samples estimated from Raman patterns are listed in table 1. The R ratio decreased with the increase of carbonization temperature suggesting the generation of the more ordered crystal structure at 950°C [17,18]. Moreover, it can be noticed that the G band shift to lower wavelength this was due to changes in the relative intensity of D band overlapping [16].

Table 1. Raman data of carbonized CB/S-PAN fibers.

Carbonization Temperature (°C)	D-mode (cm^{-1})	G-band (cm^{-1})	R
550	1354.00	1580.80	2.60
750	1340.40	1580.20	1.37
950	1340.40	1574.80	1.12

At the initial stages of carbonization, crosslinking reactions occurred in the pre-oxidized PAN. The cyclized structure began to link up in the lateral direction by denitrogenation and dehydration. A planar structure can be formed with the basal planes aligned along the fiber axis. HCN and NH_3 released at temperature up to 450°C were from an un-laddered structure while in the range of 450–900°C were from laddered structure leading to polymer crosslinking. The release of N_2 was recognized beginning from 700°C [19]. Goodhew *et al* proposed that intermolecular dehydrogenation happened in the range of between 400°C and 600°C, while denitrogenation occurred at higher temperatures. Dehydrogenation attached ladder molecules forming graphite-like ribbons, while, denitrogenation was accountable for the growth of ribbons to form sheet-like structures. The sheets could further grow at higher temperatures with the release of N_2 [20]. The chemical composition of the carbonized CB/S-PAN fibers were presented in table 2. As can be seen, a significant effect in the elements was observed. As the carbonization temperature increased the carbon content increased, while nitrogen and hydrogen content decreased by dehydrogenation and denitrogenation. This further revealed the generation of the carbon structure of the carbonized CB/S-PAN fibers.

Table 2. Elemental composition of carbonized CB/S-PAN fibers.

Carbonization	Elemental content (%)		
Temperature (°C)	C	H	N
550	68.60	2.40	18.24
750	71.99	1.48	16.50
950	87.33	0.29	9.73

3.2. Phase analysis

XRD patterns of carbonized CB/S-PAN fibers were presented in figure 2. XRD profile exhibited broad diffraction peaks at $2\Theta \approx 25°$ assigned to (002) crystalline plane, demonstrating the formation of carbon basal plane, and rearrangement of layer plane which changed the microstructure parameters such as crystal height L_c [21]. It can be pointed out that Lc increased and the FWHM values decreased as the carbonization temperature increased (table 3). This finding is correlating with Raman results suggesting the formation of a more perfect graphite structure.

Table 3. XRD data of carbonized CB/S-PAN fibers.

Carbonization Temperature (°C)	$2\Theta \approx 25°$ Peak FWHM	Crystallite height L_c (nm)
550	13.31	0.59
750	11.93	0.66
950	5.90	1.33

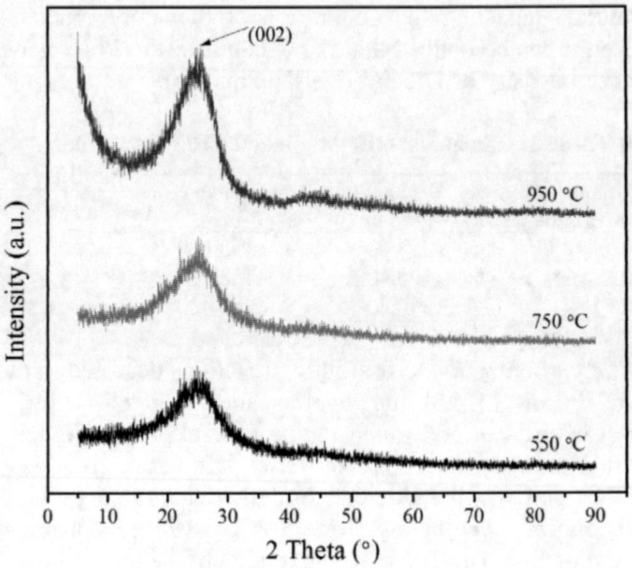

Figure 2. XRD patterns of carbonized CB/S-PAN fibers.

3.3. Thermal stability

TGA curves of carbonized CB/S-PAN fibers at different carbonization temperatures were displayed in figure 3. As shown, a primary weight loss below 200°C for carbonized fibers at 550°C and 750°C was noticed due to the removal of unstable chemical species from the fibers' surface [22]. A secondary weight loss between 400°C and 700°C was observed for carbonized fibers at 550°C, which can be ascribed to the further release of volatile products due to low carbonization temperature [23]. No obvious weight loss was noticed for carbonized fibers at 950°C indicating an improved graphite structure.

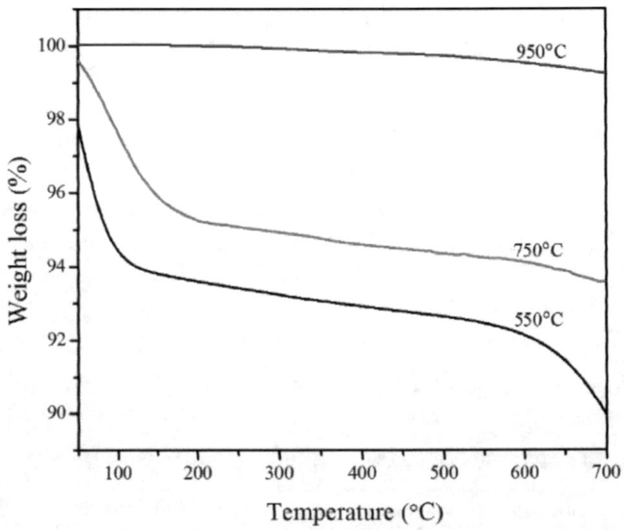

Figure 3. TGA curves in a nitrogen atmosphere of carbonized CB/S-PAN fibers.

3.4. Physical and mechanical properties

The mechanical and physical properties of carbonized CB/S-PAN fibers were listed in table 4. The tensile strength and modulus of carbonized fibers increased as the carbonization temperature increased. Tensile strength is not graphitic order and orientation dependent but is affected by defects [24] and the elementary formation of carbon basal planes [25] which could further grow at higher temperatures with the release of N_2 [20]. The modulus is dependent on the graphitic order which was affected by the increase of the crystal size, therefore in good agreement with Raman and XRD results. The carbonized fibers diameter decreased while the density increased at higher temperature interpreting that the crosslinking of ladder structure, as well as the lengthening and broadening of carbon basal planes, led to the repacking of the structure [26].

Table 4. Physical and mechanical characteristics of carbonized fibers.

Carbonization Temperature (°C)	Diameter (μm)	Density (cg/cm^3)	Tensile strength (GPa)	Tensile modulus (GPa)
550	7.40	1.47	0.64	26.93
750	7.53	1.80	0.98	53.96
950	6.96	1.80	1.00	71.06

3.5. SEM observation

Figure 4(a) displayed the surface topography of carbonized CB/S-PAN fibers. As shown, the presence of nanoparticles on the fibers' surface with aid of microwave penetration during the pre-oxidation process can enhance the surface roughness and consequently the bond strength of fibers/matrix interface during the fabrication of CFs composites. The surface of carbonized S-PAN fibers was uniformly deposited with NPs, indicating the good coupling between NPs and the PAN fibers' surface. Furthermore, the fibers' surface seemed free of defects indicating the formation of a strong graphitic structure [27]. SEM images of fracture surfaces for the carbonized CB/S-PAN fibers were presented in figure 4(b). It could be seen that almost similar cross-section regularity was observed. The fracture was seemed to begin in the specific skin area and then transferred to the core. The carbonized fibers possessed brittle fracture surface and this nature became distinct as the carbonization temperature increased. Furthermore, it exhibited somehow smooth surface and random structure at 550°C, and the roughness increased at higher carbonization temperature.

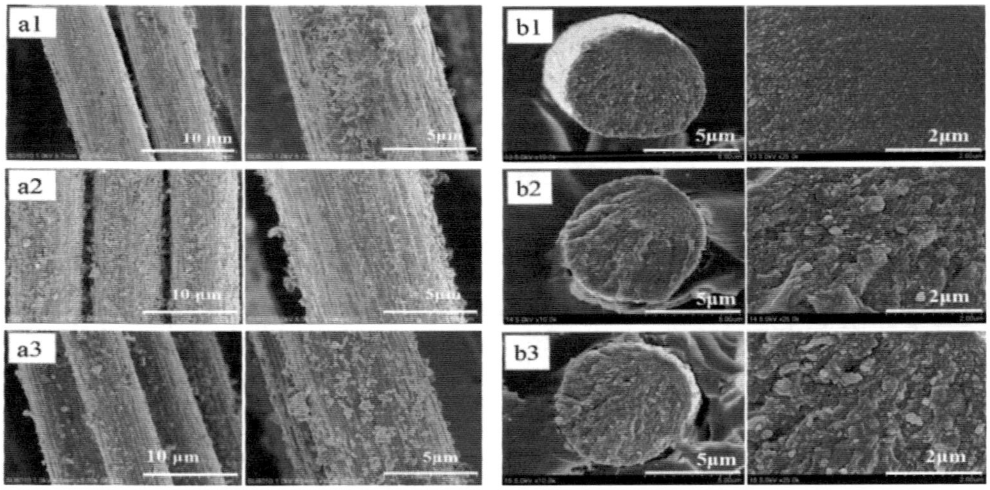

Figure 4. Different magnification SEM fracture morphology images of CB/S-PAN fibers carbonized at (a1) 550°C, (a2) 750°C, and (a3) 950°C. Different magnification SEM topography images of CB/S-PAN fibers carbonized at (b1) 550°C, (b2) 750°C, and (b3) 950°C.

4. Conclusion

The carbonization performance of microwave pre-oxidized CB/S-PAN fibers was studied. The Raman analysis revealed the formation of a graphite structure, while XRD results indicated that the ladder polymer cross-linked to enhance the growth of the carbon basal planes. The elemental analysis further revealed the generation of the carbon structure. The analysis of thermal stability confirmed the formation of a more perfect graphitic structure. The increase of Young's modulus correlated to the increase of graphitic order and crystal size. The growth of carbon basal planes and the elimination of structural defects lead to the increase of tensile strength.

Acknowledgment

This research did not receive any specific grant from funding agencies in the public, commercial, or not-for-profit sectors.

References

[1] Frank E, Hermanutz F and Buchmeiser M R 2012 Carbon fibers: Precursors, manufacturing, and properties *Macromol. Mater Eng* **297** 493-501
[2] MInus M and Kumar S 2005 The processing, properties, and structure of carbon fibers *Jom* **57** 52-8
[3] Liu Y and Kumar S 2012 Recent progress in fabrication, structure, and properties of carbon fibers *Polym Rev* **52** 234-58
[4] Kozey V V, Jiang H, Mehta V R and Kumar S 1995 Compressive behavior of materials: Part II. High performance fibers *J. Mater Res* **10** 1044-61
[5] Chand S 2000 Review carbon fibers for composites *J Mater Sci* **35** 1303-13
[6] Fitzer E 1989 Pan-based carbon fibers—present state and trend of the technology from the viewpoint of possibilities and limits to influence and to control the fiber properties by the process parameters *Carbon* **27** 621-45
[7] Rahaman M S A, Ismail A F and Mustafa A 2007 A review of heat treatment on polyacrylonitrile fiber *Polym Degrad Stab* **92** 1421-32
[8] Yu M, Xu Y, Wang C, Hu X, Zhu B, Qiao K and Yuan H 2012 Heredity and difference of multiple-scale microstructures in PAN-based carbon fibers and their precursor fibers *J Appl Polym Sci* **125** 3159-66
[9] Ko T H and Li C H 1995 The influence of pre-carbonization on the properties of PAN-based carbon fibers developed by two-stage continuous carbonization and air oxidation *Polym compos* **16** 224-32
[10] Qian X, Zou R, OuYang Q, Wang X and Zhang Y 2015 Surface structural evolvement in the conversion of polyacrylonitrile precursors to carbon fibers *Appl Surf Sci* **327** 246-52
[11] Ko T H 1991 The influence of pyrolysis on physical properties and microstructure of modified PAN fibers during carbonization *J Appl Polym Sci* **43** 589-600
[12] Elagib T H, Hassan E A, Fan C, Han K and Yu M 2019 Microwave pre-oxidation for polyacrylonitrile precursor coated with nano-carbon black *Polym Eng Sci* **59** 457-64
[13] Kim S Y, Kim S-Y, Choi J, Lee S, Jo S M, Joo J and Lee H-S 2015 Two step microwave plasma carbonization including low plasma power pre-carbonization for polyacrylonitrile based carbon fiber *Polymer* **69** 123-8
[14] Elagib T H, Hassan E A, Fan C, Han K and Yu M 2018 Single and hybrid electromagnetic absorbing coatings on polyacrylonitrile precursor to motivate the microwave pre-oxidation *Polym Degrad Stab* **158** 64-71
[15] Li D, Wang H and Wang X 2007 Effect of microstructure on the modulus of PAN-based carbon fibers during high temperature treatment and hot stretching graphitization *J Mater Sci* **42** 4642-9
[16] Huson M G, Church J S, Kafi A A, Woodhead A L, Khoo J, Kiran M, Bradby J E and Fox B L 2014 Heterogeneity of carbon fibre *Carbon* **68** 240-9

[17] Kong L, Liu H, Cao W and Xu L 2014 PAN fiber diameter effect on the structure of PAN-based carbon fibers *Fibers Polym* **15** 2480-8
[18] Liu F, Wang H, Xue L, Fan L and Zhu Z 2008 Effect of microstructure on the mechanical properties of PAN-based carbon fibers during high-temperature graphitization *J Mater Sci* **43** 4316-22
[19] Huang X 2009 Fabrication and properties of carbon fibers *Materials* **2** 2369-403
[20] Goodhew P, Clarke A and Bailey J 1975 A review of the fabrication and properties of carbon fibres *Mater Sci Eng* **17** 3-30
[21] Jing M, Wang C-G, Wang Q, Bai Y-J and Zhu B 2007 Chemical structure evolution and mechanism during pre-carbonization of PAN-based stabilized fiber in the temperature range of 350–600 C *Polym Degrad Stab* **92** 1737-42
[22] Cho C-W, Cho D-H, Ko Y-G, Kwon O-H and Kang I-K 2007 Stabilization, carbonization, and characterization of pan precursor webs processed by electrospinning technique *Carbon lett* **8** 313-20
[23] Siqueira R L, Yoshida I V P, Pardini L C and Schiavon M A 2007 Poly (borosiloxanes) as precursors for carbon fiber ceramic matrix composites *Mater Res* **10** 147-51
[24] Chae H G, Minus M L, Rasheed A and Kumar S 2007 Stabilization and carbonization of gel spun polyacrylonitrile/single wall carbon nanotube composite fibers *Polymer* **48** 3781-9
[25] Jing M, Wang C G, Zhu B, Wang Y X, Gao X P and Chen W N 2008 Effects of preoxidation and carbonization technologies on tensile strength of PAN-based carbon fiber *J Appl Polym Sci* **108** 1259-64
[26] Ko T H 1991 Influence of continuous stabilization on the physical properties and microstructure of PAN-based carbon fibers *J Appl Polym Sci* **42** 1949-57
[27] Kim S-Y, Kim S Y, Lee S, Jo S, Im Y-H and Lee H-S 2015 Microwave plasma carbonization for the fabrication of polyacrylonitrile-based carbon fiber *Polymer* **56** 590-5

IOP Conf. Series: Materials Science and Engineering **634** (2019) 012042 doi:10.1088/1757-899X/634/1/012042

Effects of dicumyl peroxide on cell formation of biopolymer blend-activated carbon composite foams

D Aussawasathien[1,3] K Ketkul[2], K Hrimchum[1] and P Threepopnatkul[2]

[1]Advanced Polymer Technology Research Group, National Metal and Materials Technology Center, 114 Thailand Science Park, Phahonyothin Rd., Khong Nueng, Khlong Luang, Pathum Thani 12120, Thailand
[2]Department of Materials Science and Engineering, Faculty of Engineering and Industrial Technology, Silpakorn University, Nakhon Pathom 73000, Thailand

E-mail: daruneea@mtec.or.th

Abstract. Immiscible biopolymer blends of polylactic acid (PLA) and polybutylene succinate (PBS) with a weight ratio of 80/20 were compatibilized by different concentrations (0, 0.1, 0.3 and 0.5 parts per hundred resin (phr)) of dicumyl peroxide (DCP) in the presence of 3 phr of activated carbon (AC). Influences of DCP on the melt rheology at foaming temperature, cell formation and morphology and foam properties of PLA-PBS-AC composites using 5 phr of azodicarbonamide (ADC) as a blowing agent were reported. It was found that the melt viscosity, relating to the melt flow index (MFI) of PLA-PBS-AC composite foams increased with increasing DCP loadings due to more compatibility between PLA and PBS. Closed cells with thick cell walls were obtained with the addition of DCP. Also, the cell size and cell density tended to decrease by DCP adding since lower cell expansion and coalescence occurred at higher polymer surface tension. The crystallinity (X_c) and thermal properties of PLA and PBS in composite foams were almost unchanged when DCP was filled in the system. The PLA-PBS-AC composite foam having 0.3 phr of DCP provided the lowest foam density (ρ_f) and the highest void fraction (V_f) of 0.5766 g cm^{-3} and 54% respectively.

1. Introduction

At these days, polymer foam packaging is seriously susceptible to environmental and disposal problems since it is not feasibly available to economically recycling process owing to high cost of the waste management. Furthermore, most of based polymers are not biodegradable, resulting in a major concern in terms of sustainability, ecology and recyclability, especially for the single-use packaging. Therefore, a requirement for the utilization of bio-based and/ or biodegradable materials in foam packaging has been extensively gained attention in order to efficiently mitigate the composting issue, nevertheless maintains all existing polymer foam properties and performances as specified. PLA is classified as a bio-based and biodegradable aliphatic polyester which has great potential to be utilized as a replacement material for typical polymers, derived from petroleum-based substances. However, it is too hard and brittle for use in packaging applications. Thus, its toughness is usually developed by mixing with PBS since PBS is another biodegradable polymer, having excellent toughness and high chemical resistance [1-3]. In addition, the compatibility between PLA and another polymer is an absolutely crucial issue to obtain the required level of properties and performances from their blends. Therefore, the compatibilizer was usually consumed in the melt blending to ameliorate the properties of polymer blends such as melt rheology, morphology and physical and mechanical properties. In the previous work, triphenyl phosphite was utilized as a reactive compatibilizer in PLA/poly (ε-caprolactone) (PCL) blends which obviously exhibited the improvement of the elasticity of blends [4]. Moreover, the PLA/poly(butylene adipate-co-terephthalate) (PBAT) blend was prepared with the filling of glycidyl methacrylate (GMA) as a compatibilizer for the reactive blending. The obtained results revealed that the toughness of such blends increased together with the improvement of other properties such as melt rheology, topology and miscibility [5]. The reactive blends of PLA and PBS were fabricated using DCP performed as a radical initiator in the extrusion process [6]. The DCP loadings of 0.2 and 0.3 phr exhibited the appropriate beneficial on mechanical properties. Furthermore, the enhancement of PLA and poly (ethylene terephthalate glycol-modified) (PETG) compatibility was reported by employing epoxy-based compatibilizer [7]. The cell structure of PLA/PETG foams was great uniformity with small pores and large expansion ratio at a suitable compatibilizer dosage in the blends.

Hence, the objective of this study was to prepare composite foams containing PLA, PBS and AC with the addition of DPC via a reactive melt extrusion process. The effect of DCP on melt rheology through MFI measurement, relating to the bubble creation and morphology and foam properties of composite foams was investigated. The X_c as well as thermal properties of PLA and PBS in composite foams were also determined.

2. Experimental procedures

2.1. Raw materials
PLA (2003D) and PBS (FZ91PM) with the MFI of 6 and 5 g (10 min)$^{-1}$ respectively were supplied from BC Polymer Marketing Co., Ltd., Thailand. AC (CGC-11A 200C) was purchased from Gigantic Carbon Co., Ltd., Thailand, possessing average surface area and particle size of about 1050 m^2 g^{-1} and 74 μm respectively. ADC (Sunfoam SB00P) with particle size around 10 μm obtained from Lautan Luas (Thailand) Co., Ltd., Thailand was used as an exothermic blowing agent in the foaming process. An antioxidant (Annox 20) and chain extender (Joncryl ADR4370) received from Optimal Tech Co., Ltd. and Merit Solution Co., Ltd, Thailand respectively were also filled in the system. DCP received from Sigma-Aldrich was used as a compatibilizing agent.

2.2. Melt mixing and sheet extrusion processes
Polymer composites containing various DCP dosages at 0, 0.1, 0.3 and 0.5 phr were parepared at selected PLA:PBS ratio and ADC and AC concentrations of 80:20 wt%, 5 phr and 3 phr respectively, based on our previous work [8]. The amount of antioxidant and chain extender were maintained at 0.5 and 0.7 phr respectively. Aforementioned components were dry blended in a plastic bag prior to the melt mixing process using twin screw extruder (LTE20-32, Labtech Engineering Co., Ltd., Thailand). The screw diameter, screw length and screw L/D ratio were 20 mm, 64 cm and 32 respectively. The mixing screw and die temperatures were set up in the range of 130-180°C. The speeds of feeding and mixing screws were 12 and 80 rpm respectively. The prepared composite was dried at 80°C for 12 h until the moisture content was about 0.02% and then kept in an aluminium packaging. Afterward, the composite foam sheet was fabricated using a single screw extruder with L/D ratio of 25 (HAAKE Rheomex, LMS Instruments Co., Ltd., Thailand) at 220°C with a screw speed of 200 rpm. The thickness of composite foam sheet was controlled by using rollers with a rolling speed of 35 rpm at room temperature.

2.3. Characterizations
The ρ_f of 1 cm × 1 cm foam sheets was analyzed by helium gas displacement method using an automatic gas pycnometer (Ultrapyc 1200e, Science Engineering and Service Co., Ltd., Thailand) at 17.0 psi and 30°C. The V_f was determined by using equation (1) from the average density of PLA and PBS resins (ρ_p) and the density of foam samples (ρ_f).

$$V_f = 100 \text{ x} \left[1 - \frac{\rho_f}{\rho_p} \right] \quad (1)$$

The cell topology was detected by using a Schottky field-emission scanning electron microscope (SEM, SU5000, Coax Group Corp. Ltd., Thailand). The specimen was immersed in liquid nitrogen for ease of fractured surface preparation. Then, the platinum was coated on the sample surface by sputtering deposition. The Image J software was utilized to determine the bubble dimension and cell density based on SEM micrographs. The cell density per unit area of composite foams was calculated by using equation (2).

$$\text{Cell density} = \left[\frac{n}{A} \right]^{\frac{3}{2}} \text{x} \left[\frac{\rho_f}{\rho_p} \right] \quad (2)$$

where A is the area of the SEM image (cm^2), n is the quantity of cells in the SEM picture, ρ_p is the average density of PLA and PBS resins and ρ_f is the density of the composite foam. The MFI was characterized using a melt flow indexer (Davenport model 10, Lloyd Instruments, Thailand), following ASTM D1238 at 220°C with an applied load of 2.16 kgf. The X_c and thermal properties of PLA, PBS in composites were obtained from thermograms using a differential scanning calorimeter (DSC, Mettler Toledo (Thailand) Ltd.) under N_2 flow, The sample was firstly heated from -60 to 200°C at a scanning rate of 10°C min^{-1} to eliminate the melt history. After that, the sample was cooled down at 10°C min^{-1} from 200 to -60°C and hold at -60°C for 5 min before the 2nd scanning to 200°C at 10°C min^{-1}. The X_c was derived from the melting enthalpy of PLA ($\Delta H_{m, PLA}$ = 93.7 J g^{-1}) and PBS ($\Delta H_{m, PBS}$ = 110.3 J g^{-1}) at 100% crystalline and the crystallization enthalpy from the melting peak (ΔH_c) (see equation (3)).

$$Xc\ (\%) = \frac{\Delta Hc}{\Delta Hm} x100 \quad (3)$$

3. Results and discussion

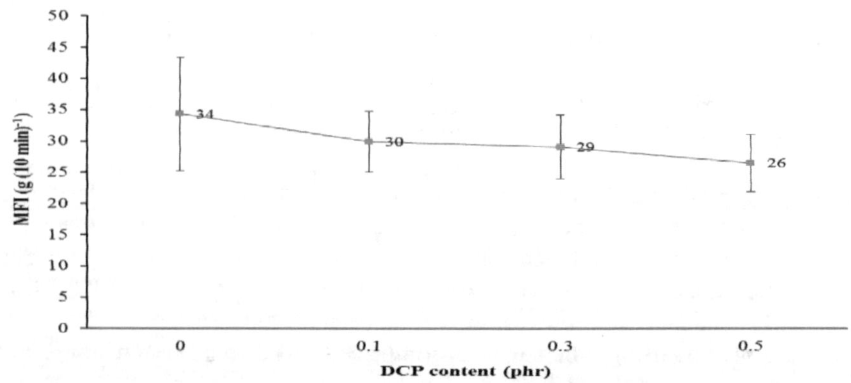

Figure 1. MFIs of composite foams at a variety of DCP concentrations.

The melt rheology of composite foams was indirectly analyzed from the MFI measurement at 220°C as shown in figure 1. The MFI of composite foams tended to decrease from 34 to 26 g (10 min)$^{-1}$) with increasing DCP contents from 0 to 0.5 phr, suggesting that the melt viscosity of polymer blends increased due to the partial enhancement of PLA and PBS compatibility. In the melt mixing process, DCP acted as a radical initiator, resulting in the formation of crosslinking and /or branching structures. Such structures performed as nucleating sites for the generation of PLA-PBS copolymers which became a compatibilizer for the PLA and PBS fractions and, thereby, attenuated the compatibility of both polymers [6]. Moreover, the variation of MFIs had a trend to decline (see error bars) with the addition of DCP in composite foams, compared with the one having 0 phr of DCP. The gas pressure generated during the bubble creation did not greatly affect the polymer melt flow when the miscibility between PLA and PBS improved.

Figures 2-5 depict the glass transition temperature (T_g), crystallization temperature (T_c), melting temperature (T_m) and X_c, of PLA and PBS in composites at various DCP loadings. The $T_{g, PBS}$ was lower than -60°C which could not be obtained because of the instrument limitation. In addition, the portion of PBS in composite foams was quite low (20 wt%), thus the effect of $T_{g, PBS}$ on the formation of cell foams could be negligible. The $T_{g, PLA}$ was around 56-58°C, almost unchanged with the DCP variation as presented in figure 2. It was expected that the T_g of both polymers did not influence the bubble formation since the foaming process was performed at 220°C after polymers were melted. The T_m of PLA (148-150°C) and PBS (112-113°C) also slightly varied with increasing DCP dosages, implying that the amount of DCP inconsiderably influenced the crystallization behavior of both polymers (see figure 3). The cold crystallization temperature of PLA ($T_{cc, PLA}$) and crystallization temperature of PBS ($T_{c, PBS}$) in the composite foam with no DCP were 120°C and 89°C respectively as indicated in figure 4. The $T_{cc, PLA}$ and $T_{c, PBS}$ of composite foams were somewhat constant with increasing DCP concentrations. The $T_{cc, PLA}$ was higher than $T_{c, PBS}$, thus the crystallization of PLA occurred before that of PBS, meaning that the crystallization of PLA had more effects on cell morphology than that of PBS did. However, the $X_{c, PLA}$ insignificantly enhanced as the DCP dosage was increased (see figure 5) since the formation of bubbles might impede the arrangement of polymer chains to be an ordered structure during polymer solidification. Hence, the $X_{c, PLA}$ might have lower impacts on bubble growth and coalescence than the polymer melt viscosity did. It should also be noted that the crystallization of polymers assists to stabilize the cell structure, nonetheless high crystallinity may obstruct the bubble expansion as a result of high proportion of hard segments in polymers [9-12].

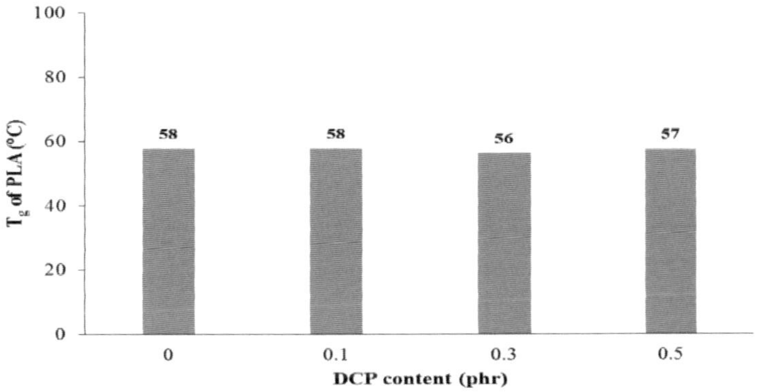

Figure 2. $T_{g, PLA}$ in PLA-PBS-AC composites at different DCP loadings.

Figures 6(a)-6(d) show SEM micrographs and cell size distributions of PLA-PBS-AC composite foams at different DCP dosages. Closed cells with thicker cell walls were observed by DCP addition because the bubble coalescence and expansion hardly happened at high polymer surface tension [13]. The average cell size of the composite foam containing no DCP was around 120-200 μm with narrower cell size distribution than that of composite foams having 0.1, 0.3 and 0.5 phr of DCP. With the addition of DCP, the average bubble size of composite foams had a trend to decrease and exhibited a wide range of variation. The composite foam possessing 0.3 phr of DCP had the broadest cell size distribution in the range of 100-120 μm to 220-240 μm. The cell density of composite foams decreased ((1.6-0.6) x 10^6 cells cm^{-3}) with DCP filling in the system from 0 to 0.5 phr, as presented in figure 7, owing to the enhancement of cell wall thickness. However, the cell density inconsiderably varied with increasing the amount of DCP from 0.1 to 0.5 phr.

IOP Conf. Series: Materials Science and Engineering **634** (2019) 012042 doi:10.1088/1757-899X/634/1/012042

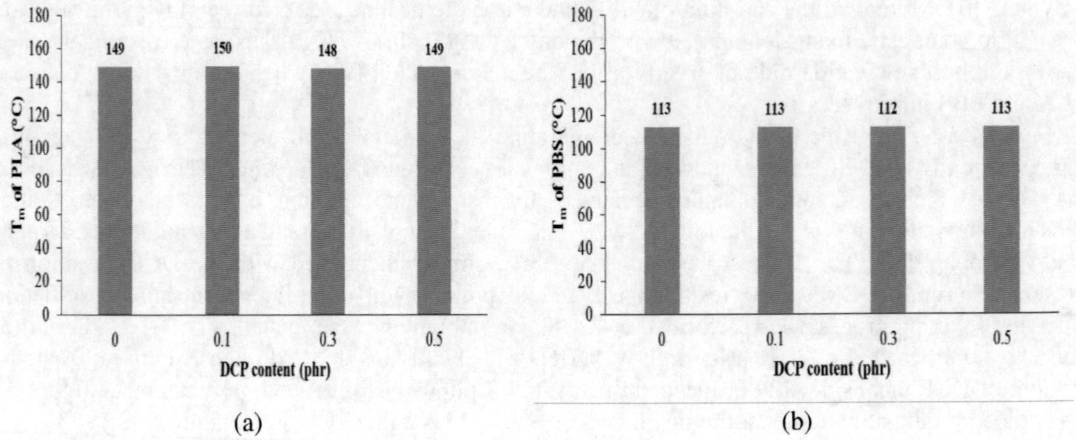

Figure 3. $T_{m,\ PLA}$ (a) and $T_{m,\ PBS}$ (b) in PLA-PBS-AC composites at various DCP dosages.

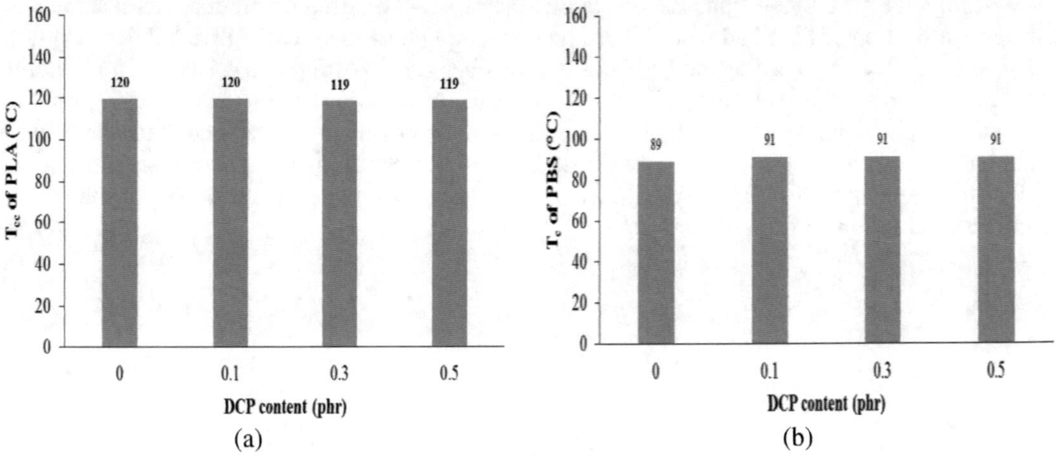

Figure 4. $T_{cc,\ PLA}$ (a) and $T_{c,\ PBS}$ (b) in PLA-PBS-AC composites at different DCP contents.

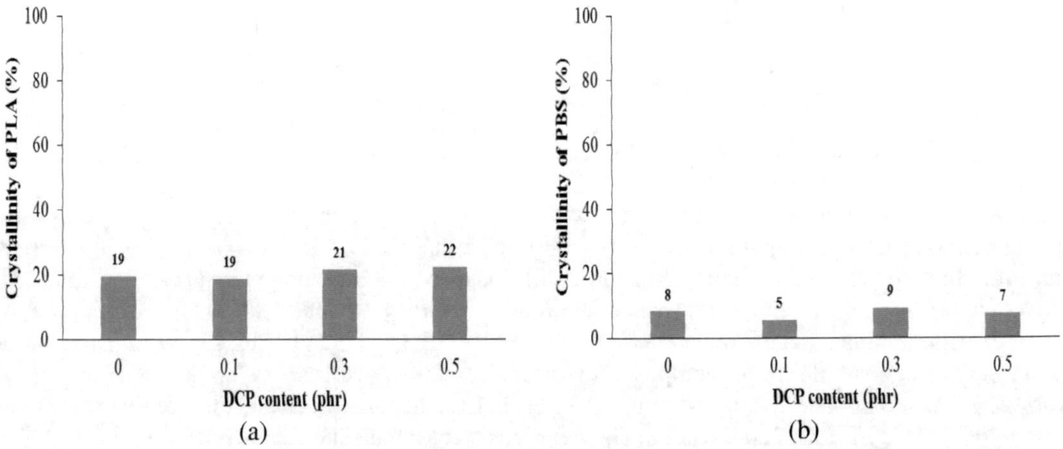

Figure 5. $X_{c,\ PLA}$ (a) and $X_{c,\ PBS}$ (b) in PLA-PBS-AC composites at different DCP concentrations.

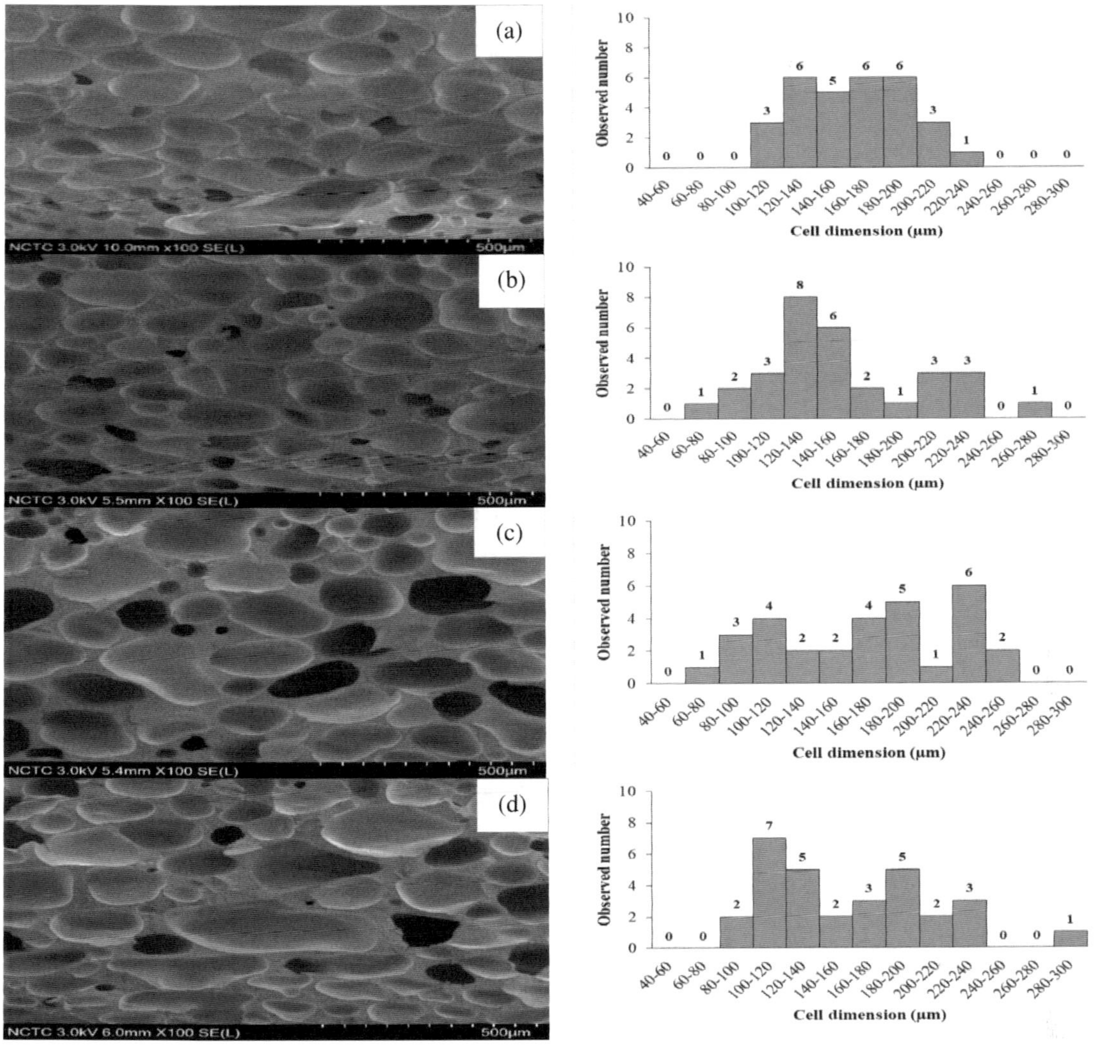

Figure 6. SEM images and cell size distributions of PLA-PBS-AC composite foams at different DCP loadings (phr): (a) 0, (b) 0.1, (c) 0.3 and (d) 0.5.

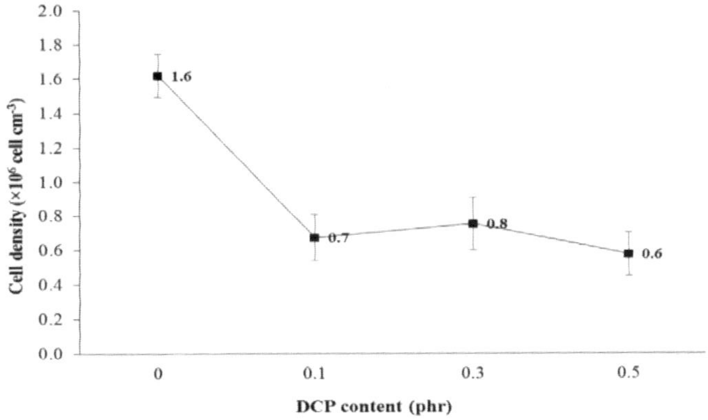

Figure 7. Bubble densities of PLA-PBS-AC composite foams at different amounts of DCP.

Figures 8 (a) and 8(b) present that the ρ_f of composite foams tended to decrease from 0.6368 to 0.5766 g cm^{-3} while the V_f enhanced from 49 to 54% with the addition of DCP in the range of 0 to 0.5 phr. The composite foam containing 3.0 phr of DCP exhibited the lowest ρ_f of about 0.5766 g cm^{-3} and the highest V_f of approximately 54% due to the balance of polymer melt viscosity and $X_{c, PLA}$ effects on the formation of cell foams.

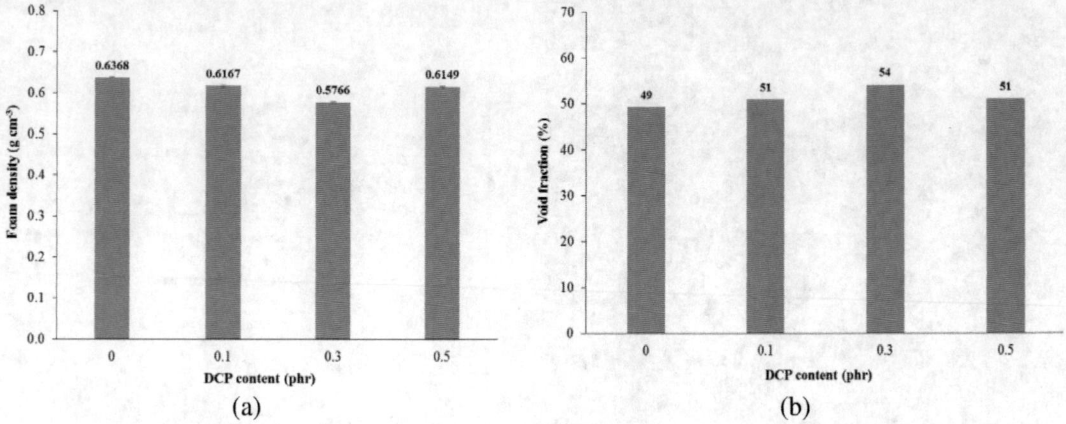

Figure 8. Foam densities (a) and void fractions (b) of composite foams at different contents of DCP.

4. Conclusions

The existence of DCP seemed to increase the melt viscosity of PLA-PBS-AC composite foams since more compatibility between PLA and PBS occurred via a reactive blend. DCP played a role as nucleating sites for the PLA-PBS copolymer formation, acting as a compatibilizing reagent for the PLA and PBS portions. The X_c and thermal properties of PLA and PBS in composites were scarcely altered with the addition of DCP, indicating that the cell creation and morphology were mainly affected by the polymer surface tension more than the polymer crystallinity. The balance of melt viscosity and $X_{c, PLA}$ is required to obtain optimum cell formation, cell dimension, cell density and foam properties. As a result, the composite foam containing the PLA/PBS proportion of 80/20 by weight, 5.0 phr of ADC, 3.0 phr of AC and 0.3 phr of DCP provided the lowest ρ_f and the highest V_f, compared with other DCP loadings.

Acknowledgments

The authors would like to thank National Metal and Materials Technology Center (MTEC) for the financial support under MTEC platform program (Project code: P1550631).

References

[1] Bhatia A, Gupta R K, Bhattacharya S N and Choi H J 2007 *Korea-Aust. Rheol. J.* **19** 125-31
[2] Harada M, Ohya T, Iida K, Hayashi H, Hirano K and Fukuda H 2007 *J. Appl. Polym. Sci.* **106** 1813-20
[3] Zhou J, Yao Z, Zhou C, Wei D and Li S 2014 *J. Appl. Polym. Sci.* **131** 40773
[4] Wang L, Ma W, Gross R A and McCarthy S P 1998 *Polym. Degrad. Stab.* **59** 161-8
[5] Zhang N, Wang Q, Re J and Wang L 2009 *J. Mater. Sci.* **44** 250-6
[6] Ji D, Liu Z, Lan X, Wu F, Xie B and Yang M 2014 *J. Appl. Polym. Sci.* doi: 10.1002/APP.39580
[7] Wang X, Liu W, Zhou H, Liu B, Li H, Du Z and Zhang C 2013 *Polymer* **54** 5839-51
[8] Ketkul K, Threepopnatkul P, Aussawasathien D and Hrimchum K 2017 *Key Eng. Mater.* **751** 344-49
[9] Kang J W, Kim J M, Kim M S, Kim Y H and Kim W N 2009 *Macromol. Res.* **17** 856-62
[10] Wang X X, Kumar V and Li W 2012 *Cell. Polym.* **31** 1-18

[11] Corre Y M, Maazouz A, Duchet J and Reigniera J 2011 *J. Supercrit. Fluids.* **58** 177-88
[12] Zhai W T, Ko Y, Zhu W L, Wong A and Park C B, 2009 *Int. J. Mol. Sci.* **12** 5381-97
[13] Liao X, Nawaby A V and Whitfield P S 2010 *Polym. Int.* **59** 1709-18

IOP Conf. Series: Materials Science and Engineering **634** (2019) 012043 doi:10.1088/1757-899X/634/1/012043

In-line rheological properties of rubber toughened Wood Polymer Composites

V Mazzanti[1,3], L Malagutti[1], M Blanchard[1], S Yi[2] and F Mollica[1]

[1]Department of Engineering, University of Ferrara, Via Saragat 1, 44122 Ferrara, Italy
[2]Department of Wood Science and Engineering, Northeast Forestry University, China

[3]E-mail: valentina.mazzanti@unife.it

Abstract. Wood Polymer Composites are materials that are more and more used and studied by both academic and industrial research. Although such materials are successfully substituting natural wood in many relevant applications, still some of their drawbacks limit their widespread diffusion. One of these is their brittleness, which can be limited by blending with a toughening agent. In this study we investigated the rheological properties, in terms of the flow curve and the wall slip characteristics, of a toughened Wood Polymer Composite. Testing was performed using an in-line slit rheometer attached directly to a single screw extruder. The material used for the experiments was a blend of polypropylene-based wood polymer composite compound, containing 35 wt.% of wood flour and a suitable thermoplastic elastomer used as a toughening agent. The results showed a shear thinning behaviour and a significant slip velocity of the material that is beneficial during processing.

1. Introduction

Wood Polymer Composites (WPCs) are materials composed of thermoplastic polymers and wood flour. They can be used as substitutes for wood, thanks to their better resistance to moisture absorption and a reduced environmental impact [1,2]. For these reasons, important applications are in outdoor structures such as decking, fencing and flooring, possibly close to wet environments. Another important benefit of these materials is that increasing the wood content, the cost of the compound decreases [3]. On the other hand, one of their main drawbacks is their brittleness, which is especially relevant at high wood content and this is often at the origin of defects, surface failure and fractures in the final component after a relatively short period of usage.

Usually, reduced toughness problems are tackled by adding a thermoplastic elastomer in blend. On the other hand, the addition of another component into the compound could introduce complexity in transformation process and raise numerous additional issues. In order to better control the many processing variables in such a way that a material with adequate mechanical properties can be obtained, it is important to characterize the composite from the rheological point of view [4].

An in-line viscometer can be used to avoid problems that often affect off-line viscometry of this type of materials [5,6]. For example, rotational rheometry in oscillatory mode in conjunction with the Cox–Merz rule is often ineffective at the processing temperatures [6,7], also due to an insufficiently large linear viscoelastic region (LVR), while the capillary rheometer is frequently difficult to use due to the WPC high viscosity, especially at high filler volume fractions [3].

The aim of this study is thus to characterize the rheology of a rubber toughened polypropylene based-WPC at 185°C for optimizing processing conditions.

2. Material

A commercial wood-polypropylene composite (PPCO68/BZ from Plasticwood, Oppeano (VR), Italy) has been mixed with Santoprene 201-55, an EPDM based thermoplastic elastomer that was purchased from ExxonMobil. Mixing was done in equal proportions, as to obtain a 35 wt.% wood content in the final composite and the compound was been obtained the same extruder used for the in-line rheological measurements, which is described next.

3. Methods

3.1. Measurement apparatus

For measuring the flow properties, an in-line slit viscometer has been used. This device consisted of an instrumented slit die attached to a single screw extruder (P.R.T. SERVICE & INNOVATION s.r.l., Sant'Agostino (FE), Italy) (Fig.1). More design details can be found in [3].

In this study, three slit thicknesses (1.33, 1.95, 4.07 mm) have been used in order to perform the Mooney wall slip correction procedure [8], which consists in performing measurements at different slit heights but at constant wall shear stress values, so that the slip contribution to flow rate could be evaluated and subtracted out. Three flush mounted pressure transducers (GEFRAN M32 type, with 100 bar full scale) allow pressure drop measurements along the die. The slit die is equipped with a thermostat for temperature control, and the polymer temperature is also checked with thermocouples. Analogic signals are conditioned and interfaced to a personal computer using the NI 9237 module for pressure and the NI 9211 module for temperature measurement (National Instruments).

3.2. Experimental protocol

In order to reduce the moisture content, the WPC pellets have been dried at 80 °C for 24 hours in an oven before testing. A uniform temperature distribution of 185°C has been fixed along the extruder as well as the slit die. The volumetric flow rate Q, controlled through the extruder screw speed, has been calculated by dividing the mass flow rate $\dot{m}$ by the density of the fluid ρ at the tested temperature.

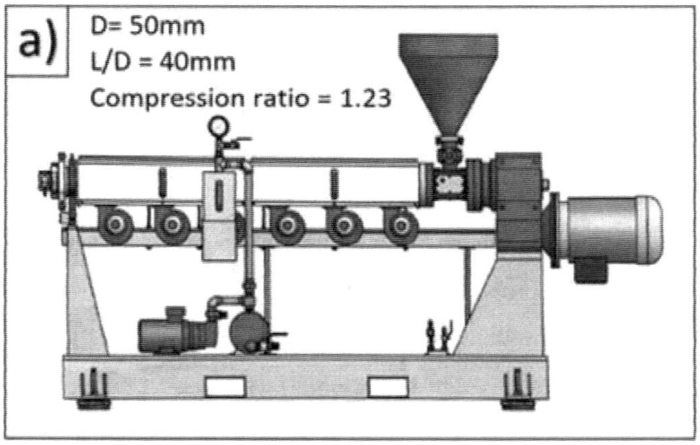

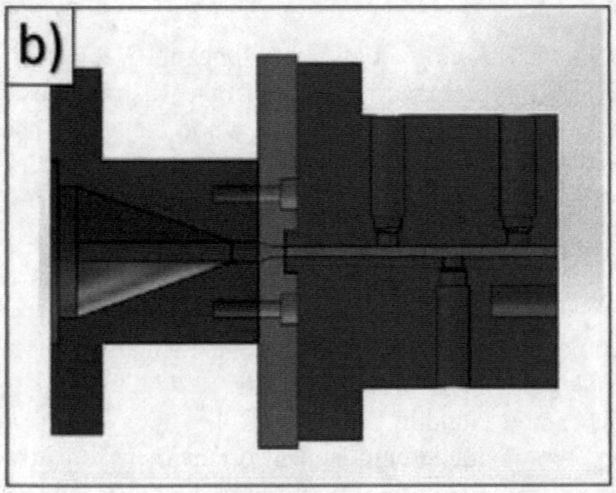

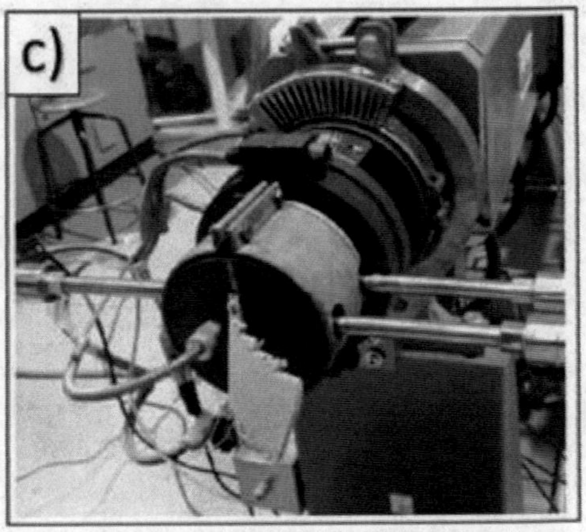

Figure 1. a) Single screw extruder, b) in-line slit viscometer scheme and c) a picture of the measurement apparatus.

4. Results

Pressure profiles measured by the transducers are linear and the pressure gradient p can be obtained from the slope of the straight line. The value of the shear stress τ_w is

$$\tau_w = -\frac{H}{2} grad\ p \tag{1}$$

where H is the slit thickness. The apparent shear rate, i.e. the shear rate the fluid would have if it were Newtonian, is defined as

$$\dot{\gamma}_{app} = \frac{6Q}{WH^2} \tag{2}$$

where Q is the volumetric flow rate and W is the width of the slit die.

In Fig. 2, τ_w vs. $\dot{\gamma}_{app}$ is shown for various slit heights. The curves can be fitted with the following power law

$$\tau_w = C\dot{\gamma}_{app}{}^n \tag{3}$$

where C and n are fitting parameters (Tab.1). It is interesting to note that both fitting parameters are almost independent of the die height.

At all slit thicknesses tested, τ_w increases with $\dot{\gamma}_{app}$ and the curves do not superimpose. Moreover, the curves place themselves from left to right in going from the largest to the smallest slit height, thus the Mooney procedure can be successful for correcting wall slip effects. This procedure works by decomposing $\dot{\gamma}_{app}$ into two contributions

$$\dot{\gamma}_{app} = \frac{6v_s}{H} + \dot{\gamma}_{app_{no\ slip}} \tag{4}$$

where v_s is the wall slip velocity and $\dot{\gamma}_{app_{no\ slip}}$ is the apparent shear rate without the contribution of wall slip.

The results obtained from the Mooney plot are shown in Fig. 3. The curves are not linear, similarly to what is reported in [5] for the case of the rheological properties of PVC. Nevertheless, an estimate of the slope can be inferred from all curves and the intercepts are positive for all levels of shear stress. In Tab. 2, the apparent shear rate corrected for slip and the wall slip velocity for the four levels of shear stress are listed. Interestingly, the wall slip velocity increases significantly with the wall shear stress.

Next, the apparent shear rate has been corrected for non-Newtonian effects using the Rabinowitsch correction [9]. The viscosity η versus true shear rate plot is pictured in Fig. 4.

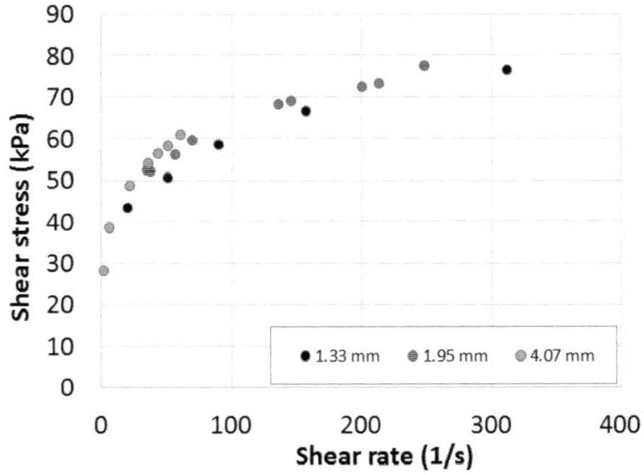

Figure 2. Shear stress vs. apparent shear rate for various slit heights.

Table 1. Fitting parameters for the power law model.

Die height (mm)	C (**kPa s^n^**)	n
1.33	22.14	0.21
1.95	26.13	0.19
4.07	23.27	0.23

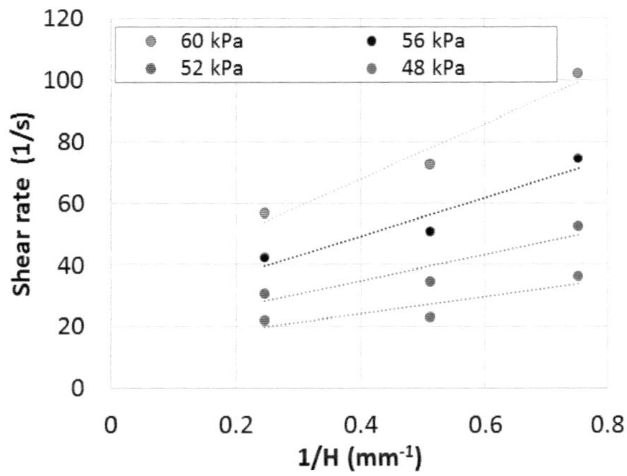

Figure 3. Apparent shear rate versus the reciprocal of slit height for four levels of shear stress.

Table 2. Apparent shear rate corrected for the slip and wall slip velocity for the four levels of shear stress.

τ_w (kPa)	$\dot{\gamma}_{app_{no\ slip}}$ (1/s)	v_s (mm/s)
48	12.95	4.69
52	17.93	7.09
56	24.24	10.43
60	32.13	14.90

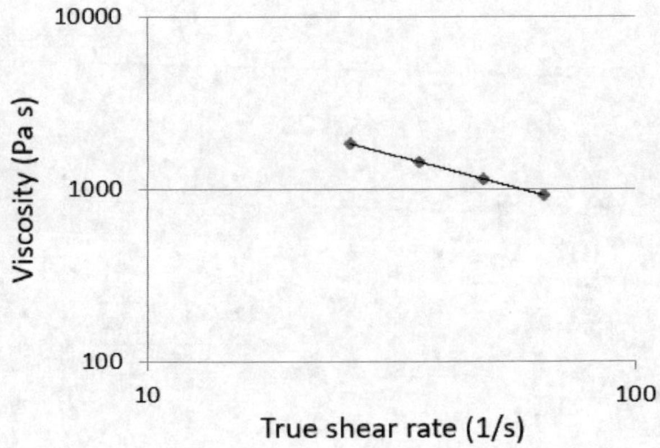

Figure 4. Shear viscosity versus true shear rate.

5. Discussion and conclusion

Mixing a thermoplastic elastomer with a wood plastic composite allows to obtain a material whose appearance is that of wood, but varying the content of elastomer in its formulation, the properties may lie in between those of a toughened plastic up to a very flexible material. This opens interesting applications in several fields, such as coatings that may resist impact, without affecting the usual WPC properties, like moisture resistance coupled to environmental friendliness.

In fact, the material at the molten state exhibits typical WPC rheological properties, such as shear thinning behavior and wall slip [2,3,10,11]. In particular, viscosity can be modeled through a power law defined as

$$\eta = K\dot{\gamma}^{(m-1)} \quad (5)$$

where K = 21.5 kPa s^m is the consistency and m = 0.25 is the power law index.

As a consequence, despite the wood content of the material is only about 35% by mass, thus typical of an injection molding grade, the rubber toughened WPC possesses a relatively high viscosity, more typical of an extrusion grade. Notice that, for this material, viscosity can hardly be lowered by significantly increasing the processing temperature. The wood flour that is used as a filler, in fact, undergoes thermo-oxidative degradation when the material is kept for too long times at temperatures close to 195°C – 200°C. Processing, though, is made easier by the relatively high slip velocity. This is possible thanks to the high content of external lubricants that are inside the Santoprene formulation [12]. The presence of external lubricants coupled to solid filler may induce pressure dependent wall slip phenomena [13] that may complicate the modeling of the transformation process.

References

[1] Adekomaya O, Jamiru T, Sadiku R, Huan Z 2016 *J. Reinf. Plast.* **35** 3
[2] Mazzanti V and Mollica F 2019 *Polym. Comp.* **40** E169
[3] Mazzanti V and Mollica F 2017 *J. Polym. Environ.* **25** 1044
[4] Li T and Wolcott M P 2004 *Compos. Part A: Appl. Sci. Manuf.* **35** 303
[5] Glomsaker T, Einar L, Thorsteinsen H P 2004 Polym. Eng. Sci. **41** 2231
[6] Hristov V, Takacs E, Vlachopoulos 2006 *J. Polym. Eng. Sci.* **46** 1204
[7] Rauwendaal C and Fernandez F 1985 *Polym. Eng. Sci.* **25** 796
[8] Mooney M 1931 *J. Rheology* **2** 221
[9] Barnes HA Hutton JF Walters K 1989 An introduction to rheology *Elsevier Science* (Amsterdam Netherland) chapter 2 pp 32
[10] Kloziński A and Jakubowska P 2019 *J. Polym. Eng.* **39** 422
[11] Barczewski M, Mysiukiewicz O, Szulc J, Kloziński A 2019 *J App. Polym. Sci.* **136** 47651

IOP Conf. Series: Materials Science and Engineering **634** (2019) 012043 doi:10.1088/1757-899X/634/1/012043

[12] Leblanc J L 2009 *Rheol Acta* **46** 1013
[13] Mazzanti V and Mollica F 2017 *J. Non Newt. Fluid Mech.* **247** 178

IOP Conf. Series: Materials Science and Engineering **634** (2019) 012044 doi:10.1088/1757-899X/634/1/012044

Study on preparation and properties of graphene reinforced epoxy resin composites

Y H Fan[1,2], S W Yu[1], H M Wang[1], Y H Yao[1], Y Wang[1] and C H Wang[1]

[1]CSIC No. 12 Research Institute, Xingping, China

E-mail: csic12@163.com

Abstract. In order to study the preparation methods of graphene reinforced epoxy resin composites and analyze the influence of graphene content on the viscosity and mechanical properties of the resin, the surface of graphene oxide was modified to make graphene oxide disperse uniformly in the matrix resin, and then composite the modified graphite oxide and resin. Three groups of graphene oxide reinforced epoxy resin composites with graphene oxide content of 0.3wt%, 0.5wt% and 0.8wt% were prepared. The results show that the modified graphene oxide keeps good dispersion uniformity in the resin matrix. With the increase of graphene oxide content, the viscosity of the composite decreases gradually. The strength of the composite is better than that of the pure resin after adding graphene oxide. When the graphene content is 0.3%, the tensile strength is the highest, which is about 16% higher than that of pure resin. With the increase of graphene content, the tensile strength decreases.

1. Background introduction

Polymer resin matrix composite is a kind of composite material which combines macromolecule resin and surface modified reinforced phase according to a certain proportion and process method to obtain excellent performance, anti-corrosion, wear resistance and other properties. It is mainly used in functional structural parts or further extended to downstream products, and can serve aviation industry, aerospace, electronics, transportation, ocean and other fields.

Graphene is a two-dimensional material with hexagonal honeycomb structure, which is composed of carbon atoms and SP2 hybrid orbital. Graphene's special monoatomic layer structure determines its rich and excellent physical properties, such as excellent barrier, anti-corrosion and wear resistance. Graphene with flexible lamellar structure can obstruct erosion, penetration and spread of corrosive media effectively [1-5]. Compound resin matrix as additive phase can improve the strength and corrosion resistance of matrix [6-8]. Compared with graphene, graphene oxide has a large number of hydroxyl, carboxyl and epoxy functional groups on its surface, which show strong hydrophilicity. On the other hand, these functional groups also provide an active structural basis for surface modification. After organic modification, graphene oxide can be stably dispersed in organic solvents and macromolecule resins [9-11].

Graphene oxide was used in this paper, using the active hydroxyl, carboxyl and epoxy groups on the surface and edge of graphene oxide, There will be a covalent reaction between graphene oxide and organic modifiers. The novelty and significance of this paper are that the mechanical properties of resin materials can be significantly improved by choosing suitable raw materials and preparation technology. In addition, this paper provides an important basic rule for the study of graphene reinforced resin matrix composites.

IOP Conf. Series: Materials Science and Engineering **634** (2019) 012044 doi:10.1088/1757-899X/634/1/012044

2. Test scheme

Three groups of graphene reinforced resin composites were prepared. The ratio of graphene oxide to resin was 0.3 wt%, 0.5 wt% and 0.8 wt%, respectively. They were recorded as 0.3GO, 0.5 GO and 0.8 GO. By comparing the viscosities and mechanical properties of three groups of samples with those of pure resin, the mechanism of graphene oxide affecting the properties of resin was summarized.

3. Material preparation

3.1. Raw material

The raw materials used in this paper include graphene oxide powder, E51 epoxy resin, amino long chain modifier, alcohol, and curing agent. Among them, graphene oxide is an important raw material. Its specifications are as follows: thickness is about 1 nm, sheet diameter is about 0.2-10 micron, carbon content is about 45 wt.%.

3.2. Preparation of composite materials

Firstly, graphene oxide and amino long chain modifiers were dissolved in alcohol solution in a certain proportion, the modifier is excessive. The reaction temperature was raised 60°C for 1-2 hours. After the reaction, the reaction products were washed with alcohol and centrifugally collected to obtain modified graphene oxide. Then, modified graphene oxide was added to the resin, and mechanical stirring for 2-3 hours to obtain modified graphene oxide reinforced epoxy.

Because graphene oxide has nano-size and can play a nano-effect, a small amount of graphene oxide added to the resin can play a significant role in improving the performance, the premise is that graphene oxide must maintain good dispersion uniformity. According to the requirement of 0.3GO, 0.5GO and 0.8GO groups of samples, modified graphene oxide was added into epoxy resin, heated to 60-80 degrees Celsius, stirred for 3-5 hours, then graphene reinforced epoxy resin composite was obtained. Amine curing agent was selected. The curing temperature was room temperature and the curing time was 48 h.

4. Result analysis

4.1. Structural analysis of modified graphene oxide

Infrared spectrum analysis of graphene oxide is shown in figure 1 and infrared spectrum analysis of modified graphene oxide, prepared by our group , is shown in figure 2. It can be seen that modified graphene oxide has more functional groups. The stretching vibration absorption peaks of - CH3 and - CH2 were found at 2955 cm^{-1} and 2892 cm^{-1}, indicating that the organic modifiers reacted successfully with graphene oxide and grafted onto the surface of graphene oxide.

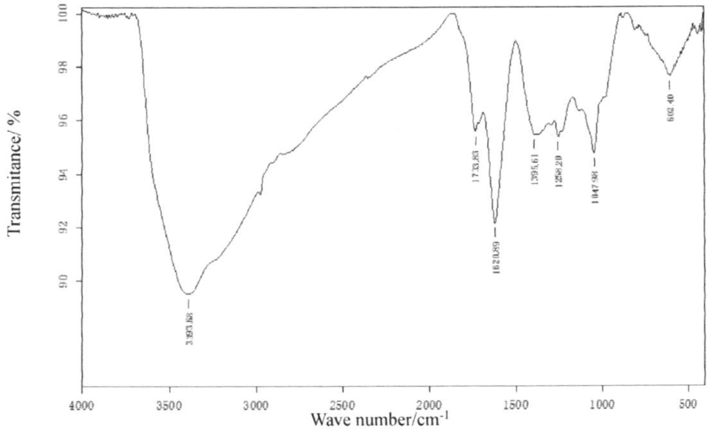

Figure 1. Infrared spectrum analysis of graphene oxide.

IOP Conf. Series: Materials Science and Engineering **634** (2019) 012044 doi:10.1088/1757-899X/634/1/012044

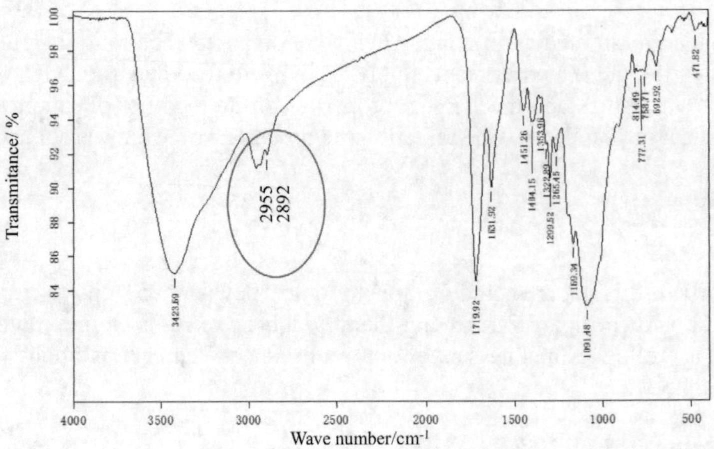

Figure 2. Infrared spectrum analysis of modified graphene oxide.

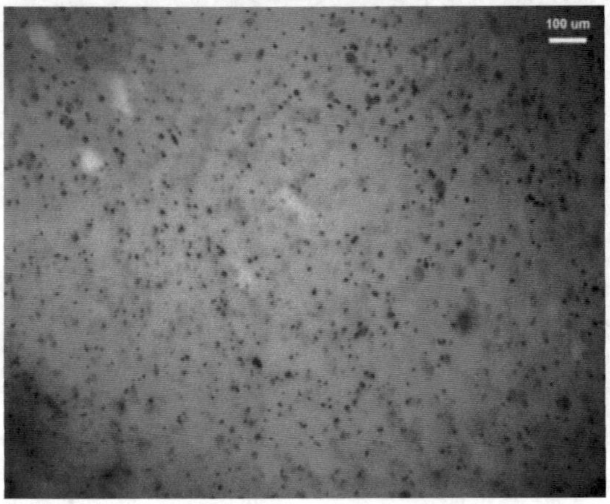

Figure 3. Distribution of modified graphene oxide in resin matrix (0.3GO).

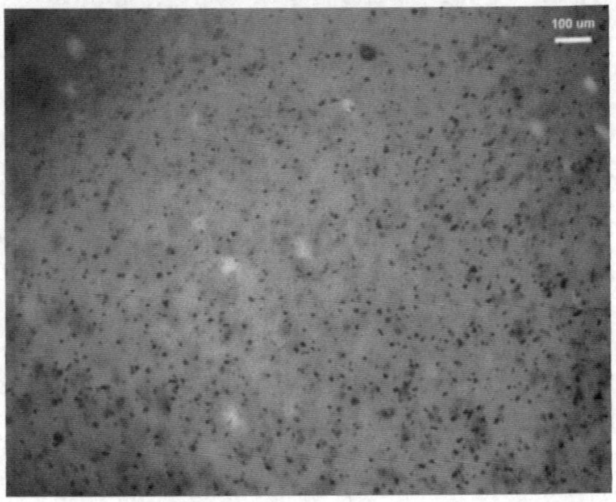

Figure 4. Distribution of modified graphene oxide in resin matrix (0.5GO).

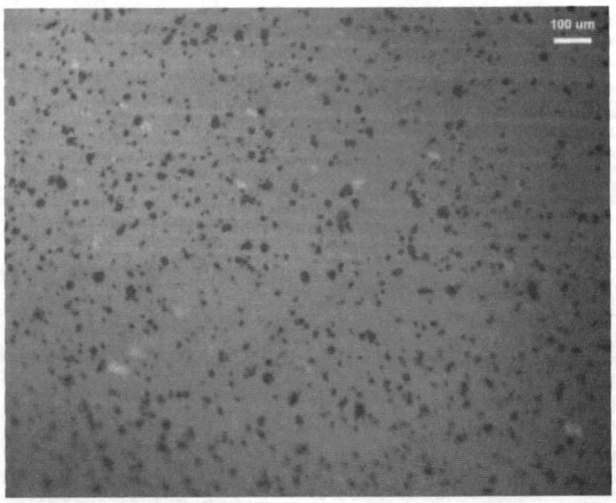

Figure 5. Distribution of modified graphene oxide in resin matrix (0.8GO).

4.2. Uniformity analysis

The uniformity of graphene reinforced epoxy resin composites with 0.3GO, 0.5GO and 0.8GO was analyzed. The distribution of graphene in epoxy resin was observed 100 times larger in three groups of composites under optical microscope. As shown in the figures 3-5, it can be seen that the distribution uniformity of modified graphene oxide in epoxy resin was good in three groups of samples. Macroscopic segregation is due to the structural changes of graphene oxide after organic modification, which has good similarity and compatibility with resins.

4.3. Viscosity analysis

The viscosities of epoxy resin, 0.3GO, 0.5GO and 0.8GO modified graphene oxide reinforced epoxy resin composites were tested by rotational rheometer. The effects of graphene content on the matrix viscosities were compared at 20 ~80°C, as shown in figure 6.

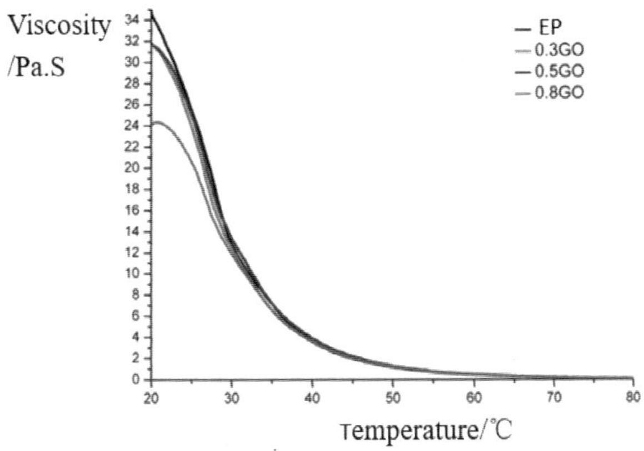

Figure 6. Viscosity analysis.

As can be seen from figure 6, with the increase of graphene content, the viscosity of resin materials tends to decrease at 20 ~30°C. The viscosity of 0.3GO and 0.5GO curves is close to each other, and the viscosity of 0.8GO decreases obviously. This is because the small size of graphene and the very thin lamellar structure of graphene play a lubricating role. When the temperature rises to about 30 ~80°C, the viscosity of the four groups of samples tends to be the same, because in the high temperature stage, the effect of temperature on the resin viscosity becomes more and more obvious, and finally the viscosity of each sample tends to be the same.

4.4. Mechanical properties analysis

Tensile specimens were prepared according to GB/T2567-2008 "Test methods for properties of resin casting body", the mechanical properties of epoxy resin, 0.3GO, 0.5GO and 0.8GO graphene reinforced epoxy resin composites were analyzed. The tensile strength was shown in table 1 and figure 7.

Table 1. Tensile strength analysis.

Sample	Tensile Strength /MPa
epoxy resin	65.7
0.3GO	76.1
0.5GO	69.6
0.8GO	66.7

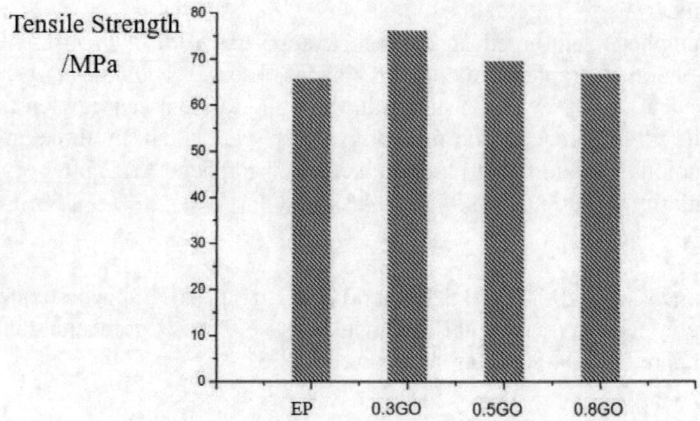

Figure 7. Tensile strength analysis contrast diagram.

From table 1 and figure 7, it can be seen that the properties of the composites with modified graphene oxide are better than those of epoxy resin. When the content of graphene is 0.3, the tensile strength reaches the highest, increasing by about 16%. With the increase of the content of graphene, the tensile strength decreases. When the content of graphene is 0.8, the tensile strength and epoxy resin are matched. This may be due to the excellent compatibility and interface structure between modified graphene oxide and resin matrix, which play a role as reinforcing phase in continuous phase resin and promote the improvement of mechanical properties [12,13]. The decrease of mechanical properties with the increase of graphene content may be attributed to the excessive introduction of modified graphite. The functional groups of olefins have an effect on the curing process of resins, resulting in changes in curing crosslinking parameters. On the other hand, with the increase of graphene oxide content, a small amount of local agglomeration will inevitably occur, which will also affect the performance [14].

5. Conclusion

- Infrared analysis showed that the surface-modified graphite oxide had more active organic functional groups.
- Microscopic optical analysis showed that the modified graphene oxide maintained good dispersion uniformity in the matrix.
- With the addition of modified graphene oxide, the viscosity of resin matrix decreases, and with the increase of graphene content, the viscosity decreases gradually.
- The tensile strength of modified graphene oxide reinforced epoxy resin composites is better than that of epoxy resin. When the content of graphene is 0.3, the tensile strength reaches the highest, increasing by about 16%. With the increase of graphene content, the tensile strength decreases.

References

[1] Hoffman A S 2012 Hydrogels for biomedical applications *Adv Drug Delivery Rev* **64 S** 18-23
[2] Geim A K and Novoselov K S 2007 The rise of grapheme *Nat Mater* **6** 83
[3] Lee C G, Wei X D, Kysar J W *et al* 2008 Measurement of the elastic properties and intrinsic strength of monolayer graphene *Sci* **321** 385-8
[4] Balandin A A,Ghosh S, Bao W *et al* 2008 Superior thermal conductivity of single-layer graphene *Nano Lett* **8** 902-7
[5] Ghosh S, Calizo I, Teweldebrhan D *et al* 2008 Extremely high thermal conductivity of graphene: Prospects for thermal management applications in nanoelectronic circuits *Appl Phys Lett* **92** 151911

[6] Chang K C, Hsu M H, Lu H *et al* 2014 Room-temperature cured hydrophobic epoxy/graphene composites as corrosion inhibitor for cold-rolled steel *Carbon* **66** 144-53
[7] Chang C H, Huang T C, Peng C W *et al* 2012 Novel anticorrosion coatings prepared from polyaniline/graphene composites *Carbon* **50** 5044-51
[8] Schluter B, Mulhaupt R and Kailer A 2014 Synthesis and tribological characterization of stable dispersions of thermally reduced graphite oxide *Tribol. Lett.* **53** 353-63
[9] He H, Riedl T, Lerf A *et al* 1996 Solid-state NMR studies of the structure of graphite oxide *J. Phys. Chem.* **100** 19954-8
[10] Lerf A, He H, Riedl T *et al* 1997 13C and 1H MAS NMR studies of graphite oxide and its chemically modified derivatives *Solid State Ionics* **101-103** Part 2 857-62
[11] He H, Klinowski J, Forster M *et al* 1998 A new structural model for graphite oxide *Chem Phys Let* **287** 53-6
[12] Wang X W, Zhang C A *et al* 2012 Enhanced of performance of biodegradable poly (butylene succinate)/graphene oxide nanocomposites via in situ polymerization *Langmuir* **28** 7091-5
[13] Zhang Q, Liu X *et al* 2012 Synthesis of acid-soluble graene and its use in producing a reduced graphene oxide-poly(benzobisoxazole)composite *J Mater Chem* **22** 12381-8
[14] Hu T, Zhao D L *et al* 2018 Preparation and mechanical properties of graphene oxide reinforced epoxy composites J *Beijing Univ Chem Technol (Nat Sci)* **45** 29-33

Effect of polymerized sewage on viscosity of polymer solution and countermeasures

W T Dong[1,3], K L Wang[1], Y Qiu[1] and W Peng[2]

[1]Key Laboratory of Enhanced Oil Recovery, Northeast Petroleum University, Ministry of Education, Daqing, Heilongjiang, 163318, China
[2]Bohai Petroleum Research Institute, Tianjin Branch, CNOOC, Tianjin, 300450, China

E-mail: dongwenting@cnooc.com.cn

Abstract. In order to better analyze the influence of sewage on the viscosity of polymer solution, this paper uses BROOKFIELD viscometer to determine the viscosity value of solution. The results show that the viscosity decreases rapidly with the increase of the of polymer solution concentration of metal cations. Especially, the concentration of Mg^{2+} and Ca^{2+} have a great influence on the viscosity and the small increase of the concentration of Mg^{2+} and Ca^{2+} can lead to the decrease of the viscosity of HPAM solution. Na_2CO_4 was chosen as the complexing agent to improve the viscosity of HPAM solution, and the viscosity increasing rate of HPAM solution with the concentration of 1500 mg/L reached 33.2%.

1. Introduction

Polymer solutions are widely used in major oil fields in China due to their good viscosity. The mechanism of enhanced oil recovery of polymer solutions is mainly through improving the oil-water mobility ratio and expanding the sweep volume [1-3]. Therefore, the viscosity is directly related to the oil displacement effect. The higher the viscosity of the polymer, the greater the recovery of crude oil [4,5]. At present, Daqing Oilfield mainly uses partially hydrolyzed polyacrylamide (HPAM) as a polymer agent for polymer flooding. The HPAM solution has a good viscosity-increasing effect, but its temperature and salt resistance is poor. In the field construction process of the oil field, sewage is usually used as the water for formulating the polymer, and the composition of the inorganic salt in the oil field sewage is very complicated, and ion concentration has a serious effect on the viscosity of polymer [6,7]. In order to investigate the effect of the polymerization wastewater on the viscosity of the polymer solution [8-10], the method of increasing the viscosity of HPAM solution by sewage polymerization method was studied. The experiment of measuring the viscosity of polymer solution by different sewage was carried out.

2. Experimental conditions

2.1. Experimental materials

Polymer: Partially hydrolyzed polyacrylamide (HPAM for short), solid content of 90%, relative molecular mass of 2500×10^4; Experimental water: 1 on-site sewage in Daqing Oilfield; 2 simulated sewage in Daqing Oilfield, mineralization and ion composition; see text; Inorganic salts: NaCl、

IOP Conf. Series: Materials Science and Engineering **634** (2019) 012045 doi:10.1088/1757-899X/634/1/012045

KCl、$MgCl_2$ and $CaCl_2$ are all produced from Yangxing Dongxing Reagent Factory, and the content is ≥96.0%.

2.2. Experimental instruments

BROOKFIELD viscometer (DVII+ type), electronic analytical balance, JJ-1 six-electromechanical stirrer, constant temperature oscillating water bath, 500 mL beaker.

2.3. Experimental methods

According to the national standard SY/T6576-2003, the vector process is to add 200 mL of simulated sewage to a 500 mL beaker, and the polymer dry powder required for mass concentration of 5 000 mg/L polymer solution is weighed with a precision balance. At a higher speed mechanical agitator, the polymer dry powder was slowly added to the simulated sewage. After the polymer dry powder was completely added to the simulated sewage for 30 s, the rotation speed was lowered to 100r/min, and stirring was continued for 3 hours. The prepared polymer solution with a mass concentration of 5000 mg/L can be used as mother liquor, which can be diluted into any concentration target solution after completely aging after sealing. Finally, the polymer solution was placed in an oven at 45 C for 12h. The BROOOKFIELD viscometer was used to measure the viscosity of the target solution at a rotating speed of 6r/min.

3. Experimental results and analysis

3.1. Effect of monovalent ion concentration on viscosity of polymer

In order to eliminate the influence of other ions and conditions in the sewage on the polymer solution, this paper first uses distilled water to prepare the polymer mother liquor. The polymer solution was diluted to the desired solution by separately adding a salt solution containing different concentrations of sodium chloride and potassium chloride. The mass concentrations of Na^+ and K^+ in the sample to be tested are 0, 500, 1000, 1500, 2000, 2500, 3000, 3500, 4000, 4500, 5000, 5500, 6000 mg/L and the polymer solution is 1500 mg/L. The results of measuring the viscosity of the polymer solution at different monovalent ion concentrations are shown in Figure 1.

The results show that the viscosity of polymer solution decreases with the increase of Na + and K + concentration. Especially when the sodium and potassium ions in the sewage are increased from 0 mg/L to 1000 mg/L, the viscosity of the polymer solution is greatly reduced. Taking the Na^+ solution as an example, when the viscosity of the solution containing no ionic component in the polycondensation water is 210.6mPa•s, the viscosity of the solution drops to 74mPa•s when the concentration of the monovalent ion is increased to 1000 mg/L, and then further increases by one. The valence ion concentration polymer solution viscosity decreases slowly. However, compared with the low concentration of Na^+ solution, the viscosity of the polymer with a monovalent ion concentration of 8000 mg/L is only 33.5mPa•s, and the oil displacement effect of the polymer solution will be greatly affected. The main reason for the above phenomenon is that the negative charge of -COO- on the polymer molecule decreases with the increase of the concentration of Na^+ and K^+, the electrostatic attraction between the molecules and the molecules decreases, and the friction between molecules decreases. It becomes curled, so the viscosity drops. Because of the small radius of Na^+ and the same charge, Na^+ is easier to enter the polymer molecular chain than K^+ to neutralize the negative charge, and Na^+ has greater influence on the viscosity of polymer solution than K^+.

3.2. Effect of divalent ion concentration on viscosity of polymer solution

The polymer solution was prepared in the same manner as above, and the mass concentrations of Ca^{2+} and Mg^{2+} in the sample to be tested were 0, 25, 50, 75, 100, 200, 300, and 400 mg/L, respectively. The results of measuring the viscosity of the polymer solution at different divalent ion concentrations are shown in Figure 2.

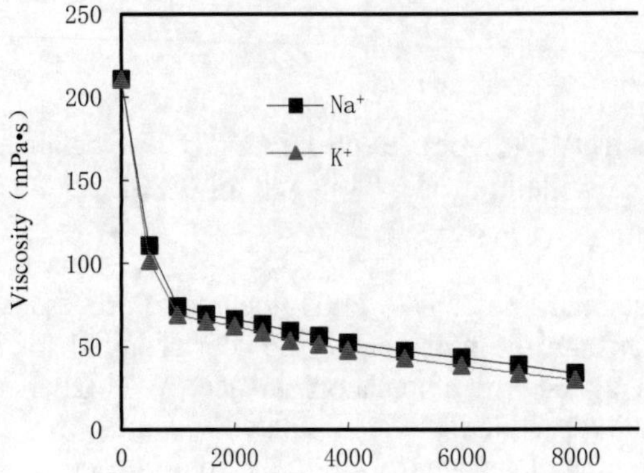

Figure 1. Effect of Na^+ and K^+ concentration on viscosity of polymer solution.

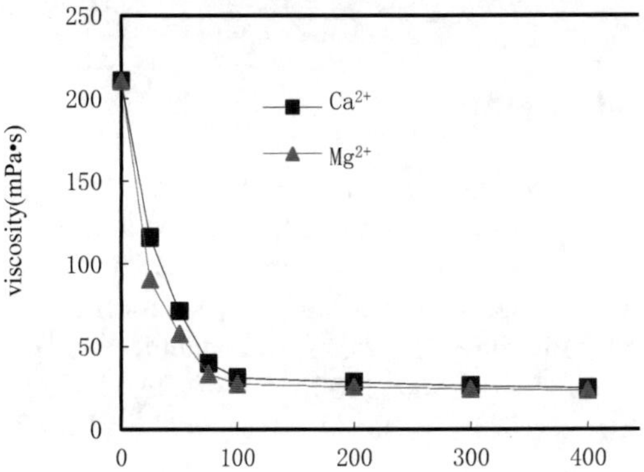

Figure 2. Effect of Ca^{2+}, Mg^{2+}concentration on viscosity of polymer solution.

The higher the concentration of Ca^{2+} and Mg^{2+}, the higher the polymer solution also shows a tendency to decrease rapidly and then gradually decrease. This rule has the same effect as the influence of monovalent ions on the viscosity of the polymer solution. Unlike monovalent ions, Ca^{2+} and Mg^{2+} have a large effect on the viscosity of the polymer at a small mass concentration change. When the Ca^{2+} concentration in the solution increased from 0mg/L to 75mg/L, the solution viscosity decreased from 210.6mPa•s to 39.9mPa•s. This is mainly because the Ca^{2+} and Mg^{2+} charge numbers are larger than Na^+ and K^+, and the ability of the -COO- to carry a negative charge on the polymer molecule is stronger, so that the polymer molecules are polycondensed, the molecular chain shrinks, and dehydration occurs. Therefore, the viscosity is reduced to a greater extent.

3.3. Effect of oilfield wastewater with different ion composition on polymer viscosity

The effects of Na^+, K^+, Ca^{2+} and Mg^{2+} on the viscosity of the solution were studied by preparing a polymer solution from a single component of simulated sewage. The composition of inorganic salts in oilfield wastewater is very complicated. In this paper, several oilfield wastewaters were used to prepare polymer solutions, and viscosity tests were carried out. The ionic composition of the sewage is shown in Table 1, and viscosities of polymer solutions with a mass concentration of 1500mg/L dissolved in wastewater are shown in Figure 3.

Table 1. Ion composition of sewage.

Total salinity (mg·L^{-1})	ion concentration (mg·L^{-1})						
	K^+ and Na^+	Ca^{2+}	Mg^{2+}	Cl^-	HCO_3^-	SO_4^{2-}	CO_3^{2-}
388	85.8	24.1	10.9	53.2	122.0	62.4	30.0
3261	1178.5	28.0	13.0	1408.9	323.3	278.4	60.6
8930	3130.8	64.9	27.8	4047.3	891.8	696.0	71.4

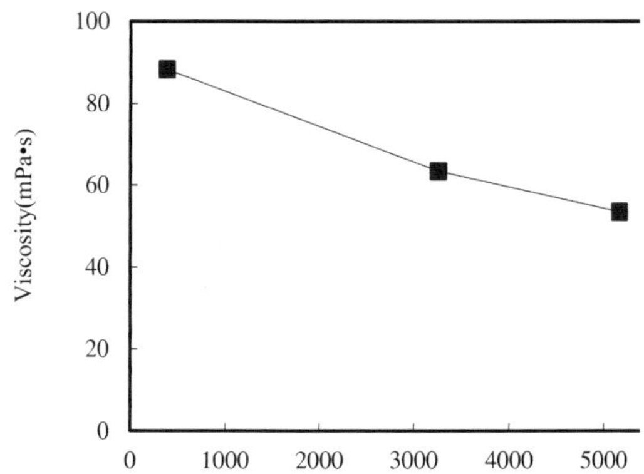

Figure 3. Viscosity of polymer solution prepared by different sewage.

The experimental results revealed that the viscosity of the polymer solution prepared using different salinities and ionic compositions varied greatly. The viscosity of HPAM decreases drastically as the degree of mineralization in the solution increases. For a wastewater with a salinity of 388.5mg/L, the viscosity of a polymer with a concentration of 1500mg/L is 88.3mPa·s; when the salinity of the sewage reaches 5176.6mg/L, the polymer solution of the same mass concentration The viscosity is 53.5mPa·s. HPAM exhibits poor salt tolerance, which is mainly due to the negative charge of COO- on the molecular chain of the polymer. The cations in the solution migrate to the vicinity of the negatively charged polymer group under the action of the opposite phase. The number of cations in the adsorption layer is increased, and the number of corresponding diffusion layer cations is reduced. The electrostatic repulsion between the polymer molecules is continuously weakened, and the polymer molecules gradually recover from the stretched state. When the molecular chain is curled, the water molecules around the cation are squeezed out. At this time, the volume of the polymer molecules in the curled state is the smallest, the contact area between the molecules is reduced, the internal friction between the molecules is minimized, and the viscosity of the solution is minimized. Reduce to the lowest value.

3.4. Increase viscosity measures

The above results indicate that the cation in the sewage is the main factor affecting the viscosity of the polymer. Therefore, the method of adding $Na_2C_2O_4$ as a metal complexing agent is used as a method for increasing the viscosity. The viscosity of the polymer solution is compared with the non-complexing agent and the addition of 210 mg/L of $Na_2C_2O_4$ as a metal complexing agent under the same sewage conditions and the same polymer mass concentration. The results are shown in Figure 4.

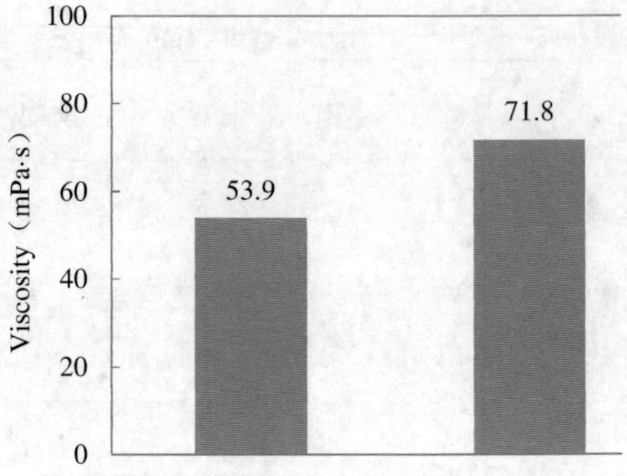

Figure 4. Viscosity enhancing effect of polymer solution.

It can be seen from the viscosity results of Figure 4 that $Na_2C_2O_4$ as a metal complexing agent can effectively reduce the influence of Na^+, K^+, Ca^{2+} and Mg^{2+} on the viscosity of the solution. After adding 210 mg/L $Na_2C_2O_4$, the viscosity of HPAM with mass concentration of 1500 mg/L increased from 53.9mPa·s to 71.8mPa·s, and viscosity growth rate reached 33.2%.

4. Conclusions

(1) The concentration of metal cations in the wastewater has a great influence on the viscosity of the polymer, and the viscosity decreases as the concentration of the metal cation increases. In particular, Ca^{2+} and Mg^{2+} have a large influence on the viscosity of the polymer solution at a small change in mass concentration.

(2) $Na_2C_2O_4$ as a metal complexing agent can effectively improve the viscosity of polymer solution prepared from sewage. After adding 210 mg/L $Na_2C_2O_4$, the viscosity increasing rate of HPAM with concentration of 1500 mg/L reached 33.2%.

References

[1] Yang C Z and Han D K 1991 Present status of EOR in the Chinese Petroleum industry and its future *J Petrol Sci Eng* **6** 175-89

[2] Sevic S and Solesa M 1999 Application, efficiency, and optimisation of chemical for injection water treatment *J Petrol Sci Eng* **15** 695-8

[3] Lü X 2010 Study on the influencing factors of preparing polymer solution by using sewage *J Southwest Petrol Univ* **32** 162-6

[4] Fernandez I J 2005 Evaluation of cationic water-soluble polymer with improved thermal stability *J Petrol Sci Eng* **9** 93-103

[5] Zhang J G 2005 Analysis of the main influencing factors of polymer solution viscosity *Fault Block Oil Gas Field* **12** 57-9

[6] Yu D S 2003 Mechanism of polyacrylamide flooding *Oil Gas Field Surf Eng* **8**

[7] He S S and Zhang X J 2002 A novel complexing agent for non-pure water polymer flooding polymer solution *Drilling Fluids Completion Fluids* **13** 13-5

[8] Ren J W, Zhou X S and Zhang D 2016 Analysis of influencing factors and viscosity effect of polymer solution in sewage *Contemp Chem Indus* **2** 272-5

[9] Luo F and Sun G 2016 Study on stability and improvement of polymer solution in sewage *Contemp Chem Indus* **4**: 661-4+692

[10] Lin C Y, Xue X S, Zhu W, Zhang X S, Sun F J and Liu H Q 2012 Study on aging law of polymer solution under reservoir flow conditions *Contemp Chem Indus* **8** 771-3

IOP Conf. Series: Materials Science and Engineering **634** (2019) 012046 doi:10.1088/1757-899X/634/1/012046

Microstructure and mechanical properties of Al/SiC surface composite with different volume fractions using friction stir process

N A El-Mahallawy[1,3,c], S H Zoalfakar[2] and A A G A Maboud[3]

[1]Faculty of Engineering and Materials Science, The German University in Cairo, Egypt
[2]Mechanical Engineering Department, Higher Technological Institute (HTI) ,10th of Ramadan city, Egypt
[3]Design and Production Engineering Department, Faculty of Engineering, Ain-Shams University, Abaseia, Cairo, Egypt

Corresponding author: E-mail: nahed.elmahallawy@guc.edu.eg

Abstract. Friction stir processing (FSP) was used to produce Al/SiC composites on the surface of Al 1050 sheets. SiC particles of 1 μm size were added on the Al sheet surface inside grooves with different sizes in order to obtain different composite volume fractions. FSP was applied with a square shape tool rotating at 1500 rpm and with a feed rate of 116 mm/min. The effect of SiC volume fraction on microstructure, and mechanical properties was investigated and the results were compared with samples treated by FSP but without SiC reinforcement. Different SiC volume fractions were obtained ranging from 7% to 16% depending on the groove size. Optical microscopic analysis (OM) in collaboration with scanning electron microscope (SEM) were used to fully discuss the microstructural changes within each sample. Mechanical properties were investigated through microhardness and tension test (including yield strength, ultimate tensile strength and percentage elongation). The results show significant improvement in microstructure changes in the FSPed composites, where equiaxed and refined grains structure were obtained due to the effect of the severe plastic deformation and continuous dynamic recrystallization caused by FSP compared to the elongated and coarse grains of the base metal. Microhardness and strength of the FSPed composites improved up to 112% and 9.7%, respectively compared to the as-received material.

1. Introduction

Aluminum is a well-known and highly used metal in several industries and applications. Due to the poor mechanical properties and low wear resistance of pure aluminum, alloying is necessary to improve these properties [1]. Preparation and fabrication of aluminum metal matrix composites can be an additional solution for such drawbacks, however, the processes used face a lot of problems [2]. On the other hand, composites may only be needed on the surface of the products in order to improve hardness and wear resistance. FSP (friction stir processing) can be applied to insert a hard phase into the alumunium on the surface, hence, increasing its hardness and wear resistance. This process is based on developing of the friction stir welding process (FSW) [3,4]. In FSP heat and pressure are generated by a rotating non-consumable tool with a special design. The tool includes a pin which penetrates into the material under an applied vertical force and a shoulder which only touches the surface [5], causing severe plastic deformation through the workpiece top layer [6]. This process offers several benefits by modifying the microstructure, improving the mechanical properties and also by offering the possibility to create a composite material on the surface without the need to melt the material [7]. Due to its effects, FSP is used in wide range of industries such as in the transportation sector, vehicles, electronics industry, defense industries and much more [8]. Composites with aluminum matrix (Al MMCs) by FSP was previously studied by adding SiC particles, which enhanced the mechanical properties through controlling the microstructural features [9-11] resulting in almost double the microhardness of the base material [12,13]. Improvement of the dispersion of SiC particles was also reached with multi FSP passes [14].

In the present study, FSP process was applied under constant process parameters which were selected based on the authors,'previous research where the process parameters were optimized using RSM [15]. Surface composite was prepared using SiC with 950 nm size with different volume fractions. Investigations include macrostructure, microstructure, hardness and tensile behavior.

2. Experimental work

Aluminum 1050 alloy sheet with 5 mm thickness was cut into square samples 200 mm x200 mm. A CNC vertical milling machine was used to perform the FSP processing using a K100 tool with square tip. In order to incorporate the SiC particles on the surface to prepare the composite, grooves with 2.1mm width were machined with depths of 2, 2.5, 3 and 3.5 mm. These grooves were then filled with 950 nm SiC powder. In order to prevent the SiC powder from escaping during FSP processing, a tool shoulder without a pin was first used to pass on the grooves to close them. This was followed by an FSP pass after adjusting the center of the tool pin to coincide on that of the groove. For comparing, the effect of adding SiC, a sample without SiC was processed (called sample c). All samples were prepared with constant process parameters: 1500 rpm, 116 mm/min, 3 passes and a tool with a square pin. Samples for microstructure investigation and for tensile test were cut and 3 samples of each condition were tested on universal testing machine as per ASTM E8M-04 procedures for tension test. The microhardness test was carried out by Vickers microhardness by load of a 200 g and a 15 s as dwelling time. The hardness distribution was obtained using five microhardness readings measured on the top surface as well as on cross sections. The microstructure was investigated on cross-sections of the samples after standard metallographic preparation and using Keller's reagent for etching. Investigation was carried out using Leco LX 31 and Ceti optical microscopes and by FEI Inspect S 50 for SEM.to

3. Results and discussions

3.1. Microstructure

In the longitudinal direction, the coarse and elongated grains of 1050 Al alloy because of the rolling process were changed to refined grains due to the FSP process, figure 1. This achieved grain refinement was attributed to a continuous dynamic recrystallization as found in many previous studies [16]. The overall macroscopic view of a composite sample, figure 2(a) illustrates the known three main zones in the FSP process: stir zone (SZ), thermo-mechanical affected zone (TMAZ), and heat affected zone (HAZ). By investigating, the microstructure in three different zones it is noticed that, the grain size in the SZ zone figure 2(b) where the smallest grain compared to TMAZ area figure 2(c) and HAZ figure 2(d). The fine grain structures in the SZ area were found because of a fully dynamic recrystallization caused by the friction heat produced by the FSP tool and the presence of SiC particles [14]. The TMAZ zone exhibited deformed structure compared to SZ zone; this grain shape is more clear in the HAZ zone where curved and elongated grains size appeared in figure 2(d). The distorted grains in TMAZ region was attributed to a shortage of the plastic deformation [17], where discontinuous dynamic recrystallization has occurred in TMAZ [18, 19] while the major amount of the recrystallization is located in the SZ. As the recrystallization effect decreased gradually, it results in large grain size in HAZ and in TMAZ zone.

IOP Conf. Series: Materials Science and Engineering **634** (2019) 012046 doi:10.1088/1757-899X/634/1/012046

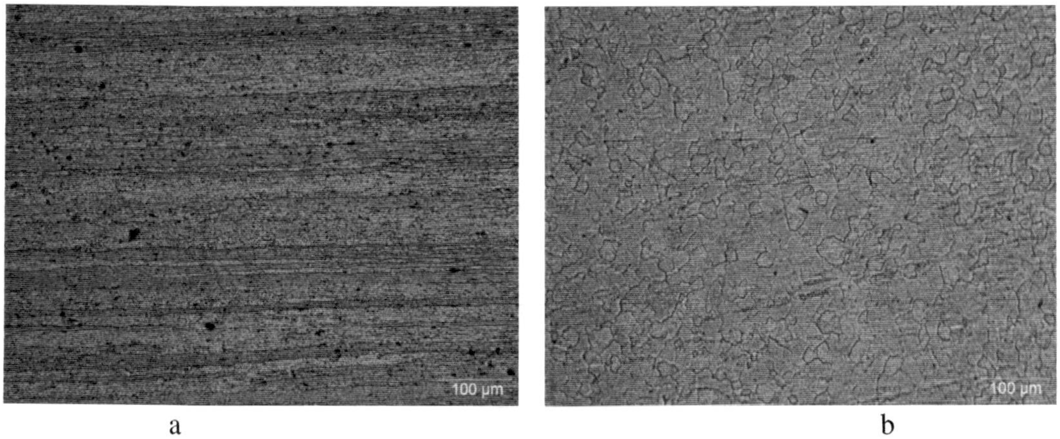

Figure 1. Metallographic investigation in FSPed sample C without SiC showing: (a) the base metal and (b) the stir zone.

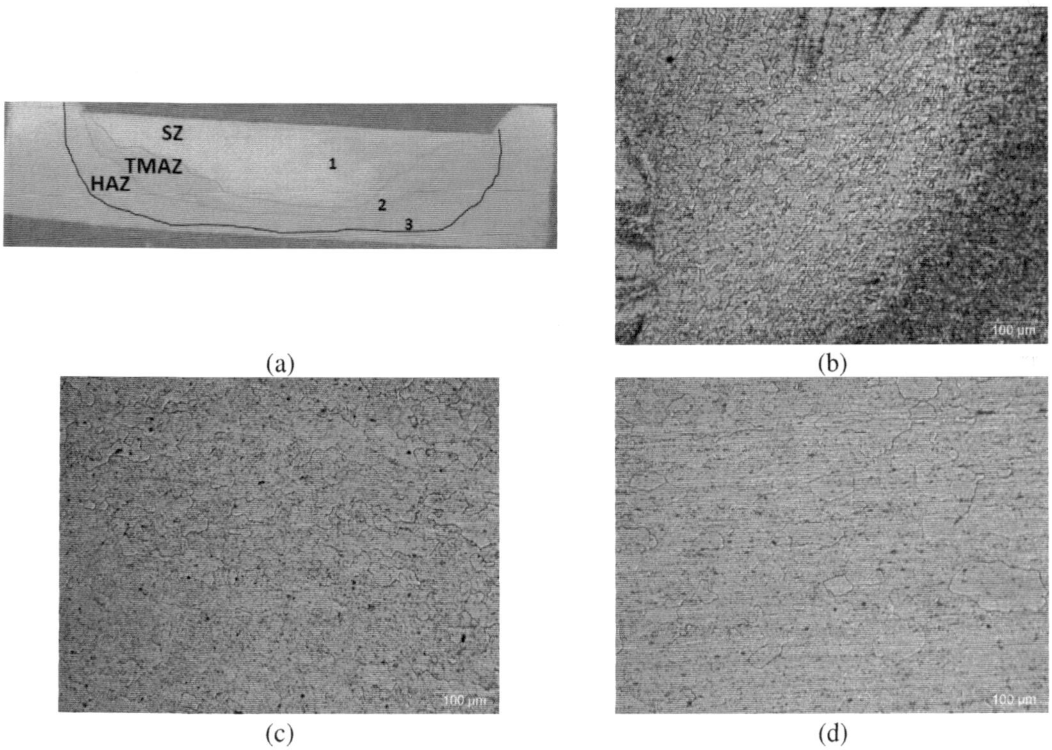

Figure 2. Microstructure of Sample 1 containing SiC particles: (a) the overall view of the FSP zones, (b) SZ: stir zone in zone 1, (c) TMAZ: thermo-mechanical affected zone 2 and (d) HAZ: Heat-affected zone 3.

The average volume fraction of SiC particles in the SZ was found to be ≈ 9 %, 11%, 7% and 16% for samples 1, 2, 3 and 4, respectively. This indicates that FSP tool shoulder was effective to block the leakage of SiC particles from the grooves. Furthermore, when the groove depth was increased from 2 mm to 2.5 mm (25%), the volume fraction of SiC was increased by 22.2%, and the increase of groove depth from 2 mm to 3.5 mm (75%), the volume fraction of SiC was increased by 77%. It is clear from the results that the groove depth can be a successful method to obtain a controllable specific volume fraction of the added particles in the surface composite using FSP.

3.2. Mechanical properties

From the values of tensile strength and elongation % ,Table 1, it can be deduced that, applying FSP only didn't improve the UTS of the as-received original material as it can be seen in sample c results, while the addition of SiC affected the strength. The UTS value was clearly increased in sample 4 where the Vf of SiC was ≈ 16% (the UTS recorded an increase of 9.7% compared to the original material). Sample 3 containing the least % of SiC (≈ 7%) exhibited the lowest UTS value among all FSPed samples with SiC particles. This strength enhancement is attributed to the presence of reinforcing SiC and to the refined grain size generated by FSP [20, 21] The percentage elongation of FSPed composites decreased strongly due to the existence of the ceramic SiC particles, that strongly reduces the material plasticity in the FSPed composites due to the non-deformable SiCp; this behavior is very common in the FSPed Al/SiC composites and mostly occurred accompanied with an increase of the strength [22]. The distribution of the microhardness around the SZ for the different SiC Vf are plotted in figures 3 and 4 together with that of the base metal. Obviously, the maximum microhardness values are found in the stir zone due to the existence of SiC particles as previously found [23] and due to formation of fully recrystallized grains attributed to dynamic recrystallization, which is a result of the FSP passes in the SZ. The stirring action has also a large effect on distributing the SiCp in the stir zone causing hardness increase as it is found in the previous work [13]. The peak VHN listed in Table 1 was 41 for sample 1, 33 for sample 2, 33 for sample 3 and 52 for sample 4. The hardness values decreased gradually in TMAZ and HAZ, which are depleted in SiC and exposed to less amount of plastic deformation compared to the SZ, figure 4. It is noticed that, FSP improved the microhardness on top of the FSPed samples by 46% for sample c and by 38% to 112% for the composites samples, Table 1. The hardness improvement is obviously related with the increase of SiC Vf, and the maximum hardness value was found in sample 4 with the maximum VF (≈ 16%) which enhanced the properties of the surface. So, the general improvement in top surface microhardness is attributed to the combined effect of plastic deformation and the existence of SiCp [24, 25].

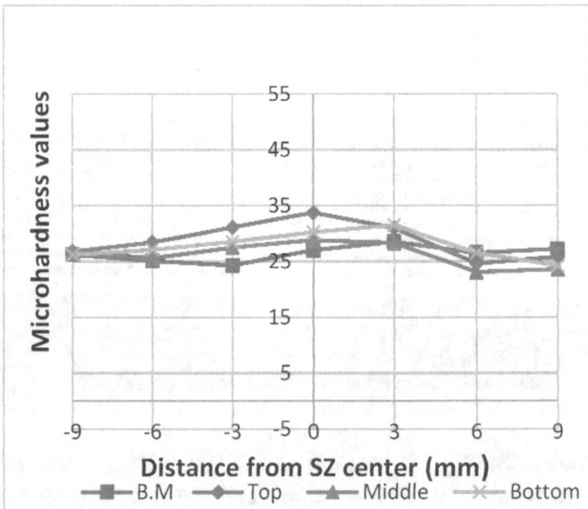

Figure 3. Micro hardness in sample C at different distances away from the SZ and at different depths from the top.

Table 1. Mechanical properties and corresponding SiC volume fraction

Sample name	Y.S (MPa)	UTS (MPa)	%El (MPa)	Microhardness (HV) (on top of the surface)	VF (%)
As-received	73.9 ± 0.1	84.3 ± 0.6	67.08 ± 2.6	27.3	0
c	37 ± 1.5	75.6 ± 0.2	72.9 ± 3.1	39.8	0
1	54.5 ± 4.9	91.22 ± 1.2	27.82 ± 5.1	41.18	9
2	62.85 ± 0.1	91.84 ± 2.1	29.8 ± 3.4	50.24	11
3	62 ± 1.4	80.5 ± 6.7	18.28 ± 1.3	37.76	7
4	50.95 ± 2.8	92.05 ± 5.8	42.16 ± 3.6	58.08	16

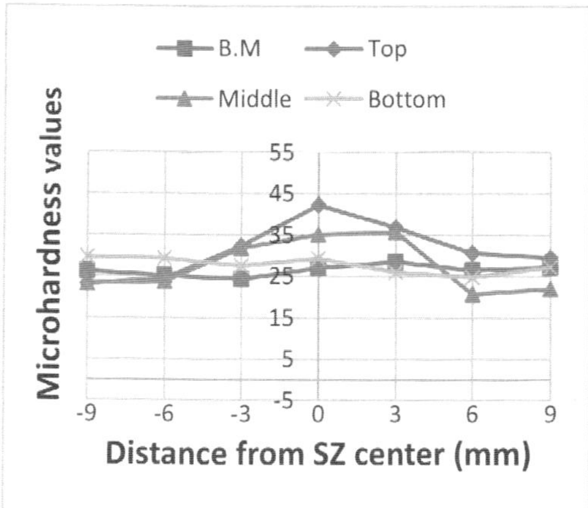

(a)

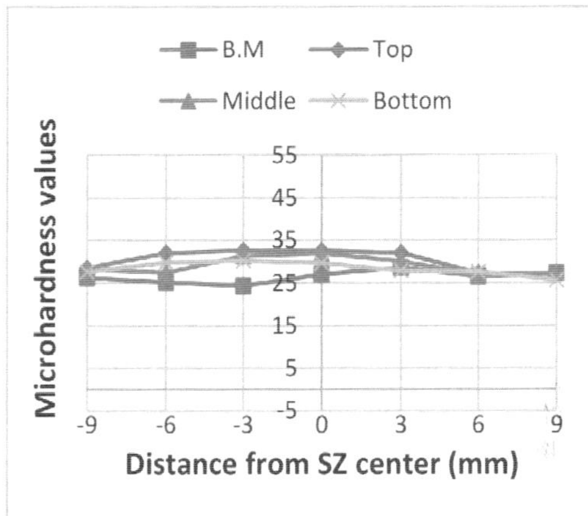

(b)

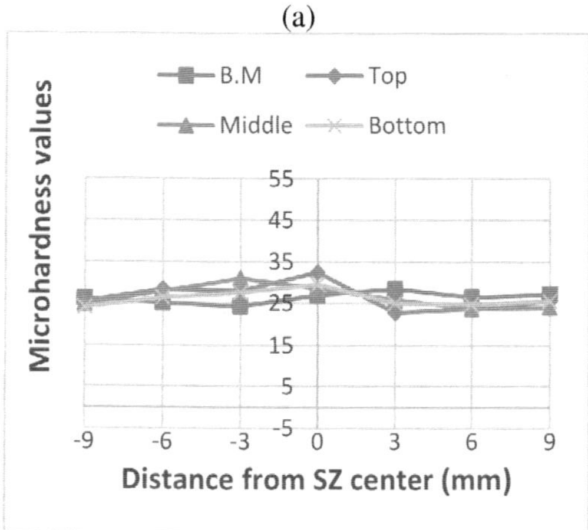

(c)

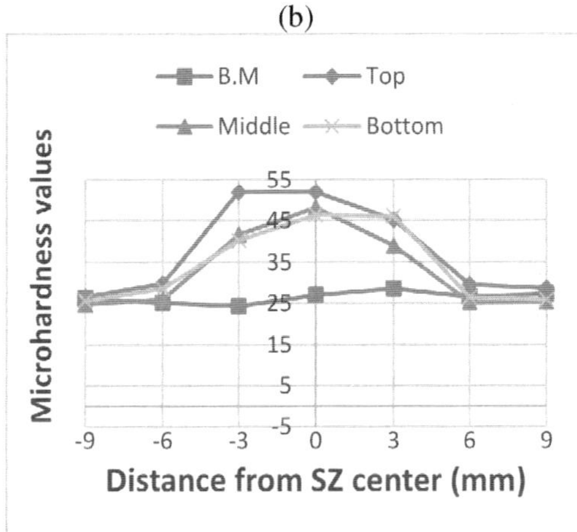

(d)

Figure 4. Micro hardness at different distances away from the SZ and at different depths from the top measured in FSPed samples together with those of original base metal, where a, b, c and d correspond to sample 1, sample 2, sample 3 and sample 4, respectively.

4. Conclusions

- Friction stir processing (FSP) can be successfully applied to create surface Al/SiC composites in addition to improving the mechanical and microstructure properties of as-received metal.
- Severe plastic deformation caused by FSP and SiC both enhanced grain refining on the surface
- The existence of SiC particles achieved 9.7% enhancement in UTS value compared to the original metal and up to 112% increase in the microhardness.

IOP Conf. Series: Materials Science and Engineering **634** (2019) 012046 doi:10.1088/1757-899X/634/1/012046

References

[1] Hehl A and Krug P 2013 *Aluminum and Aluminum Alloys* (Wiley, Hoboken, New Jersey, United States)

[2] Adesina A Y, Al-badour F A and Gasem Z M 2018 Wear resistance performance of AlCrN and TiAlN coated H13 tools during friction stir welding of A2124 / SiC composite *J. Manuf. Process.* **33** 111-25

[3] Mishra R S and Ma Z Y 2005 Friction stir welding and processing *Mater. Sci. Eng. R Reports* **50** 1-78

[4] Mishra R S and Mahoney M W 2001 Friction stir processing: A new grain refinement technique to achieve high strain rate superplasticity in commercial alloys *Materials Science Forum* **357** 507-12

[5] Bauri R and Yadav D 2017 *Metal Matrix Composites by Friction Stir Processing* 1st ed. (Elsevier, UK)

[6] Patel V V, Badheka V and Kumar A 2017 Effect of polygonal pin profiles on friction stir processed superplasticity of AA7075 alloy *J. Mater. Process. Tech.* **240** 68-76

[7] Prakrathi S, Ravikumar M, Udupa K R and Udaya Bhat K 2013 Fabrication of hybrid surface composite through friction stir processing and its impression creep behaviour *ISRN Mater. Sci.* **2013** 1-6

[8] Jana S, Mishra R S and Grant G 2015 *Friction Stir Casting Modification for Enhanced Structural Efficiency* 1st ed. (Elsevier, UK)

[9] Ghanbari D, Kasiri M, Amini K and Gharavi F 2017 Influence of heat treatment on mechanical properties and microstructure of the Al2024 / SiC composite produced by multi – pass friction stir processing *Measurement* **104** 151-8

[10] Rathee S, Maheshwari S, Noor A and Srivastava M 2017 Analysis of microstructural changes in enhancement of surface properties in sheet forming of Al alloys via friction stir processing *Mater. Today Proc.* **4** 452-8

[11] Deepak D, Sidhu R S and Gupta V K 2013 Preparation of 5083 Al-SiC surface composite by friction stir processing and its mechanical characterization *Int. J. Mech. Eng.* **3** 1-11

[12] Mishra R S, Ma Z Y and Charit I 2003 Friction stir processing : A novel technique for fabrication of surface composite *Mater. Sci. Eng. A* **341** 307-10

[13] Mahmoud E R I, Ikeuchi K and Takahashi M 2008 Fabrication of SiC particle reinforced composite on aluminium surface by friction stir processing *Sci Technol Weld JOI* **13** 607-18

[14] Abreu C M, Acuña R, Cabeza M, Cristóbal M J, Merino P and Verdera D 2017 Microstructure and mechanical properties of Al / SiC composite surface layer produced by friction stir processing *Ciência Tecnol. Dos Mater.* **29** e82-6

[15] Maboud A A G A, El-Mahallawy N A and Zoalfakar S H 2018 Process parameters optimization of friction stir processed Al 1050 aluminum alloy by response surface methodology (RSM) *Mater. Res. Express* **6** 026527

[16] Azizieh M, Iranparast D, Dezfuli M A G, Balak Z and Kim H S 2017 Fabrication of Al / Al 2 Cu in situ nanocomposite via friction stir processing *Trans. Nonferrous Met. Soc. China* **27** 779-88

[17] Iwaszko J, Kudła K, Fila K and Strzelecka M 2016 The effect of friction stir processing (FSP) on the microstructure and properties of AM60 magnesium alloy *Arch. Metall. Mater.* **61** 1555-60

[18] Komarasamy M, Kumar N, Tang Z, Mishra R S and Liaw P K 2015 Effect of microstructure on the deformation mechanism of friction stir-processed Al0.1 CoCrFeNi high entropy alloy effect of microstructure on the deformation mechanism of friction stir-processed *Mater. Res. Lett.* **3** 30-4

[19] Wang Y, Shao W Z, Zhen L and Zhang X M 2008 Microstructure evolution during dynamic recrystallization of hot deformed superalloy 718 *Mater. Sci. Eng. A* **486** 321-32

[20] Sajad S, Alipour M, Hamouda A, Matin A, Kord S, Afshari B and Koppad P 2017 Effect of multi-

pass friction stir processing on the microstructure, mechanical and wear properties of AA5083 / ZrO_2 nanocomposites *J. Alloys Compd.* **726** 1262-73
[21] Zhang Z and Chen D L 2008 Contribution of Orowan strengthening effect in particulate-reinforced metal matrix nanocomposites *Mat. Sci. Eng. A-Struct.* **484** 148-52
[22] Min S 2009 Effects of volume fraction of SiC particles on mechanical properties of SiC / Al composites *Trans. Nonferrous Met. Soc. China* **19** 1400-4
[23] Salih O S, Ou H, Wei X and Sun W 2018 Microstructure and mechanical properties of friction stir welded AA6092/SiC metal matrix composite *Mater. Sci. Eng. A* **742** 78-88
[24] Zarghani A, Bozorg S F and Hanzaki A 2009 Microstructures and mechanical properties of Al / Al_2O_3 surface nano-composite layer produced by friction stir processing *Mater. Sci. Eng. A* **500** 84-91
[25] Zhang Z, Yang R, Guo Y, Chen G, Lei Y, Cheng Y and Yue Y 2017 Microstructural evolution and mechanical properties of ZrB2/6061Al nanocomposites processed by multi-pass friction stir processing *Mater. Sci. Eng. A* **689** 411-8

IOP Conf. Series: Materials Science and Engineering **634** (2019) 012047 doi:10.1088/1757-899X/634/1/012047

Piezoelectric characteristics of inorganic 0.965(Na, K, Li)(Nb, Sb)O_3–0.035(Bi, Na)(Sr)ZrO_3 ceramics doped with Fe_2O_3 for actuator

J-H Yoo[1,4] J S Choi[1], J I Hong[2] and Y-H Jeong[3,4]

[1]*Department of Electrical Engineering, Semyung University, Jecheon 390-711, Korea*
[2]*Departmentof Electrical Information Control, Dong Seoul University, Korea*
[3]*Departmentof Electrical Engineering, Korea National University of Transportation, Korea*

E-mail: juhyun57@semyung.ac.kr (J-H Yoo)/yhjeong@ut.ac.kr (Y-H Jeong)

Abstract. In this experiment, 0.965(Na,K,Li)(Nb,Sb)O_3–0.035(Bi,Na)(Sr)ZrO_3 ceramics were fabricated with Fe_2O_3 addition, and their piezoelectric characteristics were investigated. When Fe_2O_3 was added to the ceramics, $d_{33}$and k_p were decreased. However, Q_m was increased. At the NKN-BNZ ceramics with 0.1 wt% Fe_2O_3, the d_{33}, k_p, Q_m, and ε_r were optimized as 263[pC/N], 0.450, 51.94, 1978, respectively.

1. Introduction

In recent years, Lead Zirconate Titanate (PZT) system piezoelectric ceramics have been used extensively in ultrasonic dishwashers, smartphone piezoelectric speakers, and ultrasonic surgical instruments [1]. However, they contain at least 60 wt% PbO in the composition in spite of having excellent piezoelectric properties. PbO is regulated as environmentally harmful substance because it volatilizes rapidly at 1000°C and is harmful to human body [2-3].

(Na,K)NbO_3-based piezoelectric ceramics with d_{33} = 416[pC/ N] developed by Sato et al. in 2004 have been actively studied as a substitute material for PZT system because of their excellent piezoelectric properties. However, it is hard to manufacture the dense ceramics due to volatilization of potassium and sodium at high temperature in the NKN system.

In addition, the piezoelectric properties of NKN system ceramics are very high in the orthorhombic–tetragonal phase boundary regions, and it is difficult to actually apply to the application device because the piezoelectricity changes greatly at this orthorhombic–tetragonal transition (To-t) temperature.

In recent years, Jianguo Zhu et al.have developed a composition with best d_{33} of ~350 (pC/N) and high Tc of ~315 (°C) in the NKN ceramics substituted with ($Bi_{0.5}$ $Na_{0.5}$)ZrO_3 having a rhombohedral-tetragonal(RT) phase boundary regions [4].

Here, ($Bi_{0.5}Na_{0.5}$) ZrO_3 can shift $T_{R\text{-}O}$ (rhombohedral-tetragonal transition point) and $T_{O\text{-}T}$ (orthorhombic-tetragonal transition point) of (Na, K) NbO_3 ceramics to room temperature. As the result, by forming the R and Tphases coexistence, high d_{33} and high Tc can be obtained. Originally, PZT composition ceramics with a rhombohedral-tetragonal (RT) phase regions has been extensively used because of its excellent piezoelectric properties [5]. Nowadays, smartphone piezoelectric speakers have been mounted in LG G8 phone. A high kp and high d_{33} are required for this device.

Accordingly, in this experiment, for developing NKN composition ceramics with a rhombohedral-tetragonal (RT) phase regions, $(Na,K)NbO_3$ system substituted with $(Bi,Na)(Sr)\ ZrO_3$ were fabricated as a function of Fe_2O_3 addition and their piezoelectric characteristics were analyzed for piezoelectric speakers.

2. Experiments

We manufactured the ceramic specimens as following composition formula:

$$0.965(Na,K,Li)(Nb,\ Sb)O_3 - 0.035(Bi,Na)(Sr)ZrO_3$$
$$+x\ wt\%\ Fe_2O_3\ (x=0, 0.1, 0.2, 0.3, 0.4)$$

The high purity powders of K_2CO_3, Na_2CO_3, Li_2CO_3, Sb_2O_5, $SrCO_3$, Bi_2O_3 and ZrO_2 with the purity of above 99% were weighed by mole ratio and the high purity powders were milled for 24hours. And then, they were calcined at 900℃ for 6 hours. And then a 5 wt % polyvinyl alcohol was mixed to the powders. The pressure of 17 MPa using a mold with diameter of 9mm were induced, and then sintered at 1130℃for 3h. The samples were polished to thickness 1.0mm and then electro-deposited with Ag paste. Capacitance was measured at 1 kHz using an LCR meter (ANDO AG-4034) and the dielectric constant (ε_r) was calculated for investigation of the dielectric characteristics. For analysis of the piezoelectric characteristics the fr and fa were measured by an Agilent 4294A Impedance Analyzer and then the k_p and Q_m were calculated. d_{33} was measured by APC-90-2030 piezo d_{33} meter.

3. Results and discussion

The density of the specimen with x(=Fe_2O_3) was appeared in Figure 1. The density enhanced due to the increase of the Fe_2O_3. In this experiment, Fe_2O_3 (melting point: 730℃) and Bi_2O_3 (melting point: 817℃) were added for forming the liquid phase of the samples during firing period.

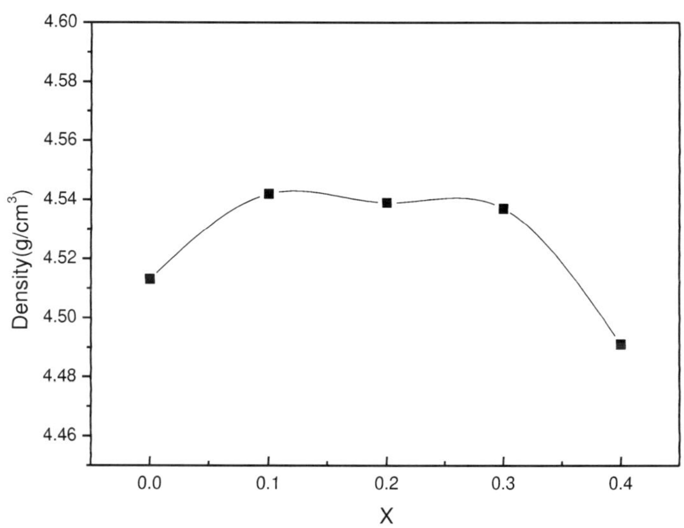

Figure 1. Density of samples with Fe_2O_3 addition.

However, these oxides firstly added before calcination process. Accordingly, except for acting as liquid phase, Fe_2O_3 can be acted as acceptor dopant. The result of the density is because the densification of the samples was increased up to 0.1 wt% as liquid phase effect and then slowly decreased owing to excess Fe_2O_3.

Figure 2 shows the kp of the specimen with Fe_2O_3 addition. A maximum value of 0.450 was obtained when the Fe_2O_3 was 0.1 because of the enhancement of sinterability. Threafter, the kp was decreased. And then, the kp increased again when the Fe_2O_3 was 0.3. The decrease of kp at the 0.2 wt% Fe_2O_3 composition is considered because Fe_2O_3 was largely acted as acceptor dopant.

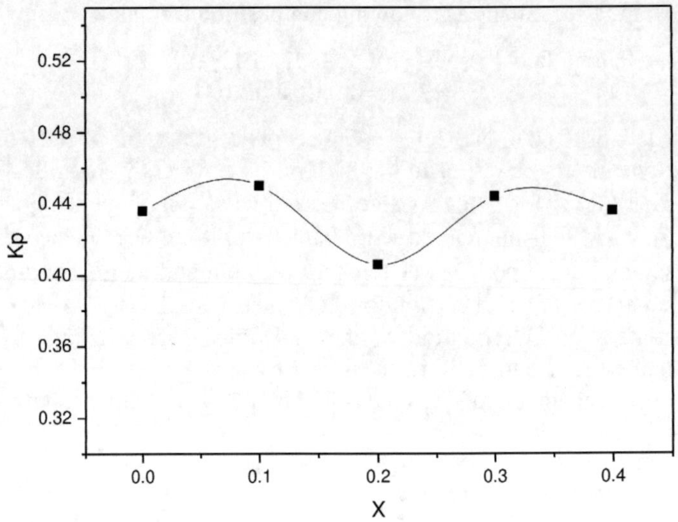

Figure 2. K_p of samples with Fe_2O_3.

The Qm with Fe_2O_3 is shown in Figure 3. When Fe_2O_3 substitution is 0.2, the highest value of 55 was appeared, and then decreased a little. This result is because Fe $^{3+}$ ion (0.64 Å) can be infiltrated at the B sites of Nb $^{5+}$ ion (0.69 Å) and Sb $^{5+}$ ion (0.62 Å). As the result, the formation of oxygen vacancies are caused, resulting in the increase of Qm through restraining the domain wall motion.

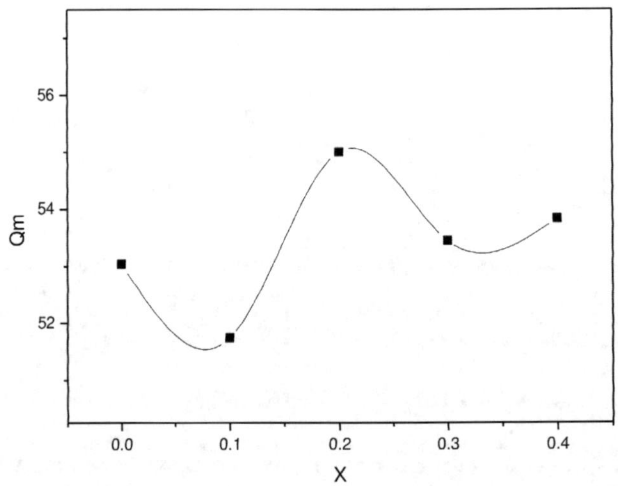

Figure 3. Q_m of samples as a function of Fe_2O_3.

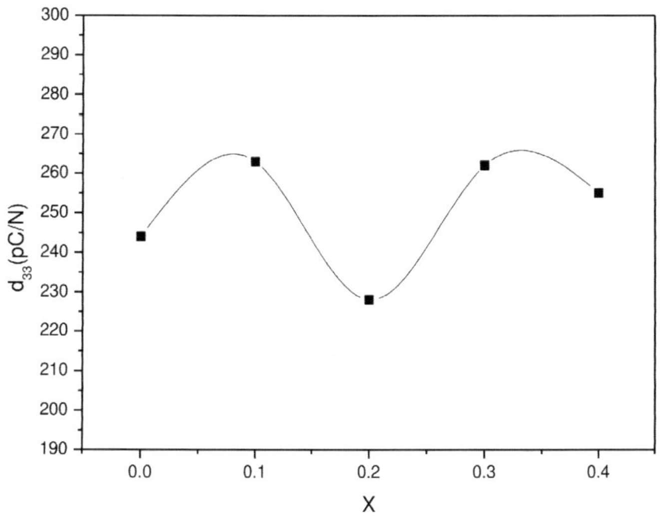

Figure 4. d_{33} of samples with Fe_2O_3.

The d_{33} piezoelectric constant with Fe_2O_3 is shown in Figure 4.

The best value of d_{33} was 263 [pC / N] when Fe_2O_3= 0.1. In this composition, perhaps, it is considered that rhombohedral and tetragonal (R-T) coexistence phases were appeared. And also, the decrease of d_{33} at the 0.2 wt% Fe_2O_3 composition is considered because Fe_2O_3 was largely acted as acceptor dopant. In general, in case of acting as acceptor dopant, Qm increases and kp and d_{33} decreases through restraining the domain wall motion.

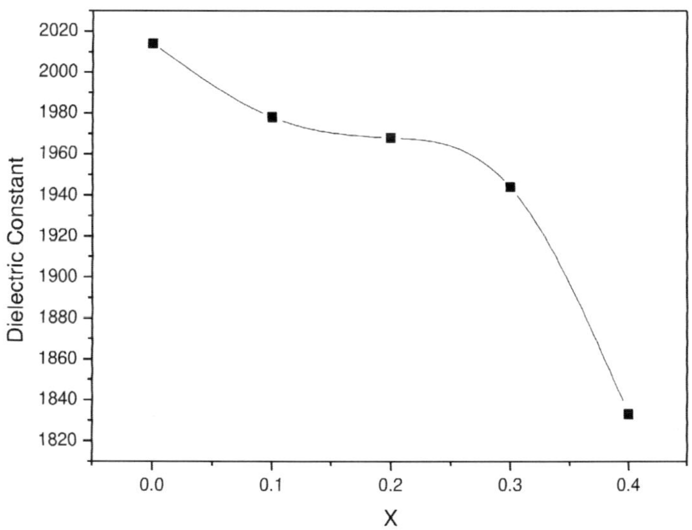

Figure 5. ε_r (dielectric constant) of samples with Fe_2O_3.

The ε_r (dielectric constant) according to Fe_2O_3 is shown Figure 5. The best value of ε_r was 2014 [pC / N] at the Fe_2O_3= 0. After x of Fe_2O_3 was more than 0.1, ε_r was abruptly decreased because Fe_2O_3 also acts as acceptor dopant. Table 1 shows electrical characteristics of samples with Fe_2O_3.

Table 1. Electrical characteristics of samples with Fe_2O_3.

X	Density [g/cm^3]	k_p	ε_r	d_{33} [pC/N]	Q_m
0	4.515	0.436	2014	244	53.04
0.1	4.542	0.450	1978	263	51.74
0.2	4.539	0.406	1968	228	55
0.3	4.537	0.444	1944	262	53.44
0.4	4.491	0.436	1833	255	53.83

4. Conclusions

In this experiment, 0.965(Na,K,Li)(Nb,Sb)O_3 –0.035(Bi ,Na)(Sr)ZrO_3 ceramics were fabricated with Fe_2O_3, and their piezoelectric characteristics were investigated.

- The $d_{33}$and k_p were decreased when Fe_2O_3 was added to the ceramics. But Q_m was increased.
- At the NKN-BNZ ceramics with 0.1 wt% Fe_2O_3, the d_{33}, k_p, Q_m, and ε_r were optimized as 263[pC/N], 0.450, 51.94, 1978, respectively, for piezo- actuators.

Acknowledgement

This paper was studied by 2016 National Research Foundation of Korea, and Korea National University of Transportation in 2018.

References

[1] Minho P and Juhyun Y 2012 *J. of Electronic Materials* **41** 3095
[2] Jungrae N and Juhyun Y 2013 *J. Electroceram* **30** 139
[3] Sang-Hoon S, Juhyun Y and Sang-Don L 2014 *Ferroelectrics Lett.* **41** 56
[4] Gou Q, Zhu J G, Wu J G, Li F L, Jiang L M and Xiao D Q 2018 *J. of Alloys and compounds* **730** 311
[5] Zhang B Y and Zhu J G 2013 *ACS Appl. Mater. Interfaces* **5** 7718

IOP Conf. Series: Materials Science and Engineering **634** (2019) 012048 doi:10.1088/1757-899X/634/1/012048

Metal-oxide nanotubes functional material tailored for membrane water/wastewater treatment

A Benamor[1], M Nasser[1], M H El-Naas[1] and H Qiblawey[2,3]

[1]Gas Processing Center, College of Engineering, Qatar University, 2713 Doha, Qatar
[2]Department of Chemical Engineering, College of Engineering Qatar University, 2713 Doha, Qatar

E-mail: hazim@qu.edu.qa

Abstract. Titanium Dioxide Nano-Tubes (TNTs) synthesized by hydrothermal method were used to prepare new polysulfone thin film nanocomposite (TFN) membranes. The TFN membranes contained different TNTs proportions (0.1, 0.3 and 0.5 wt. %). A polyamide selective layer was formed on top of the TFN membrane surface using interfacial polymerization (IP). Nanofiltration experiments were performed using NaCl and $MgSO_4$ salts solutions. The water flux was found to increase with increased TNT loading in the membrane due to high porosity of embedded nanoparticles and the formation of enlarged pores. The antifouling behaviour of the membrane was tested by bovine serum albumin (BSA) solution and found to improve with increased TNT content in the membrane.

1. Introduction

Membrane technology is considered as one of the most growing technologies for water treatment and desalination of sea and brackish water to solve fresh water scarcity and to cover the increase in drinking water demand in many parts of the world [1, 2]. Among all, TiO_2 is the most extensively studied transition-metal oxides because of its non-toxic, environmentally friendly, inexpensive, commercially available and good corrosion-resistant property. Transition metals embedded in membranes have been applied to counter a large number of environmental problems including those related to water and air purification [3, 4]. Nano materials have its inherent photocatalytic effects that decompose organic chemicals and kill bacteria [5]. Water filtration with composite membrane of TiO_2 particles and photo degradation under ultraviolet (UV) irradiation was investigated by several research groups [6, 7]. Other researchers studied TiO_2/polymer TFC reverse osmosis membrane to reduce biofouling [8, 9]. From previously mentioned outcomes, the incorporation of TiO_2 is a useful approach to improve membrane morphology and inhibit membrane fouling. Preparing TNTs from TiO_2 nanoparticles using hydrothermal method offer larger surface area and larger pore volumes compared to other methods in addition to the hydrophilic properties of TiO_2 [10-12]. These features are the most attractive properties and key features required for the development of high quality membranes [13].

Recently, a new class of nanotube-based membranes (TFN membranes) were synthesized and used for reverse osmosis (RO) operations. These membranes showed excellent performances compared to the usual polyamide (PA) TFC membranes. Based on this concept, titanate nanotubes were produced and functionalized using several amines to produce a new class of TFN RO membranes that showed water permeability up to 93 % with better anti-organic fouling properties [14]. Polyethersulfone blend membrane with incorporated TiO_2 nanotubes used in vacuum membrane distillation showed a permeate flux of 15.2 kg/m^2 h and a maximum salt rejection of 98 % for a feed salt concentration of 7000 ppm and a temperature of 65 °C [15]. The addition of TNT to PES membrane for the fabrication of asymmetric PES/TNTs blend membranes resulted in higher porosity, better hydrophilicity, increased permeate flux and higher salt rejection compared to bare PES membrane [16] which can be explained by increased membrane hydrophilicity due to the embedded hydrophilic carbon nanotubes [10, 17, 18] and inner void spaces of nanotubes [19-21]. The gaps existing between the embedded nanomaterials and the polyamide layers enhanced the water diffusion rates [5, 18, 20, 22]. Furthermore, the increased pore volumes of TNTs and their large specific surface area provide a large number of adsorption sites and diffusion channels for water movement, resulting in higher water channeling. Hence, the goal of this study is to synthesize TiO_2 nanotubes by a solution chemical method and use these nanotubes to develop TFN NF membranes for filtration of various salts.

2. Experimental

2.1. Materials

All chemicals employed in this research work are summarised in Table 1. All materials were used as received without further purification. Purified water used in all experiments. Acrylic plastic plates (11×9×0.24 in) were used to support the PSf-TNTs membrane synthesis.

Table 1. Description of materials and their source

Material description	Source
Polysulfone (PSf) Udel P-1700 in pellet	Solvay Advanced Polymers
N-Methyl pyrrolidinone (NMP) anhydrous 99.5%	Sigma-Aldrich St Louis, Ohio, USA
m-Phenylenediamine (MPD) flakes >99%	Sigma-Aldrich St Louis, Ohio, USA
n-hexane (> 99%)	Sigma-Aldrich St Louis, Ohio, USA
1, 3, 5-benzenetricarbonyl trichloride (TMC> 98%,)	Sigma-Aldrich St Louis, Ohio, USA
Sodium hydroxide (NaOH 98%)	Sigma-Aldrich St Louis, Ohio, USA
Sodium chloride (NaCl) > 9.5%	Sigma-Aldrich St Louis, Ohio, USA
Magnesium sulphate (MgSO4 99 %)	Sigma-Aldrich St Louis, Ohio, USA
Titanium dioxide nanoparticles with particle size of < 21 nm (TiO2, Anatase)	Sigma-Aldrich St Louis, Ohio, USA

2.2. Synthesis of TiO_2 nanotubes

Titanium dioxide nanotubes (TNTs) were prepared by hydrothermal method; 3 gm of TiO_2 powder were mixed with 150 ml of 10 molar NaOH solution, and stirred for 60 min at ambient temperature. The resulted suspension was then heated to 170 °C for 24 h in a Teflon-lined autoclave to ensure full dissolution of TiO_2 powder. After that the autoclave was left to cool down until it reached room temperature and a product in the form of white powder was formed. The product was initially washed with an HCl solution of 0.1 molar then washed by deionized water. When the pH of the washing solution approaches 7, cleaning with deionized water was stopped. The product in the form of white powder was then dried under vacuum at 85 °C for 4 h to make sure of complete removal of all water traces. The sample was post-treated again at 500 °C for 3 h with a temperature increase rate of 5 °C/min.

2.3. Composite membrane preparation

2.3.1. Preparation of substrate membrane. To prepare the casting solution, PVP was dissolved in NMP and the solution was stirred for 10 min following the procedure described by Elimelech et al.[23]. Weighed amount of TNT nanotubes was added to NMP/PVP mixture, and ultrasonicated for 30 min to minimize chances of TNT nanotubes agglomeration. Polysulfone (PSF) beads were then added to the solution and stirred vigorously until a homogenous solution was obtained. The obtained homogeneous casting solution was stirred for 8 h at ambiant temperature, then ultrasonicated for 1 h to remove trapped air bubbles and finally stored in a desiccator for 15 h before any use. The solution was then casted at room temperature on a non-woven polyester supported on a glass plate using a 150 μm thickness casting knife. The whole composite was immediately immersed in a distilled water coagulation bath containing 3 wt. % NMP for phase inversion to take place. The substrate was kept in the coagulation bath for 10 min where the membrane was obtained and then transferred to another water bath and kept for at least 24 h to remove the solvent and water-soluble polymer (PVP) residuals. The membrane was washed repeatedly with distilled water and stored in water. The produced substrates are designated as TFC-0, TFC-1, TFC-3, and TFC-5 corresponding to the Nanotubes loading percentage of 0, 0.1, 0.3 and 0.5% respectively.

2.3.2. Preparation of Polyamide selective layer. The polyamide (PA) selective layer was casted on top of the prepared PSF substrate surface by interfacial polymerization (IP) process. To make sure that MPD solution penetrates into the substrate pores, thirty mL of MPD aqueous solution 2 wt. % was poured and held for two minutes on top of the substrate membrane. The excess MPD solution was drained and the remaining residuals were removed by air knife. Another thirty mL of TMC 0.1 wt. % dissolved in n-hexane was poured on the substrate surface and kept for 1 min after which the excess organic solution was drained off. As a result, an ultra-thin PA layer was formed. The synthesised TFC membrane was kept at ambient condition for one min to dry then kept in an oven at a temperature of 70 °C for another eight min for complete drying. The new TFC membrane was stored in DI water at 4 °C until further use [24].

3. Results and discussions

3.1. Morphology of TiO_2 nanotubes

The morphology of TNTs are directly analyzed by SEM (Philips model XL30E, USA) using direct applying nanotubes on copper plate with sputtering at 2 mints. Also TNT were analyzed by transmission electron microscope (Phillips CM12 with motor driven stage and SIA L3C Digital Camera, The Netherlands). The TNTs are dispersed in water using an ultrasonic bath (Branson 3800 ultrasonic cleaner-M3800, USA) for 1 h. A few drops of each suspension were deposited onto 3-mm diameter carbon-coated grids. 120 kV acceleration voltage were used for the TEM analyses. The obtained images as shown in figure 1 reveal that the synthesized nanotubes are open ended nanotubes of narrow size distribution. The average inner diameter was found to be in the range of 3-5 nm while the outer diameter of 10-13 nm. The observed length of TNTs nanotubes was found to be greater than 300 nm. The TEM images revealed that the TNTs are single wall nanotubes.

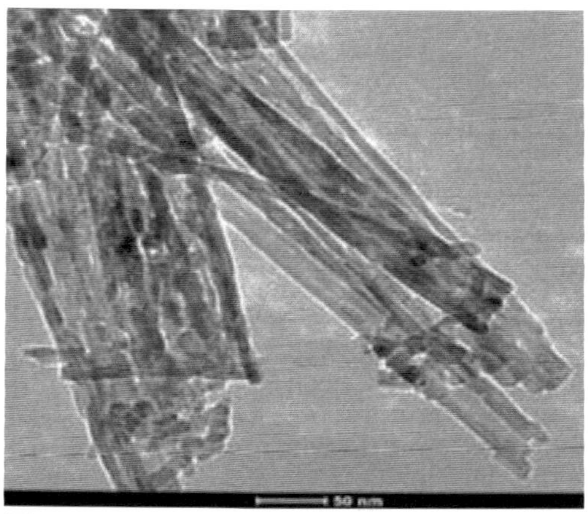

Figure 1. TEM images of synthesized TNTs.

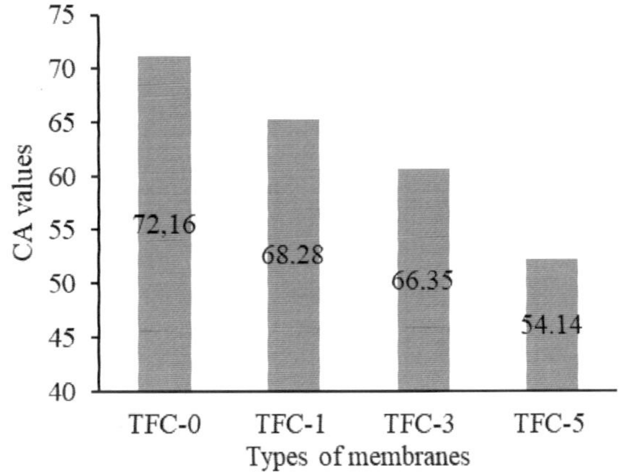

Figure 2. Contact angles of the different membranes.

3.2. Membrane contact angle

Contact angle measurement is an important parameter in membrane characterization as it indicates the membrane's hydrophilicity and flux behavior. For each membrane, the contact angles was measured at least three times and the average value was considered and shown in figure 2. The contact angle results indicate that TNTs content membranes exhibited a higher hydrophilicity than control TFN-0 membrane. In addition, the CA decreases from 72.16 ± 2◦ to 54.14 ± 2◦as TNTs percentage increased from 0.1 to 0.5 wt. % due to increase of hydrophilicity. Therefore, it was expected that the TNTs content membranes absorb more permeate into the membrane and thus enhance the flux rate. The lowest CA of TFN-5 (54.14± 2) indicates that this membrane exhibited a more hydrophilic surface due to a superior ultra-hydrophilic property of the TiO_2 nanotubes deposited, which was well distributed over the surface compared to other TFN membranes [25].

3.3. Membrane performance study

3.3.1. Permeation experiments. Two different experimental set ups were used in this work, the first one consist of a dead end cell while the second one is a cross-flow module both supplied by Sterlitech. Applying varying trans-membrane pressures (4 to 10 bars) at ambiant temperature, permeation experiments at were performed using a dead-end filtration cell l equipped with magnetic stirrers set at a constant rotation speed of 500 rpm. The effective membrane surface area was 14.6 cm^2. For each experiment around 300 millilitres of deionized water was used. To ensure reproducibility, all measurements were performed using at least three different membrane samples. Before any measurement, membrane was compacted for about 30 min under a pressure of 15 bar. The permeate flux (J_w) was calculated by Eq. (1):

$$J_w = \left(V / A.\Delta t\right) \tag{1}$$

Where,

V: volume of permeated water (L)
A: membrane area (cm^2)
Δt: permeation time (min).

The experimental permeability results of the synthesized TFN membranes are presented in figure 3.
The permeance was defined to be the ratio of the water flux over the membrane pressure difference (Δp):

$$\text{Permeance} = Jw/\Delta P = \left(V / A.t.\Delta P\right) \tag{2}$$

The average permeance of the pristine TFN-0 membrane was 1.06 L/m².h.bar. Compared to the TFN membrane, the addition of 0.3 wt. % TNT resulted in an increased permeance of 1.468 L/m².h.bar. The permeance slightly increased up to 1.518 L/m².h.bar when 0.5 wt. % TNTs was added, which is one time and half the value of the pure PA membrane (1.06 L/m².h.bar)

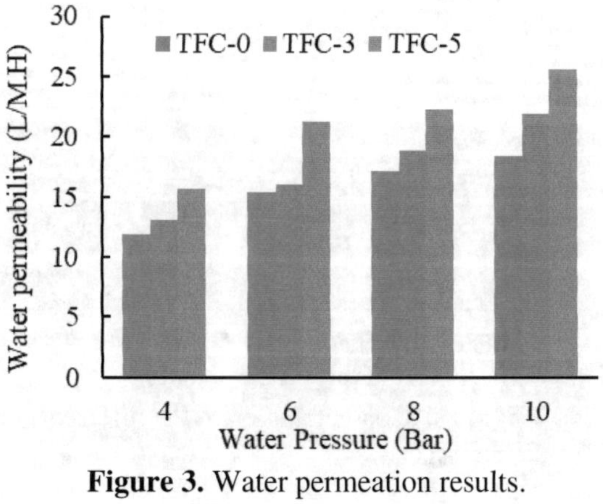

Figure 3. Water permeation results.

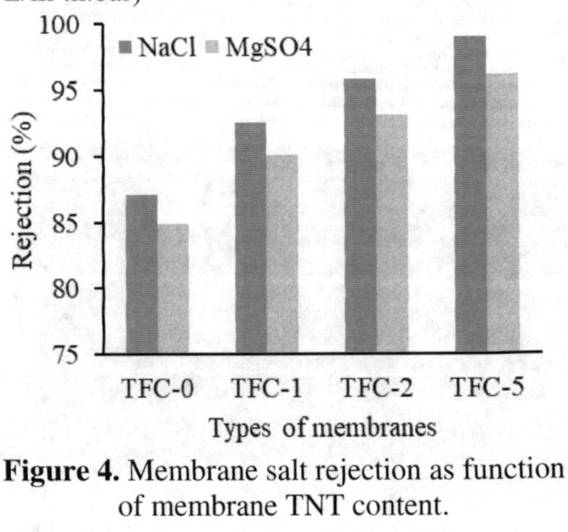

Figure 4. Membrane salt rejection as function of membrane TNT content.

3.3.2. Salt rejection. Membrane performances in term of salt rejection, as shown in figure 4 were measured in a lab scale Sterlitech™ CF042P stainless steel cross-flow module with an effective membrane area of 42 cm^2 and 18 L volume capacity tank [26]. For performance tests, 1,500 mg/L NaCl (1500 mg/L $MgSO_4$) solution were fed at a flowrate of 1.7 litre per minute at 25°C. The permeate flux was measured under 10 bar for 2 h after membrane compaction for 30 min under 15 bar. Permeate flux (Jw) was calculated by Eq. (1). Salt rejection was calculated by measuring the conductivity of the feed (C_f) and the permeate (C_p) using (HACH, HQ14 portable) conductivity meter as expressed in Eq. (3).

$$\%R = \left(1 - C_p / C_f\right) \times 100 \tag{3}$$

4. Conclusions

Titanate Nanotubes were successfully synthesized and characterized. The average inner diameter was found to be in the range of 3-5 nm while the outer diameter of 10-13 nm. The observed length of TNTs nanotubes was found to be greater than 300 nm. The TEM images revealed that the TNTs are single wall nanotubes. The TNT-incorporated polysulfone TFC membranes were developed. The prepared membranes showed improved permeation properties. The results of the performance evaluation tests recommend that the addition of TNT improved the pure water, compared to pristine PSf membrane. Both salts rejection using NaCl and $MgSO_4$ calculated by measuring the conductivity of the feed and the permeate showed improved salt rejection with increased TNT content in the membrane.

Acknowledgments

This paper was made possible by an NPRP10-0127-170270 and NPRP8-1115-2-473 from the Qatar National Research Fund (a member of Qatar Foundation). The statements made herein are solely the responsibility of the authors.

References

[1] Mjalli F, Ettouney H, Abdel-Jabbar N and Qiblawey H 2008 *Desalination* **222** 441-50
[2] Qiblawey H, Banat F and Al-Naser Q 2009 *Desalin. Water Treat.* **7** 53-9
[3] Ng L Y, Mohammad A W, Leo C P and Hilal N 2013 *Desalination* **308** 15-33
[4] Emadzadeh D, Lau W J, Matsuura T, Rahbari-Sisakht M and Ismail A F 2014 *Chem. Eng. J.* **237** 70-80
[5] Perera DHN, Nataraj S K, Thomson NM, Sepe A, Hüttner S, Steiner U, Qiblawey H and Sivaniah E 2014 *J. Membr. Sci.* **453** 212-20
[6] Rahimpour A, Madaeni S S, Taheri A H and Mansourpanah Y 2008 *J. Membr. Sci.* **313** 158-69
[7] Mansourpanah Y and Momeni Habili E 2015 *J. Membr. Sci. Res.* **1** 26-33
[8] Bae T-H and Tak T-M 2005 Effect of TiO_2 nanoparticles on fouling mitigation of ultrafiltration membranes for activated sludge filtration *J. Membr. Sci.* **249** 1-8
[9] Kwak S-Y, Kim S H and Kim S S 2001 *Environ. Sci. Technol.* **35** 2388-94
[10] Emadzadeh D, Lau W J, Rahbari-Sisakht M, Daneshfar A, Ghanbari M, Mayahi A, Matsuura T and Ismail A F 2015 *Desalination* **368** 106-13
[11] Kasuga T, Hiramatsu M, Hoson A, Sekino T and Niihara K 1998 *Langmuir* **14** 3160-3
[12] Liu W, Ni J and Yin X 2014 *Water Res.* **53** 12-25
[13] Chen Q and Peng L-M 2007 *Int. J. Nanotechnol.* **4** 44-65
[14] Emadzadeh D, Lau W, Rahbari-Sisakht M, Daneshfar A, Ghanbari M, Mayahi A, Matsuura T and Ismail A 2015 *Desalination* **368** 106-13
[15] Abdallah H, Moustafa A, AlAnezi A A and El-Sayed H 2014 *Desalination* **346** 30-6
[16] Shaban M, AbdAllah H, Said L, Hamdy H S and Khalek A A 2015 *Chem. Eng. Res. Des.* **95** 307-16
[17] Kim E-S, Hwang G, El-Din M G and Liu Y 2012 *J. Membr. Sci.* **394** 37-48
[18] Roy S, Ntim S A, Mitra S and Sirkar K K 2011 *J. Membr. Sci.* **375** 81-7
[19] Baroña G N B, Lim J, Choi M and Jung B 2013 *Desalination* **325** 138-47

[20] Chan W-F, Chen H-Y, Surapathi A, Taylor M G, Shao X, Marand E and Johnson J K 2013 *Acs Nano* **7** 5308-19
[21] Kim W-G and Nair S 2013 *Chem. Eng. Sci.* **104** 908-24
[22] Huang H, Qu X, Dong H, Zhang L and Chen H 2013 *RSC Advances* **3** 8203-7
[23] Yip N Y, Tiraferri A, Phillip W A, Schiffman J D and Elimelech M 2010 *Environ. Sci. Technol.* **44** 3812-8
[24] Perera DHN, Song Q, Qiblawey H and Sivaniah E 2015 *J. Membr. Sci.* **487** 74-82
[25] Wu G, Gan S, Cui L and Xu Y 2008 *Appl. Surf. Sci.* **254** 7080-6
[26] Ashfaq M Y, Wang T, Qiblawey H, Reesh I A and , Judd S 2017 *Desalin. Water Treat.* **60** 122-8